Lecture Notes in Physics

Volume 1003

The series Lecture Notes in Physics (LNP), founded in 1969, reports new developments in physics research and teaching - quickly and informally, but with a high quality and the explicit aim to summarize and communicate current knowledge in an accessible way. Books published in this series are conceived as bridging material between advanced graduate textbooks and the forefront of research and to serve three purposes:

- to be a compact and modern up-to-date source of reference on a well-defined topic;
- to serve as an accessible introduction to the field to postgraduate students and non-specialist researchers from related areas;
- to be a source of advanced teaching material for specialized seminars, courses and schools.

Both monographs and multi-author volumes will be considered for publication. Edited volumes should however consist of a very limited number of contributions only. Proceedings will not be considered for LNP.
Volumes published in LNP are disseminated both in print and in electronic formats, the electronic archive being available at springerlink.com. The series content is indexed, abstracted and referenced by many abstracting and information services, bibliographic networks, subscription agencies, library networks, and consortia.
Proposals should be sent to a member of the Editorial Board, or directly to the responsible editor at Springer:

Dr Lisa Scalone
Springer Nature
Physics
Tiergartenstrasse 17
69121 Heidelberg, Germany
lisa.scalone@springernature.com

Roderich Tumulka

Foundations of Quantum Mechanics

Springer

Roderich Tumulka
Department of Mathematics
University of Tübingen
Tübingen, Germany

ISSN 0075-8450 ISSN 1616-6361 (electronic)
Lecture Notes in Physics
ISBN 978-3-031-09547-4 ISBN 978-3-031-09548-1 (eBook)
https://doi.org/10.1007/978-3-031-09548-1

This Springer imprint is published by the registered company Springer Nature Switzerland AG
The registered company address is: Gewerbestrasse 11, 6330 Cham, Switzerland

Contents

Acronyms

ABL	Aharonov–Bergmann–Lebowitz
$BM_{\mathcal{F}}$	Bohmian mechanics with foliation $\mathcal{F}$
CI	Copenhagen interpretation
CSL	Continuous spontaneous localization
EPR	Einstein–Podolsky–Rosen
GONB	Generalized orthonormal basis
GRW	Ghirardi–Rimini–Weber
GRWØ	GRW theory without primitive ontology
GRWf	GRW theory with flash ontology
GRWm	GRW theory with matter density ontology
GRWp	GRW theory with particle ontology
IBC	Interior-boundary condition
LHC	Large Hadron Collider
LIGO	Laser Interferometer Gravitational-Wave Observatory
MPI	Microscopic parameter independence
NHVT	No-hidden-variables theorem
ODE	Ordinary differential equation
ONB	Orthonormal basis
OQM	Orthodox quantum mechanics
PBR	Pusey–Barrett–Rudolph
PDE	Partial differential equation
POVM	Positive-operator-valued measure
PVM	Projection-valued measure
QFT	Quantum field theory
QTWO	Quantum theory without observers
rGRWf	Relativistic GRW theory with flash ontology
rGRWm	Relativistic GRW theory with matter density ontology
SØ	Schrödinger equation without primitive ontology
Sm	Schrödinger equation with matter density ontology
$SO(n)$	Special orthogonal group in n dimensions
tr	Trace
$TRCL$	Trace class
UV	Ultraviolet

Waves and Particles

1

1.1 Overview

Quantum mechanics is the field of physics concerned with (or the post-1900 theory of) the motion of electrons, photons, quarks, and other elementary particles, inside atoms or otherwise. It is distinct from classical mechanics, the pre-1900 theory of the motion of physical objects. Quantum mechanics forms the basis of modern physics and covers most of the physics under the conditions on Earth (i.e., not-too-high temperatures or speeds, not-too-strong gravitational fields). "Foundations of quantum mechanics" is the topic concerned with what exactly quantum mechanics means and how to explain the phenomena described by quantum mechanics. It is a controversial topic, although it should not be. It is one of the goals of this book to enable readers to form their own opinion of the subject.

Here are some voices critical of the traditional, orthodox view of quantum mechanics:

> With very few exceptions (such as Einstein and Laue) [...] I was the only sane person left [in theoretical physics].
>
> (Erwin Schrödinger in a letter dated 1959)

> I think I can safely say that nobody understands quantum mechanics.
>
> (Richard Feynman, 1965)

> I think that conventional formulations of quantum theory [...] are unprofessionally vague and ambiguous.
>
> (John Bell, 1986)

In this book, we will be concerned with what kinds of reasons people have for criticizing the orthodox understanding of quantum mechanics, what the alternatives are, and which kinds of arguments have been put forward for or against important views. We will also discuss the rules of quantum mechanics for making empirical

R. Tumulka, *Foundations of Quantum Mechanics*, Lecture Notes in Physics 1003,
https://doi.org/10.1007/978-3-031-09548-1_1

predictions; they are uncontroversial. The aspects of quantum mechanics that we discuss also apply to other fields of quantum physics, in particular to quantum field theory.

We need some mathematics prerequisites: complex numbers; vectors in n dimensions and inner product; matrices and their eigenvalues and eigenvectors; multivariable calculus; and probability, continuous random variables, and the Gaussian (normal) distribution. Some mathematical topics will be introduced here when we need them: differential operators (such as the Laplace operator) and their analogy to matrices; eigenvalues and eigenvectors of differential (and other) operators; the Hilbert space of square-integrable functions, norm, and inner product; projection operators; Fourier transform of a function; positive operators and positive operator-valued measures (POVMs); tensor product of vector spaces; trace of a matrix or an operator, partial trace; certain ordinary and partial differential equations, particularly the Schrödinger equation; and exponential random variables and the Poisson process.

There is also a philosophical side to the topic of this book. Here are some philosophical questions that will come up: Is the world deterministic, or stochastic, or neither? Can and should logic be revised in response to empirical findings? Are there in principle limitations to what we can know about the world (its laws, its state)? Which theories are meaningful as fundamental physical theories? In particular: If a statement cannot be tested empirically, can it be meaningful? (Positivism versus realism.) Does a fundamental physical theory have to provide a coherent story of what happens? Does that story have to contain elements representing matter in three-dimensional space in order to be meaningful?

Physicists usually take math classes but not philosophy classes. That does not mean, though, that one does not use philosophy in physics. It rather means that physicists learn the philosophy they need in physics classes. Prior knowledge in philosophy is not assumed in this book, but we will sometimes make connections with philosophy.

1.2 The Schrödinger Equation

One of the fundamental laws of quantum mechanics is the *Schrödinger equation*

$$i\hbar \frac{\partial \psi}{\partial t} = -\sum_{i=1}^{N} \frac{\hbar^2}{2m_i} \nabla_i^2 \psi + V\,\psi\,. \tag{1.1}$$

It governs the time evolution of the *wave function* $\psi = \psi_t = \psi(t, \boldsymbol{x}_1, \boldsymbol{x}_2, \ldots, \boldsymbol{x}_N)$. (It can be expected to be valid only in the *non-relativistic* regime, i.e., when the speeds of all particles are small compared to the speed of light. In the general case (the *relativistic* case), it needs to be replaced by other equations, such as the *Klein–Gordon equation* and the *Dirac equation* described in Chap. 7.) We focus first on *spinless* particles and discuss the phenomenon of *spin* later in Sect. 2.3. I use boldface symbols such as $\boldsymbol{x}$ for three-dimensional (3d) vectors.

Equation (1.1) applies to a system of N particles in $\mathbb{R}^3$. The word "particle" is traditionally used for electrons, photons, quarks, etc. Opinions diverge about whether electrons actually are particles in the literal sense (i.e., point-shaped objects or little grains). A *system* is a subset of the set of all particles in the world. A *configuration* of N particles is a list of their positions; *configuration space* is thus, for our purposes, the Cartesian product of N copies of physical space or $\mathbb{R}^{3N}$. The wave function of quantum mechanics, at any fixed time, is a function on configuration space, either complex-valued or spinor-valued (as we will explain in Sect. 2.3); for spinless particles, it is complex-valued, so

$$\psi : \mathbb{R}_t \times \mathbb{R}^{3N}_q \to \mathbb{C}. \tag{1.2}$$

The subscript indicates the variable: t for time and $q = (\boldsymbol{x}_1, \ldots, \boldsymbol{x}_N)$ for the configuration. Note that i in (1.1) sometimes denotes $\sqrt{-1}$ and sometimes labels the particles, $i = 1, \ldots, N$; m_i are positive constants, called the *masses* of the particles; $\hbar = h/2\pi$ is a constant of nature; h is called Planck's quantum of action or Planck's constant; $h = 6.63 \times 10^{-34}\,\text{kg m}^2\text{s}^{-1}$;

$$\nabla_i = \left(\frac{\partial}{\partial x_i}, \frac{\partial}{\partial y_i}, \frac{\partial}{\partial z_i}\right) \tag{1.3}$$

is the derivative operator with respect to the variable $\boldsymbol{x}_i$, ∇_i^2 the corresponding *Laplace operator*

$$\nabla_i^2 \psi = \frac{\partial^2 \psi}{\partial x_i^2} + \frac{\partial^2 \psi}{\partial y_i^2} + \frac{\partial^2 \psi}{\partial z_i^2}. \tag{1.4}$$

V is a given real-valued function on configuration space, called the *potential energy* or just *potential.*

Fundamentally, the potential in non-relativistic physics is

$$V(\boldsymbol{x}_1, \ldots, \boldsymbol{x}_N) = \sum_{1 \le i < j \le N} \frac{e_i e_j / 4\pi\varepsilon_0}{|\boldsymbol{x}_i - \boldsymbol{x}_j|} - \sum_{1 \le i < j \le N} \frac{G m_i m_j}{|\boldsymbol{x}_i - \boldsymbol{x}_j|}, \tag{1.5}$$

where

$$|\boldsymbol{x}| = \sqrt{x^2 + y^2 + z^2} \text{ for } \boldsymbol{x} = (x, y, z) \tag{1.6}$$

denotes the Euclidean norm in $\mathbb{R}^3$, e_i are constants called the *electric charges* of the particles (which can be positive, negative, or zero), the first term is called the *Coulomb potential*, the second term is called the *Newtonian gravity potential*, ε_0 and G are constants of nature called the electric constant and Newton's constant of gravity ($\varepsilon_0 = 8.85 \times 10^{-12}\,\text{kg}^{-1}\,\text{m}^{-3}\,\text{s}^4\,\text{A}^2$ and $G = 6.67 \times 10^{-11}\,\text{kg}^{-1}\,\text{m}^3\,\text{s}^{-2}$), and m_i are again the masses. However, when the Schrödinger equation is regarded as an *effective equation* rather than as a fundamental law of nature, then the potential V may contain terms arising from particles outside the system interacting with particles belonging to the system. That is why the Schrödinger equation is often considered for rather arbitrary functions V, also time-dependent ones. The operator

$$H = -\sum_{i=1}^{N} \frac{\hbar^2}{2m_i}\nabla_i^2 + V \tag{1.7}$$

is called the *Hamiltonian operator*, so the Schrödinger equation can be summarized in the form

$$i\hbar\frac{\partial\psi}{\partial t} = H\psi \,. \tag{1.8}$$

The Schrödinger equation is a *partial differential equation* (PDE). It determines the time evolution of ψ_t in that for a given initial wave function $\psi_0 = \psi(t=0) : \mathbb{R}^{3N} \to \mathbb{C}$, it uniquely fixes ψ_t for any $t \in \mathbb{R}$. The initial time could also be taken to be any $t_0 \in \mathbb{R}$ instead of 0.

Exercise 1.1 (Plane Waves) Show that for every constant vector $\boldsymbol{k} \in \mathbb{R}^3$ (the *wave vector*), there is a unique constant $\omega \in \mathbb{R}$ so that

$$\psi(t, \boldsymbol{x}) = e^{i\boldsymbol{k}\cdot\boldsymbol{x}}e^{-i\omega t} \tag{1.9}$$

satisfies the free (i.e., $V = 0$) Schrödinger equation (1.1) for $N = 1$. Specify ω in terms of $\boldsymbol{k}$. Remark: Since for every t, $\|\psi_t\| = \infty$, this function (called a plane wave) is not square-integrable (not normalizable) and thus not physically possible as a wave function; but it is a useful toy example.

So far, I have not said anything about what this new physical object ψ has to do with the particles. One such connection is

Born's Rule *If we measure the system's configuration at time t, then the outcome is random with probability density*

$$\rho(q) = \left|\psi_t(q)\right|^2. \tag{1.10}$$

This rule refers to the concept of *probability density*, which means the following. The probability that the random outcome $X \in \mathbb{R}^{3N}$ is any particular point $x \in \mathbb{R}^{3N}$ is zero, in symbols $\mathbb{P}(X = x) = 0$. However, the probability that X lies in a set $B \subseteq \mathbb{R}^{3N}$ is given by

$$\mathbb{P}(X \in B) = \int_B \rho(q)\, d^{3N}q \tag{1.11}$$

(a $3N$-dimensional volume integral). Instead of $d^{3N}q$, we will often just write dq. A density function ρ must be non-negative and normalized,

$$\rho(x) \geq 0 \,, \quad \int_{\mathbb{R}^{3N}} \rho(q)\, dq = 1 \,. \tag{1.12}$$

A famous density function in one dimension is the *Gaussian density*

$$\rho(x) = \frac{1}{\sqrt{2\pi}\sigma} e^{-\frac{(x-\mu)^2}{2\sigma^2}} \,. \tag{1.13}$$

A random variable with Gaussian density is also called a *normal* (or *normally distributed*) random variable. It has *mean* $\mu \in \mathbb{R}$ and *standard deviation* $\sigma > 0$. The *mean value* or *expectation value* $\mathbb{E}X$ of a random variable X is its average value

$$\mathbb{E}X = \int_{\mathbb{R}} x\, \rho(x)\, dx \,. \tag{1.14}$$

The *standard deviation* of X is defined to be $\sqrt{\mathbb{E}[(X - \mathbb{E}X)^2]}$.

For the Born rule to make sense, we need that

$$\int_{\mathbb{R}^{3N}} |\psi_t(q)|^2\, dq = 1. \tag{1.15}$$

And indeed, the Schrödinger equation guarantees this relation: If it holds for $t = 0$, then it holds for any $t \in \mathbb{R}$. More generally, the Schrödinger equation implies that

$$\int dq\, |\psi_t|^2 = \int dq\, |\psi_0|^2 \tag{1.16}$$

for any ψ_0. One says that $\int dq\, |\psi_t|^2$ satisfies a *conservation law*. Indeed, the Schrödinger equation implies a *local conservation law* for $|\psi|^2$, as we will show below; this means not only that the total amount of $|\psi|^2$ is conserved but also that amounts of $|\psi|^2$ cannot disappear in one place, while the same amount appears in another place; that is, the amount of $|\psi|^2$ cannot be created or destroyed, only moved around, and in fact flows with a *current* j.

In general, a local conservation law in $\mathbb{R}^d$ gets expressed by a *continuity equation*[1]

$$\frac{\partial \rho}{\partial t} = -\sum_{\alpha=1}^{d} \frac{\partial j_\alpha}{\partial x_\alpha} \,, \tag{1.17}$$

[1] I do not know where this name comes from. It has nothing to do with being continuous. It should be called conservation equation.

where ρ is a time-dependent scalar function on $\mathbb{R}^d$ called the density and j a time-dependent vector field on $\mathbb{R}^d$ called the current. To understand why (1.17) expresses local conservation of ρ, recall the Ostrogradski–Gauss integral theorem (divergence theorem), which asserts that for a vector field $\boldsymbol{F}$ in $\mathbb{R}^n$,

$$\int_A \operatorname{div} \boldsymbol{F}(\boldsymbol{x})\, d^n\boldsymbol{x} = \int_{\partial A} \boldsymbol{F}(\boldsymbol{x}) \cdot \boldsymbol{n}(\boldsymbol{x})\, d^{n-1}\boldsymbol{x}\,, \tag{1.18}$$

where

$$\operatorname{div} \boldsymbol{F} = \partial F_1/\partial x_1 + \ldots + \partial F_n/\partial x_n \tag{1.19}$$

is called the divergence of $\boldsymbol{F}$, A is an n-dimensional region with (piecewise smooth) $n-1$-dimensional boundary ∂A, the left-hand side is a volume integral in n dimensions with volume element $d^n\boldsymbol{x}$, the right-hand side is a surface integral (flux integral) of the vector field $\boldsymbol{F}$, $\boldsymbol{n}(\boldsymbol{x})$ is the outward unit normal vector on ∂A at $\boldsymbol{x} \in \partial A$, and $d^{n-1}\boldsymbol{x}$ means the area of a surface element. The formula (1.18) implies in particular that if the vector field $\boldsymbol{F}$ has zero divergence, then its flux integral across any closed surface ∂A vanishes. Now apply this to $n = d+1$ and the vector field $\boldsymbol{F} = (\rho, j_1, \ldots, j_d)$, which has zero divergence (in $d+1$ dimensions!) according to (1.17), and consider its flux across the surface of the $d+1$-dimensional cylinder $A = [0, T] \times S$, where $S \subseteq \mathbb{R}^d$ is a ball or any set with a piecewise smooth boundary ∂S; see Fig. 1.1.

Then the surface integral of $\boldsymbol{F}$ is

$$0 = -\int_S \rho_0 + \int_S \rho_T + \int_0^T dt \int_{\partial S} d^d x\, j \cdot n_{\partial S} \tag{1.20}$$

with $n_{\partial S}$ the unit normal vector field in $\mathbb{R}^d$ on the boundary of S. That is, the amount of ρ in S at time T differs from the initial amount of ρ in S by the flux of j across the boundary of S during $[0, T]$—a local conservation law. If (and this is indeed the case with the Schrödinger equation) there is no flux to infinity, i.e., if the last

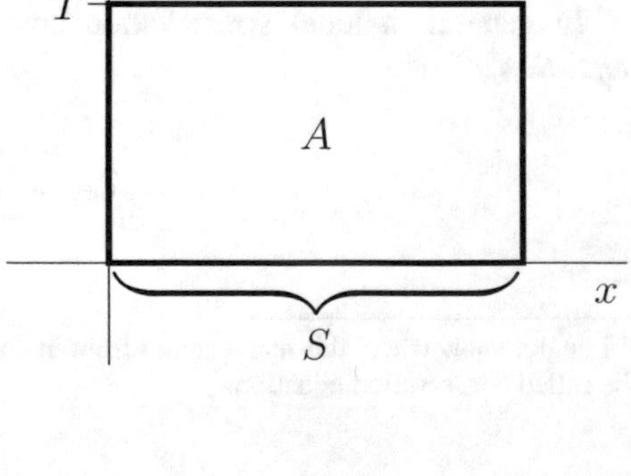

Fig. 1.1 The set A considered in the calculation, here shown for space dimension $d = 1$

integral becomes arbitrarily small by taking S to be a sufficiently big ball, then the total amount of ρ remains constant in time.

Now the Schrödinger equation implies the following continuity equation in configuration space with $d = 3N$:

$$\frac{\partial|\psi(t,q)|^2}{\partial t} = -\sum_{i=1}^{N} \nabla_i \cdot \boldsymbol{j}_i(t,q) \tag{1.21}$$

with

$$\boldsymbol{j}_i(t,q) = \frac{\hbar}{m_i}\mathrm{Im}\Big(\psi^*(t,q)\nabla_i\psi(t,q)\Big), \tag{1.22}$$

where Im means imaginary part, because

$$\frac{\partial}{\partial t}\Big(\psi^*\,\psi\Big) = 2\mathrm{Re}\Big(\psi^*\,\tfrac{-i}{\hbar}H\psi\Big) \tag{1.23}$$

$$= \tfrac{2}{\hbar}\mathrm{Im}\Big(-\sum_{i=1}^{N}\tfrac{\hbar^2}{2m_i}\psi^*\,\nabla_i^2\psi + \underbrace{V(q)|\psi|^2}_{\text{real}}\Big) \tag{1.24}$$

and

$$-\sum_{i=1}^{N}\nabla_i\cdot\boldsymbol{j}_i = -\sum_{i=1}^{N}\tfrac{\hbar}{m_i}\mathrm{Im}\Big(\psi^*\,\nabla_i^2\psi + \underbrace{(\nabla_i\psi^*)\cdot(\nabla_i\psi)}_{\text{real}}\Big), \tag{1.25}$$

so the two are equal. Thus, $|\psi|^2$ is locally conserved, and in particular its integral over all of configuration space does not change with time, as expressed in (1.16).

Since the quantity $\int dq\,|\psi|^2$ occurs frequently, it is useful to abbreviate it: The L^2 *norm* is defined to be

$$\|\psi\| = \Big(\int_{\mathbb{R}^{3N}} dq\,|\psi(q)|^2\Big)^{1/2}. \tag{1.26}$$

Thus, $\|\psi_t\| = \|\psi_0\|$, and the Born rule is consistent with the Schrödinger equation, provided the initial datum ψ_0 has norm 1, which we will henceforth assume. The wave function ψ_t will in particular be square-integrable, and this makes the space $L^2(\mathbb{R}^{3N})$ of square-integrable functions a natural arena. It is also called the *Hilbert space* and is the space of all wave functions (times finite factors).

1.3 Unitary Operators in Hilbert Space

In the following, we will often simply write L^2 for $L^2(\mathbb{R}^{3N})$. We will leave out many mathematical details.

1.3.1 Existence and Uniqueness of Solutions of the Schrödinger Equation

The Schrödinger equation defines the time evolution of the wave function ψ_t. In mathematical terms, this means that for every choice of initial wave function $\psi_0(q)$, there is a unique solution $\psi(t,q)$ of the Schrödinger equation. This leads to the question what exactly is meant by "every" wave function. Remarkably, even when ψ_0 is not differentiable, there is still a natural sense in which a "weak solution" or "L^2 solution" can be defined. This sense allows for a particularly simple statement:

Theorem 1.1 [2] *For a large class of potentials V (including Coulomb, Newton's gravity, every bounded measurable function, and linear combinations thereof) and for every $\psi_0 \in L^2$, there is a unique weak solution $\psi(t,q)$ of the Schrödinger equation with potential V and initial datum ψ_0. Moreover, at every time t, ψ_t lies again in L^2.*

1.3.2 The Time Evolution Operators

Let $U_t : L^2 \to L^2$ be the mapping defined by

$$U_t\psi_0 = \psi_t\,. \tag{1.27}$$

U_t is called the *time evolution operator* or *propagator*. Often, it is not possible to write down an explicit closed formula for U_t, but it is nevertheless useful to consider U_t. It has the following properties.

First, U_t is a *linear operator*, i.e.,

$$U_t(\psi + \phi) = (U_t\psi) + (U_t\phi) \tag{1.28}$$

$$U_t(z\psi) = z\,(U_t\psi) \tag{1.29}$$

[2] This follows from Stone's theorem and Kato's theorem together. See, e.g., Theorem VIII.8 in Reed and Simon (1980) [17] and Theorem X.16 in Reed and Simon (1975) [16].

for any $\psi, \phi \in L^2$, $z \in \mathbb{C}$. This follows from the fact that the Schrödinger equation is a *linear equation* or, equivalently, that H is a linear operator. It is common to say *operator* for linear operator.

Second, U_t preserves norms:

$$\|U_t\psi\| = \|\psi\|. \tag{1.30}$$

This is just another way of expressing Eq. (1.16). Operators with this property are called *isometric*.

Third, they obey a *composition law*:

$$U_s U_t = U_{t+s}, \quad U_0 = I, \tag{1.31}$$

for all $s, t \in \mathbb{R}$, where I denotes the *identity operator*

$$I\psi = \psi. \tag{1.32}$$

It follows from (1.31) that $U_t^{-1} = U_{-t}$. In particular, U_t is a bijection. An isometric bijection is also called a *unitary operator*; so U_t is unitary. A family of operators satisfying (1.31) is called a *one-parameter group* of operators. Thus, the propagators form a unitary one-parameter group. (The composition law (1.31) is owed to the time translation invariance of the Schrödinger equation, which depends on the time independence of the potential. If one inserted a time-dependent potential, then (1.31) and (1.32) would have to be replaced by $U_{t_2}^{t_3} U_{t_1}^{t_2} = U_{t_1}^{t_3}$ and $U_t^t = I$, where U_s^t maps ψ_s to ψ_t.)

Fourth,

$$U_t = e^{-iHt/\hbar}. \tag{1.33}$$

The exponential of an operator A can be defined by the *exponential series*

$$e^A = \sum_{n=0}^{\infty} \frac{A^n}{n!} \tag{1.34}$$

if A is a so-called bounded operator; in this case, the series converges (in the operator norm). Unfortunately, the Hamiltonian of the Schrödinger equation (1.1) is unbounded. But mathematicians agree about how to define e^A for unbounded operators (of the type that H is); we will not worry about the details of this definition.

Equation (1.33) is easy to understand: after defining

$$\phi_t := e^{-iHt/\hbar}\psi_0, \tag{1.35}$$

one would naively compute as follows:

$$i\hbar\frac{d}{dt}\phi_t = i\hbar\frac{d}{dt}e^{-iHt/\hbar}\psi_0 \tag{1.36}$$

$$= i\hbar\Big(-\frac{iH}{\hbar}\Big)e^{-iHt/\hbar}\psi_0 \tag{1.37}$$

$$= H\phi_t\,, \tag{1.38}$$

so ϕ_t is a solution of the Schrödinger equation with $\phi_0 = e^0\psi_0 = \psi_0$ and thus $\phi_t = \psi_t$. The calculation (1.36)–(1.38) can actually be justified for all ψ_0 in the domain $\mathscr{D}$ of H, a dense subspace[3] of L^2; we will not go into details here.

1.3.3 Unitary Matrices and Rotations

The space L^2 is infinite-dimensional. As a finite-dimensional analog, consider the functions on a finite set, $\psi : \{1, \ldots, n\} \to \mathbb{C}$, and the norm

$$\|\psi\| = \left(\sum_{i=1}^{n} \psi(i)\right)^{1/2} \tag{1.39}$$

instead of the L^2 norm

$$\|\psi\| = \left(\int |\psi(q)|^2\, dq\right)^{1/2}. \tag{1.40}$$

A function on $\{1, \ldots, n\}$ is always square-summable (its norm cannot be infinite). It can be written as an n-component vector

$$\big(\psi(1), \ldots, \psi(n)\big)\,, \tag{1.41}$$

and the space of these functions can be identified with $\mathbb{C}^n$.

The linear operators on $\mathbb{C}^n$ are given by the complex $n \times n$ matrices. If a matrix preserves the norm (1.39) as in (1.30), it is automatically bijective and thus unitary. A matrix U_{ij} is unitary if and only if

$$U^\dagger = U^{-1}\,, \tag{1.42}$$

[3] A subset $\mathscr{D}$ of L^2 is called *dense* if and only if for every $\psi \in L^2$ and every $\varepsilon > 0$, there is a $\phi \in \mathscr{D}$ such that $\|\phi - \psi\| < \varepsilon$.

where $U^\dagger$, the *adjoint matrix of* U, is defined by

$$U_{ij}^\dagger = (U_{ji})^* . \tag{1.43}$$

The norm (1.39) is analogous to the norm (= magnitude = length) of a vector in $\mathbb{R}^3$,

$$|\boldsymbol{u}| = \left(\sum_{i=1}^{3} u_i^2\right)^{1/2} . \tag{1.44}$$

The norm-preserving operators in $\mathbb{R}^3$ are exactly the *orthogonal matrices*, i.e., those matrices A with

$$A^t = A^{-1} , \tag{1.45}$$

where A^t denotes the transposed matrix, $A_{ij}^t = A_{ji}$. They have a geometric meaning: Each orthogonal matrix is either a rotation around some axis passing through the origin or a reflection across some plane through the origin, followed by a rotation. The set of orthogonal 3×3 matrices is denoted $O(3)$. The set of those orthogonal matrices which do not involve a reflection is denoted $SO(3)$ for "special orthogonal matrices"; they correspond to rotations and can be characterized by the condition $\det A > 0$ in addition to (1.45).

In dimension $d > 3$, one can show that the special orthogonal matrices are still compositions (i.e., products) of two-dimensional rotation matrices such as (for $d =$ 4)

$$\begin{pmatrix} \cos\alpha & \sin\alpha & & \\ -\sin\alpha & \cos\alpha & & \\ & & 1 & \\ & & & 1 \end{pmatrix} . \tag{1.46}$$

This rotation does not rotate around an axis; it rotates around a $(d-2)$-dimensional subspace (spanned by the third and fourth axes). However, in $d \geq 4$ dimensions, not every special orthogonal matrix is a rotation around a $(d-2)$-dim. subspace through a certain angle, but several such rotations can occur together, as the following example shows:

$$\begin{pmatrix} \cos\alpha & \sin\alpha & & \\ -\sin\alpha & \cos\alpha & & \\ & & \cos\beta & \sin\beta \\ & & -\sin\beta & \cos\beta \end{pmatrix} . \tag{1.47}$$

We will simply call every special orthogonal $d \times d$ matrix a "rotation."

Since $\mathbb{C}^n$ can be regarded as $\mathbb{R}^{2n}$, and the norm (1.39) then coincides with the $2n$-dimensional version of (1.44), every unitary operator then corresponds to an orthogonal operator, in fact a special orthogonal one. So if you can image $2n$-dimensional space, every unitary operator is geometrically a rotation. Also in L^2, it is appropriate to think of a unitary operator as a rotation.

1.3.4 Inner Product

In analogy to the dot product in $\mathbb{R}^3$,

$$\boldsymbol{u} \cdot \boldsymbol{v} = \sum_{i=1}^{3} u_i v_i \tag{1.48}$$

one defines the *inner product* of two functions $\psi, \phi \in L^2$ to be

$$\langle \psi | \phi \rangle = \int_{\mathbb{R}^{3N}} \psi(q)^* \, \phi(q) \, dq \, . \tag{1.49}$$

It has the following properties:

1. It is *anti-linear* (or *semi-linear* or *conjugate-linear*) in the first argument,

$$\langle \psi + \phi | \chi \rangle = \langle \psi | \chi \rangle + \langle \phi | \chi \rangle \, , \quad \langle z\psi | \phi \rangle = z^* \langle \psi | \phi \rangle \tag{1.50}$$

   for all $\psi, \phi, \chi \in L^2$ and $z \in \mathbb{C}$.
2. It is linear in the second argument,

$$\langle \psi | \phi + \chi \rangle = \langle \psi | \phi \rangle + \langle \psi | \chi \rangle \, , \quad \langle \psi | z\phi \rangle = z \langle \psi | \phi \rangle \tag{1.51}$$

   for all $\psi, \phi, \chi \in L^2$ and $z \in \mathbb{C}$. Properties 1 and 2 together are called *sesquilinear* (from Latin *sesqui* = $1\frac{1}{2}$).
3. It is *conjugate-symmetric* (or *Hermitian*),

$$\langle \phi | \psi \rangle = \langle \psi | \phi \rangle^* \tag{1.52}$$

   for all $\psi, \phi \in L^2$.
4. It is *positive definite*,[4]

$$\langle \psi | \psi \rangle > 0 \text{ for } \psi \neq 0 \, . \tag{1.53}$$

[4] Another math subtlety: This will be true only if we identify two functions ψ, ϕ whenever the set $\{q \in \mathbb{R}^{3N} : \psi(q) \neq \phi(q)\}$ has volume 0. It is part of the standard definition of L^2 to make these identifications.

Note that the dot product in $\mathbb{R}^3$ has the same properties, the *properties of an inner product*, except that the scalars involved lie in $\mathbb{R}$, not $\mathbb{C}$. Another inner product with these properties is defined on $\mathbb{C}^n$ by

$$\langle\psi|\phi\rangle = \sum_{i=1}^{n} \psi(i)^* \phi(i) . \tag{1.54}$$

The norm can be expressed in terms of the inner product according to

$$\|\psi\| = \sqrt{\langle\psi|\psi\rangle} . \tag{1.55}$$

Note that the radicand is ≥ 0.

Exercise 1.2 Show that, conversely, the inner product can be expressed in terms of the norm according to the *polarization identity*

$$\langle\psi|\phi\rangle = \tfrac{1}{4}\Big(\|\psi+\phi\|^2 - \|\psi-\phi\|^2 - i\|\psi+i\phi\|^2 + i\|\psi-i\phi\|^2\Big). \tag{1.56}$$

(Hint: Start from the right-hand side, express each norm in terms of the inner product, and watch out for cancelations.)

It follows from the polarization identity that every unitary operator U preserves inner products,

$$\langle U\psi|U\phi\rangle = \langle\psi|\phi\rangle . \tag{1.57}$$

(Likewise, every $A \in SO(3)$ preserves dot products, which has the geometrical meaning that a rotation preserves the angle between any two vectors.)

In analogy to the dot product, two functions ψ, ϕ with $\langle\psi|\phi\rangle = 0$ are said to be *orthogonal*.

1.3.5 Abstract Hilbert Space

The general and abstract definition of a *vector space* (over $\mathbb{R}$ or over $\mathbb{C}$) is that it is a set S (whose elements are called vectors) together with a prescription for how to add elements of S and a prescription for how to multiply an element of S by a scalar, such that the usual algebraic rules of addition and scalar multiplication are satisfied. Similarly, a *Hilbert space* is a vector space over $\mathbb{C}$ together with an inner product satisfying the *completeness* property: every Cauchy sequence converges. One can then prove the

Theorem 1.2 *$L^2(\mathbb{R}^d)$ is a Hilbert space.*

1.4 Classical Mechanics

Classical physics means pre-quantum (pre-1900) physics. I describe one particular version that could be called *Newtonian mechanics* (even though certain features were not discovered until after Isaac Newton's death). This version is over-simplified in that it leaves out magnetism, electromagnetic fields (which play a role for electromagnetic waves and thus the classical theory of light), and relativity theory.

1.4.1 Definition of Newtonian Mechanics

According to Newtonian mechanics, the world consists of a space, which is a three-dimensional Euclidean space, and particles moving around in space with time. Here, a *particle* means a material point—a point-shaped physical object. Let us suppose there are N particles in the world (say, $N \approx 10^{80}$), and let us fix a Cartesian coordinate system in Euclidean space. At every time t, particle number i ($i = 1, \ldots, N$) has a position $\boldsymbol{Q}_i(t) \in \mathbb{R}^3$. These positions are governed by the *equation of motion*

$$m_i \frac{d^2 \boldsymbol{Q}_i}{dt^2} = -\nabla_i V(\boldsymbol{Q}_1, \ldots, \boldsymbol{Q}_N) \tag{1.58}$$

with V the fundamental potential function of the universe as given in Eq. (1.5). This completes the definition of Newtonian mechanics.

The equation of motion (1.58) is an *ordinary differential equation* (ODE) of second order (i.e., involving second time derivatives). Once we specify, as initial conditions, the initial positions $\boldsymbol{Q}_i(0)$ and velocities $(d\boldsymbol{Q}_i/dt)(0)$ of every particle, the equation of motion (1.58) determines $\boldsymbol{Q}_i(t)$ for every i and every t.

Written explicitly, (1.58) reads

$$m_i \frac{d^2 \boldsymbol{Q}_i}{dt^2} = -\sum_{j \neq i} \frac{e_i e_j}{4\pi\varepsilon_0} \frac{\boldsymbol{Q}_j - \boldsymbol{Q}_i}{|\boldsymbol{Q}_j - \boldsymbol{Q}_i|^3} + \sum_{j \neq i} G m_i m_j \frac{\boldsymbol{Q}_j - \boldsymbol{Q}_i}{|\boldsymbol{Q}_j - \boldsymbol{Q}_i|^3}. \tag{1.59}$$

The right-hand side is called the *force* acting on particle i; the j-th term in the first sum (with the minus sign in front) is called the Coulomb force exerted by particle j on particle i; the j-th term in the second sum is called the gravitational force exerted by particle j on particle i.

Newtonian mechanics is empirically wrong. For example, it entails the absence of interference fringes in the double-slit experiment (and entails wrong predictions about everything that is considered a quantum effect). Nevertheless, it is a coherent theory, a "theory of everything," and often useful to consider as a hypothetical world to compare ours to.

Newtonian mechanics is to be understood in the following way: Physical objects such as tables, baseballs, or dogs consist of huge numbers (such as 10^{24}) of particles, and they must be regarded as just such an agglomerate of particles. Since Newtonian mechanics governs unambiguously the behavior of each particle, it also completely dictates the behavior of tables, baseballs, and dogs. Put differently, after (1.58) has been given, there is no need to specify any further laws for tables, baseballs, or dogs. Any physical law concerning tables, baseballs, or dogs is a *consequence* of (1.58). This scheme is called *reductionism.* It makes chemistry and biology subfields of physics. (This does not mean, though, that it would be of practical use to try to solve (1.58) for 10^{24} or 10^{80} particles in order to study the behavior of dogs.) Can everything be reduced to (1.58)? It seems that conscious experiences are an exception—presumably the only one.

When we consider a baseball, we are often particularly interested in the motion of its center $\boldsymbol{Q}(t)$ because we are interested in the motion of the whole ball. It is often possible to give an *effective equation* for the behavior of a variable like $\boldsymbol{Q}(t)$, for example,

$$M\frac{d^2\boldsymbol{Q}}{dt^2} = -\gamma\frac{d\boldsymbol{Q}}{dt} - Mg\begin{pmatrix}0\\0\\1\end{pmatrix}, \tag{1.60}$$

where M is the mass of the baseball, the first term on the right-hand side is called the *friction force* and the second the *gravitational force of Earth*, γ is the *friction coefficient* of the baseball, and g is the *gravitational field strength of Earth.* The effective equation (1.60) looks quite similar to the fundamental equation (1.58), but (i) it has a different status (it is not a fundamental law), (ii) it is only *approximately* valid, (iii) it contains a term that is not of the form $-\nabla V$ (the friction term), and (iv) forces that do obey the form $-\nabla V(\boldsymbol{Q})$ (such as the second force) can have other functions for V (such as $V(\boldsymbol{x}) = Mgx_3$) instead of (1.5).

The theory I call Newtonian mechanics was never actually proposed to give the correct and complete laws of physics (although we can imagine a hypothetical world where it does); for example, it leaves out magnetism. An extension of this theory, which we will not consider further but which is also considered "classical physics," includes electromagnetic fields (governed by Maxwell's field equations (7.25)) and gravitational fields (governed by Einstein's field equations, also known as the general theory of relativity).

The greatest contributions from a single person to the development of Eq. (1.58) came from Isaac Newton (1643–1727), who suggested (in his *Philosophiae Naturalis Principia Mathematica* 1687) considering ODEs, in fact of second order, suggested "forces" and the form $m\frac{d^2\boldsymbol{Q}}{dt^2} =$ force, and introduced the form of the gravitational force, now known as "Newton's law of universal gravity." Equation (1.59) was first written down, without the Coulomb term, by Leonhard Euler (1707–1783). The first term was proposed in 1784 by Charles Augustin de Coulomb (1736–1806). Nevertheless, we will call (1.58) and (1.59) "Newton's equation of motion."

1.4.2 Properties of Newtonian Mechanics

Exercise 1.3 (Time Reversal) Show that if $t \mapsto q(t) = (\boldsymbol{Q}_1(t), \ldots, \boldsymbol{Q}_N(t))$ is a solution of Newton's equation of motion (1.58), then so is $t \mapsto q(-t)$, which is called the *time reverse*.

This property is called *time reversal invariance* or *reversibility*. It is a rather surprising property, in view of the irreversibility of many phenomena. But since it has been explained in statistical mechanics, particularly by Ludwig Boltzmann (1844–1906), how reversibility of the microscopic laws and irreversibility of macroscopic phenomena can be compatible (for discussion, see, e.g., Lebowitz (2008) [14]), time reversal invariance has been widely accepted. This was also because time reversal invariance also holds in other, more refined theories after Newtonian mechanics, such as Maxwell's equations of classical electromagnetism (which we will write down in (7.25)), general relativity, and the Schrödinger equation.

Definition 1.1 Let $\boldsymbol{v}_i(t) = d\boldsymbol{Q}_i/dt$ denote the velocity of particle i. The *energy*, the *momentum*, and the *angular momentum* of the universe are defined to be, respectively,

$$E = \sum_{k=1}^{N} \frac{m_k}{2} \boldsymbol{v}_k^2 - \sum_{\substack{j,k=1 \\ j<k}}^{N} \Big(G m_j m_k - \frac{e_j e_k}{4\pi\varepsilon_0} \Big) \frac{1}{|\boldsymbol{Q}_j - \boldsymbol{Q}_k|} \tag{1.61}$$

$$\boldsymbol{p} = \sum_{k=1}^{N} m_k \boldsymbol{v}_k \tag{1.62}$$

$$\boldsymbol{L} = \sum_{k=1}^{N} m_k \boldsymbol{Q}_k \times \boldsymbol{v}_k \,, \tag{1.63}$$

where $\boldsymbol{v}^2 = \boldsymbol{v} \cdot \boldsymbol{v} = |\boldsymbol{v}|^2$ and $\times$ denotes the cross product in $\mathbb{R}^3$. The first term in (1.61) is called *kinetic energy* and the second one *potential energy*.

Exercise 1.4 (Conservation Laws) Show that E, $\boldsymbol{p}$, and $\boldsymbol{L}$ are conserved quantities, i.e., they are time independent, if the motion is governed by (1.59). Show further that, more generally, for the Newtonian motion (1.58) with arbitrary potential V that is invariant under rotations and translations,

$$V\Big(R\boldsymbol{Q}_1 + \boldsymbol{a}, \ldots, R\boldsymbol{Q}_N + \boldsymbol{a}\Big) = V(\boldsymbol{Q}_1, \ldots, \boldsymbol{Q}_N) \tag{1.64}$$

for all $\boldsymbol{a} \in \mathbb{R}^3$ and $R \in SO(3)$, E, $\boldsymbol{p}$, and $\boldsymbol{L}$ are conserved if we define

$$E = \sum_{k=1}^{N} \frac{m_k}{2} \boldsymbol{v}_k^2 + V(\boldsymbol{Q}_1, \ldots, \boldsymbol{Q}_N). \tag{1.65}$$

1.4.3 Hamiltonian Systems

A *dynamical system* is another name for an ODE. A dynamical system in $\mathbb{R}^n$ can be characterized by specifying the function $F : \Omega \to \mathbb{R}^n$ in

$$\frac{dX}{dt} = F(X, t), \tag{1.66}$$

with $\Omega \subseteq \mathbb{R}^n \times \mathbb{R}$. F can be called a time-dependent vector field on (a possibly time-dependent domain in) $\mathbb{R}^n$. (One often considers a more general concept of ODE, in which F is a time-dependent vector field on a differentiable manifold M.)

Newtonian mechanics has a time evolution that belongs to the class of dynamical systems, with $n = 6N$, $X = (\boldsymbol{Q}_1, \ldots, \boldsymbol{Q}_N, \boldsymbol{v}_1, \ldots, \boldsymbol{v}_N)$ and Ω (the *phase space*) $= \mathbb{R}^{6N}$ or rather $\Omega = \{(\boldsymbol{Q}_1, \ldots, \boldsymbol{Q}_N, \boldsymbol{v}_1, \ldots, \boldsymbol{v}_N) \in \mathbb{R}^{6N} : \boldsymbol{Q}_i \neq \boldsymbol{Q}_j \forall i \neq j\}$. The *phase point* $X(t) = (\boldsymbol{Q}_1(t), \ldots, \boldsymbol{Q}_N(t), \boldsymbol{v}_1(t), \ldots, \boldsymbol{v}_N(t))$ is determined by the equation of motion and the initial datum $X(0)$. The mapping T_t that maps any $X(0) \in \Omega$ to $X(t)$,

$$T_t(X(0)) = X(t), \tag{1.67}$$

is called the *flow map*. It satisfies a composition law analogous to that of the unitary time evolution operators for the Schrödinger equation (1.31),

$$T_s T_t = T_{s+t} \text{ and } T_0 = \mathrm{id}_\Omega \tag{1.68}$$

for all $s, t \in \mathbb{R}$, where id_Ω means the identity mapping on Ω, $\mathrm{id}_\Omega(x) = x$. In general, T_t is not a linear mapping but still a bijection.

Newtonian mechanics also belongs to a narrower class, called *Hamiltonian systems*. Simply put, these are dynamical systems for which the vector field F is a certain type of derivative of a scalar function H called the *Hamiltonian function* or simply the *Hamiltonian*. Namely, n is assumed to be even, $n = 2r$, and denoting the n components of x by $(q_1, \ldots, q_r, p_1, \ldots, p_r)$, the ODE is of the form

$$\frac{dq_i}{dt} = \frac{\partial H}{\partial p_i} \tag{1.69}$$

$$\frac{dp_i}{dt} = -\frac{\partial H}{\partial q_i}. \tag{1.70}$$

Newtonian mechanics fits this definition with $r = 3N$, $q_1, \dots, q_r$ the $3N$ components of $q = (\boldsymbol{q}_1, \dots, \boldsymbol{q}_N)$, $p_1, \dots, p_r$ the $3N$ components of $p = (\boldsymbol{p}_1, \dots, \boldsymbol{p}_N)$ (the momenta $\boldsymbol{p}_k = m_k \boldsymbol{v}_k$), and $H = H(q, p)$ the energy (1.61) expressed as a function of q and p, that is,

$$H(q, p) = \sum_{k=1}^{N} \frac{\boldsymbol{p}_k^2}{2m_k} - \sum_{\substack{j,k=1 \\ j \neq k}}^{N} \Big(G m_j m_k - \frac{e_j e_k}{4\pi\varepsilon_0} \Big) \frac{1}{|\boldsymbol{q}_j - \boldsymbol{q}_k|} . \tag{1.71}$$

For readers familiar with manifolds, I mention that the natural definition of a Hamiltonian system on a manifold M is as follows. M plays the role of phase space. Let the dimension n of M be even, $n = 2r$, and suppose we are given a symplectic form ω on M, i.e., a non-degenerate differential 2-form whose exterior derivative vanishes, $d\omega = 0$. (Non-degenerate means that it has full rank n at every point.) The equation of motion for $t \mapsto x(t) \in M$ reads

$$\omega\Big(\frac{dx}{dt}, \cdot\Big) = dH , \tag{1.72}$$

where dH means the exterior derivative of H. To make the connection with the case $M = \mathbb{R}^n$ just described, dH is then the gradient of H and ω the $n \times n$ matrix

$$\omega = \begin{pmatrix} 0 & I \\ -I & 0 \end{pmatrix} \tag{1.73}$$

with I the $r \times r$ unit matrix and 0 the $r \times r$ zero matrix; $\omega(dx/dt, \cdot)$ becomes the transpose of ω applied to the n-vector dx/dt, and (1.72) reduces to (1.69) and (1.70).

1.5 The Double-Slit Experiment

The double-slit experiment is a demonstration of fundamental features of quantum mechanics, in particular of the wave nature and the particle nature of electrons and other elementary particles. The experiment has been carried out in many variants with electrons, neutrons, photons, entire atoms, and even molecules. It involves *interference*, i.e., the constructive or destructive cooperation (i.e., addition) of waves. The word "diffraction" means more or less the same as interference.

In this experiment, an "electron gun" shoots electrons at a plate with two slits. Electrons get reflected when they hit the plate, but they can pass through the slits. Behind the plate with the slits, at a suitable distance, there is a screen capable of detecting electrons; see Fig. 1.2. Every electron leaves a (basically point-shaped) spot on the screen that we can see. The exact location of the next spot is unpredictable, but a probability distribution governing the spots is visible from a large number of spots. Before I describe the outcome of the experiment, let me

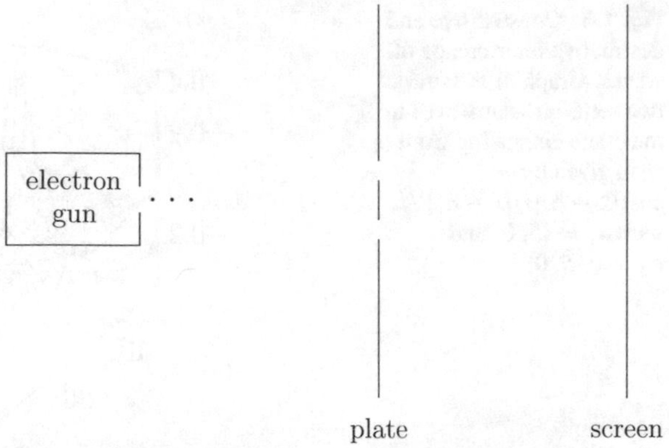

Fig. 1.2 The setup of a double-slit experiment

discuss, for the purpose of contrast, the expected outcome on the basis of Newtonian mechanics and on the basis of classical wave theory.

1.5.1 Classical Predictions for Particles and Waves

In Newtonian mechanics, a particle moves uniformly (i.e., along a straight line with constant speed) if no forces act on it. The particles hitting the plate will be scattered back and leave the setup, but let us focus on the particles that make it through the slits. Since in this experiment, gravity can usually be neglected and other forces do not matter much as the particles do not get very close to each other or to particles belonging to the plate, the particles passing the slits move along straight lines. Not all particles will arrive at the same spot, as they will be shot off at different positions and in different directions. If the particle source (the electron gun) is small and far away, then the possible spots of arrival will form two stripes corresponding to the two slits—the complement of the shadow of the plate. A particle passing through the upper (lower) slit arrives in the upper (lower) stripe. If the source is larger and less far away, then the two stripes will be blurred and may overlap.

Waves behave very differently. We may think of water waves in a basin and, playing the role of the plate, a wall with two gaps that will let waves pass into an ulterior basin; at the end of the ulterior basin, playing the role of the screen, we may measure the intensity of the arriving waves (say, their energy) as a function on the location along the rim of the basin. The first difference from Newtonian particles is that the energy does not arrive in a single spot, in a chunk carried by each particle, but arrives continuously in time and continuously distributed along the rim. A second difference shows up when the width of the slits is small enough so it becomes comparable to the wave length. Then the wave emanating from each slit will spread out in different directions; for an ideally small slit, the outcoming wave will be a semi-circular wave such as $\cos(k\sqrt{x^2 + y^2})/(x^2 + y^2)^{1/4}$ (with wave number $k = 2\pi/\text{wave length}$ and direction of propagation orthogonal to the wave fronts, while the amplitude decreases like $r^{-1/2}$ because the energy of the

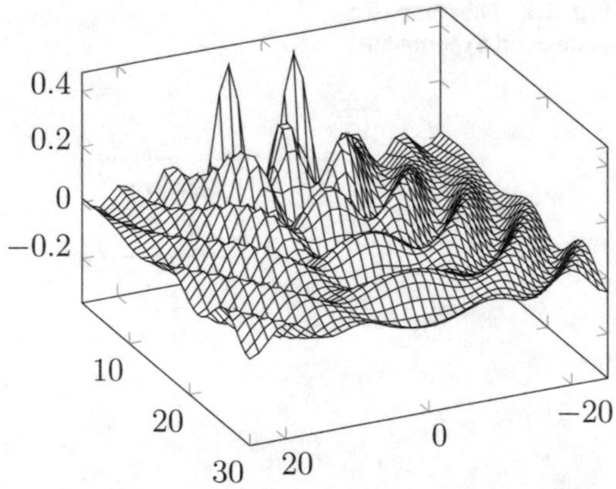

Fig. 1.3 Constructive and destructive interference of waves: Graph of the sum of two semi-circular waves in the plane emanating from $\boldsymbol{a}_i$, each given by $\cos(|\boldsymbol{x} - \boldsymbol{a}_i|)/|\boldsymbol{x} - \boldsymbol{a}_i|^{1/2}$, with $\boldsymbol{a}_1 = (5, 0)$ and $\boldsymbol{a}_2 = (-5, 0)$

wave (for water waves, seismic waves, waves in elastic materials, electromagnetic waves, and the gravitational waves of the general theory of relativity) is proportional to the square of the amplitude and the energy gets distributed over semi-circles of circumference πr). Thus, at each slit, the incoming wave propagates in the direction from the source, while the outgoing wave propagates in many directions (hence the word "diffraction," which is Latin for "breaking up," as the path of propagation suddenly changes direction).

The third difference is that, when waves emanate from both slits whose distance is not much larger than the wave length, they will cancel each other in some places (destructive interference) and add in others (constructive interference), as shown in Fig. 1.3. As a consequence, the energy arriving on the rim will vary from place to place; as a function $E(x)$ of the coordinate x along the rim, it will show ups and downs (local maxima and minima) known as an "interference pattern." In double-slit experiments done with light, these are visible as alternating bright and dark stripes known as "interference fringes."

1.5.2 Actual Outcome of the Experiment

In the quantum mechanical experiment, the energy arrives, as for Newtonian particles, in discrete localized chunks, each corresponding to one electron (or neutron, etc.) and each visible as a spot on the screen. The probability distribution $\rho(x)$ of the spots, however, features an interference pattern, indicative of the presence of waves. One speaks of *wave–particle duality*, meaning that the experiment displays aspects of particle behavior and aspects of wave behavior. But how, you may wonder, can the electron be particle and wave at the same time?

We will discuss proposals for that in the following chapters; for now, let us go into more detail about the experiment. The experiment can be carried out in such a way that, at any given time, only *one* electron (or neutron etc.) is passing the setup between the source and the screen, so we can exclude interaction between many particles as the cause of the interference pattern. Figure 1.4 shows the spots on the screen in a double-slit experiment carried out by Tonomura et al. (1989) [21]. In this experiment, 70,000 electrons were detected individually after passing through a double slit.[5] Only one electron at a time went through the setup. About 1000 electrons per second went through, at nearly half the speed of light. Each electron needed about 10^{-8} s to travel from the double slit to the screen.

The sense of surprise (or even paradox) may be further enhanced. If we place the screen directly behind the plate, the spots occur in two stripes located where the slits are (like Newtonian particles with small source). From this, it seems natural to conclude that every particle passes through one of the slits. Now it would be of interest to see how the particles passing through slit 1 behave later on. So we put the screen back at the original distance from the plate, where it shows an interference pattern, but now we close one of the slits, say slit 2. As expected, the number of particles that arrive on the screen gets (approximately) halved. Perhaps less expectedly, the interference fringes disappear. Instead of several local maxima and minima, the distribution function $\rho_1(x)$ has just one maximum in (approximately) the center and tends to 0 monotonically on both sides. Let me explain what is strange about the disappearance of the fringes. If we close slit 1 instead and keep only slit 2 open, then (if the slits are equal in size and their centers are distance a apart) the distribution function should be (and indeed is) $\rho_2(x) = \rho_1(x - a)$. Arguing that every particle passes through one of the slits and that those passing through slit $i \in \{1, 2\}$ end up with distribution ρ_i, we may expect that the distribution with both slits open, $\rho = \rho_{12}$, is given by the sum of the ρ_i. But it is not,

$$\rho_{12}(x) \neq \rho_1(x) + \rho_2(x) . \tag{1.74}$$

While the right-hand side has no minima (except perhaps for a little valley of width a in the middle, which is usually invisible since a is usually smaller than 10^{-6} m), ρ_{12} features pronounced minima, some of which even have (at least ideally) $\rho_{12}(x) = 0$. Such x are places where particles passing through slit 1 would arrive if slit 1 alone were open, but where no particles arrive if both slits are open! How does a particle passing through slit 1 even "know" whether slit 2 is open or not?

Moreover, there are detectors that can register a particle, while it passes through and moves on. If we place such a detector in each slit, while both slits are open, we will know of each particle which slit it went through. In that case, the fringes disappear again, and the observed distribution is $\rho_1 + \rho_2$. In particular, the distribution on the screen depends on whether we put detectors in the slits. It seems as if our mere knowledge of which slit each particle went through had an effect on the locations of the spots on the screen!

---

5 More precisely, electrons could pass right or left of a positively charged wire of diameter 1 μm. Those passing on the right get deflected to the left, and vice versa. Thus, the arrangement leads to the superposition of waves travelling in slightly different directions—just what is needed for interference.

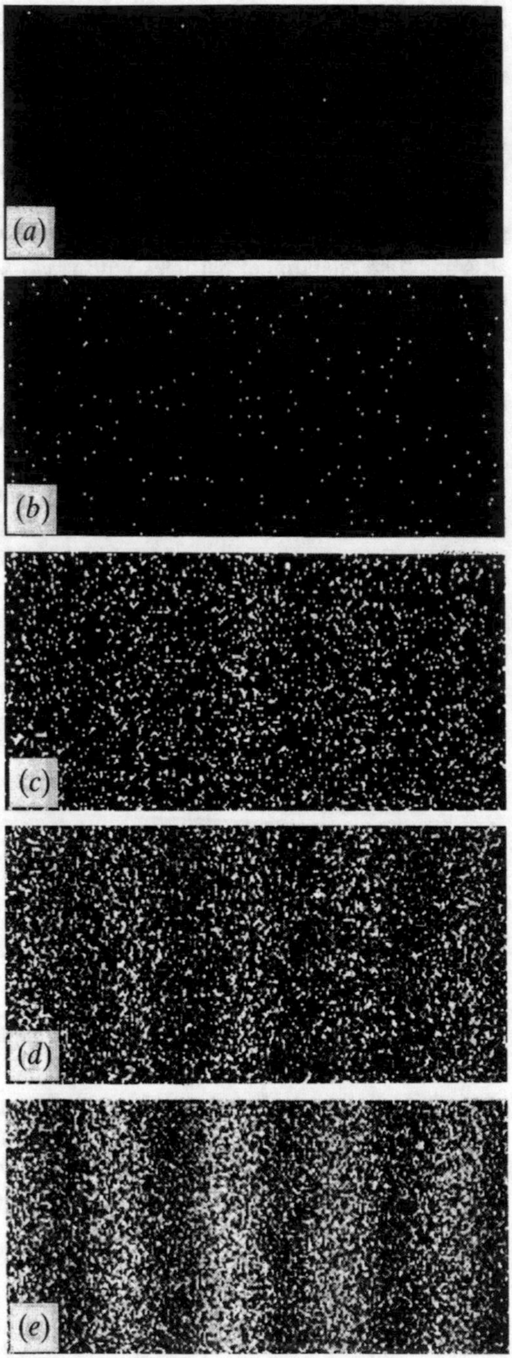

Fig. 1.4 A picture of actual results of a double-slit experiment, after (**a**) 10, (**b**) 100, (**c**) 3000, (**d**) 20,000, and (**e**) 70,000 electrons (Reprinted from [21] with permission by AIP Publishing)

The same phenomena arise when using more than two slits, except that the details of the interference pattern are different then. It is common to use dozens of slits or more (called a diffraction grating).

Note that the observations in the double-slit experiment are in agreement with, and in fact follow from, the Born rule and the Schrödinger equation: The relevant system here consists of one electron, so ψ_t is a function in just three dimensions. The potential V can be taken to be $+\infty$ (or very large) at every point of the plate, except in the slits themselves, where $V = 0$. Away from the plate, also $V = 0$. The Schrödinger equation governs the behavior of ψ_t, with the initial wave function ψ_0 being a *wave packet*, e.g., a Gaussian wave packet (as in Exercise 1.11),

$$\psi_0(\boldsymbol{x}) = (2\pi\sigma^2)^{-3/4} e^{-i\boldsymbol{k}\cdot\boldsymbol{x}} e^{-\frac{\boldsymbol{x}^2}{4\sigma^2}}, \tag{1.75}$$

moving toward the double slit. According to the Schrödinger equation, part of ψ will be reflected from the plate, and part of it will pass through the two slits. The two parts of the wave emanating from the two slits, ψ_1 and ψ_2, will overlap and thus interfere, $\psi = \psi_1 + \psi_2$.

When we detect the electron, its probability density is given, according to the Born rule, by

$$|\psi|^2 = |\psi_1 + \psi_2|^2 = |\psi_1|^2 + |\psi_2|^2 + 2\,\mathrm{Re}(\psi_1^* \psi_2)\,. \tag{1.76}$$

The third summand on the right is responsible for the minima of the interference pattern.

What if we include detectors in the slits? Then we detect the electron twice: once in a slit and once at the screen. Thus, either we have to regard it as a many-particle problem (involving the electron and the particles making up the detector), or we need a version of the Born rule suitable for repeated detection. We will study both approaches in later chapters.

1.5.3 Feynman's Discussion

Richard Feynman (1918–1988), in his widely known book ([12] Vol. 3, Ch. 1), provides a nice introduction to the double-slit experiment. I recommend that chapter as further reading and will add a few remarks about it:

- Feynman's statement on page 1

  > [The double slit experiment] has in it the heart of quantum mechanics. In reality, it contains the *only* mystery.

  is a bit too strong. Other mysteries can claim to be on equal footing with this one. Feynman weakened his statement later.

- Feynman's statements

> We cannot make the mystery go away by 'explaining' how it works. (page 1)
> Many ideas have been concocted to try to explain the curve for P_{12} [...] None of them has succeeded. (page 6)
> No one has found any machinery behind the law. No one can 'explain' any more than we have just 'explained.' No one will give you any deeper representation of the situation. We have no idea about a more basic mechanism from which these results can be deduced. (page 10)

  are too strong. We will see in Sects. 1.6 and 3.3 that Bohmian mechanics and other theories provide precisely such explanations of the double-slit experiment.
- Feynman's presentation conveys a sense of mystery and a sense of paradox about quantum mechanics. This will be a recurrent theme in this book, and one question will be whether there is any genuine, irreducible mystery or paradox in quantum mechanics.
- Feynman suggests that the mysterious character of quantum mechanics is not surprising ("perfectly reasonable") "because all of direct, human experience and of human intuition applies to large objects." This argument seems not quite on target to me. After all, the troublesome paradoxes of the double slit are not like the notions we often find hard to imagine (e.g., how big the number 6×10^{23} is, or what four-dimensional geometry looks like, or how big a light year is) but which are clearly sensible.

1.6 Bohmian Mechanics

> [Bohmian mechanics] exercises the mind in a very salutary way.
>
> John Bell (1984) [5, p. 171]

The situation in quantum mechanics is that we have a set of rules, known as the *quantum formalism*, for computing the possible outcomes and their probabilities for (more or less) any conceivable experiment, and everybody agrees (more or less) about the formalism. What the formalism does not tell us, and what is controversial, is what exactly happens during these experiments and how nature arrives at the outcomes whose probabilities the formalism predicts. There are different theories answering these questions, and Bohmian mechanics is one of them.

Let me elucidate my statements a bit. We have already encountered part of the quantum formalism: the Schrödinger equation and the Born rule. These rules have allowed us to predict the possible outcomes of the double-slit experiment with a single electron (easy here: a spot anywhere on the screen) and their probability distribution (here: a probability distribution corresponding to $|\psi|^2$ featuring a sequence of maxima and minima corresponding to interference fringes). What the rules did not tell us was what exactly happens during this experiment (e.g., how the electron moves). Bohmian mechanics fills this gap.

We have not seen all the rules of the quantum formalism yet. We will later, in Sects. 2.2 and 3.1. So far, we have formulated the Born rule only for position measurements, and we have not considered repeated detections.

1.6.1 Definition of Bohmian Mechanics

According to Bohmian mechanics (also known as pilot-wave theory or de Broglie–Bohm theory), the world consists of a space, which is a three-dimensional Euclidean space, and particles (material points) moving around in space with time. Let us suppose there are N particles in the world (say, $N \approx 10^{80}$), and let us fix a Cartesian coordinate system in Euclidean space. At every time t, particle number i ($i = 1, \ldots, N$) has a position $\boldsymbol{Q}_i(t) \in \mathbb{R}^3$. These positions are governed by *Bohm's equation of motion*

$$\frac{d\boldsymbol{Q}_i}{dt} = \frac{\hbar}{m_i} \mathrm{Im} \frac{\nabla_i \Psi}{\Psi}(t, Q(t)) \,. \tag{1.77}$$

Here, $Q(t) = (\boldsymbol{Q}_1(t), \ldots, \boldsymbol{Q}_N(t))$ is the configuration at time t, and Ψ is a wave function that is called the *wave function of the universe* and evolves according to the Schrödinger equation

$$i\hbar \frac{\partial \Psi}{\partial t} = -\sum_{i=1}^{N} \frac{\hbar^2}{2m_i} \nabla_i^2 \Psi + V\,\Psi \tag{1.78}$$

with V given by (1.5). The configuration $Q(0)$ at the initial time of the universe (say, right after the big bang) looks as if chosen randomly by nature with probability density

$$\rho_0(q) = |\Psi_0(q)|^2 \,. \tag{1.79}$$

One says that $Q(0)$ is *typical* relative to the $|\Psi_0|^2$ distribution; see Appendix A.4 for further discussion of typicality. (We write an upper case Q for the configuration of particles and a lower case q for the configuration variable in either ρ or Ψ.) This completes the definition of Bohmian mechanics.

1.6.2 Properties of Bohmian Mechanics

The central fact about Bohmian mechanics is that its predictions agree exactly with those of the quantum formalism (which so far have always been confirmed in experiment). We will understand later why this is so.

Equation (1.77) is an ordinary differential equation of first order (specifying the velocity rather than the acceleration). Thus, the initial configuration $Q(0)$

determines $Q(t)$ for all t, so Bohmian mechanics is a deterministic theory. On the other hand, $Q(t)$ is random because $Q(0)$ is. Note that this randomness does not conflict with determinism. It is a theorem, the *equivariance theorem*, that the probability distribution of $Q(t)$ is given by $|\Psi_t(q)|^2$. We will derive the equivariance theorem in Sect. 1.6.4. As a consequence, it is consistent to assume the Born distribution for every t. Note that due to the determinism, the Born distribution can be *assumed* only for one time (say $t = 0$); for any other time t, then, the distribution of $Q(t)$ is fixed by (1.77). The state of the universe at any time t is given by the pair $(Q(t), \Psi_t)$. In particular, in Bohmian mechanics, "wave–particle duality" means a very simple thing: there is a wave, and there are particles.

Let us have a closer look at Bohm's equation of motion (1.77). If we recall the formula (1.22) for the probability current, then we can rewrite Eq. (1.77) in the form

$$\frac{d\boldsymbol{Q}_i}{dt} = \frac{\boldsymbol{j}_i}{|\Psi|^2} = \frac{\text{probability current}}{\text{probability density}} . \tag{1.80}$$

This is a very plausible relation because it is a mathematical fact about any particle system with deterministic velocities that

$$\text{probability current} = \text{velocity} \times \text{probability density} . \tag{1.81}$$

We will come back to this relation when we prove equivariance.

Here is another way of rewriting (1.77). A complex number z can be characterized by its modulus $R \geq 0$ and its phase $S \in \mathbb{R}$, $z = Re^{iS}$. It will be convenient in the following to replace S by $S/\hbar$ (but we will still call S the phase of z). Then a complex-valued function $\Psi(t, q)$ can be written in terms of the two real-valued functions $R(t, q)$ and $S(t, q)$ according to

$$\Psi(t, q) = R(t, q)\, e^{iS(t,q)/\hbar} . \tag{1.82}$$

Let us plug this into (1.77): Since

$$\nabla_i \Psi = \nabla_i (Re^{iS/\hbar}) \tag{1.83}$$

$$= (\nabla_i R)e^{iS/\hbar} + R\nabla_i e^{iS/\hbar} \tag{1.84}$$

$$= (\nabla_i R)e^{iS/\hbar} + R\frac{i\nabla_i S}{\hbar}e^{iS/\hbar} , \tag{1.85}$$

we have that

$$\frac{\hbar}{m_i}\mathrm{Im}\frac{\nabla_i \Psi}{\Psi} = \frac{\hbar}{m_i}\mathrm{Im}\Bigg(\underbrace{\frac{\nabla_i R}{R}}_{\text{real}} + i\frac{\nabla_i S}{\hbar}\Bigg) \tag{1.86}$$

$$= \frac{\hbar}{m_i}\frac{\nabla_i S}{\hbar} = \frac{1}{m_i}\nabla_i S . \tag{1.87}$$

Thus, (1.77) can be rewritten as

$$\frac{d\boldsymbol{Q}_i}{dt} = \frac{1}{m_i}\nabla_i S(t, Q(t)). \tag{1.88}$$

In words, the velocity is given (up to a constant factor involving the mass) by the gradient of the phase of the wave function.

Exercise 1.5 (Current in Terms of Modulus and Phase) Show that

$$\boldsymbol{j}_i = \frac{1}{m_i} R^2 \nabla_i S. \tag{1.89}$$

Exercise 1.6 (Bohmian Trajectories for Plane Waves) Let ψ_t be a plane wave solution (1.9) of the Schrödinger equation with wave vector $\boldsymbol{k}$. Show that for every constant vector $\boldsymbol{a} \in \mathbb{R}^3$,

$$\boldsymbol{Q}(t) = \boldsymbol{a} + \frac{\hbar \boldsymbol{k}}{m} t \tag{1.90}$$

is a Bohmian trajectory with initial position $\boldsymbol{Q}(0) = \boldsymbol{a}$.

A few years before the development of the Schrödinger equation, Louis de Broglie[6] (1892–1987) had suggested a quantitative rule of thumb for wave–particle duality: A particle with momentum $\boldsymbol{p} = m\boldsymbol{v}$ should "correspond" to a wave with wave vector $\boldsymbol{k}$ according to the *de Broglie relation*

$$\boldsymbol{p} = \hbar \boldsymbol{k}. \tag{1.91}$$

The wave vector is defined by the relation $\psi = e^{i\boldsymbol{k}\cdot\boldsymbol{x}}$ (so it is defined only for plane waves); it is orthogonal to the wave fronts (surfaces of constant phase), and its magnitude is $|\boldsymbol{k}| = 2\pi/(\text{wave length})$. Now, if the wave is not a plane wave, then we can still define a *local wave vector* $\boldsymbol{k}(\boldsymbol{x})$ that is orthogonal to the surface of constant phase and whose magnitude is 1/(rate of phase change). Some thought shows that $\boldsymbol{k}(\boldsymbol{x}) = \nabla S(\boldsymbol{x})/\hbar$. If we use this expression on the right-hand side of (1.91) and interpret $\boldsymbol{p}$ as mass times the velocity of the particle, we obtain exactly Eq. (1.88), that is, Bohm's equation of motion.

[6] Although de Broglie was French, his family was originally from Italy. His last name is awkward to pronounce in French, but in Italian it is natural and sounds like "Brawl-ye."

1.6.3 Historical Overview

The idea that the wave function might determine particle trajectories as a "guiding field" was perhaps first expressed by Albert Einstein (1879–1955) around 1923 and considered in detail by John C. Slater (1900–1976) in 1924 (see [19]). Bohmian mechanics was developed by Louis de Broglie in 1927 (see [9]) but then abandoned. It was rediscovered independently by Nathan Rosen (1909–1995, known for the Einstein–Rosen bridge in general relativity and the Einstein–Podolsky–Rosen argument) in 1945 [18] and David Bohm (1917–1992) in 1952 [7, 8]. Bohm was the first to realize that it actually makes the correct predictions and the first to take it seriously as a physical theory. Several physicists mistakenly believed that Bohmian mechanics makes wrong predictions, including de Broglie, Rosen, and Einstein. Curiously, Bohm's 1952 papers [7, 8] provide a strange presentation of the theory, as Bohm insisted on writing the law of motion as an equation for the acceleration $d^2\boldsymbol{Q}_j/dt^2$, obtained by taking the time derivative of (1.77); see Appendix A.4.4.

It is widespread to call any variables that are not functions of ψ "hidden variables"; in Bohmian mechanics, the configuration Q is a variable that is not a function of ψ, so it is often called a hidden variable although the particle positions are not hidden at all in Bohmian mechanics, as they can be measured any time to any desired accuracy.

1.6.4 Equivariance

Equivariance can be expressed by saying that an ensemble of Bohmian configurations, if $|\Psi_0|^2$ distributed at time 0 (or at any one time), is $|\Psi_t|^2$ distributed at every time t. (An *ensemble* means a probability distribution ρ, thought of as a large collection of systems with statistical distribution given by ρ.) The term "equivariance" comes from the fact that the two relevant quantities, ρ_t and $|\Psi_t|^2$, vary equally with t. (Here, ρ_t is the distribution arising from ρ_0 by transport along the Bohmian trajectories.) The equivariance theorem can be expressed by means of the following diagram:

$$\begin{array}{ccc} \Psi_0 & \longrightarrow & \rho_0 \\ {\scriptstyle U_t}\downarrow & & \downarrow \\ \Psi_t & \longrightarrow & \rho_t \end{array} \tag{1.92}$$

The horizontal arrows mean taking $|\cdot|^2$, the left vertical arrow means the Schrödinger evolution from time 0 to time t, and the right vertical arrow means the transport of probability along the Bohmian trajectories. The statement about this diagram is that both paths along the arrows lead to the same result. Here is a rigorous version:

Theorem 1.3 (*Equivariance Theorem*, Berndl et al. (1995) [6], Teufel et al. (2005) [20]) *For a large class of potentials V and a dense subspace of ψ_0, the* *solution $t \mapsto Q(t)$ of Bohm's equation of motion with $Q(0) \sim |\Psi_0|^2$ exists globally (i.e., for all t) with probability 1, and equivariance holds:*

$$\mathbb{P}(Q(t) \in B) = \int_B dq\, |\Psi_t(q)|^2 \quad \forall B \subseteq \mathbb{R}^{3N}\ \forall t \in \mathbb{R}. \tag{1.93}$$

We give a non-rigorous derivation that conveys the reason why equivariance holds. As a preparation, we note that the equation of motion can be written in the form

$$\frac{dQ}{dt} = v_t(Q(t)), \tag{1.94}$$

where $v_t : \mathbb{R}^{3N} \to \mathbb{R}^{3N}$ is the vector field on configuration space $v_t = v = (\boldsymbol{v}_1, \dots, \boldsymbol{v}_N)$ whose i-th component is

$$\boldsymbol{v}_i = \frac{\hbar}{m_i}\mathrm{Im}\frac{\nabla_i \Psi}{\Psi}. \tag{1.95}$$

We now address the following question: If v_t is known for all t, and the initial probability distribution ρ_0 is known, how can we compute the probability distribution ρ_t at other times? The answer is the continuity equation

$$\frac{\partial \rho_t}{\partial t} = -\mathrm{div}\Big(\rho_t v_t\Big). \tag{1.96}$$

This follows from the fact that the *probability current* is given by $\rho_t v_t$. In fact, in any dimension d ($d = 3N$ or otherwise) and for any density (probability density or energy density or nitrogen density, etc.), it is true that

$$\text{current} = \text{density} \times \text{velocity}. \tag{1.97}$$

We are now ready to give the derivation. We first show that

$$\text{if } \rho_t = |\Psi_t|^2 \text{ then } \frac{\partial \rho_t}{\partial t} = \frac{\partial}{\partial t}|\Psi_t|^2 \tag{1.98}$$

and then conclude that if $\rho_0 = |\Psi_0|^2$, then $\rho_t = |\Psi_t|^2$ (which is the equivariance theorem). By the continuity equation (1.96) for ρ_t and the continuity equation (1.21) for $|\Psi_t|^2$, the right equation in (1.98) is equivalent to

$$-\sum_i \nabla_i \cdot \Big(\rho_t \boldsymbol{v}_i\Big) = -\sum_i \nabla_i \cdot \boldsymbol{j}_i. \tag{1.99}$$

As mentioned in (1.80), $\boldsymbol{v}_i = \boldsymbol{j}_i/|\Psi_t|^2$. Thus, if $\rho_t = |\Psi_t|^2$, then Eq. (1.99) is true, which completes the derivation.

1.6.5 The Double-Slit Experiment in Bohmian Mechanics

Let us apply what we know about Bohmian mechanics to $N = 1$ and the wave function of the double-slit experiment. We assume that the particle in the experiment moves as if it was alone in the universe, with the potential V representing the wall with two slits. We will justify that assumption in a later chapter. We know already what the wave function $\psi(t, \boldsymbol{x})$ looks like. Figure 1.5 shows possible trajectories of the particle.

We know from the equivariance theorem that the position will always have probability distribution $|\psi_t|^2$. Thus, if we detect the particle at time t, we find its distribution in agreement with the Born rule.

Note that the particle moves not along straight lines, as it would according to classical mechanics. That is because Bohm's equation of motion is different from Newton's. Note also that the wave passes through both slits, while the particle passes through one only. Note further that the particle trajectories would be different if one slit were closed: then no interference fringes would occur. How can the particle, after passing through one slit, know whether the other slit is open? Because the wave passes through both slits if both are open.

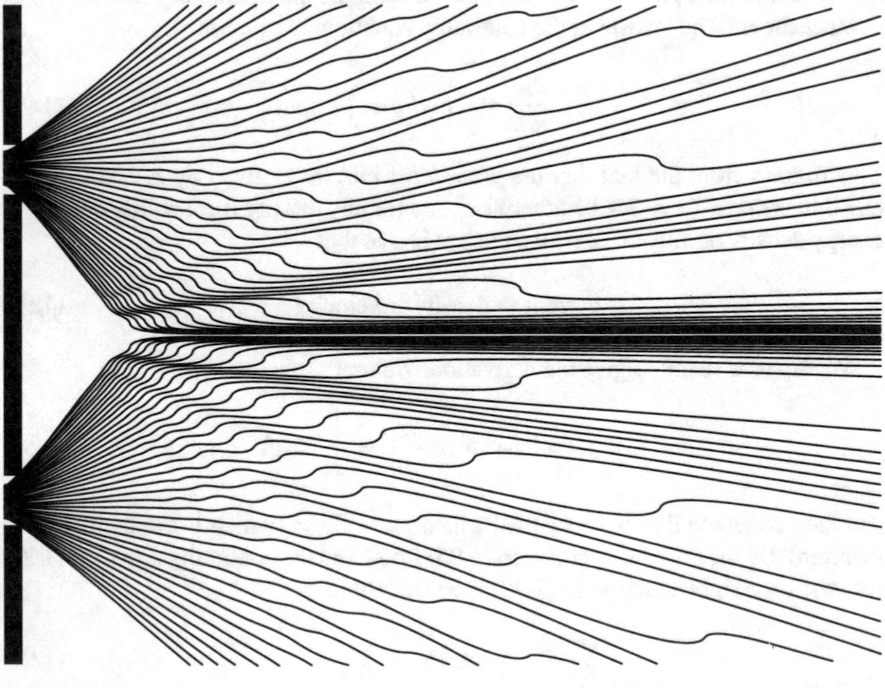

Fig. 1.5 Several alternative Bohmian trajectories of a particle in a double-slit experiment (Reprinted from Dürr and Teufel (2009) [11] with permission by Springer, based on a figure in Philippidis et al. (1979) [15])

> Is it not clear from the smallness of the scintillation on the screen that we have to do with a particle? And is it not clear, from the diffraction and interference patterns, that the motion of the particle is directed by a wave? De Broglie showed in detail how the motion of a particle, passing through just one of two holes in screen, could be influenced by waves propagating through both holes. And so influenced that the particle does not go where the waves cancel out, but is attracted to where they cooperate. This idea seems to me so natural and simple, to resolve the wave–particle dilemma in such a clear and ordinary way, that it is a great mystery to me that it was so generally ignored.
>
> John Bell (1986 [4] or [5, p. 191])

Coming back to Feynman's description of the double-slit experiment, we see that his statement that its outcome "cannot be explained" is not quite accurate. It is true that it cannot be explained in Newtonian mechanics, but it can in Bohmian mechanics.

Note also that we can find out which slit the particle went through without disturbing the interference pattern: check whether the particle arrived in the upper or lower half of the detection screen. This method takes for granted that Bohm's equation of motion is correct; in a Bohmian world, it would yield correct retrodictions.

The fact that trajectories beginning in the upper half stay in the upper half, visible from Fig. 1.5, can be understood mathematically as follows. Since the initial wave function, as well as the arrangement of the two slits, is symmetric around the horizontal middle axis, the wave function stays symmetric while evolving, $\psi_t(x, y, z) = \psi_t(x, y, -z)$ (with z the vertical axis in Fig. 1.5), and so the Bohmian velocity field is mirror symmetric,

$$\begin{aligned} v_x(x, y, z, t) &= v_x(x, y, -z, t)\,, \\ v_y(x, y, z, t) &= v_y(x, y, -z, t)\,, \\ v_z(x, y, z, t) &= -v_z(x, y, -z, t)\,. \end{aligned} \tag{1.100}$$

As a consequence, on the symmetry plane $z = 0$, the velocity field is tangent to the plane, and as a consequence of that, any trajectory with one point on the $z = 0$ plane stays on that plane (toward the future and the past), so no trajectory can cross the $z = 0$ plane. (We are using here the uniqueness of the solution of a first-order ODE for a given initial point.)

Here is an alternative reasoning. Since the velocity component in the direction perpendicular to the plate is, we may assume, constant, we can think of the horizontal axis in Fig. 1.5 as the time axis and simplify the math by pretending we are dealing with one-dimensional (1d) motion (along the z-axis). Bohmian trajectories cannot cross each other (this follows from the uniqueness of the solution of a first-order ODE for a given initial point by taking the time of a hypothetical crossing as the initial time). In one dimension, this has the consequence that alternative trajectories stay in the same order along the axis. Since, by symmetry, $Q_z(t) = 0$ is a solution, the other trajectories cannot cross it. (For comparison,

Newtonian trajectories in 1d can cross because the equation of motion is of second order. The trajectories cannot cross in phase space.)

Another alternative reasoning is based on the observation (Exercise 1.13) that, by equivariance, in 1d the α-quantile of $|\psi_0|^2$ lies on the same trajectory as the α-quantile of $|\psi_t|^2$ (i.e., the trajectories are the quantile curves). By symmetry, for $\alpha > 0.5$, the α-quantile lies in the upper half axis $\{z > 0\}$ at every t.

1.6.6 Delayed-Choice Experiments

John Archibald Wheeler (1911–2008) proposed in 1978 [22] a variant of the double-slit experiment that may increase further the sense of paradox. Since Wheeler's variant, called the *delayed-choice experiment*, uses no more than the Schrödinger equation and Born's rule, and since we know that Bohmian mechanics can account for that, it is clear that the paradox must disappear in Bohmian mechanics. Let us have a look at what Wheeler's paradox is and how Bohmian mechanics resolves it.

Wheeler considered preparing, by means of a double slit or in some other way, two wave packets moving in different directions, so that they pass through each other. After passing through each other, they continue moving in different directions and thus get separated again. Wheeler gave the experimenter two choices: either put a screen in the overlap region or put it further away, where the two wave packets have clearly separated; see Fig. 1.6. If you put the screen in the overlap region, you will see an interference pattern, which is taken to indicate that the electron is a wave and went through both slits. However, if you put the screen further away, the detection occurs in one of two clusters. If the detection occurs in the upper (lower) cluster, this is taken to indicate that the particle went through the lower (upper) slit

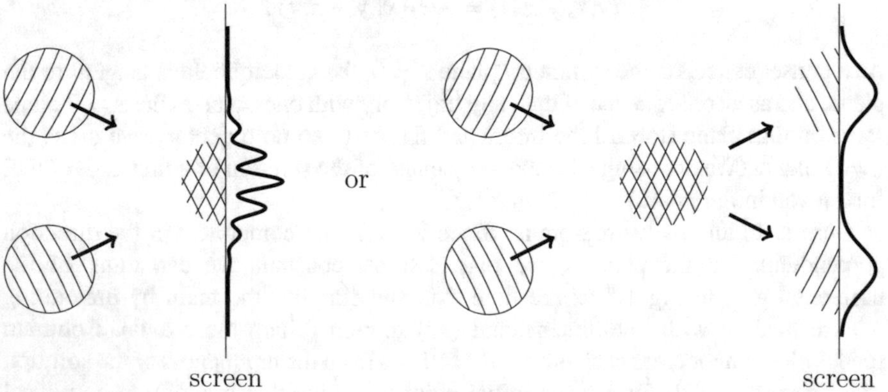

Fig. 1.6 The two options for the experimenter in the delayed-choice experiment. Left: put a screen in the interference region. Right: put a screen in the distant region where the wave packets have separated. The circles represent wave packets moving in the direction indicated by the arrows; the curve indicates the observed distribution density along the screen

because a wave packet passing through the lower (upper) slit will end up in the upper (lower) region on the screen. So, Wheeler argued, we can choose whether the electron is particle or wave: if we put the screen far away, it must be particle because we see which slit it went through; if we put the screen in the overlap, it must be wave because we see the interference pattern. Even more, we can force the electron to become wave or particle (and to go through both slits or just one) even *after* it passed through the double slit! So it seems like there must be *retrocausation*, i.e., situations in which the cause lies in the future of the effect.

Bohmian mechanics illustrates that these conclusions do not actually follow; this was first discussed by Bell (1980) [3]. To begin with, there is no retrocausation in Bohmian mechanics, as any intervention of observers will change ψ only in the future, not in the past, of the intervention and the particle trajectory will correspondingly be affected also only in the future. Another basic observation is that with the literal wave–particle dualism of Bohmian mechanics (there is a wave and there is a particle), there is nothing left of the idea that the electron is sometimes a wave and sometimes a particle and hence nothing of the idea that observers could force an electron to become a wave or to become a particle. In detail: the wave passes through both slits, the particle through one; in the overlap region, the two wave packets interfere, and the particle's $|\psi|^2$ distribution features an interference pattern; if there is no screen in the overlap region, then the particle moves on in such a way that the interference pattern disappears and two separate clusters form.

After we understand the Bohmian picture of this experiment, some steps in Wheeler's reasoning appear strange: If one assumes that there are no particle trajectories in the quantum world, as one usually does in orthodox quantum mechanics, then it would seem natural to say that there is no fact about which slit the electron went through, given that there was no attempt to detect the electron while passing a slit. Surprising it is, then, that Wheeler claims that the detection on the far-away screen reveals which slit it took! How can anything reveal which slit the electron took if the electron did not take a slit?

There is another interesting aspect to the story that I will call *Wheeler's fallacy*. When you analyze the Bohmian picture in the case of far-away screen, it turns out that the trajectories passing through the upper (lower) slit end up in the upper (lower) cluster; see Fig. 1.7. So Wheeler made the wrong retrodiction of which slit the electron passed through!

How could this happen? Wheeler noticed that if the lower (upper) slit is closed, so only one packet comes out, and it comes out of the upper (lower) slit, then only detection events in the lower (upper) region occur. This is also true in Bohmian mechanics. Wheeler concluded that when wave packets come out of both slits, and if a detection occurs in the upper cluster, then the particle must have passed through the lower slit. This is wrong in Bohmian mechanics, and once you realize this, it is obvious that Wheeler's conclusion is a *non sequitur*[7]—a fallacy.

[7] = it does not follow (Latin).

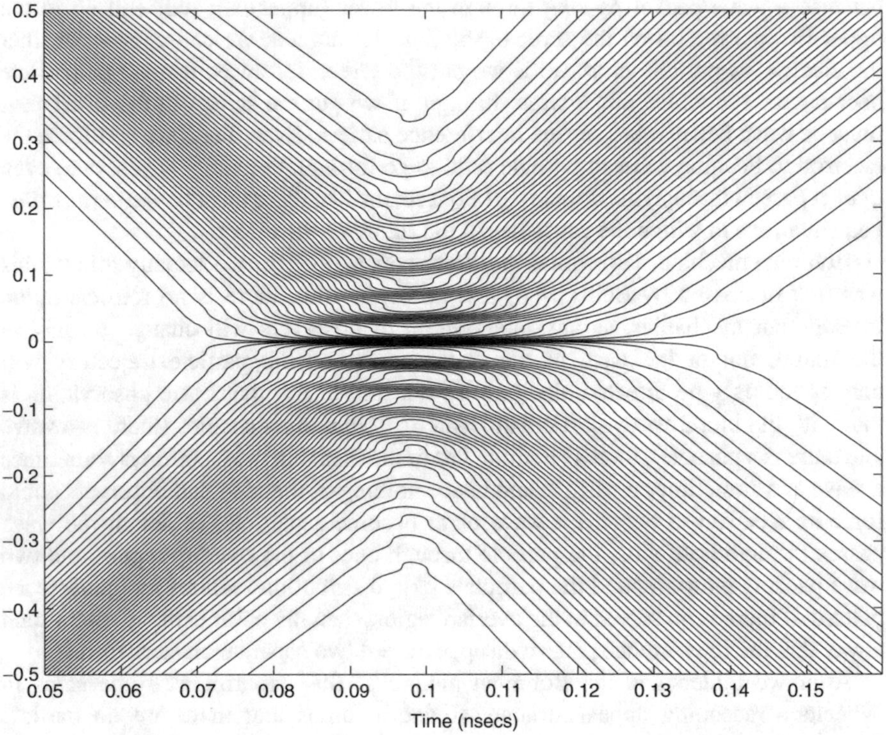

Fig. 1.7 Bohmian trajectories in the interference region of the delayed-choice experiment (Reprinted from Hiley and Callaghan (2006) [13] with permission by IOP Publishing, based on a figure in Dewdney (1985) [10])

Afshar's Experiment

Shahriar Afshar (2005) [1] proposed and carried out a further variant of the experiment, known as Afshar's experiment. In this variant, one puts the screen in the far position, but one adds obstacles (that would absorb or reflect electrons) in the overlap region, in fact in those places where the interference is destructive; see Fig. 1.8.

If an interference pattern occurs in the overlap region, even if it is not observed, then almost no particles arrive at the obstacles, and almost no particles get absorbed or reflected. Indeed, for the particular wave function we are considering, the presence of the obstacles does not significantly alter the time evolution according to the Schrödinger equation.[8] As a consequence, if all particles arrive on the far screen

[8] A way of seeing this without running a numerical simulation goes as follows. The obstacles could be represented as regions of infinite potential. In the Schrödinger equation, a region $B \subset \mathbb{R}^3$ of infinite potential is equivalent to a Dirichlet boundary condition on the boundary ∂B of B, i.e., the condition $\psi(x, t) = 0$ for all $x \in \partial B$ and all $t \in \mathbb{R}$. Imposing such a condition at places x

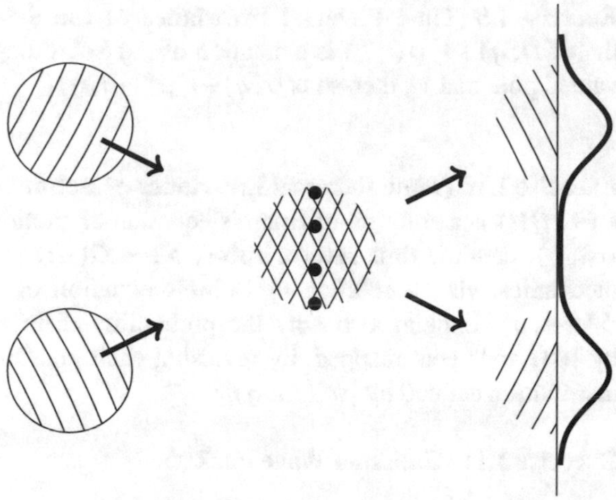

Fig. 1.8 Afshar's experiment adds obstacles in the interference region, drawn as black circles

(in either the upper or the lower region), as in fact observed in the experiment, then this indicates that no absorption or reflection occurred, so there was an interference pattern in the overlap region even though no screen was put there. Afshar argued that this experiment refutes Wheeler's view that one can *either* have an interference pattern *or* measure which slit the particle went through, but not both. (In his article, Afshar also said things based on the same fallacy as Wheeler's; but that does not make the experiment less relevant.) Again, Bohmian mechanics, having particle *and* wave, easily explains the outcome of this experiment.

Exercises

Exercise 1.1 can be found in Sect. 1.2, Exercise 1.2 in Sect. 1.3.4, Exercises 1.3 and 1.4 in Sect. 1.4.2, and Exercises 1.5 and 1.6 in Sect. 1.6.2.

Exercise 1.7 (Essay Questions) What is surprising about the double-slit experiment? How does Bohmian mechanics explain the double-slit experiment?

Exercise 1.8 (Essay Questions) Describe the delayed-choice double-slit experiment. Why does it seem paradoxical? How does the paradox get resolved in Bohmian mechanics?

---

where the solution ψ in the absence of obstacles would vanish for all t anyway does not affect the solution.

Exercise 1.9 (Time Reversal Invariance of the Schrödinger Equation) Show that if $(t, q) \mapsto \psi(t, q)$ is a solution of the Schrödinger equation (1.1) with real-valued potential V, then so is $(t, q) \mapsto \psi^*(-t, q)$.

Exercise 1.10 (Time Reversal Invariance of Bohmian Mechanics) Show that if $t \mapsto Q(t)$ is a solution of Bohm's equation of motion (1.77) with wave function $\psi(t, q)$, then the time-reverse history $t \mapsto Q(-t)$ is also a solution of Bohmian mechanics, viz., a solution of Bohm's equation of motion with wave function $\psi^*(-t, q)$. Explain also why the probability distribution over histories defined by $|\psi(t, q)|^2$ gets mapped, by reversing each possible history, to the probability distribution defined by $|\psi^*(-t, q)|^2$.

Exercise 1.11 (Gaussian Wave Packet)

(a) Show that the function

$$\psi(\boldsymbol{x}, t) = (2\pi\lambda_t^2\sigma^2)^{-3/4} e^{i\boldsymbol{k}\cdot(\boldsymbol{x}-\hbar\boldsymbol{k}t/2m)} e^{-\frac{(\boldsymbol{x}-\hbar\boldsymbol{k}t/m)^2}{4\lambda_t\sigma^2}} \tag{1.101}$$

with

$$\lambda_t = 1 + \frac{i\hbar t}{2m\sigma^2} \tag{1.102}$$

and arbitrary constants $\boldsymbol{k} \in \mathbb{R}^3$, $\sigma > 0$, is a solution of the free Schrödinger equation of a single particle in three dimensions,

$$i\hbar\frac{\partial\psi}{\partial t} = -\frac{\hbar^2}{2m}\nabla^2\psi\,. \tag{1.103}$$

(The main difficulty here is to organize the calculation so as to make it manageable.)

(b) Show that the probability density $\rho_t(\boldsymbol{x})$ is, at every t, Gaussian and specify its mean and standard deviation.

Exercise 1.12 (Bohmian Trajectories for Gaussian Wave Packets) Using Exercise 1.11, verify that for the Gaussian wave function (1.101), the velocity vector field (1.95) takes the form

$$\boldsymbol{v}(\boldsymbol{x}, t) = \frac{\hbar}{m}\boldsymbol{k} + \frac{\frac{\hbar^2}{4m^2\sigma^2}t}{\sigma^2 + \frac{\hbar^2}{4m^2\sigma^2}t^2}(\boldsymbol{x} - \hbar\boldsymbol{k}t/m) \tag{1.104}$$

and the Bohmian trajectories (shown in Fig. 1.9) are

$$\boldsymbol{Q}(t) = \frac{\hbar t}{m}\boldsymbol{k} + \frac{1}{\sigma}\sqrt{\sigma^2 + \frac{\hbar^2}{4m^2\sigma^2}t^2}\,\boldsymbol{Q}(0)\,. \tag{1.105}$$

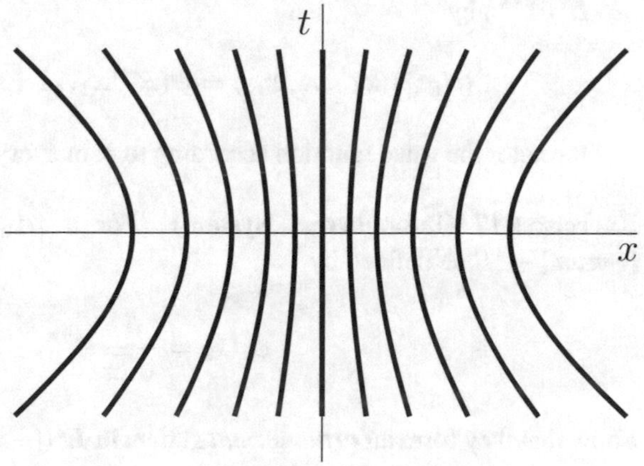

Fig. 1.9 Several alternative Bohmian trajectories in a Gaussian packet, corresponding to (1.105) with $\boldsymbol{k} = \boldsymbol{0}$

Exercise 1.13 (Quantile Rule) For a continuous probability distribution on $\mathbb{R}$ with density $\rho(x)$, the α*-quantile* for $0 < \alpha < 1$ is the point x_α where

$$\int_{-\infty}^{x_\alpha} \rho(x)\,dx = \alpha. \tag{1.106}$$

The $\frac{1}{2}$-quantile is also known as the median, the $\frac{1}{4}$-quantile as the first quartile, the $\frac{3}{4}$-quantile as the third quartile, and the $\frac{n}{100}$-quantile as the n-th percentile. Show that in Bohmian mechanics in one dimension, if Q_0 is the α-quantile of $\rho = |\psi_0|^2$, then Q_t is the α-quantile of $\rho = |\psi_t|^2$.

Exercise 1.14 (Gaussian Packets Done Differently) Use Exercises 1.13 and 1.11(b) to verify (1.105) for (1.101).

Exercise 1.15 (Computation of Trajectories) Write a computer program that generates Fig. 1.5. Instructions: To simplify the calculation, it is assumed that the x-velocity is constant, that the x-axis is thus really the time axis, and that the problem can be regarded as Bohmian motion in one dimension (the y-axis). It is assumed that the wave packet coming out of each slit is a Gaussian wave packet as in Exercise 1.11 with mean velocity $\boldsymbol{k} = \boldsymbol{0}$ and equal width σ. The trajectories are computed by computing the quantiles of $|\psi_t|^2$ at every t.

Exercise 1.16 (Unitary Operators)

(a) For any vector $\boldsymbol{a} \in \mathbb{R}^3$, the *translation operator* $T_{\boldsymbol{a}}$ is defined on $L^2(\mathbb{R}^{3N})$ by

$$(T_a\psi)(\boldsymbol{x}_1,\ldots,\boldsymbol{x}_N)=\psi(\boldsymbol{x}_1-\boldsymbol{a},\ldots,\boldsymbol{x}_N-\boldsymbol{a})\,. \tag{1.107}$$

It shifts the wave function by $\boldsymbol{a}$ in every $\boldsymbol{x}_i$. Show that T_a is unitary.

(b) For any rotation matrix $R\in SO(3)$, the *rotation operator* U_R is defined on $L^2(\mathbb{R}^{3N})$ by

$$(U_R\psi)(\boldsymbol{x}_1,\ldots,\boldsymbol{x}_N)=\psi(R^{-1}\boldsymbol{x}_1,\ldots,R^{-1}\boldsymbol{x}_N)\,. \tag{1.108}$$

It rotates the wave function according to R in every $\boldsymbol{x}_i$. Show that U_R is unitary.

Exercise 1.17 (Orthonormal System) For $n\in\mathbb{Z}$, let the function $\varphi_n:[-\pi,\pi]\to\mathbb{C}$ be defined by

$$\varphi_n(x)=\frac{1}{\sqrt{2\pi}}e^{inx}\,. \tag{1.109}$$

Show that they form an *orthonormal system* in $L^2([-\pi,\pi])$, i.e., that

$$\langle\varphi_n|\varphi_m\rangle=\delta_{nm}=\begin{cases}1 & \text{if } n=m\\ 0 & \text{if } n\neq m\end{cases} \tag{1.110}$$

with

$$\langle\psi|\phi\rangle=\int_{-\pi}^{\pi}\psi(x)^*\,\phi(x)\,dx\,. \tag{1.111}$$

Exercise 1.18 (Galilean Relativity) A Galilean change of space-time coordinates ("Galilean boost") is given by

$$\boldsymbol{x}'=\boldsymbol{x}+\boldsymbol{v}t\,,\quad t'=t \tag{1.112}$$

with a constant $\boldsymbol{v}\in\mathbb{R}^3$ called the relative velocity.

(a) Show that if V is translation invariant, then Newton's equation of motion (1.58) is invariant under Galilean boosts: If $t\mapsto(\boldsymbol{Q}_1,\ldots,\boldsymbol{Q}_N)$ is a solution, then so is $t\mapsto(\boldsymbol{Q}'_1,\ldots,\boldsymbol{Q}'_N)$.

(b) Show that if V is translation invariant and $\psi(t,\boldsymbol{x}_1,\ldots,\boldsymbol{x}_N)$ is a solution of the Schrödinger equation, then so is

$$\psi'(t',\boldsymbol{x}'_1,\ldots,\boldsymbol{x}'_N)=\exp\left[\tfrac{i}{\hbar}\sum_{i=1}^{N}m_i(\boldsymbol{x}'_i\cdot\boldsymbol{v}-\tfrac{1}{2}\boldsymbol{v}^2t')\right]\psi\Big(t',\boldsymbol{x}'_1-\boldsymbol{v}t',\ldots,\boldsymbol{x}'_N-\boldsymbol{v}t'\Big). \tag{1.113}$$

(c) Show that if V is translation invariant, then Bohmian mechanics is invariant under Galilean boosts, i.e., that if $t \mapsto (\boldsymbol{Q}_1, \ldots, \boldsymbol{Q}_N)$ is a solution of Bohm's equation of motion (1.77) for the wave function ψ, then $t \mapsto (\boldsymbol{Q}'_1, \ldots, \boldsymbol{Q}'_N)$ is a solution of (1.77) for ψ' given by (1.113). One says that Bohmian mechanics is Galilean invariant.

Exercise 1.19 (Entanglement) In a world with N particles, consider the subsystem formed by particles $1, \ldots, M$ (with $M < N$) and call it the x-system; let the y-system consist of all other particles. Let us write $x = (\boldsymbol{q}_1, \ldots, \boldsymbol{q}_M)$ and $y = (\boldsymbol{q}_{M+1}, \ldots, \boldsymbol{q}_N)$ for their respective configuration variables and $q = (x, y)$ for the full configuration variable. One says that a wave function ψ is *disentangled* if and only if it is a product of a function of x and a function of y,

$$\psi(x, y) = \psi_1(x)\,\psi_2(y)\,; \tag{1.114}$$

otherwise, ψ is called *entangled*. (Time plays no role in this consideration; all wave functions are considered at some fixed time.) For $N = 2$ particles, $M = 1$ in each system, give an example of an entangled and a disentangled wave function.

Exercise 1.20 (Interaction and Entanglement) Consider again the x-system and y-system of Exercise 1.19. If

$$V(x, y) = V_1(x) + V_2(y) \tag{1.115}$$

then one says that the two systems *do not interact*. In this exercise, we investigate the consequences of this condition.

(a) Show that in Newtonian mechanics, as given by (1.58) with the potential (1.115), the force acting on any particle belonging to the x-system is independent of the configuration of the y-system. Conclude further that the Newtonian trajectory $Q(t) = (X(t), Y(t))$ is such that $X(t)$ obeys the M-particle version of (1.58) with potential V_1 and $Y(t)$ the $(N - M)$-particle version with potential V_2.

(b) Show that if (1.115) holds and the wave function is disentangled initially,

$$\psi_0(x, y) = \psi_{1,0}(x)\,\psi_{2,0}(y) \tag{1.116}$$

then it is disentangled at all times t,

$$\psi_t(x, y) = \psi_{1,t}(x)\,\psi_{2,t}(y)\,, \tag{1.117}$$

where each factor $\psi_{i,t}$ evolves according to the (M-particle, respectively ($N - M$)-particle) Schrödinger equation with potential V_i.

(c) Show that, in the situation of (b), the Bohmian velocity of any particle belonging to the x-system is independent of the configuration of the y-system. Conclude further that the Bohmian trajectory $Q(t) = (X(t), Y(t))$ is such that $X(t)$ obeys the M-particle version of Bohm's equation of motion (1.77) with wave function ψ_1 and $Y(t)$ the $(N - M)$-particle version with wave function ψ_2. In particular, the X-trajectory does not depend on the Y-initial condition.

References

1. S.S. Afshar, Violation of the principle of complementarity and its implications. Proc. SPIE **5866**, 229–24 (2005). http://arxiv.org/abs/quant-ph/0701027
2. G. Bacciagaluppi, A. Valentini, *Quantum Theory at the Crossroads. Reconsidering the 1927 Solvay Conference* (Cambridge University Press, Cambridge, 2009). http://arxiv.org/abs/quant-ph/0609184
3. J.S. Bell, De Broglie-Bohm, delayed-choice double-slit experiment, and density matrix. Int. J. Quant. Chem. **14**, 155–159 (1980). Reprinted as chapter 14 of [5]
4. J.S. Bell, Beables for quantum field theory. Phys. Rep. **137**, 49–54 (1986). Reprinted as chapter 19 of [5]. Also reprinted in: *Quantum Implications: Essays in Honour of David Bohm*, ed. by F.D. Peat, B.J. Hiley (Routledge, London, 1987), p. 227
5. J.S. Bell, *Speakable and Unspeakable in Quantum Mechanics* (Cambridge University Press, Cambridge, 1987)
6. K. Berndl, D. Dürr, S. Goldstein, G. Peruzzi, N. Zanghì, On the global existence of Bohmian mechanics. Commun. Math. Phys. **173**, 647–673 (1995). http://arxiv.org/abs/quant-ph/9503013
7. D. Bohm, A suggested interpretation of the quantum theory in terms of "Hidden" variables I. Phys. Rev. **85**, 166–179 (1952)
8. D. Bohm, A suggested interpretation of the quantum theory in terms of "Hidden" variables II. Phys. Rev. **85**, 180–193 (1952)
9. L. de Broglie, La nouvelle dynamique des quanta, in *Electrons et Photons. Rapports et Discussions du Cinquième Conseil de Physique tenu à Bruxelles du 24 au 29 Octobre 1927 sous les Auspices de l'Institut International de Physique Solvay*, Solvay Congress 1927 (Gauthier-Villars, Paris, 1928). English translation: The new dynamics of quanta, in [2]
10. C. Dewdney, Particle trajectories and interference in a time-dependent model of neutron single crystal interferometry. Phys. Lett. A **109**, 377–384 (1985)
11. D. Dürr, S. Teufel, *Bohmian Mechanics* (Springer, Heidelberg, 2009)
12. R. Feynman, *Feynman Lectures on Physics* (Addison-Wesley, Reading, 1964)
13. B.J. Hiley, R.E. Callaghan, Delayed-choice experiments and the Bohm approach. Phys. Scripta **74**(3), 336 (2006). http://arxiv.org/abs/1602.06100
14. J.L. Lebowitz, From time-symmetric microscopic dynamics to time-asymmetric macroscopic behavior: an overview, in *Boltzmann's Legacy*, ed. by G. Gallavotti, W.L. Reiter, J. Yngvason, Zürich (European Mathematical Society, Zürich, 2008), pp. 63–88. http://arxiv.org/abs/0709.0724
15. C. Philippidis, C. Dewdney, B.J. Hiley, Quantum interference and the quantum potential. Il Nuovo Cimento Serie 11 **52B**(1), 15–28 (1979)
16. M. Reed, B. Simon, *Methods of Modern Mathematical Physics*, vol. 2 (Academic Press, Cambridge, 1975)
17. M. Reed, B. Simon, *Methods of Modern Mathematical Physics*, vol. 1 (revised edition) (Academic Press, Cambridge, 1980)
18. N. Rosen, On waves and particles. J. Elisha Mitchell Sci. Soc. **61**, 67–73 (1945)
19. J.C. Slater, *Solid-State and Molecular Theory. A Scientific Biography* (Wiley, New York, 1975)
20. S. Teufel, R. Tumulka, Simple proof for global existence of Bohmian Trajectories. Commun. Math. Phys. **258**, 349–365 (2005). http://arxiv.org/abs/math-ph/0406030

21. A. Tonomura, J. Endo, T. Matsuda, T. Kawasaki, H. Ezawa, Demonstration of single-electron buildup of an interference pattern. Am. J. Phys. **57**(2), 117–120 (1989). http://doi.org/10.1119/1.16104
22. J.A. Wheeler, The "past" and the "delayed-choice" double-slit experiment, in *Mathematical Foundations of Quantum Mechanics*, ed. by A.R. Marlow (Academic Press, Cambridge, 1978)

Some Observables 2

In this chapter, we discuss Fourier transformation and momentum measurements, the momentum operator, Heisenberg's uncertainty relation, self-adjoint operators in greater generality, the Born rule for self-adjoint operators, and spin.

2.1 Fourier Transform and Momentum

2.1.1 Fourier Transform

We know from Exercise 1.1 in Sect. 1.2 that the plane wave $e^{i\boldsymbol{k}\cdot\boldsymbol{x}}$ evolves according to the free Schrödinger equation to

$$e^{i\boldsymbol{k}\cdot\boldsymbol{x}}e^{-i\hbar k^2 t/2m}\,. \tag{2.1}$$

Since the Schrödinger equation is linear, any linear combination of plane waves with different wave vectors $\boldsymbol{k}$,

$$\sum c_{\boldsymbol{k}} e^{i\boldsymbol{k}\cdot\boldsymbol{x}} \tag{2.2}$$

with complex coefficients $c_{\boldsymbol{k}}$, will evolve to

$$\sum c_{\boldsymbol{k}} e^{i\boldsymbol{k}\cdot\boldsymbol{x}}e^{-i\hbar k^2 t/2m}\,. \tag{2.3}$$

Moreover, a "continuous linear combination"

$$\int_{\mathbb{R}^3} d^3\boldsymbol{k}\, c(\boldsymbol{k}) e^{i\boldsymbol{k}\cdot\boldsymbol{x}} \tag{2.4}$$

R. Tumulka, *Foundations of Quantum Mechanics*, Lecture Notes in Physics 1003,
https://doi.org/10.1007/978-3-031-09548-1_2

with arbitrary complex $c(\boldsymbol{k})$ will evolve to

$$\int_{\mathbb{R}^3} d^3\boldsymbol{k}\, c(\boldsymbol{k}) e^{i\boldsymbol{k}\cdot\boldsymbol{x}} e^{-i\hbar \boldsymbol{k}^2 t/2m}\,. \tag{2.5}$$

Definition 2.1 For a given function $\psi : \mathbb{R}^d \to \mathbb{C}$, the function

$$\widehat{\psi}(\boldsymbol{k}) = \frac{1}{(2\pi)^{d/2}} \int_{\mathbb{R}^d} \psi(\boldsymbol{x})\, e^{-i\boldsymbol{k}\cdot\boldsymbol{x}}\, d^d\boldsymbol{x} \tag{2.6}$$

is called the *Fourier transform* of ψ, $\widehat{\psi} = \mathscr{F}(\psi)$.

Theorem 2.1 *Inverse Fourier transformation:*

$$\psi(\boldsymbol{x}) = \frac{1}{(2\pi)^{d/2}} \int_{\mathbb{R}^d} \widehat{\psi}(\boldsymbol{k})\, e^{i\boldsymbol{k}\cdot\boldsymbol{x}}\, d^d\boldsymbol{k}\,. \tag{2.7}$$

Note the different sign in the exponent (it is crucial). If we had not put the prefactor in (2.6), we would have obtained the prefactor squared in (2.7).

We have been sloppy in the formulation of the definition and the theorem in that we have not specified the class of functions to which these formulas apply. In fact, (2.6) can be applied whenever $\psi \in L^1$ (the space of all integrable functions, i.e., those with $\|\psi\|_{L^1} = \int d\boldsymbol{x}\, |\psi| < \infty$) and then yields $\widehat{\psi} \in L^\infty$ (the space of all bounded functions) because $|\widehat{\psi}(\boldsymbol{k})| \leq (2\pi)^{-d/2} \|\psi\|_{L^1}$ by the triangle inequality, $|\int f| \leq \int |f|$. Conversely, if $\widehat{\psi} \in L^1$, then (2.7) holds, and $\psi \in L^\infty$. However, if $\psi \in L^1 \setminus L^\infty$, then $\widehat{\psi} \notin L^1$, and (2.7) is not literally applicable. For $\psi \in L^1 \cap L^\infty$, both (2.6) and (2.7) are rigorously true. Another space of interest in this context is the *Schwartz space $\mathscr{S}$ of rapidly decaying functions*, which contains the smooth functions $\psi : \mathbb{R}^d \to \mathbb{C}$ such that for every $n \in \mathbb{N}$ and every $\boldsymbol{\alpha} \in \mathbb{N}_0^d$, there is $C_{n,\boldsymbol{\alpha}} > 0$ such that $|\partial^{\boldsymbol{\alpha}} \psi(\boldsymbol{x})| < C_{n,\boldsymbol{\alpha}}\, |\boldsymbol{x}|^{-n}$ for all $\boldsymbol{x} \in \mathbb{R}^d$, where $\partial^{\boldsymbol{\alpha}} := \partial_1^{\alpha_1} \cdots \partial_d^{\alpha_d}$. For example, every Gaussian wave packet lies in $\mathscr{S}$; note that $\mathscr{S} \subset L^1 \cap L^\infty$. It turns out that Fourier transformation maps $\mathscr{S}$ bijectively to itself. Moreover, $\mathscr{S}$ is a dense subspace in L^2, and $\mathscr{F}$ can be extended in a unique way to a bounded operator $\mathscr{F} : L^2 \to L^2$, even though the integral (2.6) exists only for $\psi \in L^1 \cap L^2$.

Going back to Eq. (2.5) and taking $c(\boldsymbol{k}) = (2\pi)^{-3/2} \widehat{\psi_0}(\boldsymbol{k})$, we can express the solution of the free Schrödinger equation as

$$\psi_t(\boldsymbol{x}) = \frac{1}{(2\pi)^{3/2}} \int_{\mathbb{R}^3} d^3\boldsymbol{k} \left(e^{-i\hbar \boldsymbol{k}^2 t/2m}\, \widehat{\psi_0}(\boldsymbol{k}) \right) e^{i\boldsymbol{k}\cdot\boldsymbol{x}}\,. \tag{2.8}$$

In words, we can find ψ_t from ψ_0 by taking its Fourier transform $\widehat{\psi_0}$; multiplying by a suitable function of $\boldsymbol{k}$, viz., $e^{-i\hbar \boldsymbol{k}^2 t/2m}$; and taking the inverse Fourier transform.

The same trick can be done for N particles. Then $d = 3N$, $\psi = \psi(\boldsymbol{x}_1, \ldots, \boldsymbol{x}_N)$, $\widehat{\psi} = \widehat{\psi}(\boldsymbol{k}_1, \ldots, \boldsymbol{k}_N)$, and the factor to multiply by is

$$\exp\Big(-i \sum_{j=1}^{N} \frac{\hbar}{2m_j} \boldsymbol{k}_j^2 t\Big) \text{ instead of } \exp\Big(-i \frac{\hbar}{2m} \boldsymbol{k}^2 t\Big). \tag{2.9}$$

Note that we take the Fourier transform only in the *space* variables, not in the *time* variable. There are also applications in which it is useful to consider a Fourier transform in t, but not here.

Example 2.1 The Fourier transform of a Gauss function. Let $\sigma > 0$ and

$$\psi(\boldsymbol{x}) = C\, e^{-\frac{\boldsymbol{x}^2}{4\sigma^2}} \tag{2.10}$$

with C a constant. Then, using the substitution $\boldsymbol{y} = \boldsymbol{x}/(2\sigma)$,

$$\widehat{\psi}(\boldsymbol{k}) = \frac{C}{(2\pi)^{3/2}} \int_{\mathbb{R}^3} e^{-\boldsymbol{x}^2/4\sigma^2} e^{-i\boldsymbol{k}\cdot\boldsymbol{x}} d^3\boldsymbol{x} \tag{2.11}$$

$$= \underbrace{\frac{2^3 C \sigma^3}{(2\pi)^{3/2}}}_{=:C_2} \int_{\mathbb{R}^3} e^{-\boldsymbol{y}^2 - 2i\sigma\boldsymbol{k}\cdot\boldsymbol{y}} d^3\boldsymbol{y} \tag{2.12}$$

$$= C_2 \int_{\mathbb{R}^3} e^{-(\boldsymbol{y}+i\sigma\boldsymbol{k})^2 - \sigma^2\boldsymbol{k}^2} d^3\boldsymbol{y} \tag{2.13}$$

$$= C_2 e^{-\sigma^2\boldsymbol{k}^2} \int_{\mathbb{R}^3} e^{-(\boldsymbol{y}+i\sigma\boldsymbol{k})^2} d^3\boldsymbol{y} \tag{2.14}$$

The evaluation of the last integral involves the Cauchy integral theorem, varying the path of integration and estimating errors. Here, I just report that the outcome is the constant $\pi^{3/2}$, independent of σ and $\boldsymbol{k}$. Thus,[1]

$$\widehat{\psi}(\boldsymbol{k}) = C_3 e^{-\sigma^2\boldsymbol{k}^2} \tag{2.15}$$

with $C_3 = C_2\pi^{3/2}$. In words, the Fourier transform of a Gaussian function is another Gaussian function, but with width $1/(2\sigma)$ instead of σ. (We see here shadows of the Heisenberg uncertainty relation, which we will discuss in Sect. 2.2.1.)

For later use, I report further[2] that the formula (2.15) remains valid for complex σ with $\mathrm{Re}(\sigma^2) > 0$ (put differently, when we replace σ by $\sigma e^{i\theta}$ with $-\frac{\pi}{4} < \theta < \frac{\pi}{4}$).

---

[1] A different derivation of (2.15) is given on page 132 of Kammler (2007) [11].

[2] See pages 562 and 588 of Kammler (2007) [11].

That is,

$$\text{if } \psi(\boldsymbol{x}) = C \exp\Big(-e^{-2i\theta}\frac{\boldsymbol{x}^2}{4\sigma^2}\Big), \text{ then } \widehat{\psi}(\boldsymbol{k}) = C' \exp\Big(-e^{2i\theta}\sigma^2\boldsymbol{k}^2\Big) \tag{2.16}$$

with some constant $C' \in \mathbb{C}$.

Proposition 2.1

(a)

$$\widehat{\frac{\partial\psi}{\partial x_j}}(\boldsymbol{k}) = ik_j\,\widehat{\psi}(\boldsymbol{k})\,. \tag{2.17}$$

That is, differentiation of ψ corresponds to multiplication of $\widehat{\psi}$ by $i\boldsymbol{k}$.
(b) Conversely,

$$\widehat{-ix_j\psi} = \frac{\partial\widehat{\psi}}{\partial k_j}\,. \tag{2.18}$$

(c) If $f(\boldsymbol{x}) = e^{i\boldsymbol{k}_0\cdot\boldsymbol{x}}\,g(\boldsymbol{x})$, then $\hat{f}(\boldsymbol{k}) = \hat{g}(\boldsymbol{k}-\boldsymbol{k}_0)$.
(d) If $f(\boldsymbol{x}) = g(\boldsymbol{x}-\boldsymbol{x}_0)$, then $\hat{f}(\boldsymbol{k}) = e^{-i\boldsymbol{k}\cdot\boldsymbol{x}_0}\,\hat{g}(\boldsymbol{k})$.

Proof

(a) Indeed, using integration by parts (and assuming that the boundary terms vanish),

$$\widehat{\frac{\partial\psi}{\partial x_j}}(\boldsymbol{k}) = \frac{1}{(2\pi)^{d/2}}\int_{\mathbb{R}^d} d^d\boldsymbol{x}\,\frac{\partial\psi}{\partial x_j}(\boldsymbol{x})\,e^{-i\boldsymbol{k}\cdot\boldsymbol{x}} \tag{2.19}$$

$$= -\frac{1}{(2\pi)^{d/2}}\int_{\mathbb{R}^d} d^d\boldsymbol{x}\,\psi(\boldsymbol{x})\,\frac{\partial}{\partial x_j}e^{-i\boldsymbol{k}\cdot\boldsymbol{x}} \tag{2.20}$$

$$= -\frac{1}{(2\pi)^{d/2}}\int_{\mathbb{R}^d} d^d\boldsymbol{x}\,\psi(\boldsymbol{x})\,(-ik_j)e^{-i\boldsymbol{k}\cdot\boldsymbol{x}} \tag{2.21}$$

$$= ik_j\frac{1}{(2\pi)^{d/2}}\int_{\mathbb{R}^d} d^d\boldsymbol{x}\,\psi(\boldsymbol{x})\,e^{-i\boldsymbol{k}\cdot\boldsymbol{x}} \tag{2.22}$$

$$= ik_j\,\widehat{\psi}(\boldsymbol{k})\,. \tag{2.23}$$

(This calculation is a rigorous proof in $\mathscr{S}$.)

(b) Interchanging differentiation and integration (which again is rigorously justified in $\mathscr{S}$),

$$\frac{\partial \widehat{\psi}}{\partial k_j} = \frac{\partial}{\partial k_j} \frac{1}{(2\pi)^{d/2}} \int_{\mathbb{R}^d} \psi(\boldsymbol{x})\, e^{-i\boldsymbol{k}\cdot\boldsymbol{x}}\, d^d\boldsymbol{x} \tag{2.24}$$

$$= \frac{1}{(2\pi)^{d/2}} \int_{\mathbb{R}^d} \Big(-ix_j\, \psi(\boldsymbol{x})\Big)\, e^{-i\boldsymbol{k}\cdot\boldsymbol{x}}\, d^d\boldsymbol{x}\,. \tag{2.25}$$

(c) Indeed,

$$\hat{g}(\boldsymbol{k}-\boldsymbol{k}_0) = \frac{1}{(2\pi)^{d/2}} \int_{\mathbb{R}^d} g(\boldsymbol{x})\, e^{-i(\boldsymbol{k}-\boldsymbol{k}_0)\cdot\boldsymbol{x}}\, d^d\boldsymbol{x} \tag{2.26}$$

$$= \frac{1}{(2\pi)^{d/2}} \int_{\mathbb{R}^d} \Big(e^{i\boldsymbol{k}_0\cdot\boldsymbol{x}}\, g(\boldsymbol{x})\Big)\, e^{-i\boldsymbol{k}\cdot\boldsymbol{x}}\, d^d\boldsymbol{x}\,. \tag{2.27}$$

(d) This follows in much the same way.

□

Example 2.2 A more general Gaussian packet of the form

$$\psi(\boldsymbol{x}) = C\, e^{i\boldsymbol{k}_0\cdot\boldsymbol{x}}\, e^{-\frac{(\boldsymbol{x}-\boldsymbol{x}_0)^2}{4\sigma^2}} \tag{2.28}$$

has Fourier transform

$$\widehat{\psi}(\boldsymbol{k}) = C_3\, e^{i\boldsymbol{k}_0\cdot\boldsymbol{x}_0}\, e^{-i\boldsymbol{k}\cdot\boldsymbol{x}_0}\, e^{-\sigma^2(\boldsymbol{k}-\boldsymbol{k}_0)^2}\,, \tag{2.29}$$

which is again a Gaussian packet with center $\boldsymbol{k}_0$ and width $1/(2\sigma)$. As in (2.16), (2.29) is still valid for complex σ with $\mathrm{Re}(\sigma^2) > 0$, so it covers the evolved Gaussian as well. The most general Gauss packet is the exponential of a second-order polynomial in $\boldsymbol{x}$ for which the matrix of second-order coefficients has negative-definite self-adjoint part. Its Fourier transform is also again a Gauss packet.

Exercise 2.1 (Gaussian Wave Packets Done by Fourier Transformation) Redo Exercise 1.11 using Fourier transformation. That is, derive the formula (1.101) for ψ_t by starting from ψ_0 given by (2.28) with $\boldsymbol{x}_0 = 0$, taking the Fourier transform, evolving it up to time t, and Fourier transforming back.

Fourier transformation defines a *unitary* operator $\mathscr{F} : L^2(\mathbb{R}^d) \to L^2(\mathbb{R}^d)$, $\mathscr{F}\psi = \widehat{\psi}$. We verify that $\|\mathscr{F}\psi\|_{L^2} = \|\psi\|_{L^2}$ at least for nice ψ. Note first that, for $f, g \in L^1 \cap L^2$,

$$\int\Big(\int e^{-i\boldsymbol{k}\cdot\boldsymbol{x}} f(\boldsymbol{k})\, d^d\boldsymbol{k}\Big) g(\boldsymbol{x})\, d^d\boldsymbol{x} = \int\Big(\int e^{-i\boldsymbol{k}\cdot\boldsymbol{x}} g(\boldsymbol{x})\, d^d\boldsymbol{x}\Big) f(\boldsymbol{k})\, d^d\boldsymbol{k} \tag{2.30}$$

by changing the order of integration (which integral is done first). The theorem saying that we are allowed to change the order of integration (for an integrable integrand fg) is called *Fubini's theorem.* From Eq. (2.30), we can conclude $\langle g^* | \hat{f} \rangle = \langle \hat{g}^* | f \rangle$. Since

$$(\mathscr{F} f)(\boldsymbol{k})^* = (2\pi)^{-d/2} \int \Big(e^{-i\boldsymbol{k}\cdot\boldsymbol{x}} f(\boldsymbol{x})\Big)^* d^d\boldsymbol{x} = \mathscr{F}^{-1}(f^*)(\boldsymbol{k})\,, \tag{2.31}$$

setting $g = \mathscr{F}^{-1}(f^*) = (\mathscr{F} f)^*$ yields $\langle \hat{f} | \hat{f} \rangle = \langle f | f \rangle$, which completes the proof.

2.1.2 Momentum

"Position measurements" usually consist of detecting the particle. "Momentum measurements" usually consist of letting the particle move freely for a while and then measuring its position.[3]

We now analyze this experiment using Bohmian mechanics. We define the *asymptotic velocity* $\boldsymbol{u}$ to be

$$\boldsymbol{u} = \lim_{t\to\infty} \frac{d\boldsymbol{Q}}{dt}(t) \tag{2.32}$$

if this limit exists. It can also be expressed as

$$\boldsymbol{u} = \lim_{t\to\infty} \frac{\boldsymbol{Q}(t)}{t}\,. \tag{2.33}$$

To understand this, note that $(\boldsymbol{Q}(t) - \boldsymbol{Q}(0))/t$ is the average velocity during the time interval $[0, t]$; if an asymptotic velocity exists (i.e., if the velocity approaches a constant vector $\boldsymbol{u}$), then the average velocity over a long time t will be close to $\boldsymbol{u}$ because for most of the time, the velocity will be close to $\boldsymbol{u}$. The term $\boldsymbol{Q}(0)/t$ converges to zero as $t \to \infty$, so we obtain (2.33).

We want the momentum measurement to measure $\boldsymbol{p} := m\boldsymbol{u}$ for a free particle ($V = 0$). So we measure $\boldsymbol{Q}(t)$ for large t, divide by t, and multiply by m. We

[3] Alternatively, one lets the particle collide with another particle, makes a "momentum measurement" on the latter, and makes theoretical reasoning about what the momentum of the former must have been.

can and will also take this recipe as the *definition* of a momentum measurement, independent of whether we want to use Bohmian mechanics.

How large do we need t to be? In practice, often not very. When thinking of a particle emitted by a radioactive atom, or coming from a particle collision in an accelerator experiment (such as the Large Hadron Collider (LHC) in Geneva), a millisecond is usually enough for $d\boldsymbol{Q}/dt$ to become approximately constant.

According to the Born rule, the outcome $\boldsymbol{p}$ is random, and its distribution can be characterized by saying that, for any set $B \subset \mathbb{R}^3$,

$$\mathbb{P}(\boldsymbol{u} \in B) = \lim_{t\to\infty} \mathbb{P}(Q(t)/t \in B) \tag{2.34}$$

$$= \lim_{t\to\infty} \mathbb{P}(Q(t) \in tB) \tag{2.35}$$

$$= \lim_{t\to\infty} \int_{tB} |\psi_t(\boldsymbol{x})|^2 \, d^3\boldsymbol{x}\,, \tag{2.36}$$

where

$$tB = \{t\boldsymbol{x} : \boldsymbol{x} \in B\} \tag{2.37}$$

is the scaled set B.

Theorem 2.2 *Let $\psi(t, \boldsymbol{x})$ be a solution of the free Schrödinger equation and $B \subseteq \mathbb{R}^3$. Then*

$$\lim_{t\to\infty} \int_{tB} |\psi(t, \boldsymbol{x})|^2 \, d^3\boldsymbol{x} = \int_{mB/\hbar} |\widehat{\psi}_0(\boldsymbol{k})|^2 \, d\boldsymbol{k}\,. \tag{2.38}$$

As a consequence, the probability density of $\boldsymbol{p}$ is

$$\frac{1}{\hbar^3}\left|\widehat{\psi}_0\left(\frac{\boldsymbol{p}}{\hbar}\right)\right|^2 . \tag{2.39}$$

The theorem essentially says that when we think of ψ_0 as a linear combination of plane waves $e^{i\boldsymbol{k}\cdot\boldsymbol{x}}$ as in Eq. (2.4) or (2.7), then the contribution from a particular value of $\boldsymbol{k}$ will move at a velocity of $\hbar\boldsymbol{k}/m$ (shadows of the de Broglie relation $\boldsymbol{p} = \hbar\boldsymbol{k}$!), and in the long run, these contributions will tend to separate in space (i.e., overlap no longer), leaving the contribution from $\boldsymbol{k}$ in the region around $\hbar\boldsymbol{k}t/m$. We see the de Broglie relation again in (2.39) when we insert $\boldsymbol{p}/\hbar$ for $\boldsymbol{k}$ in $\widehat{\psi}$. The fact that the free Schrödinger evolution tends to spatially separate the contributions with different $\boldsymbol{k}$ vectors is called *dispersion*—not to be confused with diffusion (the transport of substances through thermal motion), dissipation (the conversion of usable energy to unusable heat), or diffraction (interference of waves, in particular the change of direction of propagation due to interference). A *dispersion relation* is the relation between frequency ω and wave vector $\boldsymbol{k}$, which reads

$$\omega = \frac{\hbar \boldsymbol{k}^2}{2m} \tag{2.40}$$

for the free Schrödinger equation.

The upshot of this analysis can be formulated as

Born's Rule for Momentum *If we measure the momentum of a particle with wave function ψ, then the outcome is random with probability density*

$$\rho_{\text{mom}}(\boldsymbol{p}) = \frac{1}{\hbar^3} \left| \widehat{\psi}\left(\frac{\boldsymbol{p}}{\hbar}\right) \right|^2 . \tag{2.41}$$

Likewise, if we measure the momenta of N particles with joint wave function $\psi(\boldsymbol{x}_1, \ldots, \boldsymbol{x}_N)$, then the outcomes are random with joint probability density

$$\rho_{\text{mom}}(\boldsymbol{p}_1, \ldots, \boldsymbol{p}_N) = \frac{1}{\hbar^{3N}} \left| \widehat{\psi}\left(\frac{\boldsymbol{p}_1}{\hbar}, \ldots, \frac{\boldsymbol{p}_N}{\hbar}\right) \right|^2 . \tag{2.42}$$

For this reason, the Fourier transform $\widehat{\psi}$ is also called the *momentum representation of* ψ, while ψ itself is called the *position representation* of the wave function.

Example 2.3 The Gaussian wave packet (2.28), whose Born distribution in position space is a Gaussian distribution with mean $\boldsymbol{x}_0$ and width σ, has momentum distribution

$$\rho_{\text{mom}}(\boldsymbol{p}) = (\text{const.})\, e^{-2(\sigma/\hbar)^2(\boldsymbol{p}-\hbar \boldsymbol{k}_0)^2} , \tag{2.43}$$

that is, a Gaussian distribution with mean $\hbar \boldsymbol{k}_0$ and width

$$\sigma_P = \frac{\hbar}{2\sigma} . \tag{2.44}$$

In particular, if we want a momentum distribution that is sharply peaked around some value $\boldsymbol{p}_0 = \hbar \boldsymbol{k}_0$, that is, if we want σ_P to be small, then σ must be large, so ψ must be wide, "close to a plane wave."

2.1.3 Momentum Operator

Let p_j, $j = 1, 2, 3$, be the component of the vector $\boldsymbol{p}$ in the direction of the x_j-axis. The expectation value of p_j is (using Eq. (2.17) in the fourth line and unitarity of

$\mathscr{F}$ in the sixth)

$$\langle p_j \rangle = \int_{\mathbb{R}^3} p_j \, \rho_{\text{mom}}(\boldsymbol{p}) \, d^3\boldsymbol{p} \tag{2.45a}$$

$$= \int \hbar k_j \, |\widehat{\psi}_0(\boldsymbol{k})|^2 \, d^3\boldsymbol{k} \tag{2.45b}$$

$$= \left\langle \widehat{\psi}_0 \middle| \hbar k_j \, \widehat{\psi}_0 \right\rangle \tag{2.45c}$$

$$= \left\langle \widehat{\psi}_0 \middle| (-i\hbar) \widehat{\frac{\partial \psi_0}{\partial x_j}} \right\rangle \tag{2.45d}$$

$$= -i\hbar \left\langle \widehat{\psi}_0 \middle| \widehat{\frac{\partial \psi_0}{\partial x_j}} \right\rangle \tag{2.45e}$$

$$= -i\hbar \left\langle \psi_0 \middle| \frac{\partial \psi_0}{\partial x_j} \right\rangle \tag{2.45f}$$

$$= \left\langle \psi_0 \middle| \left(-i\hbar \frac{\partial}{\partial x_j} \right) \psi_0 \right\rangle . \tag{2.45g}$$

This relation motivates calling $P_j = -i\hbar \frac{\partial}{\partial x_j}$ the *momentum operator* in the x_j-direction and (P_1, P_2, P_3) the *vector of momentum operators.*

We note for later use that, by the same reasoning,

$$\langle p_j^n \rangle = \int (\hbar k_j)^n |\widehat{\psi}_0(\boldsymbol{k})|^2 \, d\boldsymbol{k} = \left\langle \psi_0 \middle| \left(-i\hbar \frac{\partial}{\partial x_j} \right)^n \psi_0 \right\rangle \tag{2.46}$$

for every $n \in \mathbb{N}$.

2.1.4 Tunnel Effect

The *tunnel effect* is another quantum effect that is widely perceived as paradoxical. Consider the 1d Schrödinger equation with a potential V that has the shape of a *potential barrier* of height $V_0 > 0$. As an idealized example, suppose

$$V(x) = V_0 \, 1_{0 \le x \le L} \tag{2.47}$$

or a smooth approximation thereof (see Fig. 2.1).

Classically, the motion of a particle in the potential V (or any potential in one dimension) can easily be deduced from energy conservation: If the initial position is < 0 and the initial momentum is $p_0 > 0$, then the initial energy is $E = p_0^2/2m$,

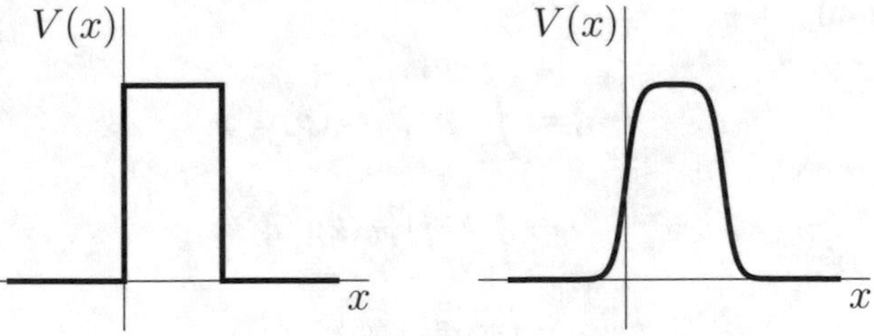

Fig. 2.1 Potential barriers in 1d. Left: Idealized "hard" barrier as in (2.47). Right: Smooth approximation thereof or "soft" barrier

and whenever the particle reaches location x, its momentum must be

$$p = \pm\sqrt{2m(E - V(x))}\,. \tag{2.48}$$

In particular, the particle can never reach a region in which $V(x) > E$; so, if $E < V_0$, then the particle will turn around at the barrier and move back to the left.

That is different in quantum mechanics. Consider a Gaussian wave packet, initially to the left of the barrier, with a rather sharp momentum distribution around a $p_0 > 0$ with $p_0^2/2m < V_0$. Then part of the packet will be reflected, and part of it will pass through the barrier! (And the part that passes through is much larger than just the tail of ρ_{mom} with $p \geq \sqrt{V_0/2m}$.) As a consequence, the Born rule predicts a substantial probability for the particle to show up on the other side of the barrier ("tunneling probability"). Figure 2.2 shows the Bohmian trajectories for such a situation (with only a small tunneling probability).

For computing the tunneling probability, an easy recipe is to assume that the initial ψ is close to a plane wave and to consider only the interior part of it that actually looks like a plane wave. One solves the Schrödinger equation for a plane wave arriving, computes the amount of probability current through the barrier, and compares it to the current associated with the arriving wave. For further discussion of why that yields a reasonable result, see Norsen (2013) [12].

What is paradoxical about tunneling? Perhaps not so much, once we give up Newtonian mechanics and accept that the equation of motion can be non-classical, such as Bohm's. Then it is only to be expected that the trajectories are different and not surprising that some barriers which Newton's trajectories cannot cross, Bohm's trajectories can. Part of the sense of paradox comes perhaps from a narrative that is often told when the tunnel effect is introduced: that the particle can "borrow" some energy for a short amount of time by virtue of an energy–time uncertainty relation. This narrative seems not very helpful.

The tunnel effect plays a crucial role in radioactive α-decay (where the α-particle leaves the nucleus by means of tunneling), beam splitters in optics (where the thickness of the barrier is adjusted so that half of the incoming wave will be reflected and half transmitted), and scanning tunneling electron microscopy (where

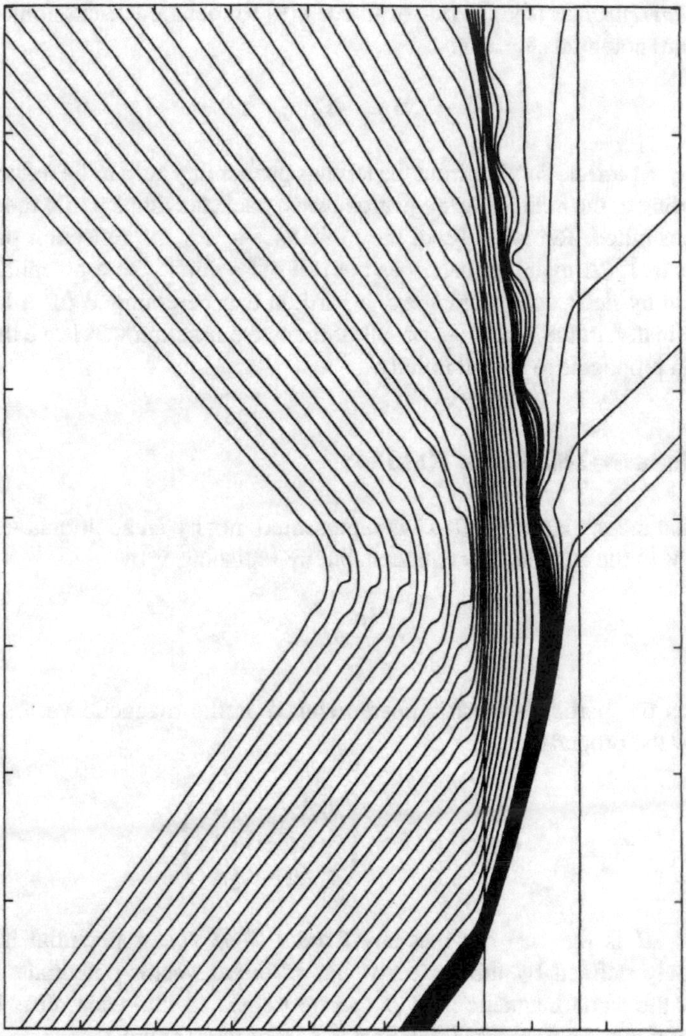

Fig. 2.2 Bohmian trajectories in a tunneling situation, with time axis upward and position axis to the right. The barrier region is delineated by two vertical lines; near the front, more trajectories are shown than in the rest of the packet (Reprinted from Bohm and Hiley (1993) [5, Fig. 5.3] with permission by Routledge)

the distance between a needle and a surface is measured by means of measuring the tunneling probability).

There are further related effects: *anti-tunneling* means that a particle gets reflected by a barrier so low that a classical particle with the same initial momentum would be certain to pass it; this happens because a solution of the Schrödinger equation will partly be reflected even at a low barrier. Another effect has been termed *paradoxical reflection* (see Garrido et al. 2011 [9] for detailed discussion): Consider a downward potential step as in

$$V(x) = -V_0\, 1_{0\leq x}\,. \tag{2.49}$$

Classically, a particle coming from the left has probability zero to be reflected back, but according to the Schrödinger equation, wave packets will be partly reflected and partly transmitted. Remarkably, in the limit $V_0 \to \infty$, the reflection probability converges to 1. "A quantum ball does not roll off a cliff!" On a potential plateau, surrounded by deep downward steps, a particle can be confined for a long time, although finally, in the limit $t \to \infty$, all of the wave function will leave the plateau region and propagate to spatial infinity.

2.1.5 External Magnetic Field

An external magnetic field $\boldsymbol{B}(t, \boldsymbol{x})$ is represented, not by an additional term in the potential V in the Schrödinger equation, but by replacing ∇ by

$$\nabla - \frac{ie}{\hbar}\boldsymbol{A}(t, \boldsymbol{x})\,, \tag{2.50}$$

where e is the particle's electric charge and $\boldsymbol{A}$ is the magnetic vector potential defined by the property

$$\boldsymbol{B} = \nabla \times \boldsymbol{A} = \begin{pmatrix} \partial_2 A_3 - \partial_3 A_2 \\ \partial_3 A_1 - \partial_1 A_3 \\ \partial_1 A_2 - \partial_2 A_1 \end{pmatrix}. \tag{2.51}$$

(In words, $\boldsymbol{B}$ is the curl of $\boldsymbol{A}$ at any fixed t. The vector potential is, in fact, not uniquely defined by this property, but different vector potentials satisfying (2.51) for the same magnetic field $\boldsymbol{B}$ can be translated into each other by *gauge transformations*; see Exercise 7.5 in Sect. 7.3.6 for more detail.)

For example, the Schrödinger equation of a single particle in a magnetic field reads

$$i\hbar\frac{\partial\psi}{\partial t} = \frac{1}{2m}\Big(-i\hbar\nabla - e\boldsymbol{A}(t, \boldsymbol{x})\Big)^2\psi(\boldsymbol{x}) + V(t, \boldsymbol{x})\psi(\boldsymbol{x}) \tag{2.52}$$

with V the electric and gravitational potential. But also in the formula for the current, which contains a derivative, the replacement (2.50) applies, yielding

$$\boldsymbol{j}(t, \boldsymbol{x}) = \tfrac{\hbar}{m}\mathrm{Im}\Big[\psi^*(t, \boldsymbol{x})\big(\nabla - \tfrac{ie}{\hbar}\boldsymbol{A}(t, \boldsymbol{x})\big)\psi(t, \boldsymbol{x})\Big]. \tag{2.53}$$

The appropriate Bohmian equation of motion reads

$$\frac{d\boldsymbol{Q}}{dt} = \frac{\hbar}{m}\text{Im}\frac{(\nabla - \frac{ie}{\hbar}\boldsymbol{A})\psi}{\psi}\bigg|_{\boldsymbol{x}=\boldsymbol{Q}_t}, \tag{2.54}$$

which is still of the form

$$\frac{d\boldsymbol{Q}}{dt} = \frac{\boldsymbol{j}}{|\psi|^2}(t, \boldsymbol{Q}_t). \tag{2.55}$$

Exercise 2.2 (Continuity Equation with *A*) Verify that in the presence of $\boldsymbol{A}$, i.e., using (2.52) and (2.53), the continuity equation

$$\frac{\partial|\psi|^2}{\partial t} = -\nabla\cdot\boldsymbol{j} \tag{2.56}$$

still holds. Conclude further that the $|\psi|^2$ distribution is still equivariant for the Bohmian trajectories.

2.2 Operators and Observables

2.2.1 Heisenberg's Uncertainty Relation

As before, $\langle X\rangle$ denotes the expectation of the random variable X. The *variance* of the momentum distribution for the initial wave function $\psi \in L^2(\mathbb{R})$ (in one dimension) is

$$\sigma_P^2 := \Big\langle\big(p - \langle p\rangle\big)^2\Big\rangle \tag{2.57a}$$

$$= \Big\langle p^2 - 2p\langle p\rangle + \langle p\rangle^2\Big\rangle \tag{2.57b}$$

$$= \langle p^2\rangle - 2\langle p\rangle^2 + \langle p\rangle^2 \tag{2.57c}$$

$$= \langle p^2\rangle - \langle p\rangle^2 \tag{2.57d}$$

$$= \langle\psi|P^2\psi\rangle - \langle\psi|P\psi\rangle^2 \tag{2.57e}$$

$$= \Big\langle\psi\Big|\big(P - \langle\psi|P\psi\rangle\big)^2\,\psi\Big\rangle. \tag{2.57f}$$

The position distribution $|\psi(x)|^2$ has expectation

$$\langle Q(0)\rangle = \int x|\psi(x)|^2\,dx = \langle\psi|X\psi\rangle \tag{2.58}$$

with the *position operator*

$$X\psi(x) = x\psi(x)\,. \tag{2.59}$$

Moreover,

$$\langle Q(0)^2\rangle = \int x^2|\psi(x)|^2\,dx = \langle\psi|X^2\psi\rangle\,, \tag{2.60}$$

so the variance of the position distribution $|\psi(x)|^2$ is

$$\sigma_X^2 := \int (x - \langle Q(0)\rangle)^2|\psi(x)|^2\,dx = \Big\langle\psi\Big|\big(X - \langle\psi|X\psi\rangle\big)^2\,\psi\Big\rangle. \tag{2.61}$$

Theorem 2.3 (Heisenberg Uncertainty Relation) *For any* $\psi \in L^2(\mathbb{R})$ *with* $\|\psi\| = 1$,

$$\sigma_X\,\sigma_P \geq \frac{\hbar}{2}\,. \tag{2.62}$$

This means that any wave function that is very narrow must have a wide Fourier transform. A generalized version will be proved later as Theorem 3.3.

Example 2.4 Consider the Gaussian wave packet (2.28), for simplicity in one dimension. The standard deviation of the position distribution is $\sigma_X = \sigma$, and we computed the width of the momentum distribution in (2.44). We thus obtain for this ψ that

$$\sigma_X\,\sigma_P = \frac{\hbar}{2}\,, \tag{2.63}$$

just the lowest value allowed by the Heisenberg uncertainty relation.

Example 2.5 Consider a wave packet passing through a slit. Let us ignore the part of the wave packet that gets reflected because it did not arrive at the slit and focus on just the part that makes it through the slit. That is a narrow wave packet, and its standard deviation in position, σ_X, is approximately the width of the slit. If that is very small, then, by the Heisenberg uncertainty relation, σ_P must be large, so the wave packet must spread quickly after passing the slit. If the slit is wider, the spreading is weaker.

2.2.2 Limitation to Knowledge

In Bohmian mechanics, the Heisenberg uncertainty relation means that whenever the wave function is such that we can know the position of a particle with (small) inaccuracy σ_X, then we are unable to know its asymptotic velocity better than with inaccuracy $\hbar/(2m\sigma_X)$; thus, we are unable to predict its future position after a large time t (for $V = 0$) better than with inaccuracy $\hbar t/(2m\sigma_X)$. This is a *limitation to knowledge* in Bohmian mechanics.

The Heisenberg uncertainty relation is often understood as excluding the possibility of particle trajectories. If the particle had a trajectory, the reasoning goes, then it would have a precise position and a precise velocity (and thus a precise momentum) at any time, so the position uncertainty would be zero and the momentum uncertainty would be zero, so $\sigma_X = 0$ and $\sigma_P = 0$, in contradiction with (2.62). We know already from Bohmian mechanics that this argument cannot be right. It goes wrong by assuming that if the particle has a precise position and a precise velocity, then they can also be precisely known and precisely controlled. Rather, inhabitants of a Bohmian universe, when they know a particle's wave function to be $\varphi(\boldsymbol{x})$, cannot know its position more precisely than the $|\varphi|^2$ distribution allows.

In the traditional, orthodox view of quantum mechanics, it is assumed that electrons do not have trajectories. It is assumed that the wave function is the *complete* description of the electron, in contrast to Bohmian mechanics, where the complete description is given by the pair $(\boldsymbol{Q}, \psi)$, and ψ alone would only be partial information and thus an incomplete description. By virtue of these assumptions, the electron *does not have a position before we attempt to detect it*. Likewise, it does not have a momentum before we attempt to measure it. Thus, in orthodox quantum mechanics, the Heisenberg uncertainty relation does *not* amount to a limitation of knowledge because there is no fact in the world that we do not know about when we do not know its position. Unfortunately, the uncertainty relation is often expressed by saying that it is impossible to measure position and momentum at the same time with arbitrary accuracy; while this would be appropriate to say in Bohmian mechanics, it is not in orthodox quantum mechanics because this formulation presumes that position and momentum have values that we could discover by measuring them.

The uncertainty relation is also involved in the double-slit experiment as follows. If it did not hold, we could make the electron move exactly orthogonal to the screen after passing through the narrow slits and arrive very near the center of the screen. Thus, the distribution on the detection screen could not have a second- or third-order maximum.

2.2.3 Self-Adjoint Operators

Certain experiments are called "quantum measurements" of "observables," with the latter mathematically represented by self-adjoint operators on Hilbert space. John

Bell (1928–1990) complained that "measurement" was a bad word because it can be misleading. In one place he wrote (1990) [4],

> On this list of bad words from good books, the worst of all is *measurement*.

I use this terminology nevertheless because it is very common, but we will keep Bell's warning in mind and come back to this point later. But first let us get acquainted with the mathematics of self-adjoint operators.

Theorem 2.4 [4] *Every bounded operator $A : \mathscr{H} \to \mathscr{H}$ on a Hilbert space $\mathscr{H}$ possesses one and only one adjoint operator $A^\dagger$, defined by the property that for all $\psi, \phi \in \mathscr{H}$,*

$$\langle \psi | A\phi \rangle = \langle A^\dagger \psi | \phi \rangle . \tag{2.64}$$

For an unbounded operator $A : \mathscr{D}(A) \to \mathscr{H}$ with dense domain $\mathscr{D}(A) \subset \mathscr{H}$, the adjoint operator $A^\dagger$ is uniquely defined by the property (2.64) *for all $\psi \in \mathscr{D}(A^\dagger)$ and $\phi \in \mathscr{D}(A)$ on the domain*

$$\mathscr{D}(A^\dagger) = \Big\{ \psi \in \mathscr{H} : \exists \chi \in \mathscr{H} \, \forall \phi \in \mathscr{D}(A) : \langle \psi | A\phi \rangle = \langle \chi | \phi \rangle \Big\} . \tag{2.65}$$

Definition 2.2 An operator A on a Hilbert space $\mathscr{H}$ is called *self-adjoint* or *Hermitian* if and only if $A = A^\dagger$. Then

$$\langle \psi | A\phi \rangle = \langle A\psi | \phi \rangle . \tag{2.66}$$

Example 2.6

- Let $\mathscr{H} = \mathbb{C}^n$. Then every operator A is bounded and corresponds to a complex $n \times n$ matrix A_{ij}. The matrix of $A^\dagger$ has entries $(A^\dagger)_{ij} = (A_{ji})^*$ ("the adjoint matrix is the conjugate transpose"). Indeed, if we define the matrix B_{ij} by $B_{ij} = (A_{ji})^*$, then we obtain, for any $\psi = (\psi_1, \ldots, \psi_n)$ and $\phi = (\phi_1, \ldots, \phi_n)$,

$$\langle \psi | A\phi \rangle = \sum_{i=1}^{n} \psi_i^* (A\phi)_i \tag{2.67a}$$

$$= \sum_i \sum_j \psi_i^* A_{ij} \phi_j \tag{2.67b}$$

[4] For the proof, see, e.g., Hall (2013) [10, p. 55–56] or Reed and Simon (1980) [13, p. 186 and 252].

$$= \sum_j \sum_i (A_{ij}^* \psi_i)^* \phi_j \tag{2.67c}$$

$$= \sum_j \Big(\sum_i B_{ji} \psi_i \Big)^* \phi_j \tag{2.67d}$$

$$= \sum_j (B\psi)_j^* \phi_j \tag{2.67e}$$

$$= \langle B\psi | \phi \rangle \,. \tag{2.67f}$$

As a consequence, a matrix A is self-adjoint if and only if $A_{ij} = A_{ji}^*$.

- A unitary operator is usually *not* self-adjoint.
- Let $\mathscr{H} = L^2(\mathbb{R}^d)$, and let A be a multiplication operator,

$$A\psi(\boldsymbol{x}) = f(\boldsymbol{x})\,\psi(\boldsymbol{x})\,, \tag{2.68}$$

such as the potential in the Hamiltonian or the position operators. Then $A^\dagger$ is the multiplication operator that multiplies by f^*. Indeed,

$$\langle \psi | A\phi \rangle = \int_{\mathbb{R}^d} \psi(\boldsymbol{x})^* f(\boldsymbol{x}) \phi(\boldsymbol{x})\, d\boldsymbol{x} \tag{2.69a}$$

$$= \int \big(f^*(\boldsymbol{x})\,\psi(\boldsymbol{x}) \big)^* \phi(\boldsymbol{x})\, d\boldsymbol{x} \tag{2.69b}$$

$$= \langle f^* \psi | \phi \rangle \,. \tag{2.69c}$$

(This calculation is rigorous if f is bounded. If it is not, then some discussion of the domains of A and $A^\dagger$ is needed.) Thus, A is self-adjoint if and only if f is real-valued.

- $(AB)^\dagger = B^\dagger A^\dagger$ and $\exp(A)^\dagger = \exp(A^\dagger)$.
- On $\mathscr{H} = L^2(\mathbb{R}^d)$, the momentum operators $P_j = -i\hbar \frac{\partial}{\partial x_j}$ are self-adjoint with the domain given by the first Sobolev space, i.e., the space of functions $\psi \in L^2$ whose Fourier transform $\widehat{\psi}$ has the property that $\boldsymbol{k} \mapsto |\boldsymbol{k}|\,\widehat{\psi}(\boldsymbol{k})$ is still square-integrable. The relation (2.66) can easily be verified on nice functions using integration by parts:

$$\langle \psi | P_j \phi \rangle = \int \psi^*(\boldsymbol{x}) (-i\hbar) \frac{\partial \phi}{\partial x_j}(\boldsymbol{x})\, d\boldsymbol{x} \tag{2.70a}$$

$$= -\int \frac{\partial \psi^*}{\partial x_j}(\boldsymbol{x}) (-i\hbar) \phi(\boldsymbol{x})\, d\boldsymbol{x} \tag{2.70b}$$

$$= \int \Big(-i\hbar \frac{\partial \psi}{\partial x_j}(\boldsymbol{x}) \Big)^* \phi(\boldsymbol{x})\, d\boldsymbol{x} \tag{2.70c}$$

$$= \langle P_j\psi|\phi\rangle\,. \tag{2.70d}$$

- In $\mathscr{H} = L^2(\mathbb{R}^d)$, the Hamiltonian is self-adjoint for suitable potentials V on a suitable domain. By formal calculation (leaving aside questions of domains), since

$$H = \sum_{j=1}^{d} \frac{1}{2m} P_j^2 + V\,, \tag{2.71}$$

we have that

$$\langle\psi|H\phi\rangle = \Big\langle\psi\Big|\Big(\sum_j \frac{1}{2m}P_j^2 + V\Big)\phi\Big\rangle \tag{2.72a}$$

$$= \sum_j \frac{1}{2m}\langle\psi|P_jP_j\phi\rangle + \langle\psi|V\phi\rangle \tag{2.72b}$$

$$= \sum_j \frac{1}{2m}\langle P_j\psi|P_j\phi\rangle + \langle V\psi|\phi\rangle \tag{2.72c}$$

$$= \sum_j \frac{1}{2m}\langle P_jP_j\psi|\phi\rangle + \langle V\psi|\phi\rangle \tag{2.72d}$$

$$= \Big\langle\Big(\sum_j \frac{P_j^2}{2m} + V\Big)\psi\Big|\phi\Big\rangle \tag{2.72e}$$

$$= \langle H\psi|\phi\rangle. \tag{2.72f}$$

2.2.4 The Spectral Theorem

Before we can formulate Born's rule for arbitrary observables, we need to learn about the spectral theorem.

Definition 2.3 If

$$A\psi = \alpha\psi\,, \tag{2.73}$$

where α is a (complex) number and $\psi \in \mathscr{H}$ with $\psi \neq 0$, then ψ is called an *eigenvector* (or *eigenfunction*) of A with *eigenvalue* α. The number α is called an eigenvalue of A if and only if there exists $\psi \neq 0$ satisfying (2.73). The set of all eigenvalues is called the *spectrum* of A. For any eigenvalue α, the set of all eigenvectors with eigenvalue α together with the zero vector forms a subspace of Hilbert space called the *eigenspace* of A with eigenvalue α. The eigenvalue α is said to be *degenerate* if and only if the dimension of its eigenspace is > 1.

If A is self-adjoint, then all eigenvalues must be real. Indeed, if ψ is an eigenvector of A with eigenvalue α, then

$$\alpha\langle\psi|\psi\rangle = \langle\psi|\alpha\psi\rangle = \langle\psi|A\psi\rangle = \langle A\psi|\psi\rangle = \langle\alpha\psi|\psi\rangle = \alpha^*\langle\psi|\psi\rangle\,, \tag{2.74}$$

so $\alpha = \alpha^*$ or $\alpha \in \mathbb{R}$.

Theorem 2.5 (Spectral Theorem[5]) *For every self-adjoint operator A in a Hilbert space $\mathscr{H}$, there is a (generalized) orthonormal basis $\{\phi_{\alpha,\lambda}\}$ consisting of eigenvectors of A,*

$$A\phi_{\alpha,\lambda} = \alpha\phi_{\alpha,\lambda}\,. \tag{2.75}$$

Such a basis is called an eigenbasis of A. ($\phi_{\alpha,\lambda}$ has two indices because for every eigenvalue α, there may be several eigenvectors, indexed by λ.)

An *orthonormal basis* (ONB) is a set $\{\phi_n\}_n$ of elements of the Hilbert space $\mathscr{H}$ such that (a) $\langle\phi_m|\phi_n\rangle = \delta_{mn}$ (i.e., it is an orthonormal system) and (b) every $\psi \in \mathscr{H}$ can be written as an (infinite) linear combination of the[6] ϕ_n,

$$\psi = \sum_n c_n\,\phi_n\,. \tag{2.76}$$

The fact that $L^2(\mathbb{R}^d)$ possesses a countable ONB means that it has, in a sense, "countably infinite" dimension, even though $\mathbb{R}^d$ is an uncountably infinite set. (For example, the orthonormal system of Exercise 1.17 is actually an ONB, called the Fourier basis.)[7]

A "generalized" orthonormal basis allows a continuous variable k instead of n,

$$\psi = \int dk\, c_k\,\phi_k\,, \tag{2.77}$$

as we have encountered with Fourier transformation, where $k = \boldsymbol{k} \in \mathbb{R}^d$, $c_k = \widehat{\psi}(\boldsymbol{k})$, and

$$\phi_{\boldsymbol{k}}(\boldsymbol{x}) = (2\pi)^{-d/2}e^{i\boldsymbol{k}\cdot\boldsymbol{x}}\,. \tag{2.78}$$

[5] For the proof, see, e.g., Hall (2013) [10, Theorem 10.10] or Reed and Simon (1980) [13, Theorem VIII.4].

[6] That is, ψ can be written as a *series* of multiples of the basis vectors ϕ_n; such a set $\{\phi_n\}$ is called a *Schauder basis*, whereas a *Hamel basis* of a vector space X is a set S of vectors from X such that every element of X can be written as a *finite* linear combination of members of S.

[7] Mathematicians call spaces with countably infinite or finite dimension *separable*; the technical definition is that a metric space is called separable if and only if there is a countable dense set.

For a generalized ONB, we do not require that the ϕ_k themselves be elements of $\mathscr{H}$; e.g., the ϕ_k of Fourier transformation are not square-integrable. We will often write a $\sum$ sign even when we mean the integral over k. The precise definition of "generalized ONB" is a unitary isomorphism $U : \mathscr{H} \to L^2(\Omega)$ with Ω the set of possible k-values indexing the generalized ONB and $U\psi(k) = c_k$. For example, for the generalized ONB (2.78), $U = \mathscr{F}$. A "non-generalized" ONB then corresponds to a unitary isomorphism $U : \mathscr{H} \to \ell^2 = L^2(\mathbb{N})$.

The big payoff of the spectral theorem is that in this ONB, it is very easy to carry out the operator A: If

$$\psi = \sum_{\alpha,\lambda} c_{\alpha,\lambda}\, \phi_{\alpha,\lambda} \tag{2.79}$$

then

$$A\psi = \sum_{\alpha,\lambda} \alpha\, c_{\alpha,\lambda}\, \phi_{\alpha,\lambda}\,. \tag{2.80}$$

Put differently, in this ONB, A is a multiplication operator, multiplying by the function $f(k) = f(\alpha, \lambda) = \alpha$. For example, in the Fourier basis (2.78), the momentum operator P_j is multiplication by $\hbar k_j$.

Put differently again, the matrix associated with the operator A in the ONB $\phi_{\alpha,\lambda}$ is a diagonal matrix. That is why one says that this ONB *diagonalizes* A.

2.2.5 Born's Rule

Born's Rule for Arbitrary Observables *In a quantum measurement of the observable* A *on a system with wave function* ψ, *then the outcome is random with probability distribution*

$$\rho_A(\alpha) = \sum_{\lambda} \big|\langle \phi_{\alpha,\lambda}|\psi\rangle\big|^2 = \sum_{\lambda} \big|U\psi(\alpha,\lambda)\big|^2\,, \tag{2.81}$$

where $\phi_{\alpha,\lambda}$ *is a (generalized) orthonormal basis diagonalizing* A; ρ_A *may mean either probability density or just probability, depending on whether* α *is a discrete or continuous variable.*

Note that the previous versions of Born's rule are contained as special cases for the position operators X (U the identity) and the momentum operator $P = -i\hbar\nabla$ (U the Fourier transformation).

We further note that the expectation value of the Born distribution is given by the simple expression $\langle\psi|A\psi\rangle$. Indeed, since the unitary isomorphism U defining the generalized ONB maps A to a multiplication operator M, $UAU^{-1} = M$, the expectation is given by

$$\int d\alpha\, \alpha\, \rho_A(\alpha) = \int_\Omega d(\alpha,\lambda)\, \alpha\, |U\psi(\alpha,\lambda)|^2 = \langle U\psi|MU\psi\rangle_\Omega = \langle\psi|U^{-1}MU\psi\rangle = \langle\psi|A\psi\rangle. \tag{2.82}$$

The spectral theorem also yields a useful perspective on the unitary time evolution operators. Since the Hamiltonian is self-adjoint, by the spectral theorem, it possesses a generalized eigenbasis diagonalizing it,

$$H\phi_{E,\lambda} = E\phi_{E,\lambda}\,. \tag{2.83}$$

As H is also called the energy operator, its (generalized) eigenvalues E are called the energy levels of H, and $\{\phi_{E,\lambda}\}$ is called the energy eigenbasis or simply the energy basis. Expressing a given vector ψ in this GONB,

$$\psi = \sum_{E,\lambda} c_{E,\lambda}\,\phi_{E,\lambda}\,, \tag{2.84}$$

one finds that

$$\psi_t = U_t\psi = e^{-iHt/\hbar}\psi = \sum_{E,\lambda} e^{-iEt/\hbar}\, c_{E,\lambda}\,\phi_{E,\lambda}\,. \tag{2.85}$$

In words, the coefficients $c_{E,\lambda}$ of ψ_t in the energy basis change with time according to

$$c_{E,\lambda}(t) = \exp(-iEt/\hbar)\, c_{E,\lambda}(0)\,, \tag{2.86}$$

which means they are rotating in the complex plane at different speeds proportional to the eigenvalues E.

2.2.6 Conservation Laws in Quantum Mechanics

As a consequence of (2.86), $|c_{E,\lambda}(t)|$ is time independent for every E and λ, i.e., is a conserved quantity. This conservation law has no classical analog. The other way around, what are the quantum analogs of the classical conservation laws of energy, momentum, and angular momentum? There are three answers.

First Answer In a certain sense, energy, momentum, and angular momentum are simply *not* conserved in quantum mechanics: that is because of the collapse of the wave function that we will discuss in detail in Chap. 3. But the point here can be understood in an elementary way as follows.

In the discussion of momentum measurements in Sect. 2.1.2, we defined the particle's momentum as mass times its asymptotic velocity. However, it is common to call the (generalized) eigenvalues of the momentum operator $P_j = -i\hbar\partial_j$ ($j = 1, 2, 3$) the momentum values in the x_j-direction. Note that the eigenfunctions are just the plane waves $e^{i\boldsymbol{k}\cdot\boldsymbol{x}}$ and the eigenvalues of P_j are $p_j = \hbar k_j$ (another version of de Broglie's relation). Now a wave function ψ is in general a superposition of plane waves with different values of p_j. As we let the wave function evolve freely, the contributions in the superposition get separated in space, and ultimately the particle is found in just one of them, corresponding to the measurement outcome p_j. So p_j as a number is not conserved, in the sense that the initial superposition may have involved also other p_j values than the outcome of the measurement. (I will comment in Sect. 3.2.6 on how that is linked to wave function collapse.)

Second Answer In quantum mechanics, energy, (the three components of) momentum, and (the three components of) angular momentum are *operators, not numbers*; they are not conserved quantities; they are conserved operators, in the following sense. With respect to any ONB $\{\phi_n\}_n$, any operator S can be represented as a matrix (possibly with infinitely many rows and columns) with entries $S_{mn} = \langle\phi_m|S\phi_n\rangle$. If we let each of the basis vectors evolve with U_t, then we obtain time-dependent matrix elements (setting, for convenience, $\hbar = 1$)

$$S_{mn}(t) = \langle\phi_m(t)|S\phi_n(t)\rangle = \langle e^{-iHt}\phi_m|Se^{-iHt}\phi_n\rangle = \langle\phi_m|e^{iHt}Se^{-iHt}\phi_n\rangle\,, \tag{2.87}$$

which are the matrix elements of $e^{iHt}Se^{-iHt}$. If S commutes with H, i.e., $SH = HS$ or $[S, H] := SH - HS = 0$, then S commutes with e^{-iHt}, so

$$e^{iHt}Se^{-iHt} = S\,, \tag{2.88}$$

and $S_{mn}(t)$ is actually time independent. One says that an operator S is *conserved* if and only if (2.88) holds, and this happens if and only if S commutes with H.[8] Examples of conserved operators include H itself, the momentum operators if V is translation invariant, and the angular momentum operators $-i\hbar\boldsymbol{x} \times \nabla$ if V is rotationally invariant.

Third Answer There is another sense in which conservation laws hold: If an operator S is conserved, then *the Born distribution associated with S is time-independent*. In particular, any eigenstate of S evolves to another eigenstate of S with the same eigenvalue.

This means, for example, for tunneling and $S = H$: At early times and at late times, the wave function hardly overlaps with the barrier region, so we can expect

[8] At least for bounded operators. For unbounded operators, this is still true if we define carefully what it means for them to commute.

that the energy distribution is (approximately) the same as for the free Hamiltonian H_0; thus, the Born distribution for H_0 is the same before and after hitting the barrier—when we take into account that the wave function after hitting the barrier comprises both the reflected and the transmitted packet. On the other hand, the Born distribution for the momentum operator P is not conserved because P does not commute with H.

For another example, if an N-particle potential V is translation invariant (such as the sum of $\binom{N}{2}$ Coulomb pair potentials), then each component of the total momentum operator is conserved (see Exercise 2.10), so the Born distribution for total momentum is time-independent. Likewise, if V is rotation invariant, then each component of the total angular momentum operator is conserved, so the Born distribution for total angular momentum is time-independent.

2.2.7 The Dirac Delta Function

Let x be a 1d variable and $g_\sigma(x)$ the Gaussian probability density,

$$g_\sigma(x) = \frac{1}{\sqrt{2\pi}\,\sigma} e^{-\frac{x^2}{2\sigma^2}}. \tag{2.89}$$

The *Dirac δ function* can be defined heuristically as

$$\delta(x) = \lim_{\sigma \to 0} g_\sigma(x). \tag{2.90}$$

Since $\delta(x) = 0$ for $x \neq 0$ and $\delta(0) = \infty$, the δ function is not a function in the ordinary sense; it is called a *distribution*. Based on the heuristic (2.90), one defines

$$\int_{\mathbb{R}} \delta(x-a)\, f(x)\, dx := \lim_{\sigma \to 0} \int_{\mathbb{R}} g_\sigma(x-a)\, f(x)\, dx. \tag{2.91}$$

It follows that if the function f is continuous at a, then

$$\int_{\mathbb{R}} \delta(x-a)\, f(x)\, dx = f(a). \tag{2.92}$$

Mathematicians take this as the definition of the δ distribution; that is, they define $\delta(\cdot - a)$ as a linear operator from some function space such as $\mathscr{S}$ (Schwartz space) to $\mathbb{C}$, $f \mapsto f(a)$.

Exercise 2.3 (Dirac Delta Function, Introductory Level)

(a) One defines the Fourier transform $\widehat{\delta_a}(k)$ of $\delta_a(x) = \delta(x-a)$ by applying the usual integral formula (2.6). Find $\widehat{\delta_a}(k)$ for arbitrary constant $a \in \mathbb{R}$, and find the function ψ whose Fourier transform is $\widehat{\psi}(k) = \delta(k-b)$ with arbitrary constant $b \in \mathbb{R}$.

(b) One defines the derivative δ' of the δ function heuristically by

$$\delta'(x) = \lim_{\sigma\to 0} g'_\sigma(x) \tag{2.93}$$

and its integrals by

$$\int_{\mathbb{R}} \delta'(x-a)\, f(x)\, dx := \lim_{\sigma\to 0} \int_{\mathbb{R}} g'_\sigma(x-a)\, f(x)\, dx\,. \tag{2.94}$$

Sketch g'_σ for small σ.

(c) Using integration by parts and (2.92), show that (for $f \in \mathscr{S}$)

$$\int_{\mathbb{R}} \delta'(x-a)\, f(x)\, dx = -f'(a)\,. \tag{2.95}$$

(d) Show that

$$\lim_{\varepsilon\to 0} \frac{1}{2\varepsilon}\big(\delta(x+\varepsilon) - \delta(x-\varepsilon)\big) = \delta'(x)\,. \tag{2.96}$$

(e) For $\alpha \in \mathbb{R} \setminus \{0\}$, show that

$$\int_{\mathbb{R}} \delta(\alpha x)\, f(x)\, dx = \frac{1}{|\alpha|} f(0)\,. \tag{2.97}$$

Exercise 2.4 (Delta Function in Higher Dimension)

(a) The d-dimensional Dirac delta function is defined by

$$\delta^d(\boldsymbol{x} - \boldsymbol{a}) = \delta(x_1 - a_1)\cdots\delta(x_d - a_d) \tag{2.98}$$

Instead of $\delta^d(\boldsymbol{x} - \boldsymbol{a})$, one sometimes simply writes $\delta(\boldsymbol{x} - \boldsymbol{a})$. Verify that

$$\int_{\mathbb{R}^d} \delta^d(\boldsymbol{x} - \boldsymbol{a})\, f(\boldsymbol{x})\, d^d\boldsymbol{x} = f(\boldsymbol{a})\,. \tag{2.99}$$

(b) For a generalized orthonormal basis (GONB) with continuous parameter, $\{\phi_{\boldsymbol{k}} : \boldsymbol{k} \in \mathbb{R}^d\}$, one requires that

$$\langle \phi_{\boldsymbol{k}_1} | \phi_{\boldsymbol{k}_2} \rangle = \delta^d(\boldsymbol{k}_1 - \boldsymbol{k}_2)\,. \tag{2.100}$$

Verify this relation for the basis functions of Fourier transformation,

$$\phi_{\boldsymbol{k}}(\boldsymbol{x}) = (2\pi)^{-d/2} e^{i\boldsymbol{k}\cdot\boldsymbol{x}}\,. \tag{2.101}$$

(c) Verify that every ϕ_k as given by (2.101) is an eigenfunction of each momentum operator $P_j = -i\hbar\partial/\partial x_j$, $j = 1, \ldots, d$. Thus, (2.101) defines a GONB that simultaneously diagonalizes all P_j. It is therefore called the *momentum basis*.

(d) Now consider another basis, given by

$$\phi_y(x) = \delta^d(x - y). \tag{2.102}$$

Verify Eq. (2.100) for these functions. Then verify that every ϕ_y is an eigenfunction of each position operator $X_j\psi(x) = x_j\psi(x)$, $j = 1, \ldots, d$. Thus, (2.102) defines a GONB that simultaneously diagonalizes all X_j. It is therefore called the *position basis*.

One can also consider *distributional solutions* of the Schrödinger equation: while we have considered so far (and will mostly consider) only solutions ψ_t of the Schrödinger equation with initial data $\psi_0 \in L^2(\mathbb{R}^d, \mathbb{C}^m)$, it is possible to define solutions for initial data that are distributions, at least for the free Schrödinger equation. For any distribution T, take its Fourier transform, multiply by the appropriate function of $\boldsymbol{k}$ and t, and Fourier transform back.

As a side remark, if one wants to make the procedure mentioned above work rigorously, one needs (i) a suitable space $\mathscr{D}$ containing "all" distributions; (ii) a definition of the Fourier transform $\widehat{T} \in \mathscr{D}$ for every $T \in \mathscr{D}$; and (iii) that a multiplication operator M_φ with a function of the form $\varphi(\boldsymbol{k}) = \exp(i\,\text{polynomial}(\boldsymbol{k}))$ maps $\mathscr{D}$ to itself. To this end, one considers the Schwartz space $\mathscr{S}$ of rapidly decaying functions, i.e., those ψ with $|\partial^{\boldsymbol{\alpha}}\psi| < C_{n,\boldsymbol{\alpha}}|\boldsymbol{x}|^{-n}$ for all $\boldsymbol{x} \in \mathbb{R}^d$, $\boldsymbol{\alpha} \in \mathbb{N}_0^d$, $n \in \mathbb{N}_0$; one endows $\mathscr{S}$ with the topology in which $\psi_n \to \psi$ if and only if

$$\left\|\psi_n - \psi\right\|_{\boldsymbol{\alpha},\boldsymbol{\beta}} := \sup_{\boldsymbol{x}\in\mathbb{R}^d} \left|\boldsymbol{x}^{\boldsymbol{\alpha}}\partial^{\boldsymbol{\beta}}(\psi_n - \psi)(\boldsymbol{x})\right| \xrightarrow{n\to\infty} 0 \text{ for all } \boldsymbol{\alpha}, \boldsymbol{\beta} \in \mathbb{N}_0^d,$$

where $\boldsymbol{x}^{\boldsymbol{\alpha}} = x_1^{\alpha_1} \cdots x_d^{\alpha_d}$ and $\partial^{\boldsymbol{\beta}} = \partial_1^{\beta_1} \cdots \partial_d^{\beta_d}$; one defines $\mathscr{D}$ as the *continuous dual space* $\mathscr{S}'$, i.e., the set of continuous $\mathbb{C}$-linear mappings $T : \mathscr{S} \to \mathbb{C}$; $\mathscr{S}'$ is called the *space of tempered distributions*; and one uses that the Fourier transformation $\mathscr{F}$ (as well as $\mathscr{F}^{-1}$) maps $\mathscr{S}$ to itself and is continuous as a mapping $\mathscr{S} \to \mathscr{S}$.

Exercise 2.5 (Continuous Dual Space)

(a) Prove that the δ distribution is a continuous mapping from $\mathscr{S}$ to $\mathbb{C}$.

(b) Let $C^\infty_{\text{poly}}(\mathbb{R}^d)$ be the space of smooth functions φ such that φ and all of its derivatives are bounded by polynomials, $|\partial^{\boldsymbol{\alpha}}\varphi(\boldsymbol{k})| \le |P_{\boldsymbol{\alpha}}(\boldsymbol{k})|$ for a suitable polynomial $P_{\boldsymbol{\alpha}}$. Show that for every such φ, the multiplication operator M_φ is a continuous operator $\mathscr{S} \to \mathscr{S}$.

2.3 Spin

The phenomenon known as *spin* does not mean that the particle is spinning around its axis, though it is in some ways similar. The simplest description of the phenomenon is to say that the wave function of an electron (at time t) is actually not of the form $\psi : \mathbb{R}^3 \to \mathbb{C}$ but instead $\psi : \mathbb{R}^3 \to \mathbb{C}^2$. The space $\mathbb{C}^2$ is called *spin space* and its elements *spinors* (short for spin vectors). We will in the following write S for spin space.

2.3.1 Spinors and Pauli Matrices

Apart from being a two-dimensional Hilbert space, spin space has the further property that with every spinor is associated a vector in physical space $\mathbb{R}^3$. This relation can be expressed as a function

$$\boldsymbol{\omega} : S \to \mathbb{R}^3, \tag{2.103}$$

given explicitly by

$$\boldsymbol{\omega}(\phi) = \left(\sum_{r,s=1}^{2} \phi_r^*(\sigma_1)_{rs}\phi_s, \sum_{r,s=1}^{2} \phi_r^*(\sigma_2)_{rs}\phi_s, \sum_{r,s=1}^{2} \phi_r^*(\sigma_3)_{rs}\phi_s \right), \tag{2.104}$$

where σ_i are the three *Pauli matrices*

$$\sigma_1 = \begin{pmatrix} 0 & 1 \\ 1 & 0 \end{pmatrix}, \quad \sigma_2 = \begin{pmatrix} 0 & -i \\ i & 0 \end{pmatrix}, \quad \sigma_3 = \begin{pmatrix} 1 & 0 \\ 0 & -1 \end{pmatrix}. \tag{2.105}$$

Obviously, they are self-adjoint complex 2×2 matrices.

Exercise 2.6 (Pauli Matrices)

(a) For each of σ_1 and σ_2, find an orthonormal basis of eigenvectors in $\mathbb{C}^2$.
(b) Show that for every unit vector $\boldsymbol{n} \in \mathbb{R}^3$, the Pauli matrix in direction $\boldsymbol{n}$, $\sigma_{\boldsymbol{n}} := \boldsymbol{n} \cdot \boldsymbol{\sigma} = n_1\sigma_1 + n_2\sigma_2 + n_3\sigma_3$, has eigenvalues ± 1. (Hint: compute the determinant and the trace.)
(c) Show that every self-adjoint complex 2×2 matrix A is of the form $A = cI + \boldsymbol{u} \cdot \boldsymbol{\sigma}$ with $c \in \mathbb{R}$ and $\boldsymbol{u} \in \mathbb{R}^3$.

It is common to write $\boldsymbol{\sigma} = (\sigma_1, \sigma_2, \sigma_3)$ for the *vector of Pauli matrices*. With this notation, and writing

$$\phi^*\chi = \sum_{s=1}^{2} (\phi_s)^* \chi_s \tag{2.106}$$

for the inner product in spin space, Eq. (2.104) can be expressed more succinctly as

$$\boldsymbol{\omega}(\phi) = \phi^* \boldsymbol{\sigma} \phi . \tag{2.107}$$

For example, the spinor $\phi = (1, 0)$ has $\boldsymbol{\omega}(\phi) = (0, 0, 1)$, which points in the $+z$-direction; $(1, 0)$ is therefore called a *spin-up spinor*. The spinor $(0, 1)$ has $\boldsymbol{\omega}(0, 1) = (0, 0, -1)$, which points in the $-z$-direction; $(0, 1)$ is therefore called a *spin-down spinor*. $\boldsymbol{\omega}$ has the properties

$$\boldsymbol{\omega}(z\phi) = |z|^2 \boldsymbol{\omega}(\phi) \tag{2.108}$$

and that unit spinors are associated with unit vectors:

Exercise 2.7 (Spinor Norm) Show that

$$|\boldsymbol{\omega}(\phi)| = \|\phi\|_S^2 = \phi^* \phi , \tag{2.109}$$

where $\| \cdot \|_S$ means the norm in the spin space $S = \mathbb{C}^2$.[9]

(This way of mapping unit elements of $\mathbb{C}^2$ to unit vectors in $\mathbb{R}^3$ is also sometimes called the *Bloch sphere*.)

Spinors have the curious property that if spinors ϕ, χ have angle θ in spin space, then the corresponding vectors $\boldsymbol{\omega}(\phi), \boldsymbol{\omega}(\chi)$ in real space have angle 2θ. Here, we define the angle θ between two vectors ϕ, χ in any complex Hilbert space by the relation

$$\cos\theta = \frac{|\langle \phi | \chi \rangle|}{\|\phi\| \|\chi\|} . \tag{2.110}$$

Exercise 2.8 (Half Angle)

(a) Show that for unit vectors ϕ, χ in spin space S,

$$2 \left| \langle \phi | \chi \rangle \right|^2 = 1 + \sum_{a=1}^{3} \langle \phi | \sigma_a \phi \rangle \langle \chi | \sigma_a \chi \rangle .$$

(b) Conclude that if ϕ and χ have angle $\theta = \arccos \left| \langle \phi | \chi \rangle \right|$ in S, then $\boldsymbol{\omega}(\phi)$ and $\boldsymbol{\omega}(\chi)$ have angle 2θ in $\mathbb{R}^3$.

[9] Instructions: By (2.108), $\boldsymbol{\omega}(z\phi) = |z|^2 \boldsymbol{\omega}(\phi)$, it suffices to show that unit spinors are associated with unit vectors. By (2.108) again, it suffices to consider ϕ with $\phi_1 \in \mathbb{R}$ (else replace ϕ by $e^{i\theta}\phi$ with appropriate θ). So we can assume, without loss of generality, $\phi = (\cos\alpha, e^{i\beta} \sin\alpha)$ with $\alpha, \beta \in \mathbb{R}$. Evaluate $\phi^* \boldsymbol{\sigma} \phi$ explicitly in terms of α and β, using the explicit formulas (2.105) for $\boldsymbol{\sigma}$. Then check that it is a unit vector.

For example, $(0, 1)$ is orthogonal in S to $(1, 0)$, while the corresponding 3-vectors point in the $+z$- and $-z$-direction, differing by 180°. Expressed the other way around, spinors differ by half the angle of their associated vectors. That is why one says that electrons have *spin one half*.

There are also other types of spinors, other than spin-$\frac{1}{2}$: spin-1, spin-$\frac{3}{2}$, spin-2, spin-$\frac{5}{2}$, etc. The space of spin-s spinors has complex dimension $2s + 1$, and the analogs of the Pauli matrices are $(2s+1)\times(2s+1)$ matrices; we note in passing what their entries are[10] ($r, r' \in \{1, 2, \ldots, 2s + 1\}$):

$$(\tilde{\sigma}_1)_{rr'} = (\delta_{r,r'+1} + \delta_{r+1,r'})\sqrt{(s+1)(r+r'-1) - rr'} \tag{2.111a}$$

$$(\tilde{\sigma}_2)_{rr'} = (i\delta_{r+1,r'} - i\delta_{r,r'+1})\sqrt{(s+1)(r+r'-1) - rr'} \tag{2.111b}$$

$$(\tilde{\sigma}_3)_{rr'} = 2(s+1-r)\delta_{rr'}\,. \tag{2.111c}$$

In this context, wave functions $\psi : \mathbb{R}^3 \to \mathbb{C}$ are said to have *spin 0*. Electrons, quarks, and all known species of elementary *matter particles* have spin $\frac{1}{2}$; all known species of elementary *force particles* (photons, W and Z bosons, and gluons) have spin 1; the predicted but not yet confirmed graviton has spin 2; the only particle species with spin 0 in the standard model of particle physics is the *Higgs particle* or *Higgs boson*, which was experimentally confirmed in 2012 at LHC at the European nuclear research center CERN in Geneva, Switzerland (but it is possible that the Higgs boson consists of several particles).

2.3.2 The Pauli Equation

When spin is taken into account, the Schrödinger equation reads a little differently. The appropriate version is known as the *Pauli equation*. We will not study this equation in detail; we write it down mainly for the sake of completeness:

$$i\hbar\frac{\partial\psi}{\partial t} = \frac{1}{2m}\Big(-i\hbar\nabla - e\boldsymbol{A}(\boldsymbol{x})\Big)^2\psi(\boldsymbol{x}) - \frac{\hbar e}{2m}\boldsymbol{\sigma}\cdot\boldsymbol{B}(\boldsymbol{x})\psi(\boldsymbol{x}) + V(\boldsymbol{x})\psi(\boldsymbol{x}) \tag{2.112}$$

with $\boldsymbol{B}$ the magnetic field, V the electric and gravitational potential, and $\boldsymbol{A}$ the magnetic vector potential. So, in comparison to (2.52), it has an additional term proportional to $\boldsymbol{\sigma}\cdot\boldsymbol{B}\,\psi$.

The Hilbert space of wave functions with spin is denoted by $L^2(\mathbb{R}^3, \mathbb{C}^2)$ and contains the square-integrable functions $\mathbb{R}^3 \to \mathbb{C}^2$. The inner product is

$$\langle\psi|\phi\rangle = \int_{\mathbb{R}^3} d^3\boldsymbol{x}\,\psi^*(\boldsymbol{x})\,\phi(\boldsymbol{x}) = \int_{\mathbb{R}^3} d^3\boldsymbol{x}\sum_{s=1}^{2}\psi_s^*(\boldsymbol{x})\,\phi_s(\boldsymbol{x})\,. \tag{2.113}$$

[10] Relations (2.111) are quoted from http://en.wikipedia.org/wiki/3D_rotation_group.

Born Rule for Position, Given a Spinor-Valued Wave Function.

$$\rho(\boldsymbol{x}) = |\psi(\boldsymbol{x})|^2 := \psi^*(\boldsymbol{x})\,\psi(\boldsymbol{x}) = \|\psi(\boldsymbol{x})\|_S^2 = \sum_{s=1}^{2} |\psi_s(\boldsymbol{x})|^2\,. \tag{2.114}$$

Note that this is a special case of the general Born rule (2.81) for the position operators X_j. In the following, we will simply write $|\cdot|$ instead of $\|\cdot\|_S$.

2.3.3 The Stern–Gerlach Experiment

Let us write

$$\psi(\boldsymbol{x}) = \begin{pmatrix} \psi_1(\boldsymbol{x}) \\ \psi_2(\boldsymbol{x}) \end{pmatrix}. \tag{2.115}$$

In the first half of a Stern–Gerlach experiment (first done in 1922 with silver atoms), a wave packet moves through a magnetic field that is carefully designed so as to deflect $\psi_1(\boldsymbol{x})$ in a different direction than $\psi_2(\boldsymbol{x})$ and thus to separate the two components in space (Fig. 2.3). That is, while the wave packet traverses the magnetic field in the x-direction, the ψ_1 part will be deflected in the z-direction and ψ_2 in the $-z$-direction. Put differently, if the initial wave function $\psi(t = 0)$ has support[11] in the ball $B_r(\boldsymbol{y})$ of radius r around the center $\boldsymbol{y}$, then the final wave function $\psi(t = 1)$ (i.e., the wave function after passing through the magnetic field) is such that $\psi_1(\boldsymbol{x}, t = 1)$ has support in $B_+ := B_r\big(\boldsymbol{y} + (1, 0, d)\big)$ and $\psi_2(\boldsymbol{x}, t = 1)$ in $B_- := B_r\big(\boldsymbol{y} + (1, 0, -d)\big)$ with deflection distance $d > r$ (so that ψ_1 and ψ_2 do not overlap). The arrangement creating this magnetic field is called a Stern–Gerlach magnet. In the second half of the Stern–Gerlach experiment, one applies detectors to the regions $B_\pm$. If the electron is found in B_+, then the outcome of the experiment is said to be *up* and if in B_-, then *down*.

A case of particular interest is that the initial wave function satisfies

$$\psi_s(\boldsymbol{x}) = \phi_s\,\chi(\boldsymbol{x})\,, \tag{2.116}$$

where $\phi \in S$, $\|\phi\|_S = 1$, and $\chi : \mathbb{R}^3 \to \mathbb{C}$, $\|\chi\| = 1$. One says that for such a ψ, the spin degree of freedom is disentangled from the spatial degrees of freedom. (Before, we have considered many-particle wave functions for which some particles were disentangled from others. We may also consider a single particle and say that

[11] The *support* of a function is the set of points where the function is nonzero. (In the math literature, it is defined as the *closure* of the set of points where the function is nonzero.) In this book, we take the *support* to mean the set of points where the function is *significantly* nonzero. That is, if a wave packet has thin tails, then we will not include them in the support; as a consequence, of course, the notion of support is not precisely defined.

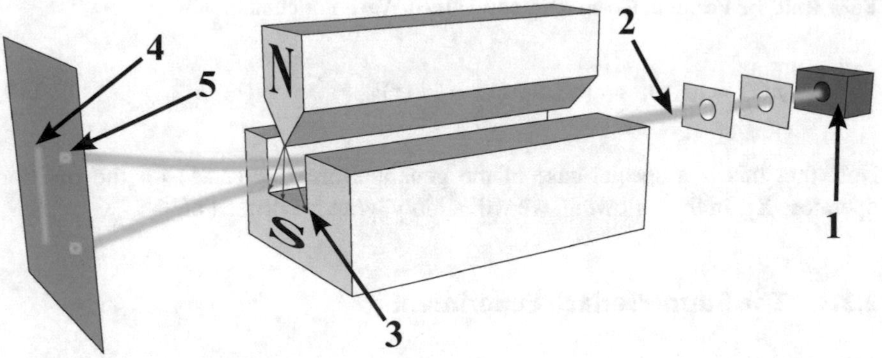

Fig. 2.3 Setup of the Stern–Gerlach experiment: (1) furnace, (2) beam of silver atoms, (3) inhomogeneous magnetic field, (4) classically expected result, and (5) observed result. Picture credit: wikipedia/Theresa Knott under CC BY-SA 3.0 license

the x variable is disentangled from the y and z variables if and only if $\psi(x, y, z) = f(x)\, g(y, z)$.)

In the case (2.116), assuming that χ has support in $B_r(\boldsymbol{y})$, the wave function after passing the magnet is

$$\begin{pmatrix} \phi_1\, \chi\big(\boldsymbol{x} - (1, 0, d)\big) \\ \phi_2\, \chi\big(\boldsymbol{x} - (1, 0, -d)\big) \end{pmatrix}, \tag{2.117}$$

and it follows from the Born rule (2.114) for position that the probability of outcome "up" is $|\phi_1|^2$ and that of "down" is $|\phi_2|^2$.

These probabilities agree with what we would have obtained from the general Born rule (2.81) for the observable $A = \sigma_3$ and the vector ϕ in the Hilbert space $\mathscr{H} = S$. The spinors $\phi_{+1} = (1, 0)$ and $\phi_{-1} = (0, 1)$ form an orthonormal basis of S consisting of eigenvectors of σ_3 (with eigenvalues $+1$ and -1, respectively); ϕ plays the role of ψ in (2.81); its coefficients in the ONB referred to in Eq. (2.81) are $\langle\phi_{+1}|\phi\rangle = \phi_1$ and $\langle\phi_{-1}|\phi\rangle = \phi_2$. That is why the Stern–Gerlach experiment is often called a "quantum measurement of σ_3" or "of the z-component of spin."

The Stern–Gerlach magnet can be rotated into any direction. For example, by rotating by 90° around the x-axis (a rotation that will map the z-axis to the y-axis), we obtain an arrangement that will deflect part of the initial wave packet ψ in the $+y$-direction and another part in the $-y$-direction. However, these parts are not ϕ_1 and ϕ_2. Instead, they are the parts along a different ONB of S:

$$\phi^{(+)} = \frac{1}{\sqrt{2}}(1, i) \text{ and } \phi^{(-)} = \frac{1}{\sqrt{2}}(1, -i) \text{ form an ONB of } S \text{ with } \boldsymbol{\omega}(\phi^{(\pm)}) = (0, \pm 1, 0). \tag{2.118}$$

That is, any $\psi : \mathbb{R}^3 \to S$ can be written as $\psi(\boldsymbol{x}) = c_+(\boldsymbol{x})\phi^{(+)} + c_-(\boldsymbol{x})\phi^{(-)}$, and these two terms will get spatially separated (in the $\pm y$-direction, in fact). The probabilities of outcomes "up" and "down" are then $\int d\boldsymbol{x}\,|c_\pm(\boldsymbol{x})|^2$. In the special case (2.116), the probabilities are just $|c_\pm|^2$, where $\phi = c_+\phi^{(+)} + c_-\phi^{(-)}$. Equivalently, the probabilities are $|\langle\phi^{(\pm)}|\phi\rangle|^2$. These values are in agreement with the general Born rule for $A = \sigma_2$ because $\phi^{(\pm)}$ are eigenvectors of σ_2 with eigenvalues ± 1.

Generally, if the Stern–Gerlach magnet is rotated from the z-direction to direction $\boldsymbol{n}$, where $\boldsymbol{n}$ is any unit vector in $\mathbb{R}^3$, then the probabilities of its outcomes are governed by the Born rule (2.81) for $A = \boldsymbol{n} \cdot \boldsymbol{\sigma}$, which for any $\boldsymbol{n}$ is a self-adjoint 2×2 matrix with eigenvalues ± 1.

2.3.4 Bohmian Mechanics with Spin

Bell (1966) [2] realized how to do Bohmian mechanics for particles with spin. It is surprisingly simple. Here is the single-particle version. Replace the Schrödinger equation by the Pauli equation and Bohm's equation of motion (2.54) by

$$\frac{d\boldsymbol{Q}}{dt} = \frac{\hbar}{m}\mathrm{Im}\frac{\psi^*(\nabla - \frac{ie}{\hbar}\boldsymbol{A})\psi}{\psi^*\psi}(t, \boldsymbol{Q}(t))\,. \tag{2.119}$$

In the case without magnetic field, $\boldsymbol{A} = \boldsymbol{0}$, this reduces to

$$\frac{d\boldsymbol{Q}}{dt} = \frac{\hbar}{m}\mathrm{Im}\frac{\psi^*\nabla\psi}{\psi^*\psi}(t, \boldsymbol{Q}(t))\,. \tag{2.120}$$

(An alternative to this equation of motion is suggested by relativistic considerations and will be discussed in Sect. 7.3.7. For the time being, we will focus on this equation of motion.)

Recall that $\psi^*\psi$ means the inner product in spin space, so the denominator means

$$\psi^*(\boldsymbol{x})\psi(\boldsymbol{x}) = |\psi_1(\boldsymbol{x})|^2 + |\psi_2(\boldsymbol{x})|^2. \tag{2.121}$$

Likewise, the numerator means

$$\psi^*(\boldsymbol{x})(\nabla - \tfrac{ie}{\hbar}\boldsymbol{A})\psi(\boldsymbol{x}) = \psi_1^*(\boldsymbol{x})(\nabla - \tfrac{ie}{\hbar}\boldsymbol{A})\psi_1(\boldsymbol{x}) + \psi_2^*(\boldsymbol{x})(\nabla - \tfrac{ie}{\hbar}\boldsymbol{A})\psi_2(\boldsymbol{x})\,. \tag{2.122}$$

The initial position $\boldsymbol{Q}(0)$ is assumed to be random with probability density

$$\rho_0(\boldsymbol{x}) = |\psi_0(\boldsymbol{x})|^2\,. \tag{2.123}$$

It follows that $\boldsymbol{Q}(t)$ has probability density $|\psi_t|^2$ at every t. This version of the *equivariance theorem* can be obtained by a very similar computation as in the spinless case, involving the following variant of the continuity equation:

$$\frac{\partial|\psi|^2}{\partial t} = -\nabla \cdot \Big(\tfrac{\hbar}{m}\mathrm{Im}\big[\psi^*(\nabla - \tfrac{ie}{\hbar}\boldsymbol{A})\psi\big]\Big). \tag{2.124}$$

As a consequence of the equivariance theorem, Bohmian mechanics leads to the correct probabilities for the Stern–Gerlach experiment.

2.3.5 Is an Electron a Spinning Ball?

If it were, then the following paradox would arise. According to classical electrodynamics (which of course is well confirmed for macroscopic objects), a spinning, electrically charged object behaves like a magnet in two ways: it creates its own magnetic field, and it reacts to an external magnetic field. Just as the strength of the electric charge can be expressed by a number, the charge e, the strength of the magnet can be expressed by a vector, the *magnetic dipole moment* or just *magnetic moment* $\boldsymbol{\mu}$. Its direction points from the south pole to the north pole, and its magnitude is the strength of the magnet. The magnetic moment of a charge e spinning at angular frequency ω around the axis along the unit vector $\boldsymbol{u}$ is, according to classical electrodynamics,

$$\boldsymbol{\mu} = \gamma e \omega \boldsymbol{u}\,, \tag{2.125}$$

where the factor γ depends on the size and shape of the object. Furthermore, if such an object flies through a Stern–Gerlach magnet oriented in direction $\boldsymbol{n}$, then, still according to classical electrodynamics, it gets deflected by an amount proportional to $\boldsymbol{\mu}\cdot\boldsymbol{n}$. Put differently, the Stern–Gerlach experiment for a classical object measures μ_z or the component of $\boldsymbol{\mu}$ in the direction of $\boldsymbol{n}$. The vector $\omega\boldsymbol{u}$ is called the *spin vector*.

Where is the paradox? It is that different choices of $\boldsymbol{n}$, when applied to objects with the same $\boldsymbol{\mu}$, would lead to a continuous interval of deflections $[-\gamma|e|\omega, +\gamma|e|\omega]$, whereas the Stern–Gerlach experiment, for whichever choice of $\boldsymbol{n}$, leads to a discrete set $\{+d, -d\}$ of two possible deflections.

The latter fact was called by Wolfgang Pauli (1900–1958) the "non-classical two-valuedness of spin." This makes it hard to come up with a theory in which the outcome of a Stern–Gerlach experiment has anything to do with a spinning motion. While Feynman went too far when claiming that the double-slit experiment does not permit any deeper explanation, it seems safe to say that the Stern–Gerlach experiment does not permit an explanation in terms of spinning balls.

2.3.6 Are There Actual Spin Values?

Here is another perspective on the question whether the electron is a spinning ball, from a Bohmian angle. We have seen that Bohmian mechanics does not involve any spinning motion to account for (what has come to be called) spin; electrons have actual positions but not an actual spin vector. However, some authors felt electrons *should* have an actual spin vector and have made proposals in this direction; let me explain why the most natural proposal in this direction, due to Bohm, Schiller, and Tiomno (1955) [6], is unconvincing.

Consider a single electron. Since ψ_t is a function from $\mathbb{R}^3$ to spin space $\mathbb{C}^2$, $\psi_t(\boldsymbol{Q}_t)$ is a vector in $\mathbb{C}^2$ and thus associated with a direction in $\mathbb{R}^3$, i.e., that of $\boldsymbol{\omega}(\psi_t(\boldsymbol{Q}_t))$. The proposal is to regard the real 3-vector

$$\boldsymbol{S}_t := \frac{\boldsymbol{\omega}(\psi_t(\boldsymbol{Q}_t))}{|\boldsymbol{\omega}(\psi_t(\boldsymbol{Q}_t))|} = \frac{\psi^*\boldsymbol{\sigma}\psi}{\psi^*\psi}(t, \boldsymbol{Q}_t) \tag{2.126}$$

as a further fundamental variable representing the actual spin vector of the particle, so that the full state is given by the triple $(\psi_t, \boldsymbol{Q}_t, \boldsymbol{S}_t)$. It is tempting to imagine the electron as a little ball spinning at angular velocity proportional to $\boldsymbol{S}_t$ (i.e., at a fixed angular speed around the axis in the direction of $\boldsymbol{S}_t$), while its center moves according to $\boldsymbol{Q}_t$.

There are two problems with this picture and with $\boldsymbol{S}_t$ as a further fundamental variable. They become visible when we consider a Stern–Gerlach experiment, say in the z-direction, starting at time τ. Since Stern–Gerlach experiments are often called "measurements of z-spin," one might expect that the outcome of the experiment is the z-component of $\boldsymbol{S}_\tau$. But that is not the case. This is the first problem. Second, the outcome of the experiment is read off from the final position of the particle, and that position depends on the initial wave function ψ_τ and the initial position $\boldsymbol{Q}_\tau$, but the equation of motion for $\boldsymbol{Q}_t$ does not involve $\boldsymbol{S}_t$, so one can say that the further fundamental variable actually has no influence on the outcome! The variable $\boldsymbol{S}_t$ is superfluous, and we have already discussed how the Stern–Gerlach experiment and indeed all phenomena involving spin are naturally explained with just ψ_t and Q_t as fundamental variables. There is no phenomenon whose explanation would require the introduction of $\boldsymbol{S}_t$ or would merely be made simpler by the introduction of $\boldsymbol{S}_t$. An actual spin vector is neither useful nor needed in Bohmian mechanics.

Let me return once more to the thought that, in view of the Bohmian particles having actual positions, they should also have actual spin values. I am arguing against this thought. Let us have a look at another natural possibility for implementing this thought and see why it fails. The fact that $L^2(\mathbb{R}^3, \mathbb{C}^2)$ is canonically isomorphic to $L^2(\mathbb{R}^3 \times \{1, 2\}, \mathbb{C})$ (i.e., that the index s of $\psi_s(\boldsymbol{x})$ could be thought of as a discrete variable with the same status as $\boldsymbol{x}$) may suggest that in a Bohm-style approach, one should have an actual value not just of $\boldsymbol{Q}$ but of $(\boldsymbol{Q}, \Sigma)$ with $\Sigma \in \{1, 2\}$ and that Born's rule should say $\mathbb{P}(\boldsymbol{Q} \in d^3\boldsymbol{x}, \Sigma = s) = |\psi_s(\boldsymbol{x})|^2$. Leaving aside the question how Σ should evolve, the main problem with this proposal is that it would break rotational invariance: it would prefer σ_3 over σ_1 and σ_2.[12] The upshot

[12] I mention in passing that this point becomes even more severe in curved space-time, where it is not necessarily possible to define what counts as the "same" direction of space at different

is that the idea of introducing actual spin values is not as plausible as it may have seemed at first.

2.3.7 Many-Particle Systems

The wave function of N electrons is of the form

$$\psi_{s_1,s_2,\ldots,s_N}(\boldsymbol{x}_1, \boldsymbol{x}_2, \ldots, \boldsymbol{x}_N)\,, \tag{2.127}$$

where each $\boldsymbol{x}_j$ varies in $\mathbb{R}^3$ and each index s_j in $\{1, 2\}$. Thus, at any configuration, ψ has 2^N complex components, or $\psi : \mathbb{R}^{3N} \to \mathbb{C}^{2^N}$. Note that while $\mathbb{R}^{3N}$ is the Cartesian product of N copies of $\mathbb{R}^3$, $\mathbb{C}^{2^N}$ is *not* the Cartesian product of N copies of $\mathbb{C}^2$ (which would have dimension $2N$) but the *tensor product* of N copies of $\mathbb{C}^2$. Equivalently, we could write ψ as a function $\mathbb{R}^{3N} \times \{1, 2\}^N \to \mathbb{C}$, where the set $\{1, 2\}^N$ of possible index values $(s_1, \ldots, s_N)$ is a Cartesian product of N copies of $\{1, 2\}$; but in the following, it will be convenient to write ψ as a function $\mathbb{R}^{3N} \to \mathbb{C}^{2^N}$.

The Pauli equation then reads

$$i\hbar\frac{\partial\psi}{\partial t} = \frac{1}{2m}\sum_{k=1}^{N}\big(-i\hbar\nabla_k - \boldsymbol{A}(\boldsymbol{x}_k)\big)^2\psi - \sum_{k=1}^{N}\frac{\hbar}{2m}\boldsymbol{\sigma}_{(k)}\cdot\boldsymbol{B}(\boldsymbol{x}_k)\psi + V\psi\,, \tag{2.128}$$

where $\boldsymbol{\sigma}_{(k)}$ means $\boldsymbol{\sigma}$ acting on the index s_k of ψ. Change the definition (2.106) of the spin inner product $\phi^*\psi$, and Born's rule (2.114), so as to sum over all spin indices s_j. Moreover, in Bohm's equation of motion (2.119), replace $\boldsymbol{Q} \in \mathbb{R}^3$ by $Q \in \mathbb{R}^{3N}$.

2.3.8 Representations of $SO(3)$

A deeper understanding of spinors comes from group representations.[13] Let us start easily. Consider the wave function of a single particle. Suppose it were, instead of a complex scalar field, a vector field, so $\boldsymbol{\psi} : \mathbb{R}^3 \to \mathbb{R}^3$. Well, it should be complex, so we complexify the vector field, $\boldsymbol{\psi} : \mathbb{R}^3 \to \mathbb{C}^3$. Now rotate your coordinate system according to $R \in SO(3)$. Then in the new coordinates, the same physical wave function is represented by a different mathematical function,

$$\tilde{\boldsymbol{\psi}}(\boldsymbol{x}) = R\boldsymbol{\psi}(R^{-1}\boldsymbol{x})\,. \tag{2.129}$$

locations. We are thus led away from thinking of s as a discrete variable and toward thinking of the wave function as spinor-valued, possibly with different spin spaces at different locations.

[13] More details about the topic of this section can be found in Sexl and Urbantke (2001) [14].

Instead of real-valued potentials, the Schrödinger equation could then include matrix-valued potentials, provided the matrices are always self-adjoint:

$$i\hbar\frac{\partial\psi}{\partial t} = -\frac{\hbar^2}{2m}\Delta\psi + V\psi\,. \tag{2.130}$$

Now consider another possibility: that the wave function is tensor-valued, ψ_{ab} with $a, b = 1, 2, 3$. Then in a rotated coordinate system,

$$\tilde{\psi}_{ab}(\boldsymbol{x}) = \sum_{c,d=1}^{3} R_{ac} R_{bd} \psi_{cd}(R^{-1}\boldsymbol{x})\,. \tag{2.131}$$

What the two examples have in common is that the components of the wave function get transformed as well according to the scheme, for $\psi : \mathbb{R}^3 \to \mathbb{C}^d$,

$$\tilde{\psi}_r(\boldsymbol{x}) = \sum_{s=1}^{d} M_{rs}(R)\, \psi_s(R^{-1}\boldsymbol{x})\,. \tag{2.132}$$

The matrices $M(R)$ satisfy the composition law

$$M(R_1)\, M(R_2) = M(R_1 R_2) \quad \text{and} \quad M(I) = I\,, \tag{2.133}$$

which means that they form a *representation* of the group $SO(3)$ of rotations—in other words, a group homomorphism from $SO(3)$ to $GL(\mathbb{C}^d)$, the "general linear group" comprising all invertible operators on $\mathbb{C}^d$. Further representations of $SO(3)$ provide further possible value spaces for wave functions ψ.

Spin space S for spin-$\frac{1}{2}$ is almost of this kind, but there is one more complication: $SO(3)$ is represented, not by linear mappings $S \to S$, but by *projective mappings* $P(S) \to P(S)$. Here, $P(S)$ is the set of all one-dimensional subspaces of S, called the *projective space* of S. The use of projective space seems fitting as two wave functions that differ only by a phase factor, $\phi(x) = e^{i\theta}\psi(x)$, are usually regarded as representing the same physical quantum state (they yield the same Born distribution, at all times and for all observables, and the same Bohmian trajectories for all times). That is, one can say that a wave function is really an element of $P(\mathscr{H})$ rather than $\mathscr{H}$ because every normalized element of $\mathbb{C}\psi$ is as good as ψ.

A *projective mapping* $F : P(S) \to P(S)$ is one that is consistent with a linear mapping, i.e., is such that there is a linear mapping $M : S \to S$ with $F(\mathbb{C}\psi) = \mathbb{C}M\psi$. While M determines F uniquely, F does not determine M, as zM with any $z \in \mathbb{C}\backslash\{0\}$ leads to the same F. In particular, if we are given $F(R)$ and want an $M(R)$, then $-M(R)$ is always another possible candidate. For spin-$\frac{1}{2}$, it turns out that while $F(R_1)\, F(R_2) = F(R_1 R_2)$ as it should, $M(R)$ can at best be found in such a way that

$$M(R_1)\, M(R_2) = \pm M(R_1 R_2)\,. \tag{2.134}$$

This sign mismatch has something to do with the halved angles. The M are elements of $SU(2)$ (the "special unitary" group of unitary 2×2 matrices with determinant 1) and with every element R of $SO(3)$ are associated two elements of $SU(2)$ that differ by a sign.

This association can actually be regarded as a mapping in the opposite direction,

$$\varphi : SU(2) \to SO(3)\,, \; M \mapsto R\,. \tag{2.135}$$

This mapping φ is a group homomorphism (i.e., $\varphi(M_1)\varphi(M_2) = \varphi(M_1 M_2)$ and $\varphi(I) = I$) and is smooth, two-to-one $[\varphi(-M) = \varphi(M)]$, and locally a diffeomorphism. The situation is similar to the group homomorphism $\chi : \mathbb{R} \to U(1)$, $\theta \mapsto e^{i\theta}$, which is also smooth, many-to-one, and locally a diffeomorphism; just like $\mathbb{R}$ is what you get from the circle $U(1)$ when you unfold it, $SU(2)$ is what you get from $SO(3)$ when you "unfold" it. (The unfolding of a manifold $\mathcal{Q}$ is called the *universal covering space* $\widehat{\mathcal{Q}}$, so $\widehat{SO(3)} = SU(2)$.) For every continuous curve γ in $SO(3)$ starting in $I \in SO(3)$, there is a unique continuous curve $\hat{\gamma}$ in $SU(2)$ starting at $I \in SU(2)$ with $\varphi \circ \hat{\gamma} = \gamma$, called the *lift* of γ. An interesting special case is a continuous curve γ in $SO(3)$ starting at I that keeps the axis of rotation fixed but continuously increases the rotation angle from 0 up to a final value θ. Its lift $\hat{\gamma}$ in $SU(2)$ will have a fixed eigenbasis (i.e., a simultaneous eigenbasis of $\hat{\gamma}(s)$ for every s) but eigenvalues that vary continuously from 1 to $e^{\pm i\theta/2}$; we see again the phenomenon that angles in spin space are only half as large as in physical space. In this sense, a rotation in physical space by 360° will correspond to one by 180° in spin space and carry ϕ to $-\phi$, whereas a rotation in physical space by 720° will carry ϕ to itself.[14]

The upshot of all this is that S is equipped with a particular projective representation of $SO(3)$; in other words, spinors are one of the various types of mathematical objects (besides vectors and tensors) that react to rotations in a well-defined way, and that is why they qualify as possible values of a wave function.

2.3.9 Inverted Stern–Gerlach Magnet and Contextuality

Consider again the Stern–Gerlach experiment in the $+z$-direction on an initial wave function of the form $\psi_s(\boldsymbol{x}) = \phi_s\, \chi(\boldsymbol{x})$ as in (2.116), with χ a fixed scalar packet, while we consider different $\phi \in S$. As mentioned, Bohmian mechanics leads to outcome Z = "up" with probability $|\phi_1|^2$ and Z = "down" with $|\phi_2|^2$, in agreement with the Born rule for σ_3.

[14] This behavior of spinors is similar to complex square roots: on the one hand, every nonzero complex number has two square roots differing by a sign; on the other, for any continuous curve γ in $\mathbb{C}\setminus\{0\}$ starting in 1, there is a unique continuous curve $\hat{\gamma}$ in $\mathbb{C}\setminus\{0\}$ starting in 1 with $\hat{\gamma}(t)^2 = \gamma(t)$ for all t. (The sign ambiguity is resolved by continuity.) When γ is closed and winds around 0 once, then $\hat{\gamma}$ will not be closed. But when γ is closed and winds around 0 twice, then $\hat{\gamma}$ will be closed. Note that the $\boldsymbol{\omega}$ mapping provides a sense in which spinors are square roots of vectors. In relativity (see Sect. 7.2.2), we will encounter a deepened sense of spinors as square roots of vectors.

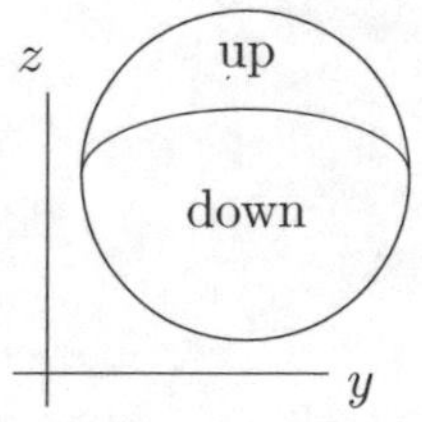

Fig. 2.4 Outcome of the experiment as a function of the y- and z-components of the initial position $Q(\tau) = (x, y, z)$ of the Bohmian particle. The curve separating the two regions indicates the critical surface mentioned in the text; the outer circle encloses the support of χ

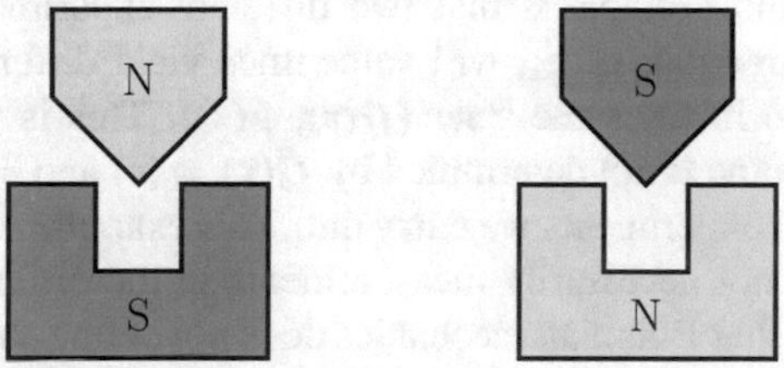

Fig. 2.5 Left: Schematic picture of a Stern–Gerlach magnet. Right: Modified Stern–Gerlach magnet with inverted polarity (north and south exchanged roles) while keeping the shape

Moreover, the outcome Z is determined by ϕ and the initial position $Q(\tau)$ of the Bohmian particle at the time τ at which the experiment begins. In fact, suppose for simplicity that the approximate time evolution of ψ described in Sect. 2.3.3 including (2.117) is valid and all the Bohmian trajectories have equal x-velocities and vanishing y-velocity; then the topmost $Q(\tau)$ (above a certain surface) as in Fig. 2.4 will end up in the "up" packet in B_+, and those below the critical surface in the "down" packet in B_-.

Now comes a subtle point: The outcome Z is *not* determined by $Q(\tau)$, $\psi(\tau)$, and σ_3 alone. To understand this statement, let us consider an example (due to David Albert (1992) [1]) consisting of a modified Stern–Gerlach experiment in which the polarity of the magnet has been changed as in Fig. 2.5.

It follows that the spin-up part of the wave function, ψ_1, gets deflected downward and the spin-down part, ψ_2, deflected upward. For this reason, let us decide that if the particle gets detected in the upward deflected location, B_+, then we say that the outcome Z is "down," and if detected in B_-, then the outcome Z is "up." With this convention, the probability of "up" is $|\phi_1|^2$ and that of "down" is $|\phi_2|^2$, in agreement with the Born rule for σ_3. That is, the modified experiment (flipped in two ways) is again a quantum measurement of the observable σ_3.

In the modified experiment, the Bohmian particle will again end up in B_+ if the initial position $Q(\tau)$ lies above a certain critical surface (possibly different from before) and in B_- if initially below the surface; see Fig. 2.6. But now, ending up in B_+ means outcome "down." Thus, if $|\phi_1|^2$ is neither 0 nor 1, then an initial position

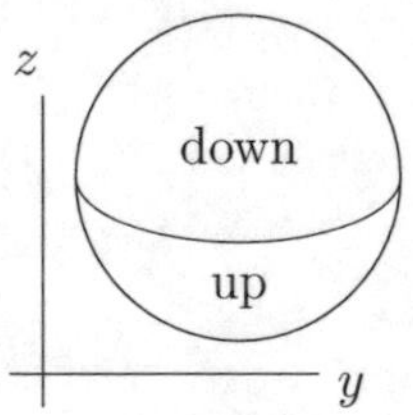

Fig. 2.6 Outcome of the modified experiment as a function of the initial position

$Q(\tau)$ near the top will lead to outcome "up" in the original experiment but "down" in the modified experiment.

The upshot of this example is that two different experiments, both of which are quantum measurements of σ_3, will sometimes yield different outcomes when applied to a particle in the same state $(Q(\tau), \psi(\tau))$. This is what I meant when saying that the outcome is not determined by $Q(\tau)$, $\psi(\tau)$ and σ_3 alone—it depends on which of the two experiments we carry out. This example shows that "quantum measurements" are not necessarily measurements in the ordinary meaning of the word. It also shows that Bohmian mechanics does not define an actual value of σ_3.

The fact that different ways of "measuring σ_3" can yield different outcomes is called *contextuality*. In the literature, contextuality is sometimes presented as the weird, mysterious trait (of the quantum world or of Bohmian mechanics) that a measurement outcome may depend on the "context" of the measurement. But really the sense of paradox arises only from taking the word "measurement" too literally, and the trivial essence of the perceived mystery has been nicely formulated by Detlef Dürr, Sheldon Goldstein, Nino Zanghì (2004) [7, Sec. 8.4]:

> The result of an experiment depends upon the experiment.

The idea that there should be an actual value of σ_3 has led to a lot of discussion in the literature associated with the key words "non-contextual hidden variables." It turns out that they are mathematically impossible, as we will see later in the section on no-hidden-variables theorems, Sect. 5.7.

Exercises

Exercise 2.1 can be found in Sect. 2.1.1, Exercise 2.2 in Sect. 2.1.5, Exercises 2.3–2.5 in Sect. 2.2.7, and Exercises 2.6–2.8 in Sect. 2.3.1.

Exercise 2.9 (Cannot Distinguish Non-orthogonal State Vectors)

(a) Alice gives to Bob a single particle whose spin state ψ is either $(1, 0)$ or $(0, 1)$ or $\frac{1}{\sqrt{2}}(1, 1)$. Bob can carry out a quantum measurement of an arbitrary self-adjoint operator. Show that it is impossible for Bob to decide with certainty which of the three states ψ is.

(b) The same with only $(1, 0)$ and $\frac{1}{\sqrt{2}}(1, 1)$.

Exercise 2.10 (Conserved Operator) Show that if the potential $V(\boldsymbol{x}_1, \ldots, \boldsymbol{x}_N)$ is translation invariant [as in (1.64) with $R = I$], then each component of the total momentum operator

$$P_a = -i\hbar \sum_{j=1}^{N} \frac{\partial}{\partial x_{ja}} \tag{2.136}$$

($a = 1, 2, 3$) commutes with the Hamiltonian

$$H = -\sum_{j=1}^{N} \frac{\hbar^2}{2m_j} \nabla_j^2 + V\,. \tag{2.137}$$

Show further that if V is rotation invariant [as in (1.64) with $\boldsymbol{a} = \boldsymbol{0}$], then each component of the total angular momentum operator

$$\boldsymbol{L} = -i\hbar \sum_{j=1}^{N} \boldsymbol{x}_j \times \nabla_j \tag{2.138}$$

commutes with H.

Exercise 2.11 (Pauli Matrices) We know from Exercise 2.6(b) that for every unit vector $\boldsymbol{n} \in \mathbb{R}^3$, $\sigma_{\boldsymbol{n}} = \boldsymbol{n} \cdot \boldsymbol{\sigma}$ has eigenvalues ± 1. Derive from this fact that, for any $\alpha \in \mathbb{R}$ and $\boldsymbol{\beta} \in \mathbb{R}^3$, $A = \alpha I + \boldsymbol{\beta} \cdot \boldsymbol{\sigma}$ has eigenvalues $\alpha \pm |\boldsymbol{\beta}|$.

References

1. D.Z. Albert, *Quantum Mechanics and Experience* (Harvard University Press, Cambridge, 1992)
2. J.S. Bell, On the problem of hidden variables in quantum mechanics. Rev. Mod. Phys. **38**, 447–452 (1966). Reprinted as chapter 1 of [3]
3. J.S. Bell, *Speakable and Unspeakable in Quantum Mechanics* (Cambridge University Press, Cambridge, 1987)
4. J.S. Bell, Against "measurement", in *Sixty-Two Years of Uncertainty. Historical, Philosophical, and Physical Inquiries into the Foundations of Quantum Physics*, ed. by A.I. Miller. NATO ASI Series B, vol. 226 (Plenum Press, New York, 1990). Reprinted in: Physics World **3**(8), 33–40 (1990)
5. D. Bohm, B.J. Hiley, *The Undivided Universe. An Ontological Interpretation of Quantum Theory* (Routledge, London, 1993)
6. D. Bohm, R. Schiller, J. Tiomno, A causal interpretation of the Pauli equation (A). Il Nuovo Cimento Supplementi **1**, 48–66 (1955)
7. D. Dürr, S. Goldstein, N. Zanghì, Quantum equilibrium and the role of operators as observables in quantum theory. J. Stat. Phys. **116**, 959–1055 (2004). http://arxiv.org/abs/quant-ph/0308038. Reprinted in [8]

8. D. Dürr, S. Goldstein, N. Zanghì, *Quantum Physics Without Quantum Philosophy* (Springer, Heidelberg, 2013)
9. P.L. Garrido, S. Goldstein, J. Lukkarinen, R. Tumulka, Paradoxical reflection in quantum mechanics. Am. J. Phys. **79**(12), 1218–1231 (2011). http://arxiv.org/abs/0808.0610
10. B.C. Hall, *Quantum Theory for Mathematicians* (Springer, New York, 2013)
11. D. Kammler, *A First Course in Fourier Analysis*, 2nd edn. (Cambridge University Press, Cambridge, 2007)
12. T. Norsen, The pilot-wave perspective on quantum scattering and tunneling. Am. J. Phys. **81**, 258 (2013). http://arxiv.org/abs/1210.7265
13. M. Reed, B. Simon, *Methods of Modern Mathematical Physics*, vol. 1 (revised edition) (Academic Press, Cambridge, 1980)
14. R.U. Sexl, H.K. Urbantke, *Relativity, Groups, Particles* (Springer, Heidelberg, 2001)

Collapse and Measurement 3

3.1 The Projection Postulate

3.1.1 Notation

In the *Dirac notation*, one writes $|\psi\rangle$ for ψ. This may seem like a waste of symbols at first, but often it is the opposite, as it allows us to replace a notation such as $\phi_1, \phi_2, \ldots$ by $|1\rangle, |2\rangle, \ldots$. Of course, a definition is needed for what $|n\rangle$ means, just as one would be needed for ϕ_n. It is also convenient when using long subscripts, such as replacing $\psi_{\text{left slit}}$ by $|\text{left slit}\rangle$. In spin space S, one commonly writes

$$|z\text{-up}\rangle = |\uparrow\rangle = \begin{pmatrix}1\\0\end{pmatrix}, \quad |z\text{-down}\rangle = |\downarrow\rangle = \begin{pmatrix}0\\1\end{pmatrix} \tag{3.1}$$

$$|y\text{-up}\rangle = \frac{1}{\sqrt{2}}\begin{pmatrix}1\\i\end{pmatrix}, \quad |y\text{-down}\rangle = \frac{1}{\sqrt{2}}\begin{pmatrix}1\\-i\end{pmatrix} \tag{3.2}$$

$$|x\text{-up}\rangle = \frac{1}{\sqrt{2}}\begin{pmatrix}1\\1\end{pmatrix}, \quad |x\text{-down}\rangle = \frac{1}{\sqrt{2}}\begin{pmatrix}1\\-1\end{pmatrix} \tag{3.3}$$

(Compare to Eq. (2.118) and Exercise 2.6 in Sect. 2.3.1.)

Furthermore, in the Dirac notation, one writes $\langle\phi|$ for the mapping $\mathscr{H} \to \mathbb{C}$ given by $\psi \mapsto \langle\phi|\psi\rangle$. Obviously, $\langle\phi|$ applied to $|\psi\rangle$ gives $\langle\phi|\psi\rangle$, which suggested the notation. Paul Dirac (1902–1984) called $\langle\phi|$ a *bra* and $|\psi\rangle$ a *ket*. Obviously, $\langle\phi|A|\psi\rangle$ means the same as $\langle\phi|A\psi\rangle$. Dirac suggested that for self-adjoint A, the notation $\langle\phi|A|\psi\rangle$ conveys better that A can be applied equally well to either ϕ or ψ. $|\phi\rangle\langle\phi|$ is an operator that maps ψ to $|\phi\rangle\langle\phi|\psi\rangle = \langle\phi|\psi\rangle\phi$. If ϕ is a unit vector, then this is the part of ψ parallel to ϕ or the *projection of ψ to ϕ*.

R. Tumulka, *Foundations of Quantum Mechanics*, Lecture Notes in Physics 1003,
https://doi.org/10.1007/978-3-031-09548-1_3

Another common and useful notation is $\otimes$, called the *tensor product*. For

$$\Psi(x, y) = \psi(x)\,\phi(y) \tag{3.4}$$

one writes

$$\Psi = \psi \otimes \phi\,. \tag{3.5}$$

Likewise, for Eq. (2.116), one writes $\psi = \phi \otimes \chi$.

The symbol $\otimes$ also has meaning when applied to Hilbert spaces.

$$L^2(x, y) = L^2(x) \otimes L^2(y)\,, \tag{3.6}$$

where $L^2(x)$ means the square-integrable functions of x, etc. Note that not all elements of $L^2(x) \otimes L^2(y)$ are of the form $\psi \otimes \phi$—only a minority are. A general element of $L^2(x) \otimes L^2(y)$ is an infinite linear combination of tensor products such as $\psi \otimes \phi$. Likewise, when we replace the continuous variable y by the discrete index s for spin, the tensor product of the Hilbert space $\mathbb{C}^2$ of vectors ϕ_s and the Hilbert space $L^2(\mathbb{R}^3, \mathbb{C})$ of wave functions $\chi(\boldsymbol{x})$ is the Hilbert space $L^2(\mathbb{R}^3, \mathbb{C}^2)$ of wave functions $\psi_s(\boldsymbol{x})$:

$$\mathbb{C}^2 \otimes L^2(\mathbb{R}^3, \mathbb{C}) = L^2(\mathbb{R}^3, \mathbb{C}^2)\,. \tag{3.7}$$

Another notation we use is

$$f(t-) = \lim_{s \nearrow t} f(s)\,, \quad f(t+) = \lim_{s \searrow t} f(s) \tag{3.8}$$

for the left and right limits of a function f at a jump.

3.1.2 The Projection Postulate

Here is the last rule of the quantum formalism:

Projection Postulate *If we carry out a quantum measurement of the observable A at time t on a system with wave function ψ_{t-} and obtain the outcome α, then the system's wave function ψ_{t+} right after the experiment is the eigenfunction of A with eigenvalue α. If there are several mutually orthogonal eigenfunctions, then*

$$\psi_{t+} = C \sum_{\lambda} |\phi_{\alpha,\lambda}\rangle\langle\phi_{\alpha,\lambda}|\psi_{t-}\rangle\,, \tag{3.9}$$

where $C > 0$ is the normalizing constant.

If λ is a continuous variable, then $\sum_\lambda$ should be $\int d\lambda$. The value of C is, explicitly,

$$C = \left\| \sum_\lambda |\phi_{\alpha,\lambda}\rangle\langle\phi_{\alpha,\lambda}|\psi_{t-}\rangle \right\|^{-1} . \tag{3.10}$$

According to the projection postulate (also known as the measurement postulate or the collapse postulate), the wave function changes dramatically in a quantum measurement. The change is known as the *reduction of the wave packet* or the *collapse of the wave function*.

Example 3.1 In a quantum measurement of spin-z (or σ_3), the wave function before the measurement is an arbitrary spinor $(\phi_1, \phi_2) \in S$ with $|\phi_1|^2 + |\phi_2|^2 = 1$ (assuming Eq. (2.116) and ignoring the space dependence). With probability $|\phi_1|^2$, we obtain outcome "up" and the collapsed spinor $(\phi_1/|\phi_1|, 0)$ after the measurement. The term $\phi_1/|\phi_1|$ is just the phase of ϕ_1. With probability $|\phi_2|^2$, we obtain "down" and the collapsed spinor $(0, \phi_2/|\phi_2|)$.

3.1.3 Projection and Eigenspace

To get a better feeling for what the expression on the right-hand side of (3.9) means, consider a vector $\psi = \psi_{t-}$ and an orthonormal basis $\phi_n = \phi_{\alpha,\lambda}$, and expand ψ in that basis:

$$\psi = \sum_n c_n \phi_n . \tag{3.11}$$

The coefficients are then given by

$$c_m = \langle\phi_m|\psi\rangle \tag{3.12}$$

because

$$\langle\phi_m|\psi\rangle = \left\langle\phi_m \middle| \sum_n c_n\phi_n\right\rangle = \sum_n c_n\langle\phi_m|\phi_n\rangle = \sum_n c_n\delta_{mn} = c_m . \tag{3.13}$$

Now change ψ by replacing some of the coefficients c_n by zero while retaining the others unchanged:

$$\tilde{\psi} = \sum_{n\in J} c_n\phi_n , \tag{3.14}$$

where J is the set of those indices retained. This procedure is called *projection* to the subspace spanned by $\{\phi_n : n \in J\}$, and the projection operator is

$$P = \sum_{n \in J} |\phi_n\rangle\langle\phi_n| . \tag{3.15}$$

(The only projections we consider are *orthogonal projections*.)

Exercise 3.1 (Projections)

(a) Show that an operator P is a projection (i.e., is diagonal in some ONB with diagonal entries 0 or 1) if and only if it is self-adjoint $[P = P^\dagger]$ and idempotent $[P^2 = P]$.
(b) Suppose that $P : \mathscr{H} \to \mathscr{H}$ is a projection with range $\mathscr{K}$; one says that P is the projection to $\mathscr{K}$. Show that $I - P$ is the projection to the orthogonal complement $\mathscr{K}^\perp = \{\phi \in \mathscr{H} : \langle\phi|\psi\rangle = 0\, \forall \psi \in \mathscr{K}\}$ of $\mathscr{K}$.
(c) Suppose that P is the projection to $\mathscr{K}$. Show that the element in $\mathscr{K}$ closest to a given vector $\psi \in \mathscr{H}$ is $P\psi$.
(d) Show that if $\|\psi\| = 1$, then $|\psi\rangle\langle\psi|$ is the projection to $\mathbb{C}\psi$.

In Eq. (3.9), the index n numbers the index pairs (α, λ), and the subset J corresponds to those pairs that have a given α and arbitrary λ. Except for the factor C, the right-hand side of (3.9) is the corresponding projection of ψ_{t-}, which gives the projection postulate its name. The subspace of Hilbert space spanned by the $\phi_{\alpha,\lambda}$ with given α is the eigenspace of A with eigenvalue α. Thus, the projection postulate can be equivalently rewritten as

$$\psi_{t+} = \frac{P_\alpha \psi_{t-}}{\|P_\alpha \psi_{t-}\|}, \tag{3.16}$$

where P_α denotes the projection to the eigenspace of A with eigenvalue α.

For every closed subspace, there is a projection operator that projects to this subspace. For example, for any region $B \subseteq \mathbb{R}^{3N}$ in configuration space, the functions whose (precise) support lies in B (i.e., which vanish outside B) form an ∞-dimensional closed subspace of $L^2(\mathbb{R}^{3N})$. The projection to this subspace is

$$(P_B\psi)(q) = \begin{cases} \psi(q) & q \in B \\ 0 & q \notin B, \end{cases} \tag{3.17}$$

that is, multiplication by the characteristic function 1_B of B.

3.1.4 Position Measurements

For a position measurement, the projection postulate implies that the wave function collapses to a Dirac delta function (which we introduced in Sect. 2.2.7). This is not realistic; it is over-idealized. A delta function is not a square-integrable function, and it contains in a sense an infinite amount of energy.[1] More realistically, a position measurement has a finite inaccuracy ε and could be expected to collapse the wave function to one of width ε, such as

$$\psi_{t+}(\boldsymbol{x}) = C e^{-\frac{(\boldsymbol{x}-\boldsymbol{\alpha})^2}{4\varepsilon^2}}\, \psi_{t-}(\boldsymbol{x})\,. \tag{3.18}$$

However, this operator (multiplication by a Gaussian) is not a projection because its spectrum is more than just 0 and 1.

Another simple model of position measurement, still highly idealized but less so than collapse to $\delta(\boldsymbol{x}-\boldsymbol{\alpha})$, considers a region $B \subset \mathbb{R}^3$ and assumes that a detector either finds the particle in B or not. The corresponding observable is $A = P_B$ as defined in (3.17), and the probability of outcome 1 is

$$\int_B d^3\boldsymbol{x}\, |\psi_{t-}(\boldsymbol{x})|^2\,. \tag{3.19}$$

In case of outcome 1, ψ_{t-} collapses to

$$\psi_{t+} = \frac{P_B \psi_{t-}}{\|P_B \psi_{t-}\|}\,. \tag{3.20}$$

3.1.5 Consecutive Quantum Measurements

With the projection postulate, the formalism provides a prediction of probabilities for any sequence of measurements. If we prepare the initial wave function ψ_0 and make a measurement of A_1 at time t_1, then the Schrödinger equation determines what ψ_{t_1-} is, the general Born rule (2.81) determines the probabilities of the outcome α_1, and the projection postulate determines the wave function after the measurement. The latter is the initial wave function for the Schrödinger equation, which governs the evolution of ψ until the time t_2 at which the second measurement, of observable A_2, occurs. The probability distribution of the outcome α_2 is given by the Born rule again and depends on α_1 because the initial wave function in

[1] Still, we can note for the sake of completeness that the free Schrödinger evolution can also be defined for an initial wave function in $\mathbb{R}^3$ that is $\psi_0(\boldsymbol{x}) = \delta^3(\boldsymbol{x})$: the solution is $\psi_t(\boldsymbol{x}) = (-im/2\pi\hbar t)^{3/2} \exp(im\boldsymbol{x}^2/2\hbar t)$ for $t \neq 0$. The corresponding Bohmian trajectories are of the form $\boldsymbol{Q}(t) = t\,\boldsymbol{Q}(1)$; all trajectories pass through $\boldsymbol{x} = \boldsymbol{0}$ at $t = 0$ (in violation of the rule, valid for regular wave functions, that trajectories cannot cross).

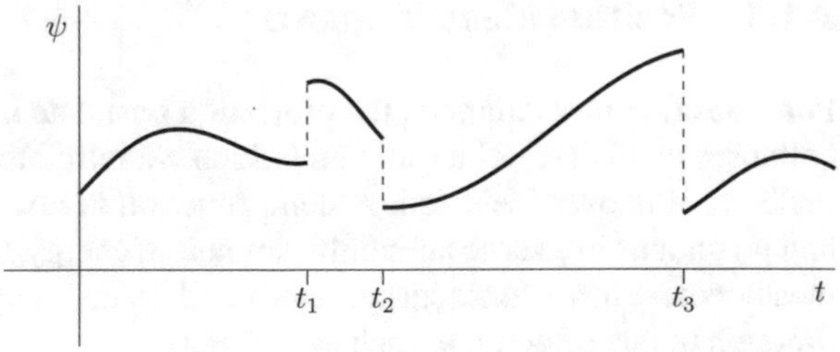

Fig. 3.1 Schematic representation of the time evolution of the wave function under consecutive quantum measurements at times t_1, t_2, t_3: episodes of unitary evolution are interrupted by discontinuous jumps

the Schrödinger equation, ψ_{t_1+}, did. And so on; see Fig. 3.1. This scheme is the *quantum formalism*. Note that the observer can choose t_2 and A_2 after the first measurement and thus make this choice depending on the first outcome α_1.

The projection postulate implies that if we make another measurement of A right after the first one, then with probability 1, we will obtain the same outcome α.

You may feel a sense of paradox about the two different laws for how ψ changes with time: the unitary Schrödinger evolution and the collapse rule. Already at first sight, the two seem rather incompatible: the former is deterministic, while the latter stochastic; the former is continuous, while the latter not; and the former is linear, while the latter not. It seems strange that time evolution is governed not by a single law but by two and even stranger that the criterion for when the collapse rule takes over is something as vague as an observer making a measurement. Upon scrutiny, the sense of paradox will persist and even deepen in the form of what is known as the *measurement problem of quantum mechanics*, which we will discuss in detail in Sect. 3.2.

Let us now turn to the situation of consecutive quantum measurements for non-commuting operators. Compare two experiments, each consisting of two quantum measurements: (a) first measure σ_2 and then σ_3, and (b) first measure σ_3 and then σ_2. (That is, we take the limit in which the time between the two experiments tends to 0; we take the collapsed wave function immediately after the first measurement as the wave function immediately before the second measurement.)

Exercise 3.2 (Iterated Stern–Gerlach Experiment) Consider the following experiment on a single electron. Suppose it has a wave function of the product form $\psi_s(\boldsymbol{x}) = \phi_s\, \chi(\boldsymbol{x})$, and we focus only on the spinor. The initial spinor is $\phi = (1, 0)$.

(a) A Stern–Gerlach experiment in the y-direction (or σ_2-measurement) is carried out and then a Stern–Gerlach experiment in the z-direction (or σ_3-measurement). Both measurements taken together have four possible outcomes: up-up, up-down, down-up, and down-down. Find the probabilities of the four outcomes.

(b) As in (a), but now the z-experiment comes first and the y-experiment afterward.

More generally, suppose the two self-adjoint operators A and B in $\mathscr{H}$ have spectral decomposition $A = \sum_\alpha \alpha P_\alpha$ and $B = \sum_\beta \beta Q_\beta$ with purely discrete (i.e., countable) spectra. (So P_α is the projection to A's eigenspace with eigenvalue α, and Q_β the projection to B's eigenspace with eigenvalue β.) Suppose that A is measured first on a system with wave function $\psi \in \mathscr{H}$ and B right afterward. In the first stage, with probability $\|P_\alpha\psi\|^2$, the outcome Z_A of the A-measurement has the value α, and ψ collapses to $\psi' = P_\alpha\psi/\|P_\alpha\psi\|$. Then, in the second stage, the outcome Z_B of the B-measurement has the value β with probability $\|Q_\beta\psi'\|^2$. Thus, the conditional probability of $Z_B = \beta$, given that $Z_A = \alpha$, is given by

$$\mathbb{P}_{A\,\text{first}}(Z_B = \beta | Z_A = \alpha) = \left\| Q_\beta \frac{P_\alpha\psi}{\|P_\alpha\psi\|} \right\|^2 . \tag{3.21}$$

Since conditional probabilities can be expressed as $\mathbb{P}(E|F) = \mathbb{P}(E \text{ and } F)/\mathbb{P}(F)$ for events E, F with $\mathbb{P}(F) \neq 0$, the total probability (i.e., the joint distribution of the outcomes) is

$$\mathbb{P}_{A\,\text{first}}(Z_B = \beta, Z_A = \alpha) = \mathbb{P}_{A\,\text{first}}(Z_B = \beta | Z_A = \alpha)\, \mathbb{P}_{A\,\text{first}}(Z_A = \alpha) \tag{3.22}$$

$$= \left\| Q_\beta \frac{P_\alpha\psi}{\|P_\alpha\psi\|} \right\|^2 \|P_\alpha\psi\|^2 \tag{3.23}$$

$$= \left\| Q_\beta P_\alpha \psi \right\|^2 . \tag{3.24}$$

If, however, we measure B first and A right afterward, then, by the same consideration with A and B changing roles, the joint distribution is given by

$$\mathbb{P}_{B\,\text{first}}(Z_B = \beta, Z_A = \alpha) = \left\| P_\alpha Q_\beta \psi \right\|^2 . \tag{3.25}$$

Now in general, operators do not commute, and also P_α and Q_β cannot be expected to commute. So in general,

$$\mathbb{P}_{B\,\text{first}} \neq \mathbb{P}_{A\,\text{first}} . \tag{3.26}$$

Definition 3.1 One says that two self-adjoint operators A, B *can be measured simultaneously* if the joint distribution of outcomes does not depend on which is measured first, that is, if (3.24) agrees with (3.25) for all α and β and $\psi \in \mathscr{H}$.

So generically, observables cannot be measured simultaneously, and Exercise 3.2 showed that σ_2 and σ_3 cannot be measured simultaneously. Here are examples of observables that can: X_2 and X_3, the y-component of position and the z-component, or σ_2 of particle 1 and σ_3 of particle 2. We now move toward a proof that this happens if and only if the two operators commute with each other.

Theorem 3.1 (Extension of the Spectral Theorem to Several Commuting Operators) *If and only if A and B commute, then there exists a generalized ONB* $\{\phi_n\}$ *whose elements are eigenvectors of both operators A and B,* $A\phi_n = \alpha_n\phi_n$ *and* $B\phi_n = \beta_n\phi_n$.

Sketch of Proof If A commutes with B, then B maps the eigenspace of A with eigenvalue α to itself, so B is block diagonal in any eigen ONB of A; now diagonalize each block. □

Theorem 3.2 *Of two observables A and B with discrete spectrum, one is measured after the other. The joint probability distribution of the outcomes* (α, β) *is independent of the order of the two measurements for every wave function if and only if the operators A and B commute,* $AB = BA$.

Proof Let A have spectral decomposition $A = \sum_\alpha \alpha P_\alpha$, likewise $B = \sum_\beta \beta Q_\beta$ with Q_β the projection to the eigenspace of B with eigenvalue β. The joint distribution, if A is measured first and B thereafter, is given by (3.24).

"if": Suppose $AB = BA$. By Theorem 3.1, they can be simultaneously diagonalized, so $P_\alpha Q_\beta = Q_\beta P_\alpha$ for all α, β, leading to the same probability of (α, β).

"only if": Fix α, β. We have that $\|PQ\psi\| = \|QP\psi\|$ for all ψ and show that $\|(QP-PQ)\psi\| = 0$. Since every ψ can be decomposed as $\psi = u+v$ with $Qu = u$ and $Qv = 0$,

$$\|(QP - PQ)(u+v)\|^2 = \langle u+v|(QP - PQ)(PQ - QP)|u+v\rangle \tag{3.27a}$$

$$= \langle u+v|(QPQ - QPQP - PQPQ + PQP)|u+v\rangle \tag{3.27b}$$

$$\begin{aligned} &= \langle u|(P - PQP)|u\rangle \\ &\quad + \langle u|(-PQP + PQP)|v\rangle \\ &\quad + \langle v|(-PQP + PQP)|u\rangle \\ &\quad + \langle v|PQP|v\rangle \end{aligned} \tag{3.27c}$$

$$= \|Pu\|^2 - \|QPu\|^2 + \|QPv\|^2 \tag{3.27d}$$

$$= \|Pu\|^2 - \|PQu\|^2 + \|PQv\|^2 \tag{3.27e}$$

$$= \|Pu\|^2 - \|Pu\|^2 + 0 = 0. \tag{3.27f}$$

□

Example 3.2 Indeed, σ_2 and σ_3 do not commute:

$$\sigma_2\sigma_3 = \begin{pmatrix} 0 & i \\ i & 0 \end{pmatrix}, \quad \sigma_3\sigma_2 = \begin{pmatrix} 0 & -i \\ -i & 0 \end{pmatrix}. \tag{3.28}$$

Any two multiplication operators commute. In particular, the position operators X_i and X_j commute with each other. The momentum operators $P_j = -i\hbar\partial/\partial x_j$ commute with each other. X_i commutes with P_j for $i \neq j$, but

$$[X_j, P_j] = i\hbar I, \tag{3.29}$$

with I the identity operator. Equation (3.29) is called *Heisenberg's canonical commutation relation*. To verify it, it suffices to consider a function ψ of a one-dimensional variable x. Using the product rule,

$$[X, P]\psi(x) = XP\psi(x) - PX\psi(x) \tag{3.30a}$$

$$= x(-i\hbar)\frac{\partial\psi}{\partial x} - (-i\hbar)\frac{\partial}{\partial x}\Big(x\psi(x)\Big) \tag{3.30b}$$

$$= -i\hbar x\frac{\partial\psi}{\partial x} + i\hbar\psi(x) + i\hbar x\frac{\partial\psi}{\partial x} \tag{3.30c}$$

$$= i\hbar\psi(x). \tag{3.30d}$$

So, for two commuting observables, the quantum formalism provides a *joint probability distribution*. For non-commuting observables, it provides *two* joint probability distributions, one for each order. Moreover,

two non-commuting observables typically do not both have sharp values at the same time. (3.31)

For example, there is no quantum state that is an eigenvector to both σ_2 and σ_3. As a further consequence, a measurement of B must disturb the value of A if $AB \neq BA$. (Think of Exercise 3.2(a): After the σ_2-measurement, the particle was not certain any more to yield "up" in the σ_3-measurement.)

Also the Heisenberg uncertainty relation is connected to (3.31), as it expresses that position and momentum cannot both have sharp values (i.e., $\sigma_X = 0$ and $\sigma_P = 0$) at the same time. In fact, the following generalized version of Heisenberg's uncertainty relation applies to observables A and B instead of X and P:

Theorem 3.3 (Robertson–Schrödinger Inequality, Robertson (1929) [41] and Schrödinger (1930) [46]) *For any bounded self-adjoint operators A, B and any $\psi \in \mathscr{H}$ with $\|\psi\| = 1$,*

$$\sigma_A\,\sigma_B \geq \frac{1}{2}\Big|\langle\psi|[A, B]|\psi\rangle\Big|. \tag{3.32}$$

Note that the inequality is so much the stronger as the commutator $[A, B]$ is bigger and becomes vacuous when $[A, B] = 0$.

Proof Recall that the distribution over the spectrum of A defined by ψ has expectation value $\langle A\rangle := \langle\psi|A|\psi\rangle$ and variance

$$\sigma_A^2 = \langle\psi|(A - \langle A\rangle)^2|\psi\rangle = \|\phi_A\|^2 \tag{3.33}$$

with

$$\phi_A := (A - \langle A\rangle)\psi\,, \tag{3.34}$$

where we simply wrote $\langle A\rangle$ for $\langle A\rangle I$. By the Cauchy–Schwarz inequality,

$$\sigma_A^2\,\sigma_B^2 = \|\phi_A\|^2\,\|\phi_B\|^2 \geq \big|\langle\phi_A|\phi_B\rangle\big|^2\,. \tag{3.35}$$

Since

$$\langle\phi_A|\phi_B\rangle = \langle\psi|(A - \langle A\rangle)(B - \langle B\rangle)|\psi\rangle \tag{3.36a}$$
$$= \langle\psi|(AB - \langle A\rangle B - A\langle B\rangle + \langle A\rangle\langle B\rangle)|\psi\rangle \tag{3.36b}$$
$$= \langle AB\rangle - \langle A\rangle\langle B\rangle\,, \tag{3.36c}$$

we obtain that

$$\big|\langle\phi_A|\phi_B\rangle\big|^2 \geq \Big(\mathrm{Im}\langle\phi_A|\phi_B\rangle\Big)^2 \tag{3.37a}$$
$$= \left|\frac{\langle\phi_A|\phi_B\rangle - \langle\phi_B|\phi_A\rangle}{2i}\right|^2 \tag{3.37b}$$
$$= \left|\frac{\langle AB\rangle - \langle A\rangle\langle B\rangle - \langle BA\rangle + \langle B\rangle\langle A\rangle}{2}\right|^2 \tag{3.37c}$$
$$= \frac{1}{4}\big|\langle\psi|[A, B]|\psi\rangle\big|^2\,. \tag{3.37d}$$

□

For unbounded operators, a statement analogous to Theorem 3.3 holds, but one must be careful about the exact hypotheses because $A\psi$ will only be defined if $\psi \in \mathrm{domain}(A)$. The correct statement (Hall (2013) [32] Chap. 12) is: (3.32) holds for self-adjoint A, B if $\psi \in \mathrm{domain}(AB) \cap \mathrm{domain}(BA)$. In the special case of position and momentum, $A = X$ and $B = P$; however, one can prove the relation

$$\sigma_X\sigma_P \geq \hbar/2 \tag{3.38}$$

even for *all* unit vectors $\psi \in \mathscr{H}$ (as we stated in Theorem 2.3). (But note that $\sigma_X = \infty$ when $\psi \notin \text{domain}(X)$ and likewise for P, and since σ_P and σ_X are never zero, it follows that whenever $\psi \notin \text{domain}(X) \cap \text{domain}(P)$, then the left-hand side of (3.38) is infinite and the relation (3.38) is trivially true.)

3.2 The Measurement Problem

3.2.1 What the Problem Is

This is a problem about orthodox quantum mechanics. It is solved in Bohmian mechanics and several other theories. Because of this problem, the orthodox view is in trouble when it comes to analyzing the process of measurement.

Consider a "quantum measurement of the observable A." Realistically, there are only finitely many possible outcomes, so A should have finite spectrum. Consider the system formed by the object together with the apparatus. Since the apparatus consists of electrons and quarks, too, it should itself be governed by quantum mechanics. (That is reductionism at work.) So I write Ψ for the wave function of the system (object and apparatus). Suppose for simplicity that the system is isolated (i.e., there is no interaction with the rest of the universe), so Ψ evolves according to the Schrödinger equation during the experiment (recall Exercise 1.20), which begins at the initial time t_i and ends at the final time t_f. It is reasonable to assume that

$$\Psi(t_\text{i}) = \psi(t_\text{i}) \otimes \phi \tag{3.39}$$

with $\psi = \psi(t_\text{i})$ the wave function of the object before the experiment and ϕ a wave function representing a "ready" state of the apparatus. By the spectral theorem, ψ can be written as a linear combination (superposition) of eigenfunctions of A,

$$\psi = \sum_\alpha c_\alpha \psi_\alpha \quad \text{with} \quad A\psi_\alpha = \alpha\psi_\alpha \text{ and } \|\psi_\alpha\| = 1\,. \tag{3.40}$$

If the object's wave function is an eigenfunction ψ_α, then, by Born's rule (2.81), the outcome is certain to be α. Set $\Psi_\alpha(t_\text{i}) = \psi_\alpha \otimes \phi$. Then $\Psi_\alpha(t_\text{f})$ must represent a state in which the apparatus displays the outcome α (e.g., by a pointer pointing to the appropriate position on a scale).

Now consider again a general ψ as in Eq. (3.40). Since the Schrödinger equation is linear, the wave function of object and apparatus together at t_f is

$$\Psi(t_\text{f}) = \sum_\alpha c_\alpha \Psi_\alpha(t_\text{f})\,, \tag{3.41}$$

a superposition of states corresponding to different outcomes (see Fig. 3.2) and not a random state corresponding to a unique outcome, as one might have expected from the projection postulate.

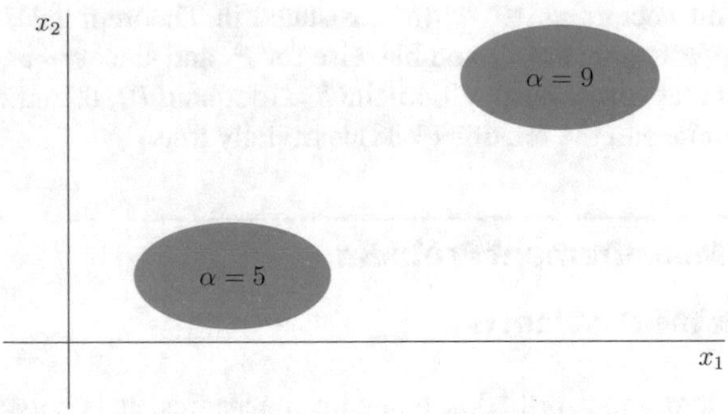

Fig. 3.2 The post-measurement wave function $\Psi(t_f)$. Of the configuration space, only 2 of the $>10^{23}$ dimensions are shown; of the wave function, only the significant support (i.e., the region where $\Psi(t_f)$ is significantly nonzero) is indicated; of the several wave packets, only two are shown, corresponding to different values of the outcome α

This is the measurement problem. The upshot is that there is a conflict between the following assumptions:

- In each run of the experiment, there is a unique outcome.
- The wave function is a complete description of a system's physical state.
- The evolution of the wave function of an isolated system is always given by the Schrödinger equation.

Thus, we have to drop one of these assumptions. The first is dropped in the many-worlds picture, in which all outcomes are realized, albeit in parallel worlds. If we drop the second, we opt for additional ("hidden") variables as in Bohmian mechanics, where the state at time t is described by the pair (Q_t, ψ_t). If we drop the third, we opt for replacing the Schrödinger equation by a non-linear evolution (as in the GRW = Ghirardi–Rimini–Weber approach). Of course, a theory might also drop several of these assumptions. Orthodox quantum mechanics insists on all three assumptions, and that is why it has a problem.

For further reading on the measurement problem, I recommend Maudlin (1995) [36].

3.2.2 How Bohmian Mechanics Solves the Problem

Since it is assumed that the Schrödinger equation is valid for a closed system, the after-measurement wave function of object and apparatus together is

$$\Psi = \sum_\alpha c_\alpha \Psi_\alpha \,. \tag{3.42}$$

Since the Ψ_α have disjoint supports[2] in the configuration space (of object and apparatus together), and since the particle configuration Q has distribution $|\Psi|^2$, the probability that Q lies in the support of Ψ_α is

$$\mathbb{P}\big(Q \in \text{support}(\Psi_\alpha)\big) = \int\limits_{\text{support}(\Psi_\alpha)} d^{3N}q\,|\Psi(q)|^2 = \int\limits_{\mathbb{R}^{3N}} d^{3N}q\,|c_\alpha \Psi_\alpha(q)|^2 = |c_\alpha|^2, \tag{3.43}$$

which agrees with the prediction of the quantum formalism for the probability of the outcome α. And indeed, when $Q \in \text{support}(\Psi_\alpha)$, then the particle positions (including the particles of both the object and the apparatus!) are such that the pointer of the apparatus points to the value α. Thus, the way out of the measurement problem is that although the wave function is a superposition of terms corresponding to different outcomes, the actual particle positions define the actual outcome.

As a consequence of the above consideration, we also see that the predictions of Bohmian mechanics for the probabilities of the outcomes of experiments agree with those of the quantum formalism. In particular, there is no experiment that could empirically distinguish between Bohmian mechanics and the quantum formalism, while there are (in principle) experiments that distinguish the two from a GRW world.

If Bohmian mechanics and the quantum formalism agree about all probabilities, then where do we find the collapse of the wave function in Bohmian mechanics? There are two parts to the answer, depending on which wave function we are talking about.

The first part of the answer is, if the Ψ_α are macroscopically different, then they will never overlap again (until the time when the universe reaches thermal equilibrium, perhaps in $10^{10^{10}}$ years). This fact is independent of Bohmian mechanics; it is a trait of the Schrödinger equation called *decoherence*.[3] If Q lies in the support of one among several disjoint packets, then only the packet containing Q is relevant, by Bohm's law of motion (1.77), to determining dQ/dt. Thus, as long as the packets stay disjoint, only the packet containing Q is relevant to the trajectories of the particles, and all other packets could be replaced by zero without affecting the trajectories. That is why we can replace Ψ by $c_\alpha \Psi_\alpha$, with α the actual outcome. Furthermore, the factor c_α cancels out in Bohm's law of motion (1.77) and thus can be dropped as well.

The second part of the answer is, the quantum formalism does not, in fact, talk about the wave function Ψ of object and apparatus but about the wave function ψ

[2] As described in Footnote 11, we mean by "support" the set where a function is significantly nonzero.

[3] "Coherence" originally meant the ability to interfere; "decoherence" the loss thereof. Another, related but inequivalent, widespread meaning of "decoherence" is that the reduced density matrix is (approximately) diagonal in the eigenbasis of A; see Sect. 5.5.5.

of the object alone. This leads us to the question what is meant by the wave function of a subsystem. If

$$\Psi(x, y) = \psi(x)\phi(y) \tag{3.44}$$

then it is appropriate to call ψ the wave function of the x-system, but in general Ψ does not factorize as in (3.44). In Bohmian mechanics, a natural general definition for the wave function of a subsystem is the *conditional wave function*

$$\psi(x) = \mathcal{N}\,\Psi(x, Y)\,, \tag{3.45}$$

where Y is the actual configuration of the y-system (while x is not the actual configuration X but any configuration of the x-system) and

$$\mathcal{N} = \left(\int |\Psi(x, Y)|^2 dx\right)^{-1/2} \tag{3.46}$$

is the normalizing factor. The conditional wave function does not, in general, evolve according to a Schrödinger equation, but in a complicated way depending on Ψ, Y, and X. There are special situations in which the conditional wave function does evolve according to a Schrödinger equation, in particular when the x-system and the y-system do not interact *and* the wave packet in Ψ containing $Q = (X, Y)$ is of a product form as in (3.44). Indeed, this is the case for the object *before*, but not *during*, the measurement; as a consequence, the wave function of the object (i.e., its conditional wave function) evolves according to the Schrödinger equation before, but not during, the measurement—in agreement with the quantum formalism. To determine the conditional wave function after the quantum measurement, suppose that Ψ_α is of the form

$$\Psi_\alpha = \psi_\alpha \otimes \phi_\alpha \tag{3.47}$$

with ϕ_α a wave function of the apparatus with the pointer pointing to the value α. Let α be the actual outcome, i.e., $Q \in \text{support}(\Psi_\alpha)$. Then $Y \in \text{support}(\phi_\alpha)$, and the conditional wave function is indeed

$$\psi = \psi_\alpha \tag{3.48}$$

up to a constant phase factor. That is, under suitable assumptions on the wave function of the apparatus and the interaction between object and apparatus, the projection postulate follows from Bohmian mechanics as a theorem.

3.2.3 Decoherence

People sometimes say that decoherence solves the measurement problem. We have seen that decoherence (i.e., the fact that the Ψ_α stay disjoint practically forever) plays a role in how Bohmian mechanics solves the measurement problem. But it is clear that mere disjointness of the packets Ψ_α does not make any of the packets go away. So the problem remains unless we drop one of the three assumptions mentioned in Sect. 3.2.1.

It is striking that $\sum c_\alpha \Psi_\alpha$ is the kind of wave function for which, if we applied the Born rule to a quantum measurement of the pointer position, we would get $\|c_\alpha \Psi_\alpha\|^2 = |c_\alpha|^2$ as the probability that the pointer points to α, and that is exactly the value we wanted. So a "super-measurement" of the pointer position would seem to help. But if we apply the reasoning of the measurement problem to the "super-apparatus" used for the super-measurement, we obtain again a non-trivial superposition of terms associated with different outcomes α. So which idea wins? When we push this thought further and iterate it further, we have to stop at the point where the system is the whole universe and includes all types of apparatus used. Thus, we end up with a superposition, and the measurement problem remains.

It is also striking that the super-observer, applying her super-apparatus to measure the pointer position of the first apparatus, cannot distinguish between decoherence and collapse; that is, she cannot decide whether the wave function of the system was $\sum c_\alpha \Psi_\alpha$ (a superposition) or one of the Ψ_α with probability $|c_\alpha|^2$ (a "mixture"). This means two things: first, that both will yield to the super-observer the result α with probability $|c_\alpha|^2$; and second, that even if she tried to manipulate the system (i.e., act on it with external forces, etc.), she would not be able to find out whether it is a superposition or a mixture. That is because the crucial difference between a superposition and a mixture is that a superposition is capable of interference. However, if the packets Ψ_α are macroscopically disjoint (i.e., if decoherence has occurred), then it becomes so extraordinarily difficult as to be practically impossible to make these packets overlap again, which would be a necessary condition for interference.

In particular, the probability of the outcome obtained by the super-observer does not depend on whether we treat the first apparatus as a quantum mechanical system (as we did in the measurement problem) or simply as triggering a collapse of the wave function. As Heisenberg (1930) [33] put it, it does not matter to the super-observer whether to put the cut (often called the "Heisenberg cut") between the quantum mechanical system (governed by the Schrödinger equation) and the measurement apparatus (that requires use of the projection postulate) before or after the first apparatus. This fact is a consistency property of the rules for making predictions. But this fact does not mean the measurement problem did not exist.

That is because the measurement problem is not about what the outcome will be; it is about what happens *in reality*. If the three assumptions about reality are true, then it follows that reality will not agree with the prediction of the quantum formalism. That is the problem.

This point also makes clear that right from the start, the super-apparatus did not actually help. If we are talking about what happens in reality, we expect that already the first apparatus produces an actual outcome, not merely a superposition. But it did not without the super-apparatus, and that is why there is a problem.

3.2.4 Schrödinger's Cat

Often referred to in the literature, this is Schrödinger's (1935) [47] formulation of the measurement problem:

> One can even set up quite ridiculous cases. A cat is penned up in a steel chamber, along with the following diabolical device (which must be secured against direct interference by the cat): in a Geiger counter there is a tiny bit of radioactive substance, so small, that *perhaps* in the course of one hour one of the atoms decays, but also, with equal probability, perhaps none; if it happens, the counter tube discharges and through a relay releases a hammer which shatters a small flask of hydrocyanic acid. If one has left this entire system to itself for an hour, one would say that the cat still lives *if* meanwhile no atom has decayed. The first atomic decay would have poisoned it. The ψ-function of the entire system would express this by having in it the living and dead cat (pardon the expression) mixed or smeared out in equal parts.
>
> It is typical of these cases that an indeterminacy originally restricted to the atomic domain becomes transformed into macroscopic indeterminacy, which can then be *resolved* by direct observation. That prevents us from so naively accepting as valid a 'blurred model' for representing reality. In itself it would not embody anything unclear or contradictory. There is a difference between a shaky or out-of-focus photograph and a snapshot of clouds and fog banks.

3.2.5 Positivism and Realism

Positivism is the view that a statement is unscientific or even meaningless if it cannot be tested experimentally, that an object is not real if it cannot be observed, and that a variable is not well-defined if it cannot be measured. For example, the statement in Bohmian mechanics that an electron went through the upper slit of a double slit if and only if it arrived in the upper half of the screen cannot be tested in experiment. After all, if you try to check which slit the electron went through by detecting every electron at the slit, then the statement is no longer true in Bohmian mechanics (and in fact, no correlation with the location of arrival is found). So a positivist thinks that this statement is meaningless (although you may feel you understand its meaning) and that Bohmian mechanics is unscientific. Good statements for a positivist are *operational statements*, i.e., statements of the form "if we set up an experiment in this way, the outcome has such-and-such a probability distribution." Positivists think that the quantum formalism (thought of as a summary of all true operational statements of quantum mechanics) is the only scientific formulation of quantum mechanics. They also tend to think that ψ is the complete description of a system, as it is the only information about the system that can be found experimentally without disturbing ψ. They tend not to understand the measurement problem, or not to take it seriously, because it requires thinking about reality.

Realism is the view that a fundamental physical theory needs to provide a coherent story of what happens. Bohmian mechanics, GRW theory, and many-worlds are examples of realist theories. For a realist, the quantum formalism by itself does not qualify as a fundamental physical theory. The story provided by Bohmian mechanics, for example, is that particles have trajectories, that there is a physical object that is mathematically represented by the wave function, and that the two evolve according to certain equations. For a realist, the measurement problem is serious and can only be solved by denying one of the three conflicting premises.

Feynman (1967) [27, p. 169] had a nice example for expressing his reservations about positivism:

> For those people who insist that the only thing that is important is that the theory agrees with experiment, I would like to imagine a discussion between a Mayan astronomer and his student. The Mayans were able to calculate with great precision predictions, for example, for eclipses and for the position of the moon in the sky, the position of Venus, etc. It was all done by arithmetic. They counted a certain number and subtracted some numbers, and so on. There was no discussion of what the moon was. There was no discussion even of the idea that it went around. They just calculated the time when there would be an eclipse, or when the moon would rise at the full, and so on. Suppose that a young man went to the astronomer and said, 'I have an idea. Maybe those things are going around, and they are balls of something like rocks out there, and we could calculate how they move in a completely different way from just calculating what time they appear in the sky.' 'Yes,' says the astronomer, 'and how accurately can you predict the eclipses?' He says, 'I haven't developed the thing very far yet.' Then says the astronomer, 'Well, we can calculate eclipses more accurately than you can with your model, so you must not pay any attention to your idea because obviously the mathematical scheme is better'.

The point is that positivism, if taken too far (as the imaginary ancient astronomer did), will stifle efforts to understand the world around us. People often say that the goal of physics is to make predictions that can be compared to experiment. I would not say that. I think that the goals of physics include to understand how the world works and to figure out what its fundamental laws are. In fact, often we do not make theories to compute predictions for experiments but make experiments to investigate our theories. Physics also has further goals, including to make use of the laws of nature for our purposes (such as technical applications), or to determine what happens in remote places or in the distant past or future, or to find explanations of phenomena, or to study remarkable behavior of special physical systems.

Positivism may appear as particularly safe and modest. After all, operational statements may appear as safe statements, and it may seem modest to refrain from speculation about the nature of things and the explanation of the phenomena we observe. However, often this appearance is an *illusion of modesty*, and Feynman's example suggests why. Someone who refrains too much from speculation may miss out on understanding the nature of things and explanation of phenomena. Here is another example: Wheeler's fallacy (see Sect. 1.6.6). Wheeler took for granted that a particle will reach the lower detector if and only if it went through the upper slit. Positivists like to call this an "operational definition" of what it means for a particle to have gone through the upper slit in a situation in which no attempt was made at detection during the passage through the slits. But such a "definition" is actually neither safe nor modest: it goes far beyond what is within our choice to define, and, as described in Sect. 1.6.6, it conflicts with where the particle actually went in that theory in which it makes sense to ask which slit the particle went through—i.e., in Bohmian mechanics.[4]

Positivism and realism play a role not only in the foundations of quantum mechanics but widely in philosophy. As a side remark, I mention that they also play a role in mathematics. According to the positivist view of mathematics (also known as formalism), a mathematical statement that can neither be proved nor disproved is meaningless, whereas a realist about mathematics (also called a Platonist) would object that if we can understand the content of a mathematical statement, then it must be meaningful and have a truth value (i.e., be either true or false), regardless of whether it can be proven. Perhaps the most prominent positivist in mathematics was David Hilbert (1862–1943), and perhaps the most prominent realist Kurt Gödel (1906–1978) who, in his famous incompleteness theorem (1931) [30], gave an explicit example of an obviously meaningful mathematical statement that can neither be proven nor disproven from standard axioms using standard rules such as *Principia Mathematica* (Russell and Whitehead 1913 [43]), but, for curious reasons I will not discuss here, can actually be known to be true.[5]

3.2.6 Experiments and Operators

After our analysis of the measurement process, we now have a clearer picture of what a "quantum measurement" actually is. It can be summarized in the following:

Definition 3.2 Let $A = \sum_\alpha \alpha P_\alpha$ be a self-adjoint operator on $\mathscr{H}$ with finite spectrum $\sigma(A)$. An experiment is a *quantum measurement of A* if and only if there are mutually disjoint regions B_α in the apparatus configuration space such that for

[4] One could argue that Wheeler's definition is not strictly positivistic because "if the particle triggers the lower detector then it went through the upper slit" goes beyond operational statements. But in practice, positivists often feel they need operational definitions for statements like "the particle went through the upper slit" or "the atom got ionized" or "an α particle was emitted," statements which physicists cannot avoid making.

[5] Another illuminating example of (presumably) unprovable mathematical statements is due to Tim Maudlin and involves "reasonless truths": Define that any two real numbers x, y *match* if their decimal expansions have equal digit 1, or digits 2 and 3, or digits 4–6, or digits 7–10, etc. Random numbers match with probability $1/10 + 1/100 + 1/1000 + \ldots = 1/9$ (geometric series). Let P be the statement "$54^{1/18}$ and π^{78} do not match." If P is false, it can be disproven, but if true, then presumably it cannot be proven from standard axiom systems of mathematics such as *Principia Mathematica* [43] because there is no deeper reason behind it—it happens to be true "by coincidence." Now generate many similar statements; 8/9 of them should be true and unprovable. Let P' be the statement "For all $n \in \mathbb{N}$, $\sqrt{2}$ matches $\cos n$." It is presumably unprovable no matter if true or false and presumably false because it has probability 0.

all unit vectors $\psi \in \mathscr{H}$, $\psi \otimes \phi$ at the initial time t_i of the experiment evolves to $\sum_\alpha (P_\alpha \psi) \otimes \phi_\alpha$ at the final time t_f, where ϕ_α has support in B_α.

At this point, we may realize that so far a statement like this was missing from the quantum formalism. It is missing from orthodox textbooks. We were supposed to somehow know already which experiments are quantum measurements and which experiment corresponds to which operator! Well, there is an obvious candidate for the reason why it was missing: because orthodox quantum mechanics cannot achieve a clean analysis of the measurement process.

In Definition 3.2, an *experiment* is understood as specified by specifying the ready state ϕ, the initial and final times t_i and t_f, the Hamiltonian governing the interaction between object and apparatus during the time interval $[t_\mathrm{i}, t_\mathrm{f}]$, and the regions B_α that define how to read off the outcome from the configuration of the apparatus.

Definition 3.2 can straightforwardly be extended to the concept of an *approximate quantum measurement of* A: that is an experiment for which the final wave function is *approximately* $\sum_\alpha (P_\alpha \psi) \otimes \phi_\alpha$ and the ϕ_α are *approximately* concentrated in B_α.

Another remark concerns the conservation of energy (or momentum, or angular momentum). As pointed out already in Sect. 2.2.6, energy as a number is usually not defined, not even in Bohmian mechanics, so we cannot ask whether it changes over time. Energy as an operator is defined. The energy operator H is conserved, and its associated Born distribution is conserved, for the full wave function Ψ of object and apparatus but not for the conditional wave function (3.45) of the object. Likewise, they are not conserved when we replace $\Psi(t_\mathrm{f})$ by $\Psi_\alpha(t_\mathrm{f})$. So we can now say very easily and briefly in what sense energy is or is not conserved.

3.3 The GRW Theory

Bohmian mechanics is not the only possible explanation of quantum mechanics. Another one is provided by the GRW theory, named after GianCarlo Ghirardi (1935–2018), Alberto Rimini, and Tullio Weber, who proposed it in 1986 [29]. Similar theories were proposed by Philip Pearle (1989) [40] under the name CSL (for *continuous spontaneous localization*) and by Lajos Diósi (1989) [21], but our discussion will focus on the GRW theory, which is mathematically the simplest one.

In all of these theories, the key idea is that the Schrödinger equation is only an approximation, valid for systems with few particles (say, $N < 10^4$) but not for macroscopic systems (say, $N > 10^{23}$). The true evolution law for the wave function, these theories say, involves something like collapses and thereby avoids superpositions (such as Schrödinger's cat) of macroscopically different contributions. These theories, called *dynamical state reduction theories* or *collapse theories*, regard the collapse of ψ as a physical process governed by laws of nature, for which they propose explicit equations. The wave function collapses, not when an observer makes a measurement, but *spontaneously*, even in the absence of observers.

Collapse theories deserve particular interest in connection with relativity, as we will discuss in Chap. 7.

In collapse theories, Ψ_t does not evolve according to the Schrödinger equation, but according to a modified evolution law. This evolution law is *stochastic*, as opposed to deterministic. That is, for any fixed Ψ_0, it is random what Ψ_t is, and the theory provides a probability distribution over Hilbert space. A family of random variables X_t, with one variable for every time t, is called a *stochastic process*. Thus, the family $(\Psi_t)_{t\geq 0}$ is a stochastic process in Hilbert space. In the *GRW process*, periods governed by the Schrödinger equation are interrupted by random jumps. Such a jump occurs, within any infinitesimal time interval dt, with probability $\lambda\, dt$, where λ is a constant called the *jump rate*. Let us call the random jump times $T_1, T_2, \ldots$; the sequence $T_1, T_2, \ldots$ is known as the *Poisson process with rate* λ; it has widespread applications in probability theory. Let us have a closer look.

3.3.1 The Poisson Process

Think of $T_1, T_2, \ldots$ as the times at which a certain type of random event occurs; standard examples include the times when an earthquake (of a certain strength) occurs, or when the phone rings, or when the price of a certain share falls below a certain value. We take for granted that the ordering is chosen such that $0 < T_1 < T_2 < \ldots$; see Fig. 3.3 an example of what a Poisson process looks like.

Let us figure out the probability density function of T_1. The probability that T_1 occurs between 0 and dt is $\lambda\, dt$. Thus, the probability that it does not occur is $1 - \lambda\, dt$. Suppose that it did not occur between 0 and dt. Then the probability that it does not occur between dt and $2\, dt$ is again $1 - \lambda\, dt$. Thus, the total probability that no event occurs between 0 and $2\, dt$ is $(1 - \lambda\, dt)^2$. Proceeding in the same way, the total probability that no event occurs between 0 and $n\, dt$ is $(1 - \lambda\, dt)^n$. Thus, the total probability that no event occurs between 0 and t, $\mathbb{P}(T_1 > t)$, can be approximated by setting $dt = t/n$ and letting $n \to \infty$. That is,

$$\mathbb{P}(T_1 > t) = \lim_{n\to\infty} \left(1 - \frac{\lambda t}{n}\right)^n = e^{-\lambda t}. \tag{3.49}$$

Let us write $\rho(t)$ for the probability density function of T_1. By definition,

$$\rho(t)\, dt = \mathbb{P}(t < T_1 < t + dt). \tag{3.50}$$

To compute this quantity, we reason as follows. If T_1 has not occurred until t, then the probability that it will occur within the next dt is $\lambda\, dt$. Thus, (3.50) differs from

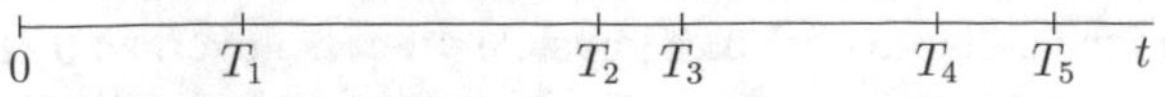

Fig. 3.3 A realization of the Poisson process

(3.49) by a factor $\lambda\, dt$, or, as the factor dt cancels out,

$$\rho(t) = 1_{t>0} e^{-\lambda t} \lambda\,, \tag{3.51}$$

where the expression 1_C is 1 whenever the condition C is satisfied and 0 otherwise. The distribution (3.51) is known as the *exponential distribution with parameter* λ, $\mathrm{Exp}(\lambda)$. We have thus found that *the waiting time for the first event has distribution* $\mathrm{Exp}(\lambda)$.

After T_1, the next dt has again probability $\lambda\, dt$ for the next event to occur. The above reasoning can be repeated, with the upshot that *the waiting time* $T_2 - T_1$ *for the next event has distribution* $\mathrm{Exp}(\lambda)$ *and is independent of what happened up to time* T_1. The same applies to the other waiting times $T_{n+1} - T_n$. In fact, at any time t_0, the waiting time until the next event has distribution $\mathrm{Exp}(\lambda)$.

Exercise 3.3 (Mean of the Exponential Distribution) Verify through integration that $\mathrm{Exp}(\lambda)$ has expectation value (mean)

$$\int_0^\infty t\, \rho(t)\, dt = \frac{1}{\lambda}\,. \tag{3.52}$$

This fact is very plausible if you think of it this way: If in every second the probability of an earthquake is, say, 10^{-8}, then you would guess that an earthquake occurs on average every 10^8 seconds. The constant λ, whose dimension is 1/time, is thus the average frequency of the earthquakes (or whichever events).

Another way of representing the Poisson process is by means of the random variables

$$X_t = \#\{i \in \mathbb{N} : T_i < t\}\,, \tag{3.53}$$

the number of earthquakes up to time t. (The symbol $\#S$ means the number of elements of the set S.)

Theorem 3.4 *If* $\mathcal{T} = \{T_1, T_2, \ldots\}$ *is a Poisson process with rate* λ *and* $\mathcal{T}'$ *is an independent Poisson process with rate* λ', *then* $\mathcal{T} \cup \mathcal{T}'$ *is a Poisson process with rate* $\lambda + \lambda'$.

For example, suppose earthquakes in Australia occur with rate λ and are independent of those in Africa, which occur with rate λ'; then the earthquakes in Africa and Australia together occur with rate $\lambda + \lambda'$. See, e.g., Ross (1996) [42] for the proof of this and the subsequent theorem.

Theorem 3.5 *If we choose n points at random in the interval* $[0, n/\lambda]$, *independently with uniform distribution, then the joint distribution of these points converges, as* $n \to \infty$, *to the Poisson process with parameter* λ.

3.3.2 Definition of the GRW Process

Now let us get back to the definition of the GRW process. To begin with, set the particle number $N = 1$, so that $\Psi_t : \mathbb{R}^3 \to \mathbb{C}$. The random events are, instead of earthquakes, spontaneous collapses of the wave function. That is, suppose that the random variables $T_1, T_2, T_3, \ldots$, are governed by a Poisson process with parameter λ; suppose that between T_{k-1} and T_k, the wave function Ψ_t evolves according to the Schrödinger equation (where $T_0 = 0$); at every T_k, the wave function changes discontinuously ("collapses") as if an outside observer made an unsharp position measurement with inaccuracy $\sigma > 0$. I will give the formula below.

The constants λ and σ are thought of as new constants of nature, for which GRW suggested the values

$$\lambda \approx 10^{-16}\,\text{sec}^{-1}\,, \quad \sigma \approx 10^{-7}\,\text{m}\,. \tag{3.54}$$

Alternatively, Stephen Adler (2007) [2] suggested

$$\lambda \approx 3 \times 10^{-8}\,\text{sec}^{-1}\,, \quad \sigma \approx 10^{-6}\,\text{m}\,. \tag{3.55}$$

This completes the definition of the GRW process for $N = 1$.

Now consider arbitrary $N \in \mathbb{N}$, and let Ψ_0 be (what is normally called) an N-particle wave function $\Psi_0 = \Psi_0(\boldsymbol{x}_1, \ldots, \boldsymbol{x}_N)$. Consider N independent Poisson processes with rate λ, $T_{i,1}, T_{i,2}, \ldots$ for every $i \in \{1, \ldots, N\}$. Let T_1 be the smallest of all these random times, T_2 the second smallest, etc., and let I_1 be the index associated with T_1, I_2 the index associated with T_2, etc. Equivalently, $T_1, T_2, \ldots$ is a Poisson process with rate $N\lambda$, and along with every T_k, we choose a random index I_k from $\{1, \ldots, N\}$ with uniform distribution (i.e., each i has probability $1/N$), independent of each other and of the T_k. Equivalently, a collapse with index i occurs with rate λ for each $i \in \{1, \ldots, N\}$. Between T_{k-1} and T_k, Ψ_t evolves according to the Schrödinger equation. At T_k, Ψ changes as if an observer outside of the system[6] made an unsharp position measurement with inaccuracy σ on particle number I_k.

3.3.3 Definition of the GRW Process in Formulas

Let us begin with $N = 1$.

$$\Psi_{T_k+} = \frac{C(X_k)\Psi_{T_k-}}{\|C(X_k)\Psi_{T_k-}\|}\,, \tag{3.56}$$

[6] Or, rather, outside of the universe, as the idea is that the entire universe is governed by GRW theory.

where the *collapse operator* $C(\boldsymbol{X})$ is a multiplication operator multiplying by the square root of a 3d Gaussian function centered at $\boldsymbol{X}$:

$$C(\boldsymbol{X})\Psi(\boldsymbol{x}) = \sqrt{g_{\boldsymbol{X},\sigma}(\boldsymbol{x})}\,\Psi(\boldsymbol{x}) \tag{3.57}$$

with

$$g_{\boldsymbol{X},\sigma}(\boldsymbol{x}) = \frac{1}{(2\pi\sigma^2)^{3/2}} e^{-(\boldsymbol{X}-\boldsymbol{x})^2/2\sigma^2} . \tag{3.58}$$

The point $\boldsymbol{X}_k \in \mathbb{R}^3$ is chosen at random with probability density

$$\rho(\boldsymbol{X}_k = \boldsymbol{y}|T_1, \ldots, T_k, \boldsymbol{X}_1, \ldots, \boldsymbol{X}_{k-1}) = \|C(\boldsymbol{y})\Psi_{T_k-}\|^2 , \tag{3.59}$$

where $\rho(\cdots|\cdots)$ means the probability density, *given* the values of $T_1, \ldots, T_k$, $\boldsymbol{X}_1, \ldots, \boldsymbol{X}_{k-1}$. The right-hand side of (3.59) is indeed a probability density because it is nonnegative and

$$\int d^3\boldsymbol{y}\, \rho(\boldsymbol{X}_k = \boldsymbol{y}|\cdots) = \int d^3\boldsymbol{y}\, \|C(\boldsymbol{y})\Psi\|^2 = \int d^3\boldsymbol{y} \int d^3\boldsymbol{x}\, |C(\boldsymbol{y})\Psi(\boldsymbol{x})|^2 = \tag{3.60}$$

$$= \int d^3\boldsymbol{x} \int d^3\boldsymbol{y}\, g_{\boldsymbol{y},\sigma}(\boldsymbol{x})\, |\Psi(\boldsymbol{x})|^2 = \int d^3\boldsymbol{x}\, |\Psi(\boldsymbol{x})|^2 = 1 . \tag{3.61}$$

For arbitrary $N \in \mathbb{N}$ and $\Psi_t = \Psi_t(\boldsymbol{x}_1, \ldots, \boldsymbol{x}_N)$,

$$\Psi_{T_k+} = \frac{C_{I_k}(\boldsymbol{X}_k)\Psi_{T_k-}}{\|C_{I_k}(\boldsymbol{X}_k)\Psi_{T_k-}\|} \tag{3.62}$$

where the collapse operator $C_I(\boldsymbol{X})$ is the following multiplication operator:

$$C_I(\boldsymbol{X})\Psi(\boldsymbol{x}_1, \ldots, \boldsymbol{x}_N) = \sqrt{g_{\boldsymbol{X},\sigma}(\boldsymbol{x}_I)}\,\Psi(\boldsymbol{x}_1, \ldots, \boldsymbol{x}_N) . \tag{3.63}$$

The random point $\boldsymbol{X}_k$ is chosen at random with probability density

$$\rho(\boldsymbol{X}_k = \boldsymbol{y}|T_1, \ldots, T_k, I_1, \ldots, I_k, \boldsymbol{X}_1, \ldots, \boldsymbol{X}_{k-1}) = \|C_{I_k}(\boldsymbol{y})\Psi_{T_k-}\|^2 . \tag{3.64}$$

See Fig. 3.4 for an example of the effect of a GRW collapse on a wave function.

Let us examine the probability distribution (3.59) of the center $\boldsymbol{X}$ of a collapse. For a one-particle wave function Ψ, it is essentially $|\Psi|^2$; more precisely, it is the Born distribution $|\Psi|^2$ convolved with g_σ, that is, smeared out (or blurred, or coarse-grained) over a distance σ that is smaller than the macroscopic scale. The formula

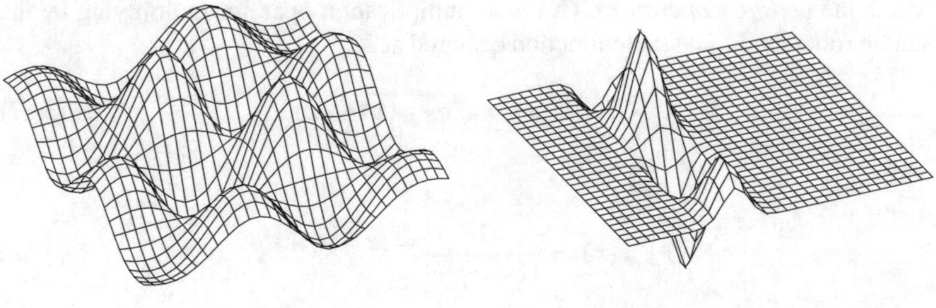

Fig. 3.4 Plot of the real part of a wave function of two particles in one space dimension: Left, before the GRW collapse; Right, after the collapse (i.e., $C_1(X)\Psi$). Note that the Gaussian profile is visible at the rim of the post-collapse surface and that the post-collapse wave function is almost zero outside a strip in configuration space corresponding to, say, the interval $[X - 3\sigma, X + 3\sigma]$ on the x_1-axis

for the convolution $f * g$ ("f convolved with g") of two functions $f, g : \mathbb{R}^d \to \mathbb{C}$ reads

$$(f * g)(\boldsymbol{x}) = \int_{\mathbb{R}^d} d^d\boldsymbol{y}\, f(\boldsymbol{y})\, g(\boldsymbol{x} - \boldsymbol{y})\,. \tag{3.65}$$

For an N-particle wave function Ψ, $\rho(\boldsymbol{X} = \boldsymbol{y})$ is essentially the *marginal* of $|\Psi|^2$ connected to the $\boldsymbol{x}_I$-variable, i.e., the distribution on 3-space obtained from the $|\Psi|^2$ distribution on $3N$-space by integrating out $3N - 3$ variables—more precisely, convolved with g_σ, i.e., smeared over width σ. Thus, again, on the macroscopic scale, the distribution of $\boldsymbol{X}$ is the same as the quantum mechanical probability distribution for the position of the I-th particle. For many purposes, it suffices to think of $\boldsymbol{X}$ as $|\Psi|^2$-distributed; the reason why GRW chose its distribution not as exactly $|\Psi|^2$ is that the definition (3.59) and (3.64) above will lead to a *no-signaling theorem*, i.e., to the property of the theory that the collapse distribution of one system cannot be influenced faster than light by collapses acting on another system, as we will show in Sect. 5.5.9 using density matrices.

This completes the definition of the GRW process. But not yet the definition of the GRW theory.

3.3.4 Primitive Ontology

There is a further law in GRW theory, concerning matter in 3-space. There are two different versions of this law and, accordingly, two different versions of the GRW theory, abbreviated as GRWm (m for *matter density ontology*) and GRWf (f for *flash ontology*). For comparison, in Bohmian mechanics, the matter in 3-space consists of particles with trajectories.

In GRWm, it is a law that, at every time t, matter is continuously distributed in space with density function $m(\boldsymbol{x}, t)$ for every location $\boldsymbol{x} \in \mathbb{R}^3$, given by

$$m(\boldsymbol{x}, t) = \sum_{i=1}^{N} m_i \int_{\mathbb{R}^{3N}} d^3\boldsymbol{x}_1 \cdots d^3\boldsymbol{x}_N \, \delta^3(\boldsymbol{x}_i - \boldsymbol{x}) \left|\psi_t(\boldsymbol{x}_1, \ldots, \boldsymbol{x}_N)\right|^2 . \tag{3.66}$$

In words, one starts with the $|\psi|^2$ distribution in configuration space $\mathbb{R}^{3N}$, then obtains the marginal distribution of the i-th degree of freedom $\boldsymbol{x}_i \in \mathbb{R}^3$ by integrating out all other variables $\boldsymbol{x}_j$, $j \neq i$, multiplies by the mass associated with $\boldsymbol{x}_i$, and sums over i.

In GRWf, it is a law that matter consists of material points in space-time called flashes. That is, matter is made neither of particles following world lines nor of a continuous distribution of matter such as in GRWm, but rather of discrete points in space-time. According to GRWf, the space-time locations of the flashes can be read off from the history of the wave function: every flash corresponds to one of the spontaneous collapses of the wave function, and its space-time location is just the space-time location of that collapse. The flashes form the set

$$F = \{(\boldsymbol{X}_1, T_1, I_1), \ldots, (\boldsymbol{X}_k, T_k, I_k), \ldots\} . \tag{3.67}$$

Note that if the number N of the degrees of freedom in the wave function is large, as in the case of a macroscopic object, the number of flashes is also large (if $\lambda = 10^{-16}\ \mathrm{s}^{-1}$ and $N = 10^{23}$, we obtain 10^7 flashes per second). Therefore, for a reasonable choice of the parameters of the GRWf theory, a cubic centimeter of solid matter contains more than 10^7 flashes per second. That is to say that large numbers of flashes can form macroscopic shapes, such as tables and chairs. "A piece of matter then is a galaxy of [flashes]" (Bell 1987 [12, Sec. 3]). That is how we find an image of our world in GRWf.

A few remarks. The m function of GRWm and the flashes of GRWf are called the *primitive ontology* of the theory. *Ontology* means what exists according to a theory; for example, in Bohmian mechanics ψ and Q, in GRWm ψ and m, and in GRWf ψ and F. The "primitive" ontology is the part of the ontology representing matter in 3d space (or 4d space-time): Q in Bohmian mechanics (a "particle ontology"), m in GRWm, and F in GRWf.

Bell coined the word *beables* (pronounced bee-abbles) for variables representing the ontology. The word is a counterpart to "observables"; in contrast to the observed outcomes of experiments, the beables represent what is real. The suffix "able" can be understood as reflecting the fact that the ontology can be different for different theories.

It may seem that a continuous distribution of matter should conflict with the evidence for the existence of atoms, electrons, and quarks and should thus make wrong predictions. We will see below why that is not the case—why GRWm makes nearly the same predictions as the quantum formalism.

3.3.5 How GRW Theory Solves the Measurement Problem

We will now look at why the GRW process succeeds in solving the measurement problem, specifically in collapsing macroscopic (but not microscopic) superpositions, and why the deviations from quantum mechanics are in a sense small.

First, the collapses are supposed to occur *spontaneously*, just at random, without the intervention of an outside observer, indeed without any physical cause described by the theory; GRW is a stochastic theory. Let us look at the number of collapses. The average waiting time between two collapses is $1/N\lambda$. For a single particle, $N = 1$, this time is $\approx 10^{16}$ sec $\approx 10^8$ years. That is, for a single particle, the wave function collapses only every 100 million years. So we should not expect to see any of these spontaneous collapses when doing an experiment with a single particle or even with hundreds of particles. If, however, we consider a macroscopic system, consisting perhaps of 10^{23} particles, then the average waiting time is 10^{-7} sec, so we have a rather dense shower of collapses.

A collapse amounts to multiplication by a Gaussian with width $\sigma \approx 10^{-7}$ m, which is large on the atomic scale (recall that the size of an atom is about one Angstrom = 10^{-10} m) but small on the macroscopic scale. So, if an electron is in a superposition of being Paris and being in Tokyo, and if the center $\boldsymbol{X}$ of the collapse lies in Paris, then the collapse operator has the effect of damping the wave function in Tokyo (which is roughly 10^7 m away from Paris) by a factor of $\exp(10^{28})$. Thus, after the collapse, the wave function in Tokyo is very near zero. On the other hand, if a collapse hits an electron in a bound state in an atom, the collapse will not much affect the electron's wave function.

A wave function like the one we encountered in the measurement problem,

$$\Psi = \sum_\alpha c_\alpha \Psi_\alpha \,, \tag{3.68}$$

where Ψ_α is a wave function corresponding to the pointer pointing to the value α, would behave in the following way. Assuming the pointer contains 10^{23} particles, then every 10^{-7} sec a collapse would occur connected to one of the pointer particles. Since Ψ_α is concentrated in a region in configuration space where all of the pointer particles are at some location $\boldsymbol{y}_\alpha$, and assuming that the $\boldsymbol{y}_\alpha$ are sufficiently distant for different values of α (namely, much more than σ), a single collapse connected to any of the pointer particles will suffice for essentially removing all contributions Ψ_α except one. Indeed, suppose the collapse is connected to the particle $\boldsymbol{x}_i$, which is one of the pointer particles. Then the random center $\boldsymbol{X}$ of the collapse will be distributed according to a coarse-grained version of the i-th marginal of $|\Psi|^2$; since the separation between the $\boldsymbol{y}_\alpha$ is greater than σ, we can neglect the coarse graining, and we can just take the i-th marginal of the $|\Psi|^2$ distribution. Thus, $\boldsymbol{X}$ will be close to one of the $\boldsymbol{y}_\alpha$, and the probability that $\boldsymbol{X}$ is close to $\boldsymbol{y}_{\alpha'}$ is $|c_{\alpha'}|^2$. Then, the multiplication by a Gaussian centered at $\boldsymbol{X}$ will shrink all other packets Ψ_α by big factors, of the order $\exp(-(\boldsymbol{y}_\alpha - \boldsymbol{y}_{\alpha'})^2/2\sigma^2)$, effectively collapsing them away; see Fig. 3.5.

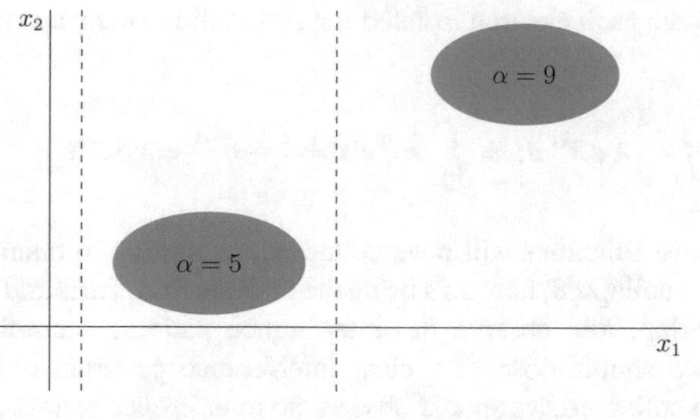

Fig. 3.5 Schematic representation of how a GRW collapse may act on a post-measurement wave function as in Fig. 3.2. If a collapse acts on the x_1 variable and sets the wave function outside the strip bounded by the dashed lines to almost zero, then the packet corresponding to $\alpha = 9$ gets effectively collapsed away

Thus, within a fraction of a second, a superposition such as (3.68) would decay into one of the packets Ψ_α (times a normalization factor) and indeed into $\Psi_{\alpha'}$ with probability $|c_{\alpha'}|^2$, the same probability as attributed by quantum mechanics to the outcome α'.

Let us make explicit how GRW succeeded in setting up the laws in such a way that they are effectively different laws for microscopic and macroscopic objects: (1) We realize that a few collapses (or even a single collapse) acting on a few (or one) of the pointer particles will collapse the entire wave function Ψ of object and apparatus together to essentially just one of the contributions Ψ_α. (2) The frequency of the collapses is proportional to the number of particles (which serves as a quantitative measure of "being macroscopic"). (3) We cannot ensure that microscopic systems experience *no collapses at all*, but we can ensure the collapses are *very infrequent*. (4) We cannot ensure that macroscopic superpositions such as $\Psi = \sum c_\alpha \Psi_\alpha$ collapse *immediately*, but we can ensure they collapse *within a fraction of a second*.

3.3.6 Empirical Tests

I have pointed out why GRW theory leads to essentially the same probabilities as prescribed by the quantum formalism. Yet, it is obvious that there are some experiments for which GRW theory predicts different outcomes than the quantum formalism. Here is an example. GRW theory predicts that if we keep a particle isolated, it will spontaneously collapse after about 100 million years, and quantum mechanics predicts it will not collapse. So let's take 10^4 electrons, for each of them prepare its wave function to be a superposition of a packet in Paris and a packet in

Tokyo; let's keep each electron isolated for 100 million years; according to GRW, a fraction of

$$\int_0^{1/\lambda} \lambda\, e^{-\lambda t}\, dt = \int_0^1 e^{-s}\, ds = 1 - e^{-1} = 63.2\% \tag{3.69}$$

of the 10^4 wave functions will have collapsed; according to quantum mechanics, none will have collapsed; now let's bring the packets from Paris and Tokyo together, let them overlap, and observe the interference pattern; according to quantum mechanics, we should observe a clear interference patterns; if all of the wave functions had collapsed, we should observe no interference pattern at all; according to GRW, we should observe only a faint interference pattern, damped (relative to the quantum prediction) by a factor of e. Ten thousand points should be enough to decide whether the damping factor is there or not. This example illustrates two things: that in principle GRW makes different predictions and that in practice these differences may be difficult to observe (because of the need to wait for 100 million years and because of the difficulty with keeping the electrons isolated for a long time, in particular avoiding decoherence).

Another testable consequence of the GRW process is *universal warming*. Since the GRW collapse usually makes wave packets narrower, their Fourier transforms (momentum representation) become wider, by the Heisenberg uncertainty relation. As a tendency, this leads to a long-run increase in energy. This effect amounts to a spontaneous warming at a rate of the order of 10^{-15} K per year. For the computation of this rate, see Exercise 3.15.

No empirical test of GRW theory against the quantum formalism can presently be carried out, but experimental techniques are progressing; see Fig. 3.6. Adler's parameters have in the meantime been empirically refuted as a byproduct of the LIGO experiment that detects gravitational waves (Carlesso et al. 2016 [18]) and other experiments. A test of GRW's parameters seems feasible using a planned interferometer on a satellite in outer space. Interferometers are disturbed by the presence of air, temperatures far from absolute zero, vibrations of the apparatus, and the presence of gravity; that is why being in outer space is an advantage for an interferometer and allows for heavier objects shot through the double slit and longer flight times. Such an interferometer has been proposed to the European Space Agency (ESA) (Kaltenbaek 2020 [35]).

3.3.7 The Need for a Primitive Ontology

Primitive ontology is a subtle philosophical topic.

We may wonder whether, instead of GRWf or GRWm, we could assume that only ψ exists and no primitive ontology; let us call this view GRWØ. To illustrate the difference between GRWf/GRWm and GRWØ, let me make up a creation myth (as a metaphorical way of speaking): Suppose God wants to create a universe governed by GRW theory. He creates a wave function ψ of the universe that starts out as

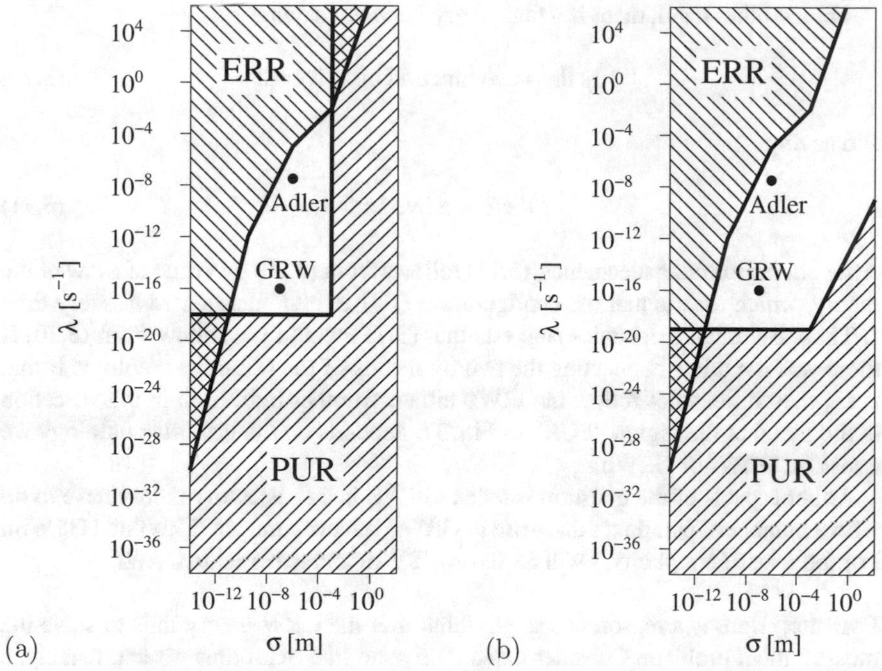

Fig. 3.6 Parameter diagram (log-log-scale) of the GRW theory with the primitive ontology given by (**a**) flashes and (**b**) the matter density function. ERR, empirically refuted region as of 2012 (equal in (**a**) and (**b**)); PUR, philosophically unsatisfactory region. GRW's and Adler's choice of parameters are marked. (Reprinted from Feldmann and Tumulka (2012) [25] with permission by the Institute of Physics)

a particular ψ_0 that he chose and evolves stochastically according to a particular version of the GRW time evolution law. According to GRWØ, God is now done. According to GRWf or GRWm, however, a second act of creation is necessary, in which he creates the matter, i.e., either the flashes or continuously distributed matter with density m, in both cases coupled to ψ by the appropriate laws.

There are several motivations for considering GRWØ. First, it seems more parsimonious than GRWm or GRWf. Second, it was part of the motivation behind GRW theory to avoid introducing an ontology in addition to ψ. In fact, much of the motivation came from the measurement problem, which requires that we either modify the Schrödinger equation or introduce additional ontology (such as Q in Bohmian mechanics), and GRW theory was intended to choose the first option, not the second.

Furthermore, there is a sense in which GRWØ clearly works: The GRW wave function ψ_t is, at almost all times, concentrated, except for tiny tails, on a set of configurations that are macroscopically equivalent to each other. So we can read off from the post-measurement wave function, e.g., what the actual outcome of a quantum measurement was.

On the other hand, there is a logical gap between saying

$$\text{“}\psi \text{ is the wave function of a live cat”} \tag{3.70}$$

and saying

$$\text{“there is a live cat.”} \tag{3.71}$$

After all, in Bohmian mechanics, (3.71) follows from (3.70) by virtue of a law of the theory, which asserts that the configuration $Q(t)$ is $|\psi_t|^2$ distributed at every time t. Thus, Bohmian mechanics suggests that (3.71) would not follow from (3.70) if there was not a law connecting the two by means of the primitive ontology. If that is so, then it does not follow in GRWØ either. Another indication in this direction is the fact that the region "PUR" in Fig. 3.6 depends on the primitive ontology we consider, GRWf or GRWm.

Other aspects of the question whether GRWØ is a satisfactory theory have to do with a number of paradoxes that arise in GRWØ but evaporate in GRWf and GRWm. For the sake of simplicity, I will focus on GRWm and leave aside GRWf.

Paradox Here is a reason one might think that the GRW theory fails to solve the measurement problem. Consider a quantum state like Schrödinger's cat, namely, a superposition

$$\psi = c_1\psi_1 + c_2\psi_2 \tag{3.72}$$

of two macroscopically distinct states ψ_i with $\|\psi_1\| = 1 = \|\psi_2\|$, such that both contributions have nonzero coefficients c_i. Given that there is a problem—the measurement problem—in the case in which the coefficients are equal, one should also think that there is a problem in the case in which the coefficients are not exactly equal, but roughly of the same size. One might say that the reason there is a problem is that, according to quantum mechanics, there is a superposition, whereas according to our intuition, there should be a definite state. But then it is hard to see how this problem should go away just because c_2 is much smaller than c_1. How small would c_2 have to be for the problem to disappear? No matter if $c_2 = c_1$ or $c_2 = c_1/100$ or $c_2 = 10^{-100}c_1$, in each case, both contributions are there. But the only relevant effect of the GRW process replacing the unitary evolution, as far as Schrödinger's cat is concerned, is to randomly make one of the coefficients much smaller than the other (although it also affects the shape of the suppressed contribution).

Answer From the point of view of GRWm, the reasoning misses the primitive ontology. Yes, the wave function is still a superposition, but the definite facts that our intuition wants can be found in the primitive ontology. The cat is made of m, not of ψ. If ψ is close to $|\text{dead}\rangle$, then m equals $m_{|\text{dead}\rangle}$ up to a small perturbation, and that can reasonably be accepted as the m function of a dead cat. While the wave function is a superposition of two packets ψ_1, ψ_2 that correspond to *two very different* kinds of (particle) configurations in orthodox quantum mechanics or Bohmian mechanics, there is only *one* configuration of the matter density m—the definite fact that our intuition wants.[7]

Paradox As a variant of the first paradox, one might say that even after the GRW collapses have pushed $|c_1|^2$ near 1 and $|c_2|^2$ near 0 in the state vector (3.72), there is still a positive probability $|c_2|^2$ that if we make a quantum measurement of the macro-state—of whether the cat is dead or alive—we will find the state ψ_2, even though the GRW state vector has collapsed to a state vector near ψ_1, a state vector that might be taken to indicate that the cat is really dead (assuming $\psi_1 = |\text{dead}\rangle$). Thus, it seems not justified to say that, when ψ is close to $|\text{dead}\rangle$, the cat is really dead.

Answer In GRWm, what we mean when saying that the cat is dead is that the m function looks and behaves like a dead cat. In orthodox quantum mechanics, one might mean instead that a quantum measurement of the macro-state would yield $|\text{dead}\rangle$ with probability 1. These two meanings are not exactly equivalent in GRWm: that is because, if $m \approx m_{|\text{dead}\rangle}$ (so we should say that the cat is dead) and if ψ is close but not exactly equal to $|\text{dead}\rangle$, then there is still a tiny but nonzero probability that within the next millisecond, the collapses occur in such a way that the cat is suddenly alive! But that does not contradict the claim that a millisecond before, the cat was dead; it only means that GRWm allows resurrections to occur—with tiny probability! In particular, if we observe the cat after that millisecond, there is a positive probability that we find it alive (simply because it *is* alive) even though before the millisecond, it actually was dead.

[7] Here is an idea for an alternative way out of the paradox suggested to me by Ilmar Bürck: Suppose we change GRW theory so that we use, instead of a Gaussian, a collapse profile with compact support, i.e., a function that vanishes *exactly* outside a bounded region (of diameter of order σ). Would that have the consequence that the disfavored packet disappears completely, $c_2 = 0$? Yes, immediately after the collapse. However, it is a mathematical property of the Schrödinger equation that wave functions with compact support develop tails to infinity in arbitrarily short time, so the amplitude in the region $R_2 \subset \mathbb{R}^{3N}$ associated with ψ_2 quickly becomes nonzero again. Then again, the Schrödinger equation is non-relativistic, and the Dirac equation, a relativistic version of the Schrödinger equation, has the property that wave functions can propagate at most at the speed of light c. (In configuration space, the set $M = \{q \in \mathbb{R}^{3N} : \psi(q) \neq 0\}$ grows at speed c in each $\boldsymbol{x}_j$ variable, i.e., $M(t) \subseteq \bigcup_{(\boldsymbol{x}_1 \ldots \boldsymbol{x}_N) \in M} \overline{B_{c|t|}(\boldsymbol{x}_1)} \times \cdots \times \overline{B_{c|t|}(\boldsymbol{x}_N)}$ where $\overline{B_r(\boldsymbol{x})} = \{\boldsymbol{y} \in \mathbb{R}^3 : |\boldsymbol{x} - \boldsymbol{y}| \leq r\}$ means the closed ball of radius r around $\boldsymbol{x}$.) If the two relevant regions R_1, R_2 of configuration space (say, corresponding to a dead and to a live cat) differ by (order of magnitude) centimeters in some particle coordinates, then it should take the wave function (order of magnitude) 10^{-10} seconds to propagate from R_1 to R_2. So whether this way out works depends on the frequency of collapses. Let us say a cat consists of (order of magnitude) 3×10^{27} nucleons (which are the "particles" GRW had in mind). Then it should undergo (order of magnitude) 3×10^{11} collapses per second, suggesting the amplitude in R_2 may indeed be exactly 0 all of the time after the first collapse. Then again, if we consider, instead of a cat, a needle weighing 1 gram (6×10^{23} nucleons), the amplitude would be nonzero most of the time. So this way out apparently does not work in all circumstances.

Paradox Let ψ_1 be the state "the marble is inside the box" and ψ_2 the state "the marble is outside the box"; these wave functions have disjoint supports S_1, S_2 in configuration space (i.e., wherever one is nonzero, the other is zero). Let ψ be given by (3.72) with $0 < |c_2|^2 \ll |c_1|^2 < 1$; finally, consider a system of n (non-interacting) marbles at time t_0, each with wave function ψ, so that the wave function of the system is $\psi^{\otimes n}$. Then for each of the marbles, we would feel entitled to say that it is inside the box, but on the other hand, the probability that all marbles be found inside the box is $|c_1|^{2n}$, which can be made arbitrarily small by making n sufficiently large.

Answer According to the m function, each of the marbles is inside the box at the initial time t_0. However, it is known that, if we assume $H = 0$ for simplicity, a superposition like (3.72) of macroscopically distinct states ψ_i will converge as $t \to \infty$ under the GRW evolution with probability $|c_1|^2$ to a function $\psi_1(\infty)$ concentrated in S_1 and with probability $|c_2|^2$ to a function $\psi_2(\infty)$ concentrated in S_2.[8] Thus, as $t \to \infty$, the initial wave function $\psi^{\otimes n}$ will evolve toward one consisting of approximately $n|c_1|^2$ factors $\psi_1(\infty)$ and $n|c_2|^2$ factors $\psi_2(\infty)$ for large n, so that ultimately about $n|c_1|^2$ of the marbles will be inside and about $n|c_2|^2$ outside the box—independent of whether anybody observes them or not. The occurrence of some factors $\psi_2(\infty)$ at a later time provides another example of the resurrection-type events mentioned earlier; they are unlikely but do occur, of course, if we make n large enough.

The act of observation plays no role in the argument and can be taken to merely record pre-existing macroscopic facts. To be sure, the physical interaction involved in the act of observation may have an effect on the system, such as speeding up the evolution from ψ toward either $\psi_1(\infty)$ or $\psi_2(\infty)$; but GRWm provides unambiguous facts about the marbles also in the absence of observers.

This concludes my discussion of these paradoxes; further detail is given in my article [48]. As a final remark concerning the primitive ontology, I want to mention an example of an unreasonable choice of primitive ontology that was considered (but not seriously proposed!) by Allori et al. (2014) [7]: We set up a theory GRWp combining the GRW wave function ψ_t with a particle ontology governed by Bohm's equation of motion. Nobody seriously proposed this theory, and it makes completely wrong predictions. For example, suppose that ψ_{t-} is the wave function of Schrödinger's cat, the Bohmian configuration Q lies in the support of $|\text{alive}\rangle$, and a GRW collapse occurs; since the collapse center is chosen randomly (and independent of Q), it may well collapse the wave function to near $|\text{dead}\rangle$. Should we say then that the cat is really alive, as suggested by Q, or really dead, as suggested by ψ_{t+}? If

[8] As an idealization, consider instead of Gaussian factors the characteristic functions of S_1 and S_2, so that the coefficients of the superposition will change with every collapse but not the shape of the two contributions, $\psi_1(\infty) = \psi_1$ and $\psi_2(\infty) = \psi_2$. Although both coefficients will still be nonzero after any finite number of collapses, one of them will tend to zero in the limit $t \to \infty$.

we take the primitive ontology seriously, then we should conclude the cat is alive. However, the collapse has deformed the wave packet of the live cat due to the slopes of the tails of the Gaussian, and the packet will from now on evolve in a way very different from a usual live cat. Despite its wrong predictions, this theory is useful to consider because it illustrates the role of the primitive ontology and that of laws linking the wave function to the matter.

3.4 The Copenhagen Interpretation

A very influential view, almost synonymous with the orthodox view of quantum mechanics, is the *Copenhagen interpretation* (CI), named after the research group headed by Niels Bohr (1885–1962), who was the director of the Institute for Theoretical Physics at the University of Copenhagen, Denmark. Further famous defenders of this view and members of Bohr's group (temporarily also working in Copenhagen) include Werner Heisenberg (1901–1976), Wolfgang Pauli, and Leon Rosenfeld (1904–1974). Bohr and Einstein were antagonists in a debate about the foundations of quantum mechanics that began around 1925 and continued until Einstein's death in 1955. Here is a description of the main elements of CI.

3.4.1 Two Realms

In CI, the world is separated into two realms: macroscopic and microscopic. In the macroscopic realm, there are no superpositions. Pointers always point in definite directions. The macroscopic realm is described by the classical positions and momenta of objects. In the microscopic realm, there are no definite facts. For example, an electron does not have a definite position. The microscopic realm is described by wave functions. One could say that the primitive ontology of CI consists of the macroscopic matter (described by its classical positions and momenta). In CI terminology, the macroscopic realm is called *classical* and the microscopic realm *quantum*.[9] Instead of classical and quantum, Bell called them *speakable* and *unspeakable*. (The macroscopic realm hosts the objects with definite properties, of which one can speak. Since in ordinary English, something "unspeakable" is not something nice, you may have gotten the sense that Bell is not a supporter of the idea of two separate realms.)

[9] This is a somewhat unfortunate terminology because the word classical suggests not only definite positions but also particular laws (say, Newton's equation of motion) which may actually not apply. The word quantum is somewhat unfortunate as well because in a reductionist view, all laws (also those governing macroscopic objects) should be consequences of the quantum laws applying to the individual electrons, quarks, etc.

The microscopic realm, when isolated, is governed by the Schrödinger equation. The macroscopic realm, when isolated, is governed by classical mechanics. The two realms interact whenever a measurement is made; then the macro realm records the measurement outcome, and the micro realm undergoes a collapse of the wave function.

I see a number of problems with the concept of two separate realms.

- It is not precisely defined where the border between micro and macro lies. That lies in the nature of the word "macroscopic." Clearly, an atom is micro and a table is macro, but what is the exact number of particles required for an object to be "macroscopic"? The vagueness inherent in the concept of "macroscopic" is unproblematical in Bohmian mechanics, GRW theory, or classical mechanics, but it is problematical here because it is involved in the formulation of the laws of nature. Laws of nature should not be vague.
- Likewise, what counts as a measurement and what does not? This ambiguity is unproblematical when we only want to compute the probabilities of outcomes of a given experiment because it will not affect the computed probabilities. But an ambiguity is problematical when it enters the laws of nature.
- The special role played by measurements in the laws according to CI is also implausible and artificial. Even if a precise definition of what counts as a measurement were given, that definition would seem arbitrary, and it would not seem believable that during measurement, other laws than normal are in place.
- The separation of the two realms, without the formulation of laws that apply to both, is against reductionism. If we think that macro objects are made out of micro objects, then the separation is problematical.

3.4.2 Elements of the Copenhagen View

Positivism

CI leans toward positivism. In the words of Werner Heisenberg (1958) [34]:

> We can no longer speak of the behavior of the particle independently of the process of observation.

Feynman (1962) [28, p. 14] did not like that:

> Does this mean that my observations become real only when I observe an observer observing something as it happens? This is a horrible viewpoint. Do you seriously entertain the thought that without observer there is no reality? Which observer? Any observer? Is a fly an observer? Is a star an observer? Was there no reality before 10^9 B.C. before life began? Or are you the observer? Then there is no reality to the world after you are dead? I know a number of otherwise respectable physicists who have bought life insurance.

Purported Impossibility of Non-paradoxical Theories

Another traditional part of CI is the claim that it is impossible to provide any coherent (non-paradoxical) realist theory of what happens in the micro realm. Heisenberg (1958) [34] again:

> The idea of an objective real world whose smallest parts exist objectively in the same sense as stones or trees exist, independently of whether or not we observe them [...], is impossible.

We know from Bohmian mechanics that this claim is, in fact, wrong.

Completeness of the Wave Function

In CI, a microscopic system is completely described by its wave function. That is, there are no further variables (such as Bohm's particle positions) whose values nature knows and we do not. For this reason, the wave function is also called the *quantum state* or the *state vector*.

Language of Measurement

CI introduced (and established) the words "measurement" and "observable" and emphasized the analogy suggested by these words: For example, that the momentum operator is analogous to the momentum variable in classical mechanics and that the spin observable $\boldsymbol{\sigma} = (\sigma_1, \sigma_2, \sigma_3)$ is analogous to the spin vector of classical mechanics (which points along the axis of spinning and whose magnitude is proportional to the angular frequency).

I have already mentioned that these two words are quite inappropriate because they suggest that there was a pre-existing value of the observable A that was merely discovered (i.e., made known to us) in the experiment, whereas in fact the outcome is often only created during the experiment. Think, for example, of a Stern–Gerlach experiment in Bohmian mechanics: The particle does not have a value of z-spin before we carry out the experiment. And in CI, since it insists that wave functions are complete, it is true in spades that A does not have a pre-existing, well-defined value before the experiment. So this terminology is *even less* appropriate in CI—and yet, it is a cornerstone of CI! Well, CI leans toward paradoxes.

Narratives, But No Serious Ones

When calculating predictions that can be compared to experimental data, adherents of CI often tell a story about the physical meaning of the mathematical elements of the calculation. This story may involve particles or waves, may be imprecise, may conflict with other stories told on another occasion, or may contain several parts conflicting with each other. But this story is not intended to describe what actually happens. On the contrary, CI insists that such narratives should not be taken seriously. They are just metaphor, or allegory, or analogy; they just serve as a mnemonic for the calculation, as a help for remembering the correct formulas or for setting up the corresponding formulas in similar calculations. In contrast, theories such as Bohmian mechanics or GRW or many-worlds are hypothesizing about what actually happens in nature and correspondingly aim at providing a single, coherent story that fits all experiments and situations.

3.4.3 Complementarity

Another idea of CI, called *complementarity*, is that in the micro realm, reality is paradoxical (contradictory) but the contradictions can never be seen (and are therefore not problematical) because of the Heisenberg uncertainty relation. I would describe the idea as follows. In order to compute a quantity of interest (e.g., the wave length of light scattered off an electron), we use both Theory A (e.g., classical theory of billiard balls) and Theory B (e.g., classical theory of waves) although A and B contradict each other.[10] It is impossible to find one Theory C that replaces both A and B and explains the entire physical process. (Here, we meet again the impossibility claim mentioned in Section "Purported Impossibility of Non-paradoxical Theories".) Instead, we should leave the conflict between A and B unresolved and accept the idea that reality is paradoxical.

Bell (1986, [11] or [13, p. 190]) wrote the following about complementarity:

> It seems to me that Bohr used this word with the reverse of its usual meaning. Consider for example the elephant. From the front she is head, trunk and two legs. From the back she is bottom, tail, and two legs. From the sides she is otherwise, and from the top and bottom different again. These various views are complementary in the usual sense of the word. They supplement one another, they are consistent with one another, and they are all entailed by the unifying concept 'elephant.' It is my impression that to suppose Bohr used the word 'complementary' in this ordinary way would have been regarded by him as missing his point and trivializing his thought. He seems to insist rather that we must use in our analysis elements which *contradict* one another, which do not add up to, or derive from, a whole. By 'complementarity' he meant, it seems to me, the reverse: contradictoriness.

Einstein (1949) [22]:

> Despite much effort which I have expended on it, I have been unable to achieve a sharp formulation of Bohr's principle of complementarity.

Bell commented (1986) [11]:

> What hope then for the rest of us?

In a way, the Copenhagen belief in contradictions fits together with the use of narratives or intuitive pictures that we are advised not to take seriously. If we did take

[10] In fact, before 1926, many successful theoretical considerations for predicting the results of experiments proceeded in this way. For example, people made a calculation about the collision between an electron and a photon as if they were classical billiard balls, then converted the momenta into wave lengths using de Broglie's relation $p = \hbar k$, and then made another calculation about waves with wave number k.

them seriously, contradictions would be a problem, but if we do not, then why worry about contradictions! Of course, it remains unclear what *can* be taken seriously. Correspondingly, Copenhagenists often speak of the "interpretation" of certain formulas, which they say is not easy to obtain. For comparison, in theories such as Bohm's and GRW's, the interpretation of formulas is mostly obvious because these theories provide clear hypotheses of what happens in reality.

Another version of complementarity concerns observables that cannot be simultaneously measured. We have discussed this situation in Sect. 3.1.5: their joint distribution depends on the order of the experiments, and their distributions are not both sharply peaked at the same time. According to CI, the existence of incompatible observables is a paradoxical trait of the micro realm that we are forced to accept. The connection between paradox and non-commutativity fits nicely with the analogy between operators in quantum mechanics and quantities in classical mechanics (as described in Section "Language of Measurement"): In classical mechanics, which is free of paradoxes, all physical quantities (e.g., components of positions, momenta, rotation vectors) are just numbers and therefore commute with each other. At the same time, the connection between paradox and non-commutativity suggests that we can never observe a contradiction in experiment because we can only observe one of two non-commuting observables. And again, since we cannot observe contradictions, the contradictions are somehow regarded as unproblematical. That is, according to Copenhagen, the situation is like in the cartoon shown in Fig. 3.7.

Here is a sentence from Bohr (1934) [14, p. 10] sometimes quoted as a description of complementarity:

> Any given application of classical concepts precludes the simultaneous use of other classical concepts which in a different connection are equally necessary for the elucidation of the phenomena.

This sounds like saying that paradox is inevitable.

Let me dwell a bit on this quote. It is not very clear here what counts as "classical concepts." Particles and waves? But why would particles and waves preclude each other? Bohmian mechanics illustrates the obvious possibility that there might be particles *and* waves. Or are position and momentum meant by "classical concepts"? But why would they be necessary? GRWm illustrates the obvious possibility that electrons may not always be concentrated in a point and thus may not always *have* a position.[11]

[11] Also Bohr's reference to "elucidation," which after all is a mental act, is striking. The idea that certain concepts are necessary for our minds to consider a real-world situation sounds reminiscent of the doctrine of German philosopher Immanuel Kant (1724–1804) that the reason why the laws of Euclidean geometry can be proved with mathematical certainty is that they are not part of the outside world but part of how our minds work—that our minds cannot think about the outside world differently than according to 3d Euclidean geometry. The modern view about Euclidean geometry, which goes back in particular to Carl Friedrich Gauss (1777–1855) and Einstein, is rather different: Our minds can very well think about curved spaces (i.e., geometries that differ from Euclidean), and physical space has a geometry that may very well be curved (in fact it is, according to the

Fig. 3.7 According to the Copenhagen view, we never see the paradoxical thing happen. But we see traces showing that it must have happened. (Drawing © 1940 Charles Addams, renewed 1967, reprinted with permission by the Tee and Charles Addams Foundation)

3.4.4 Reactions to the Measurement Problem

While Bohmian mechanics, GRW theory, and many-worlds theories have clear answers to the measurement problem, this is not so with Copenhagen. I report some answers that I heard Copenhagenists give (with some comments of mine in brackets); I must say I do not see how these answers would resolve or remove the problem.

- Nobody can actually solve the Schrödinger equation for 10^{23} interacting particles. (Sure, and we do not need to. If Ψ_α looks like a state including a pointer pointing to α, then we know by linearity that Ψ_{t_i} evolves to $\Psi_{t_\mathrm{f}} = \sum c_\alpha \Psi_\alpha$, a superposition of macroscopically different states.)
- Systems are never isolated. (If we cannot solve the problem for an isolated system, what hope can we have to treat a non-isolated one? The way you usually treat a non-isolated system is by regarding it as a subsystem of a bigger, isolated system, maybe the entire universe.)

empirical facts supporting general relativity), and Euclidean geometry is merely a good and simple approximation to (and idealization of) the geometry of physical space.

- Maybe there is no wave function of the universe. (It is up to Copenhagenists to propose a formulation that applies to the entire universe. Bohm, GRW, and many-worlds can do that. If, according to Copenhagen, there is nothing other than the wave function, and if even the wave function does not exist for the universe, then what is the complete description of the universe?)
- Who knows whether the initial wave function is really a product as in $\Psi_{t_i} = \psi \otimes \phi$. (It is not so important that it is precisely a product, as long as we can perform a quantum measurement on ψs that are non-trivial superpositions of eigenvectors. Note that if $\Psi(t_i)$ is approximately but not exactly equal to $\psi \otimes \phi$ with $\psi = \sum c_\alpha \psi_\alpha$, then $\Psi(t_f)$ is still approximately equal to $\sum c_\alpha \Psi_\alpha(t_f)$, so it is still a non-trivial superposition of contributions corresponding to different outcomes.)
- The goal of the measurement problem is to derive the collapse rule from the unitary evolution, but why should we burden ourselves with trying that? Why can we not take the collapse rule as an independent axiom? (I do not agree that the measurement problem arises only if we set our goal to derive the collapse rule. We ask what happens in reality, and we allow any number of axioms, but we derive that if ψ is the only reality and is governed by the Schrödinger equation, then a conflict with the collapse rule arises.)
- The collapse of the wave function is like the collapse of a probability distribution: as soon as I have more information, such as $X \in B$, I have to update my probability distribution ρ_{t-} for X accordingly, namely, to

$$\rho_{t+}(x) = 1_{x \in B}\, \rho_{t-}(x)\,. \tag{3.73}$$

  (The parallel is indeed striking. However, if we insist that the wave function is complete, then there never is any new information, as there is nothing that we are ignorant of.)
- Decoherence makes sure that you can replace the superposition $\Psi = \sum c_\alpha \Psi_\alpha$ by a mixture [i.e., a random one of the Ψ_α]. (A super-observer cannot distinguish between the superposition and the mixture, but we are asking whether in reality it is a superposition or a mixture; see Sect. 3.2.3.)

As Angelo Bassi once put it,

> The measurement problem is not going to go away.

3.4.5 The Transactional Interpretation

Under the name "transactional interpretation," John G. Cramer (1986) [19] published some ideas about the foundations of quantum mechanics. In 2005, I met John Cramer at a conference in Sydney, and since we had both arrived a day early, we spent a pleasant afternoon at the zoo chatting about, among other things, the foundations of quantum mechanics. Let me describe my take on his view.

I have already pointed out in Section "Narratives, But No Serious Ones" that adherents to the Copenhagen interpretation tend to tell stories along with their calculations and insist at the same time that such stories should not be taken seriously. Instead, the stories serve as something like a mnemonic to make the formulas more vividly intuitive.

In my conversation with Cramer, it became clear that he was following the same spirit. That is, with the "transactional interpretation," he did not intend to replace orthodox quantum mechanics with something else, he wanted to flesh it out further, and he wanted to provide further ones of these stories that help you remember the formulas and that attach some picture to each symbol in the calculation. For example, he thought of $\psi(x)$ or $|\psi\rangle$ as a wave propagating from the particle source to the particle detector and of $\psi^*(x)$ or $\langle\psi|$ as a wave propagating from the detector to the source. If you read Cramer's paper expecting a fundamental physical theory, i.e., a hypothesis about what happens in nature comparable to Bohmian mechanics, GRW, and many-worlds, then you will find a mess. It will remain unclear, for example, what the ontology is, what the unexplained notion of "transaction" means in terms of the ontology, or how those of his stories should be understood that involve several rounds of revision of space-time histories (such as a wave propagating from space-time point x to a space-time point y in the future of x, in the next round nature making a random decision at y and creating another wave that propagates back in time to x, where nature makes another random decision and sends another wave to y). But if the stories serve more as a visualization of the formulas, then they do not have to provide an unambiguous, coherent picture of events.

3.5 Many Worlds

It is important to be familiar with another type of theory of quantum mechanics, the many-worlds theories, which I describe in this section. After our discussion of GRW theories, I could summarize two of these theories very briefly by saying that Everett's many-worlds theory is GRWØ with $\lambda = 0$ and Schrödinger's many-worlds theory is GRWm with $\lambda = 0$.

The motivation for the many-worlds view comes from the wave function (3.41), $\Psi(t_f) = \sum_\alpha c_\alpha \Psi_\alpha$, of object and apparatus together after a quantum measurement. It is a superposition of macroscopically different terms. If we insist that the Schrödinger equation is correct (and thus reject non-linear modifications such as GRW), and if we insist that the wave function is complete, then we must conclude that there are different parts of reality, each looking like our world but with a different measurement outcome and without any interaction between the different parts. They are parallel worlds. This view was suggested by Hugh Everett III (1930–1982) in 1955 [23] and 1957 [24].

Everett's is not the only many-worlds theory, though. It is less well known that also Erwin Schrödinger (1887–1961) proposed a many-worlds theory in 1926 [44] (see also Allori et al. 2011 [6]), and it is useful to compare the two. Schrödinger, however, did not realize that his proposal was a many-worlds theory. He thought of it as a single-world theory. He came to the conclusion that it was empirically inadequate and abandoned it. Let us first try to get a good understanding of this theory.

3.5.1 Schrödinger's Many-Worlds Theory

According to Schrödinger's 1926 theory, matter is distributed continuously in space with density

$$m(\boldsymbol{x},t) = \sum_{i=1}^{N} m_i \int_{\mathbb{R}^{3N}} d^3\boldsymbol{x}_1 \cdots d^3\boldsymbol{x}_N\, \delta^3(\boldsymbol{x}_i - \boldsymbol{x}) \left|\psi_t(\boldsymbol{x}_1,\ldots,\boldsymbol{x}_N)\right|^2, \tag{3.74}$$

and ψ_t evolves according to the Schrödinger equation. The equation for m is exactly the same as in GRWm, except that ψ is not the same wave function. (Actually, Schrödinger replaced the mass factor m_i by the electric charge e_i, but this difference is not crucial. It amounts to a different choice of weights in the weighted average over i. In fact, Schrödinger's choice has the disadvantage that the different signs of charges will lead to partial cancellations and thus to an m function that looks less plausible as the density of matter. Nevertheless, the two choices turn out to be empirically equivalent, i.e., lead to the same predictions.)

In analogy to GRWm, we may call this theory Sm (where S is for the Schrödinger equation and m for the matter density ontology). Consider a double-slit experiment in this theory. Before the arrival at the detection screen, the contribution to the m function coming from the electron sent through the double slit (which is the only contribution in the region of space between the double slit and the detection screen) is a lump of matter smeared out over rather large distances (as large as the interference pattern). This lump is not homogeneous; it has interference fringes. And the overall amount of matter in this lump is tiny: If you integrate $m(\boldsymbol{x},t)$ over $\boldsymbol{x}$ in the region between the double slit and the detection screen, the result is 10^{-30} kg, the mass of an electron. But focus now on the fact that the matter is spread out. Schrödinger incorrectly thought that this fact must lead to the wrong prediction that the entire detection screen should glow faintly instead of yielding one bright spot, and that was why he thought Sm was empirically inadequate.

To understand why this reasoning was incorrect, consider a post-measurement situation (e.g., Schrödinger's cat). The wave function is a superposition of macroscopically different terms, $\Psi = \sum_\alpha c_\alpha \Psi_\alpha$. The Ψ_α do not overlap; i.e., where one Ψ_α is significantly nonzero, the others are near zero. Thus, when we compute $|\Psi|^2$, there are no (significant) cross-terms; that is, for each q, there is at most one α contributing, so

$$|\Psi(q)|^2 = |c_\alpha|^2\, |\Psi_\alpha(q)|^2 . \tag{3.75}$$

Define $m_\alpha(\boldsymbol{x})$ as what m would be according to (3.74) with $\psi = \Psi_\alpha$. Then we obtain (to an excellent degree of approximation)

$$m(\boldsymbol{x}) = \sum_\alpha |c_\alpha|^2 \, m_\alpha(\boldsymbol{x}) \,. \tag{3.76}$$

In words, the m function is a linear combination of m functions corresponding to the macroscopically different terms in Ψ. So, for Schrödinger's cat in Sm, there is a dead cat, and there is a live cat, each with half the mass. However, they do not notice they have only half the mass, and they do not notice the presence of the other cat. That is because, if we let the time evolve, then each $\Psi_\alpha(t)$ evolves in a way that corresponds to a reasonable story of just one cat; after all, it is how the wave function would evolve according to the projection postulate after a measurement of the cat had collapsed the superposition to one of the Ψ_α. Furthermore, $\Psi(t) = \sum_\alpha c_\alpha \Psi_\alpha(t)$ by linearity, and since the $\Psi_\alpha(t)$ remain non-overlapping, we have that (3.76) applies to every t from now on, that is,

$$m(\boldsymbol{x}, t) = \sum_\alpha |c_\alpha|^2 \, m_\alpha(\boldsymbol{x}, t) \,. \tag{3.77}$$

Each $m_\alpha(t)$ looks like the reasonable story of just one cat that $\Psi_\alpha(t)$ corresponds to. Thus, the two cats do not interact with each other; they are causally disconnected. After all, the two contributions m_α come from Ψ_α that are normally thought of as *alternative* outcomes of the experiment. So the two cats are like ghosts to each other: they can see and walk through each other.

And not just the cat has split in two. If a camera takes a photograph of the cat, then Ψ must be taken to be a wave function of the cat and the camera together (among other things). Ψ_1 may then correspond to a dead cat and a photo of a dead cat and Ψ_2 to a live cat and a photo of a live cat. If a human being interacts with the cat (say, looks at it), then Ψ_1 will correspond to a brain state of seeing a dead cat and Ψ_2 to one of seeing a live cat. That is, there are two copies of the cat, two copies of the photo, two copies of the human being, and two copies of the entire world. That is why I said that Sm has a many-worlds character. In each world, though, things seem rather ordinary: Like a single cat in an ordinary (though possibly pitiful) state, and all records and memories are consistent with each other and in agreement with the state of the cat.

3.5.2 Everett's Many-Worlds Theory

Everett's many-worlds theory, which could be called SØ (S for the Schrödinger equation and Ø for the empty primitive ontology), is based on the idea that the same picture would arise if we dispense with the m function. Frankly, I do not see how it would; I actually cannot make sense of SØ as a physical theory. Some authors argue that SØ has a problem with how to obtain probabilities, but I would say the more basic problem is how to obtain *things* such as cats, chairs, and pointers. I understand why the thought can be attractive that the ontology consists only of the wave function and the wave function evolves according to the Schrödinger equation. That would make for a very simple theory. It just does not make sense.

Let me elaborate a bit on this point. For SØ, we would have to assume a re-interpretation of ordinary language, that the statement "there is a live cat" does not actually refer to a thing called a cat but is really a statement about the wave function, really expressing that there is the kind of wave packet that a live cat would have. And this re-interpretation seems like an unconvincing contortion to me.

Here is another angle on this point. Everettians seem to conclude, whenever $\Psi(q_0) \neq 0$ at some point $q_0 \in \mathbb{R}^{3N}$ that is a configuration of a live cat, then there is a live cat. But how would that follow? Consider, for example, another function on configuration space $\mathbb{R}^{3N}$, the potential function $V(q)$. Often, we allow for an arbitrary function $V(q)$ when considering Newtonian mechanics or the Schrödinger equation. If this function were nonzero only in a small neighborhood of q_0, would we say that therefore there exists a live cat? Of course not. In Newtonian mechanics, there is a configuration $Q(t)$ formed by the particles at every time t, but it can *of course* be a configuration at which V vanishes, and *of course* V can be nonzero at a configuration different from $Q(t)$. In short, the primitive ontology is missing in SØ. And that problem is solved in Sm. Further discussion of the need for a primitive ontology can be found in Maudlin (2010) [37] and Allori et al. (2014) [7].

I have some remarks on the psychology of SØ. A person who believes that SØ makes sense will naturally conclude that a many-worlds view is inevitable. After all, they think that the presence of a wave function of a live cat somehow implies (or even means) the presence of a live cat, and since the Schrödinger evolution inevitably leads to macroscopic superpositions such as the $\Psi = \sum_\alpha c_\alpha \Psi_\alpha$ above, to wave functions containing a packet for each outcome of an experiment, and in the situation of Schrödinger's cat to a wave packet of a live cat and a wave packet of a dead cat, it seems inevitable to them that in this situation there is a live cat and there is a dead cat (e.g., Brown and Wallace 2005 [17]). I call this the *illusion of necessity* of many worlds. Of course, Bohmian mechanics is a counterexample to this idea of necessity. A person who believes that Sm makes sense will not have this sense of necessity, as the many-worlds character of the theory does not come from Ψ but from m, and if we had postulated a different primitive ontology such as Bohmian particles, then no many-worlds character would have arisen.

Let me now address some common objections to Everett's many-worlds theory. While there is disagreement in the literature about the relevance of a primitive ontology, many authors argue that SØ has a *preferred basis problem*: If there exists nothing more than Ψ, and if Ψ is just a vector in Hilbert space $\mathscr{H}$, then how do we know which basis to choose in $\mathscr{H}$ to obtain the different worlds? For example, if

$$\Psi = \tfrac{1}{\sqrt{2}}|\text{dead}\rangle + \tfrac{1}{\sqrt{2}}|\text{alive}\rangle\,, \tag{3.78}$$

then we could also write

$$\Psi = \tfrac{1-i}{2}|+\rangle + \tfrac{1+i}{2}|-\rangle\,, \tag{3.79}$$

where

$$|+\rangle = \tfrac{1}{\sqrt{2}}|\text{dead}\rangle + \tfrac{i}{\sqrt{2}}|\text{alive}\rangle\,, \qquad |-\rangle = \tfrac{1}{\sqrt{2}}|\text{dead}\rangle - \tfrac{i}{\sqrt{2}}|\text{alive}\rangle \tag{3.80}$$

form another ONB of the subspace spanned by $|\text{dead}\rangle$ and $|\text{alive}\rangle$. So how do we know that the two worlds correspond to $|\text{dead}\rangle$ and $|\text{alive}\rangle$ rather than to $|+\rangle$ and $|-\rangle$? Obviously, in Sm, there is no such problem because a preferred basis (the position basis) is built into the law (3.74) for m.

It is sometimes objected against many-worlds hypotheses that one cannot observe the other worlds. I must admit I do not see why that could be an objection. Sm correctly predicts (and so does SØ, if it works at all) that any inhabitant of one world cannot observe the other worlds, so it is not a question of making a wrong prediction. The existence of other worlds, whether we can see them or not, is a mathematical consequence of the Schrödinger equation and the law (3.74) for the matter density. In fact, the existence of wave packets $c_\alpha \Psi_\alpha$ in Ψ that in some sense look like these other worlds is a consequence of the Schrödinger equation alone, so it seems inevitable unless we modify the Schrödinger equation as in collapse theories. By decoherence, macroscopically disjoint packets stay macroscopically disjoint, so it seems we will have to accept that the wave function of the universe or of macroscopic systems tends to split more and more, resulting in a tree-like shape of its significant support in time × configuration space as shown schematically in Fig. 3.8. The macroscopically different packets Ψ_α are therefore also often called *branches* of Ψ.

So the objection that many-worlds hypotheses are bad because the other worlds cannot be observed seems like exaggerated positivism. A related objection against many worlds appeals to a rule from the philosophy of science known as *Ockham's razor*. This rule, whose idea is attributed to English philosopher and theologian William of Ockham (circa 1287–1347), is often summarized with the Latin words

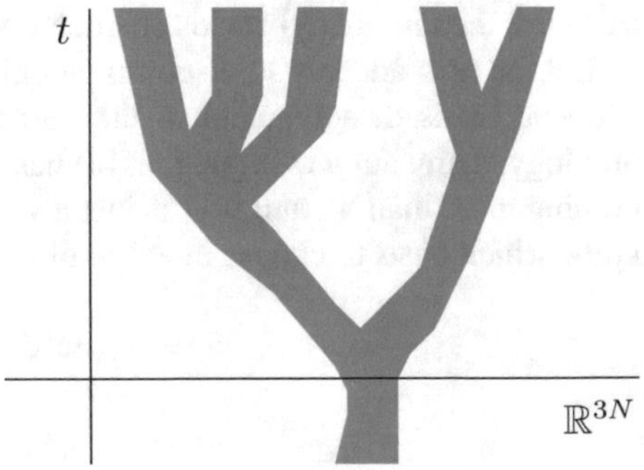

Fig. 3.8 Qualitative picture of the tree-like structure typically featured by wave functions of macroscopic objects such as measurement apparatus. Of the $3N$ dimensions of configuration space, only one is shown. The shaded area consists of those (q, t) where $\Psi(q, t)$ is significantly nonzero

"Non sunt multiplicanda entia sine necessitate," which means entities are not to be multiplied without necessity. I would express the rule as *the simpler explanation seems more believable.* As an objection against many worlds, it is sometimes invoked as follows: since any many-worlds theory has to posit the existence of far more physical objects than a single-world theory, it is less parsimonious than the latter, and as Ockham's razor recommends, we should prefer the more parsimonious theory. In Sm, however, it would actually not be possible to remove the other worlds in a simple and clean way, in part because it is not precisely defined where to draw the boundary between worlds (and which contribution to m to regard as belonging to our world), and in part because it would destroy the simplicity of the defining equations of Sm, i.e., the Schrödinger equation and (3.74).

In fact, once we grant the existence of a wave function ψ_t and a matter density function $m(t, \boldsymbol{x})$, we do not have to "multiply entities," to introduce further objects or to posit the existence of further things, for the many worlds to show up. On the contrary, one could argue that in view of the great simplicity of the two defining equations of Sm, the Schrödinger equation and (3.74), Ockham's razor works *in favor* of Sm, as it favors simple theories. In this spirit, advocates of S∅ have argued that Ockham's razor favors S∅ over all other theories of quantum mechanics. If I thought S∅ was an acceptable theory, then I would agree, as its simplicity cannot be denied.

Another consideration arises if we want to weigh a many-worlds theory (such as Sm) against a single-world theory (such as Bohmian mechanics): then one of the aspects to consider is that, other things being equal, the more obvious, down-to-earth, intuitively plausible character of a single-world theory may be more believable than the more eccentric, dramatic, unexpected character of a many-worlds theory. While this thought is not an instance of Ockham's razor, it is similar in spirit.

3.5.3 Bell's First Many-Worlds Theory

Bell also made a proposal (first formulated in 1971, published in 1981 [10]) adding a primitive ontology to Everett's S∅; Bell did not seriously propose or defend the resulting theory; he just regarded it as an ontological clarification of Everett's theory. According to this theory, at every time t, there exists an uncountably infinite collection of universes, each of which consists of N material points in Euclidean 3-space. Thus, each world has its own configuration Q, but some configurations are more frequent in the ensemble of worlds than others, with $|\Psi_t|^2$ distribution across the ensemble. At every other time t', there is again an infinite collection of worlds, but there is no fact about which world at t' is the same as which world at t.

3.5.4 Bell's Second Many-Worlds Theory

Another variant of this theory, also considered by Bell (1976) [9], supposes that there is really a single world at every time t consisting of N material points in Euclidean 3-space. The configuration Q_t is chosen with $|\Psi_t|^2$ distribution independently at every time. Although this theory has a definite Q_t at every t, it also has a many-worlds character because in every arbitrarily short time interval, configurations from all over configuration space are realized, in fact with distribution roughly equal to $|\Psi_t|^2$ (if the interval is short enough and Ψ_t depends continuously on t) across the ensemble of worlds existing at different times. This theory seems rather implausible compared to Bohmian mechanics, as it implies that our memories are completely wrong: after all, it implies that 1 minute ago, the world was not at all like what we remember it to be like a minute ago. Given that all of our reasons for believing in the Schrödinger equation and the Born rule are based on memories of reported outcomes of experiments, it seems that this theory undercuts itself: if we believe it is true, then we should conclude that our belief is not justified.

It is not very clear to me whether the same objection applies to Bell's first many-worlds theory. But certainly, both theories have, due to their radically unusual idea of what reality is like, a flavor of skeptical scenarios (such as the brain in the vat; see Sect. 8.3), in fact a stronger such flavor than Sm.

3.5.5 Probabilities in Many-Worlds Theories

Here is another common objection to the many-worlds view: If every outcome α of an experiment is realized, what could it mean to say that outcome α has probability $|c_\alpha|^2$ to occur? If, as in Sm and in SØ, all the equations are deterministic, then there is nothing random; and in the situation of the measurement problem, there is nothing that we are ignorant of. So what could talk of probability mean?

Here is what it could mean in Sm (Allori et al. 2011 [6]): Suppose we have a way of *counting worlds*. And suppose we repeat a quantum experiment (say, a Stern–Gerlach experiment with $|c_{\mathrm{up}}|^2 = |c_{\mathrm{down}}|^2 = 1/2$) many times (say, a thousand times). Then we obtain in each world a sequence of 1000 ups and downs such as

$$\uparrow\downarrow\uparrow\uparrow\downarrow\uparrow\downarrow\downarrow\downarrow \ldots. \tag{3.81}$$

Note that there are $2^{1000} \approx 10^{300}$ such sequences. The statement that the fraction of ups lies between 47 and 53% is true in some worlds and false in others. Now count the worlds in which the statement is true. Suppose that the statement is true in *the overwhelming majority of worlds*. Then that would explain why we find ourselves in such a world. And that, in turn, would explain why we observe a relative frequency of ups of about 50%. And *that* is what we needed to explain for justifying the use of probabilities.

Now consider $|c_{\text{up}}|^2 = 1/3$, $|c_{\text{down}}|^2 = 2/3$. Then the argument might seem to break down, because it is then still true that in the overwhelming majority of sequences such as (3.81), the frequency of ups is about 50%. But consider the following:

Rule for Counting Worlds *The "fraction of worlds" $f(P)$ with property P in the splitting given by $\Psi = \sum_\alpha c_\alpha \Psi_\alpha$ and $m(\boldsymbol{x}) = \sum_\alpha |c_\alpha|^2 m_\alpha(\boldsymbol{x})$ is*

$$f(P) = \sum_{\alpha \in M} |c_\alpha|^2 , \tag{3.82}$$

where M is the set of worlds α with property P.

Note that $f(P)$ lies between 0 and 1 because $\sum_\alpha |c_\alpha|^2 = 1$. It is not so clear whether this rule makes sense—whether there is room in physics for such a law. But let us accept it for the moment and see what follows. Consider the property P that the relative frequency of ups lies between 30 and 36%. Then $f(P)$ is actually the same value as the probability of obtaining a frequency of ups between 30 and 36% in 1000 consecutive independent random tossings of a biased coin with $\mathbb{P}$(up) = 1/3 and $\mathbb{P}$(down) = 2/3. And in fact, this value is very close to 1. Thus, the above rule for counting worlds implies the frequency of ups lies between 30 and 36% in the overwhelming majority of worlds. This reasoning was essentially developed by Everett (1955 [23] and 1957 [24]).

A comparison with Bohmian mechanics is useful. The initial configuration of the lab determines the precise sequence such as (3.81). If the initial configuration is chosen with $|\Psi_0|^2$ distribution, then with overwhelming probability, the sequence will have a fraction of ups between 30 and 36%. That is, if we *count initial conditions* with the $|\Psi_0|^2$ distribution, that is, if we say that the fraction of initial conditions lying in a set $B \subseteq \mathbb{R}^{3N}$ is $\int_B |\Psi_0|^2$, then we can say that for *the overwhelming majority of Bohmian worlds*, the observed frequency is about 33%. Now to make the connection with many-worlds, note that the reasoning does not depend, in fact, on whether all of the worlds are realized or just one. That is, imagine many Bohmian worlds with the same initial wave function Ψ_0 but different initial configurations, distributed across the ensemble according to $|\Psi_0|^2$. Then there is an explanation for why inhabitants should see a frequency of about 33%. Also, the rule above is robust against re-drawing the boundaries of branches (i.e., does not depend sensitively on where exactly we draw the boundary of a wave packet or on whether we regard two packets as macroscopically different).

The problem that remains is whether there is room for a rule for counting worlds. In terms of a creation myth, suppose God created the wave function Ψ and made it a law that Ψ evolves according to the Schrödinger equation; then He created matter in 3-space distributed with density $m(\boldsymbol{x}, t)$ and made it a law that m is given by (3.74). Now what would God need to do in order to make the rule for counting worlds a law? He does not create anything further, so in which way would two universes with equal Ψ and m but different rules for counting worlds differ? That is a reason for thinking that ultimately, Sm fails to work (though in quite a subtle way).

Various authors have proposed other reasonings for justifying probabilities in many-worlds theories; they seem less relevant to me, but let me mention a few. David Deutsch (1999) [20] proposed that it is rational for inhabitants of a universe governed by a many-worlds theory (a "multiverse," as it is often called) to behave as if the events they perceive were random with probabilities given by the Born rule; he proposed certain principles of rational behavior from which he derived this. (Of course, this reasoning does not provide an explanation of why we observe frequencies in agreement with Born's rule.[12]) Lev Vaidman (1998) [49] proposed that in a many-worlds scenario, I can be ignorant of which world I am in: before the measurement, I know that there will be a copy of me in each post-measurement world, and afterward, I do not know which world I am in until I look at the pointer position. And I could try to express my ignorance through a probability distribution, although it is not clear why (and in what sense) the Born distribution would be "correct" and other distributions would not.

For comparison, in Bell's many-worlds theories, it is not hard to make sense of probabilities. In Bell's first theory, there is an ensemble of worlds at every time t, and clearly most of the worlds have configurations that look as if randomly chosen with $|\Psi|^2$ distribution, in particular with a frequency of ups near 33% in the example described earlier. In Bell's second theory, Q_t is actually random with $|\Psi_t|^2$ distribution, and although the recorded sequence of outcomes fluctuates within every fraction of a second, the sequence in our memories and records at time t has, with probability near 1, a frequency of ups near 33%.

3.6 Some Morals

"Quantum theories without observers" is a label for theories of quantum mechanics that propose a possibility for what reality might be like. We have seen three types of such theories: Bohmian mechanics, collapse theories, and many-worlds theories. The main divide in the foundations of quantum mechanics lies between quantum theories without observers on one side and the orthodox or Copenhagen-style views on the other.

It is remarkable how small the modification to ordinary quantum mechanics is that is needed to turn it into a clear theory, a quantum theory without observers: we can keep the entire quantum formalism, and all calculations done in the spirit of ordinary quantum mechanics (e.g., binding energies in molecules, tunneling probabilities, etc.) remain valid. We only need to add a primitive ontology (particle

[12] And that concerns not only lab experiments: The wave function of planet Earth branches all the time and did already before humans existed. Born's rule governs also radioactive decay, the formation of stars, and the magnetization of macroscopic solids, for example.

world lines or flashes or matter density) and a suitable law governing it. The main difference to the calculations is that now they have a clear justification.

It is also worth mentioning that once we use a clear theory, we not only obtain a clear justification of the Born rule for arbitrary observables and of the projection postulate. Once we use a clear theory, we can give a clear and justified answer to whether the theory is deterministic, whether it is time reversal invariant, and so on. Moreover, we are liberated from the paradoxes that burden us in orthodox quantum theory, as well as from the need for excuses whenever we analyze a measurement process or even just consider the Schrödinger equation for a macroscopic object.

For further reading on comparing orthodox quantum mechanics to quantum theories without observers, I recommend Bricmont (2016) [16] and Norsen (2017) [38].

Another remarkable and perhaps surprising observation is that we found we needed, in order to arrive at a satisfactory and acceptable theory that can be considered as a fundamental physical theory, to introduce variables representing the distribution of matter in space and time—what is called a *primitive ontology*. They can take the form of particle trajectories or flashes or a matter density, and perhaps there are further possibilities. It seems to me that having a primitive ontology is a general trait of and a general necessity for fundamental physical theories. Remarkable also that we needed to introduce such variables *although*, among the three ways out of the quantum measurement problem, collapse theories (such as GRW) and many-worlds theories were intended *not* to choose hidden variables. How can this be? Because a satisfactory theory of quantum physics ultimately has to achieve more than just solve the measurement problem.

3.7 Special Topics

3.7.1 Einstein's View

For a volume collecting contributions from various famous scientists in honor of Einstein's 70th birthday, also Einstein himself wrote a paper (1949) [22] in which he reacted to some of the contributions on the subject of quantum mechanics. I recommend this paper as further reading; it conveys his view of the foundational issues of quantum mechanics toward the end of his life.

Here, I would like to quote a few passages from this paper, particularly with regard to the following questions: What was Einstein's criticism of orthodox quantum mechanics? Was the criticism reasonable? Was he obsessed with determinism? Einstein opposed Copenhagenism, and it is sometimes suggested that that was because he was stubborn, conservative, and unwilling to accept new ideas,[13] or

[13] For example, Heisenberg wrote [15]: "[I]t can happen that a new range of empirical data can be completely understood only when the enormous effort is made to enlarge this framework and to change the very structure of the thought process. In the case of quantum mechanics, Einstein was apparently no longer willing to take this step, or perhaps no longer able to do so."

because he was fixated on locality (which later turned out wrong; see Chap. 4), or because he was fixated on determinism.

In this paper, Einstein wrote (p. 666, 2nd paragraph):

> In what follows, I wish to adduce reasons which keep me from falling in line with the opinion of almost all contemporary theoretical physicists. I am, in fact, firmly convinced that the essentially statistical character of contemporary quantum theory is solely to be ascribed to the fact that this theory operates with an incomplete description of physical systems.

That is, Einstein said he disagrees with orthodox quantum mechanics and believes that the true theory of quantum mechanics will be a deterministic hidden variables theory. Later on p. 666:

> [Q]uantum theory [...] correctly describes the empirical relations between statable phenomena[.]

That is, he thought that the empirical predictions of the quantum formalism are correct. On p. 667:

> What does not satisfy me in [orthodox quantum mechanics] is its attitude towards [...] the complete description of any (individual) real situation (as it supposedly exists irrespective of any act of observation [...]).

So, when Einstein tried to summarize his criticism in one sentence, he said the problem with orthodox quantum mechanics was that it does not say what happens in reality. In particular, while he said he believed that the true theory is deterministic, his criticism of orthodox quantum mechanics was not its lack of determinism. On p. 669, he confirmed again:

> What I dislike [about Copenhagen-ish ways of reasoning] is the basic positivistic attitude, which from my point of view is untenable[.]

On p. 667–670, he outlined an experiment similar to Schrödinger's cat, but with the cat replaced by a machine that puts a mark on a paper strip in a place depending on the time when the Geiger counter was triggered. He commented:

> [T]here is in principle no objection to treating this entire system from the standpoint of quantum mechanics. [...] [F]or a time large compared with the average decay-time of the radioactive atom, there will be (at most) *one* such registration-mark on the paper strip. [...] If we attempt to work with the interpretation that the quantum theoretical description is to be understood as a complete description of the individual system, we are forced to the interpretation that the location of the mark on the strip is nothing which belongs to the system *per se*, but that the existence of that location is essentially dependent upon the carrying out of an observation made on the registration strip. Such an interpretation is certainly by no means absurd from a purely logical standpoint; yet there is hardly likely to be anyone who would be inclined to consider it seriously.

That is, Einstein outlined a variant of the measurement problem, pointed out that there is no outcome unless a super-observer is invoked, and judged this conclusion unacceptable. That was his main argument against orthodox quantum mechanics. In particular, his criticism was neither unreasonable nor based on taking determinism or locality for granted.

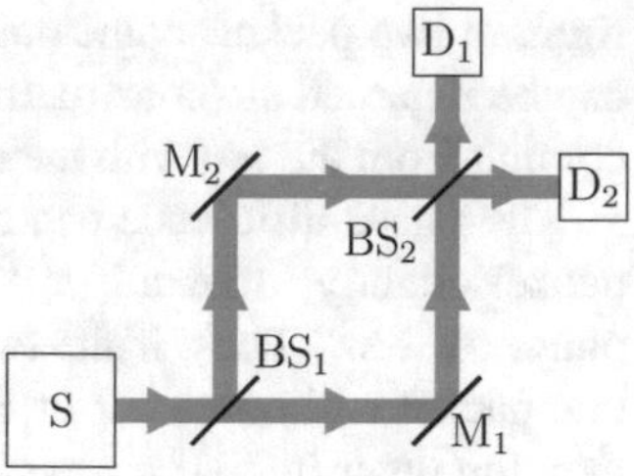

Fig. 3.9 Design of a Mach–Zehnder experiment. Wave packets travel along gray paths in the direction indicated. S, source; BS, beam splitter; M, mirror; D, detector

3.7.2 The Mach–Zehnder Interferometer

A variant of the double-slit experiment is the Mach–Zehnder interferometer, developed around 1892, an arrangement used for experiments with photons, although in principle it could also be set up for use with electrons. Like the double slit, it involves splitting the wave packet in two, having the two packets travel along different paths, and then making them overlap again and interfere; see Fig. 3.9; together with the double slit, such experiments are called *two-way* or *which-way* experiments. In practice, the paths are usually between millimeters and meters long.

A *beam splitter* is a potential barrier with height and width so adjusted that, by the tunnel effect (see Sect. 2.1.4), an incoming wave packet with approximate wave number k will be half reflected and half transmitted; that is, a normalized incoming wave packet ψ_{in} evolves to $c_{\text{refl}}\psi_{\text{refl}} + c_{\text{transm}}\psi_{\text{transm}}$, and the reflection coefficient c_{refl} and the transmission coefficient c_{transm} both have modulus $1/\sqrt{2}$. For photons, beam splitters are realized as thin layers of metal ("half-silvered mirror") or resin, usually on top of a glass plate or in between two glass plates (the potential is higher inside the metal than in glass or air).

The following further properties of a beam splitter, which can be regarded as consequences of the Schrödinger equation for a 1d potential barrier, are relevant to the Mach–Zehnder experiment. First, the reflected and transmitted packet have, up to the sign, the same pretty sharp wave number k as the incoming packet. Second, by symmetry of the potential $V(-x) = V(x)$ (taking the middle of the barrier as the origin), the behavior is symmetric under the *parity transformation* $x \to -x$ represented by the *parity operator* $P\psi(x) = \psi(-x)$: specifically, $P\psi_{\text{in}}$ evolves to $c_{\text{refl}}P\psi_{\text{refl}} + c_{\text{transm}}P\psi_{\text{transm}}$. Since ψ_{refl} and ψ_{transm} have the same wave number up to the sign, move at the same speed, and were generated during the same time interval, $\psi_{\text{refl}} = P\psi_{\text{transm}}$ to a good degree of approximation; likewise, $\psi_{\text{in}} = \psi^*_{\text{refl}}$, provided the shape of ψ_{in} is symmetric and provided we consider the right instant of time. Third, by time reversal, $c^*_{\text{refl}}\psi^*_{\text{refl}} + c^*_{\text{transm}}\psi^*_{\text{transm}}$ evolves to ψ^*_{in}. That is, if we send in two packets, one from each side, that have the same absolute wave number and arrive at the same time, then only one packet comes out, leaving to the left—at least, if the phases of the coefficients c_i are prepared in the right way. After all, $c^*_{\text{refl}}P\psi^*_{\text{refl}} + c^*_{\text{transm}}P\psi^*_{\text{transm}}$ evolves to $P\psi^*_{\text{in}}$, a packet leaving to the right. So, depending on the phase difference of two wave packets arriving at a beam splitter, either only one packet comes out toward the left, or only one packet comes out to the right, or two packets come out (one to the left and one to the right). And this effect can be regarded as (constructive or destructive) interference of the transmitted part coming from the left with the reflected part coming from the right.

The phase difference can be influenced in a practical way by shifting a wave packet slightly; after all, $e^{ik(x+\Delta x)} = e^{ik\Delta x}e^{ikx}$, corresponding to a change of phase by $k\Delta x$. Thus, if the two paths (via M_1 or M_2) have the same length, only one packet will come out of BS_2 leaving toward D_1. If one of the paths is longer than the other by half a wave length (phase change π), only one packet will come out of BS_2 leaving toward D_2. If one of the paths is longer than the other by a quarter wave length (phase change $\pi/2$), then two packets of equal magnitude will come out of BS_2, one toward D_1 and one toward D_2. In this way, the setup can be used for detecting small changes in the path lengths (similar principles are used for detecting gravitational waves) or changes in potentials (or, for photons, refraction index) somewhere along one of the paths.

If one of the paths is blocked, then only one packet arrives at BS_2, and two packets of equal magnitude come out of it. The situation is analogous to the double slit, with D_1 corresponding to a maximum and D_2 to a minimum of the interference pattern. If none of the paths is blocked (and they have the same length and no further potentials), then D_2 never clicks.

3.7.3 Path Integrals

Path integrals are a way of computing the time evolution operators U_t; they arise as follows. Consider a unitary one-parameter group $U_t = \exp(-iHt/\hbar)$, for simplicity on a finite-dimensional Hilbert space $\mathscr{H} = \mathbb{C}^d$ and just at times t that are multiples of a (small) time step $\tau > 0$, $t = n\tau$ with $n \in \mathbb{Z}$. Set $U := U_\tau$. Then $U_t = U_{n\tau} = U^n$, and the matrix elements of this power can be expressed through repeated matrix multiplication as in $(A^2)_{ik} = \sum_j A_{ij}A_{jk}$:

$$(U^n)_{i_n i_0} = \sum_{i_1 \ldots i_{n-1}=1}^{d} U_{i_n i_{n-1}} \cdots U_{i_2 i_1} U_{i_1 i_0} \,. \tag{3.83}$$

Now think of the sequence $(i_0, i_1, \ldots, i_{n-1}, i_n)$ as a path in the set $\{1, \ldots, d\}$ as in Fig. 3.10.

To think of it as a path will be particularly natural if U_{ij} is small unless i and j are "close" to each other (in a sense to be determined). Basically, Eq. (3.83) is already a path integral, except that the integral is in this case a sum: it is a sum over all possible paths connecting a given value of i_0 to a given value of i_n. Now, we want to let $\tau \to 0$ while keeping t fixed, and we want to let $\mathscr{H}$ approach $L^2(\mathbb{R}^{3N})$, say $\mathscr{H} = L^2(\Omega)$ with Ω a finite set approaching $\mathbb{R}^{3N}$, for example, $\Omega = \varepsilon\mathbb{Z}^{3N} \cap B_R(\mathbf{0})$ in the limit $\varepsilon \to 0$, $R \to \infty$. Then i_0 and i_n get replaced by two points $q_1, q_2 \in \mathbb{R}^{3N}$, $(U^n)_{i_0 i_n}$ becomes $\langle q_2|U_t|q_1\rangle$, and the sum becomes an integral over all smooth functions $q : [0, t] \to \mathbb{R}^{3N}$ with $q(0) = q_1$ and $q(t) = q_2$;

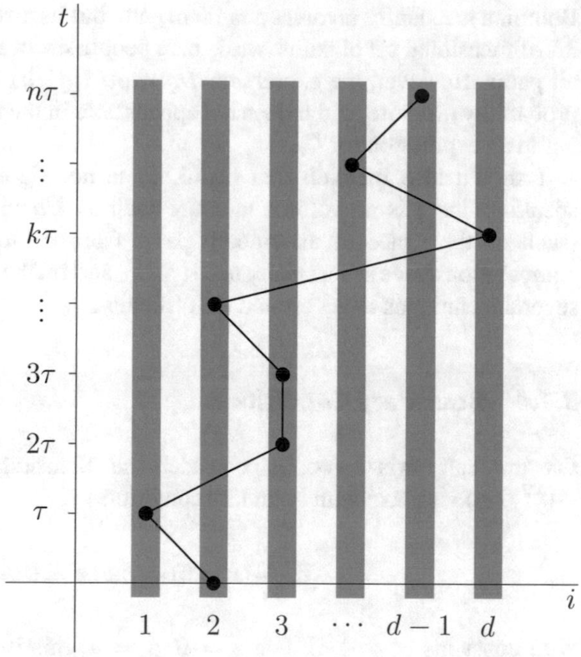

Fig. 3.10 Example of a path in $\{1, \ldots, d\}$

let us write $\int Dq$ for this kind of integral. I report that, although less obviously, the integrand also converges to a simple expression, as discovered by Feynman in 1942 [26]: to $\exp(iS[q]/\hbar)$ with the so-called classical action functional

$$S[q] := \int_0^t dt' \left[\tfrac{m}{2} \dot{q}(t')^2 - V\big(q(t')\big) \right], \tag{3.84}$$

where $\dot{q}$ means the derivative of $t \mapsto q(t)$. Thus,

$$\langle q_2 | U_t | q_1 \rangle = \int Dq \, e^{iS[q]/\hbar} . \tag{3.85}$$

People sometimes say that this formula shows or suggests that "the particle takes all paths from q_1 to q_2," but I find this statement incomprehensible and without basis. To begin with, the path $t \mapsto q(t)$ is not the path of the particle configuration, as would be the $t \mapsto Q(t)$ in Bohmian mechanics. In fact, since the formula expresses U_t and thus the time evolution of ψ, it is the *wave* ψ, not the particle, that follows $q(t)$. And for a wave, it is not strange to follow many paths. Note also that one can just as well re-write Maxwell's equations of classical electrodynamics, as in fact any linear field equation, in terms of path integrals, but nobody would claim that in classical electrodynamics, any particle "takes all paths." People sometimes also talk as if path integrals meant that the particle took a random path. Of course,

Bohmian mechanics involves a random path, but its distribution is concentrated on a $3N$-dimensional set of paths, while here people mean a distribution spread out over all paths. However, the expression $Dq \exp(iS[q]/\hbar)$ is a *complex* measure, not a probability measure, and it does not appear here in the role of a probability measure but in an expression for U_t.

I should also mention that Eq. (3.85) is not rigorously true because, strictly speaking, there is no volume measure such as Dq in infinite-dimensional spaces (such as the space of all smooth paths from q_1 to q_2). Nevertheless, various computations have successfully used (3.85), and mathematicians have come up with several techniques to get around the difficulty.

3.7.4 Boundary Conditions

On the half axis $(-\infty, 0]$, consider the Schrödinger equation $i\hbar\partial\psi/\partial t = -(\hbar^2/2m)\partial^2\psi/\partial x^2$ with boundary condition

$$\alpha\frac{\partial\psi}{\partial x}(x=0) + \beta\psi(x=0) = 0 \tag{3.86}$$

with constants $\alpha, \beta \in \mathbb{C}$. For $\alpha = 0, \beta = 1$, this is called a *Dirichlet boundary condition* and for $\alpha = 1$ and $\beta = 0$, a *Neumann boundary condition*. [This is Carl Neumann (1832–1925), not John von Neumann (1903–1957).] For general $(\alpha, \beta) \neq (0, 0)$, it is called a *Robin boundary condition* after French mathematician Victor Gustave Robin (1855–1897).

Exercise 3.4 (Boundary Conditions) Show that if α and β are real, then $j(x = 0) = 0$. (They are reflecting boundary conditions and lead to a unitary time evolution.) Show that if neither α nor β vanishes and α/β is imaginary, then $j(x = 0) > 0$ whenever $\psi(x = 0) \neq 0$. (They are absorbing boundary conditions, lead to loss of probability, and are discussed further in Sect. 5.2.4.)

For the use of boundary conditions, the concept of the *domain* of an operator plays a role: Strictly speaking, an operator in a Hilbert space $\mathscr{H}$ consists of a dense subspace $\mathscr{D} \subseteq \mathscr{H}$ and a linear mapping $H : \mathscr{D} \to \mathscr{H}$. When the operator is bounded (i.e., when there is $C > 0$ such that $\|H\psi\| \leq C\|\psi\|$ for all $\psi \in \mathscr{D}$), then we can always assume $\mathscr{D} = \mathscr{H}$; more precisely, then H possesses a unique extension to a bounded linear mapping $\mathscr{H} \to \mathscr{H}$. But Hamiltonians are often unbounded.

When we use a boundary condition, then the domain contains only functions obeying the boundary condition. For example, let us compare the negative Laplacian (i.e., the free Schrödinger Hamiltonian with $\hbar^2/2m = 1$) with and without boundary. Consider first the case without boundary: then $\mathscr{H}_1 = L^2(\mathbb{R}, \mathbb{C})$, $\mathscr{D}_1$ is the second Sobolev space (i.e., the set of functions $\psi \in \mathscr{H}$ such that $\int dk\, |k^2\widehat{\psi}(k)|^2 < \infty$ with $\widehat{\psi} = \mathscr{F}\psi$ the Fourier transform of ψ), and H_1 can be expressed as

$H_1 = \mathscr{F}^{-1} M_{k^2} \mathscr{F}$ with M_{k^2} the multiplication operator by k^2, $M_{k^2}\varphi(k) = k^2\varphi(k)$. Second, in the case with the boundary condition (3.86), we have that $\mathscr{H}_2 = L^2\big((-\infty, 0], \mathbb{C}\big)$; $\mathscr{D}_2$ comprises, among the restrictions of functions from the second Sobolev space $\mathscr{D}_1$ to $(-\infty, 0]$, those satisfying the boundary condition (3.86); and H_2 can be expressed as follows: For any $\psi \in \mathscr{D}_2$, let f be a function from $\mathscr{D}_1$ that agrees with ψ on $(-\infty, 0]$, and define $H_2\psi$ as the restriction of $H_1 f$ to $(-\infty, 0]$.[14] It is a further consequence that all L^2 eigenfunctions of H_2 satisfy the boundary condition since they lie in the domain $\mathscr{D}_2$ of H_2.

If we had defined the domain differently, say as the set $\mathscr{D}_2'$ comprising the restrictions of all functions from the second Sobolev space $\mathscr{D}_1$ to $(-\infty, 0]$, then it would still have been possible to define a linear mapping $H_2' : \mathscr{D}_2' \to \mathscr{H}$ in the same way: for any $\psi \in \mathscr{D}_2$, let f be a function from $\mathscr{D}_1$ that agrees with ψ on $(-\infty, 0]$, and define $H_2'\psi$ as the restriction of $H_1 f$ to $(-\infty, 0]$. However, H_2' is not self-adjoint, whereas H_2 is. This fact reflects the need for a boundary condition when we want to use the negative Laplacian on a configuration space with boundary as the Hamiltonian.

3.7.5 Point Interaction

A potential given by a Dirac delta function, for example,

$$V(x) = \lambda \delta^d(x) \tag{3.87}$$

for a single particle in d dimensions with real (positive or negative) prefactor λ, would represent an interaction, of the quantum particle with another particle fixed at the origin, that occurs only at contact between the two particles. It is called *point interaction* or *zero-range interaction*. It is not obvious that a potential like that makes mathematical sense, i.e., that a self-adjoint operator H exists in $L^2(\mathbb{R}^d)$ that corresponds to $-(\hbar^2/2m)\Delta + V$ with (3.87). It turns out (see, e.g., Albeverio et al. 1988 [5]) that no such operator exists in dimension $d \geq 4$.

In dimension $d = 1$, we can reason as follows. Suppose ψ is an eigenfunction of H, $H\psi = E\psi$, that is,

$$-\frac{\hbar^2}{2m}\Delta\psi(x) + \lambda\,\delta(x)\,\psi(x) = E\,\psi(x)\,. \tag{3.88}$$

[14] It can be proved (e.g., Adams and Fournier 2003 [1]) that $H_1 f$ does not depend on f as long as f agrees with ψ on $(-\infty, 0]$. Moreover, we have made use of the fact [1] that for f from the second Sobolev space, it can be defined uniquely what $f(0)$ and $\partial f/\partial x(0)$ mean despite the fact that a general $g \in L^2(\mathbb{R}, \mathbb{C})$ is an equivalence class of functions modulo change on a set of volume 0 and thus does not possess a well-defined value at a point, such as $g(0)$.

Integrate this relation over x from $-\varepsilon$ to $+\varepsilon$ for small $\varepsilon > 0$ to obtain

$$-\frac{\hbar^2}{2m}\Big[\psi'(\varepsilon)-\psi'(-\varepsilon)\Big]+\lambda\,\psi(0)=E\int_{-\varepsilon}^{\varepsilon}dx\,\psi(x) \tag{3.89}$$

with ψ' the derivative of ψ. Taking the limit $\varepsilon \to 0$ and assuming that ψ is bounded near 0, the right-hand side vanishes, so

$$\psi'(0+)-\psi'(0-)=\frac{2m\lambda}{\hbar^2}\,\psi(0)\,. \tag{3.90}$$

That is, ψ' has a jump discontinuity at 0 of height given by the right-hand side. Conversely, assuming (3.90) while ψ'' exists everywhere except at 0, then $-\frac{\hbar^2}{2m}\Delta\psi$ consists of a Dirac delta peak $-\frac{\hbar^2}{2m}[\psi'(0+)-\psi'(0-)]\delta(x)=-\lambda\psi(0)\,\delta(x)$ at the origin and a regular function everywhere else, with the consequences that $H\psi(x)=-\frac{\hbar^2}{2m}\Delta\psi(x)+\lambda\,\delta(x)\,\psi(x)$ is a regular function (and, for suitable ψ, a square-integrable one). Mathematicians say that the *domain* $\mathscr{D}\subset L^2(\mathbb{R})$ of the Hamiltonian consists of functions obeying (3.90) and H maps $\mathscr{D}$ to $L^2(\mathbb{R})$; $\mathscr{D}$ is a dense subspace. Condition (3.90) can be called a boundary condition if we regard 0 as the common boundary of the positive and negative half axis.

Without giving details, I report that in $d = 3$ dimensions, the potential (3.87) makes sense (i.e., admits a self-adjoint Hamiltonian) for λ of the form $\lambda = \eta + \alpha\eta^2$ with infinitesimal η and finite $\alpha \in \mathbb{R}$. The domain of H then consists of functions satisfying the *Bethe–Peierls boundary condition* at the origin,

$$\lim_{r\searrow 0}\Big[\partial_r\big(r\psi(r\boldsymbol{\omega})\big)+\alpha r\psi(r\boldsymbol{\omega})\Big]=0 \tag{3.91}$$

for all unit vectors $\boldsymbol{\omega}\in\mathbb{R}^3$. Put differently, if we can expand ψ in powers of $r=|\boldsymbol{x}|$ as

$$\psi(r\boldsymbol{\omega})=c_{-1}(\boldsymbol{\omega})\,r^{-1}+\sum_{n=0}^{\infty}c_n(\boldsymbol{\omega})\,r^n\,, \tag{3.92}$$

then (3.91) demands that

$$c_0(\boldsymbol{\omega})+\alpha c_{-1}(\boldsymbol{\omega})=0\,. \tag{3.93}$$

In particular, for $\alpha \neq 0$ and $c_0 \neq 0$, it follows that $c_{-1} \neq 0$, so ψ is forced to diverge at the origin like $1/r$.

3.7.6 No-Cloning Theorem

We show that it is impossible to duplicate the quantum state of an object without destroying the original quantum state. Let

$$\mathbb{S}(\mathscr{H}) = \{\psi \in \mathscr{H} : \|\psi\| = 1\} \tag{3.94}$$

denote the unit sphere in $\mathscr{H}$. A *cloning mechanism* for the Hilbert space $\mathscr{H}_{\text{obj}}$ would consist of a Hilbert space $\mathscr{H}_{\text{app}}$, a ready state $\phi_0 \in \mathbb{S}(\mathscr{H}_{\text{app}})$ of the apparatus, a ready state $\psi_0 \in \mathbb{S}(\mathscr{H}_{\text{obj}})$ of the copy, and a unitary time evolution U on $\mathscr{H}_{\text{obj}} \otimes \mathscr{H}_{\text{obj}} \otimes \mathscr{H}_{\text{app}}$ such that, for all $\psi \in \mathbb{S}(\mathscr{H}_{\text{obj}})$,

$$U(\psi \otimes \psi_0 \otimes \phi_0) = \psi \otimes \psi \otimes \phi_\psi \tag{3.95}$$

with some $\phi_\psi \in \mathbb{S}(\mathscr{H}_{\text{app}})$ that may depend on ψ.

Theorem 3.6 *If* $\dim \mathscr{H}_{\text{obj}} \geq 2$, *then no cloning mechanism exists.*

Proof Assume that such a U existed. Consider unit vectors $\psi_1 \perp \psi_2 \in \mathbb{S}(\mathscr{H}_{\text{obj}})$ and $\psi_3 = \frac{1}{\sqrt{2}}\psi_1 + \frac{1}{\sqrt{2}}\psi_2$. Then

$$\langle \psi_1 | \psi_3 \rangle = \frac{1}{\sqrt{2}} \tag{3.96}$$

and

$$\begin{aligned} \langle U\psi_1 | U\psi_3 \rangle &= \langle \psi_1 \otimes \psi_1 \otimes \phi_{\psi_1} | \psi_3 \otimes \psi_3 \otimes \phi_{\psi_3} \rangle & (3.97)\\ &= \langle \psi_1 | \psi_3 \rangle_{\text{obj}} \, \langle \psi_1 | \psi_3 \rangle_{\text{obj}} \, \langle \phi_{\psi_1} | \phi_{\psi_3} \rangle_{\text{app}} & (3.98)\\ &= \frac{1}{\sqrt{2}} \frac{1}{\sqrt{2}} \langle \phi_{\psi_1} | \phi_{\psi_3} \rangle_{\text{app}} \,. & (3.99) \end{aligned}$$

By the Cauchy–Schwarz inequality,

$$\left| \langle \phi_{\psi_1} | \phi_{\psi_3} \rangle_{\text{app}} \right| \leq \|\phi_{\psi_1}\| \, \|\phi_{\psi_3}\| = 1, \tag{3.100}$$

so

$$\left| \langle U\psi_1 | U\psi_3 \rangle \right| \leq \frac{1}{2} \tag{3.101}$$

in contradiction to

$$\langle U\psi_1|U\psi_3\rangle = \frac{1}{\sqrt{2}}\,, \tag{3.102}$$

which follows from (3.96) and the unitarity of U. □

The no-cloning theorem is often attributed to Wootters and Zurek (1982) [51] but was proved before by GianCarlo Ghirardi and even earlier by Park (1970) [39].

3.7.7 Aharonov–Bergmann–Lebowitz Time Reversal Symmetry

We have seen that Bohmian mechanics is invariant under time reversal. In this section, we look at a different kind of time reversal symmetry. On the face of it, it may appear that every stochastic theory breaks time reversal symmetry because random decisions are made during the course of events, so the outcomes of these random decisions are fixed afterward but not before. A more careful consideration, however, reveals that stochastic processes can indeed be time reversal invariant as follows. A stochastic process can be characterized by means of a probability distribution μ over a space of histories $t \mapsto x(t)$. Now for each history, we can consider the reversed history $t \mapsto \overline{x}(t) := x(-t)$ (the overbar has nothing to do with complex conjugation); the distribution of $\overline{x}$ we call $\overline{\mu}$. If μ depends on a wave function Ψ (let us write $\mu = \mu^\Psi$ to indicate this dependence), then time reversal symmetry means that for every Ψ, there exists a $\overline{\Psi}$ such that

$$\overline{\mu^\Psi} = \mu^{\overline{\Psi}}\,. \tag{3.103}$$

For example, Bohmian mechanics, when regarded as a stochastic process, is symmetric in this sense (see Exercise 1.10); another example will be provided by Bell's jump process in Sect. 6.3.1.

Now consider several quantum measurements, carried out consecutively on the same system at times $t_1 < t_2 < \ldots < t_n$. The random outcomes $Z_1, \ldots, Z_n$ can be regarded as analogous to a stochastic process. Yakir Aharonov, Peter Bergmann, and Joel Lebowitz (1964) [3] (ABL) first pointed out that under certain circumstances, this process is time symmetric.

Let us formulate this symmetry explicitly. Let $A_1, \ldots, A_n$ be self-adjoint operators in $\mathscr{H}$ whose spectra $\sigma(A_k)$ are purely discrete (i.e., countable), so that

$$A_k = \sum_{\alpha\in\sigma(A_k)} \alpha\, P_{k,\alpha} \tag{3.104}$$

with $P_{k,\alpha}$ the projection to the eigenspace of A_k with eigenvalue α. Consider a quantum system with initial wave function $\psi_0 \in \mathscr{H}$ with $\|\psi_0\| = 1$ at time t_0.

Suppose that at times $t_1 < t_2 < \ldots < t_n$, quantum measurements of $A_1, \ldots, A_n$ (respectively) are carried out with outcomes $Z_1, \ldots, Z_n \in \mathbb{R}$ $(t_0 < t_1)$.

Exercise 3.5 (Consecutive Quantum Measurements) Show that

$$\mathbb{P}\Big(Z_1 = z_1, \ldots, Z_n = z_n\Big) = \Big\| P_{nz_n} U_{t_n - t_{n-1}} \cdots P_{1z_1} U_{t_1 - t_0} \psi_0 \Big\|^2 \tag{3.105}$$

with

$$U_t = e^{-iHt/\hbar} \tag{3.106}$$

the unitary propagator of the system.

Now consider $n+1$ measurements instead of n, and suppose that $A_{n+1} = |\phi_0\rangle\langle\phi_0|$ is a one-dimensional projection, $\|\phi_0\| = 1$, and that we conditionalize on $Z_{n+1} = 1$. The resulting conditional distribution can then be written as the *ABL formula*

$$\begin{aligned} &\mathbb{P}\big(Z_1 = z_1, \ldots, Z_n = z_n \big| Z_{n+1} = 1\big) \\ &\quad = \mathcal{N} \Big| \langle \phi_0 | U_{t_{n+1} - t_n} P_{n,z_n} \cdots U_{t_2 - t_1} P_{1z_1} U_{t_1 - t_0} | \psi_0 \rangle \Big|^2 \end{aligned} \tag{3.107}$$

with normalizing constant $\mathcal{N}$. Now consider another experiment, in which we start the system with initial wave function ϕ_0, then evolve it with Hamiltonian $-H$ instead of H for the time $t_{n+1} - t_n$, then measure A_n and call the outcome Y_n, then evolve the collapsed wave function with $-H$ for $t_n - t_{n-1}$, then measure A_{n-1} and call the outcome Y_{n-1} and so on, then evolve the collapsed wave function after the A_1 measurement with $-H$ for $t_1 - t_0$, and finally measure $A_0 := |\psi_0\rangle\langle\psi_0|$ and conditionalize on the outcome $Y_0 = 1$. Then

$$\begin{aligned} &\mathbb{P}\big(Y_1 = z_1, \ldots, Y_n = z_n \big| Y_0 = 1\big) \\ &\quad = \mathcal{N} \Big| \langle \psi_0 | U_{t_0 - t_1} P_{1z_1} U_{t_1 - t_2} \cdots P_{n,z_n} U_{t_n - t_{n+1}} | \phi_0 \rangle \Big|^2 . \end{aligned} \tag{3.108}$$

The crucial observation is that (3.107) and (3.108) are equal! That is, another way of computing the distribution of $Z_1, \ldots, Z_n$ is to start from ϕ_0 as if it were the initial wave function, to apply the whole procedure backward (i.e., unitary evolution with $-H$ and measurements in the opposite order), and finally to measure $|\psi_0\rangle\langle\psi_0|$. This is the *ABL symmetry*.

This kind of reversal symmetry is quite surprising because we would not expect it when we consider the physical details of the apparatus that have been left out of the description in terms of projections: For example, in each measurement, further wave packets $\Psi_{\alpha'}$ are generated corresponding to other possible outcomes than the one realized, and these packets exist in reality except in collapse theories;

if each measurement has (say) two possible outcomes, then the wave function has 1 branch at time t_0 and 2^n branches after t_n. These branches are completely left out of the consideration, they are not part of the symmetry, and when they are kept in the picture, then the experiment is not at all the time reverse of the second experiment (with outcomes Y and initial wave function ϕ_0). Likewise, there will usually be memory devices that store the value of Z_k until time t_{n+1}; these devices are completely left out of the consideration. Likewise, when we carry out a measurement in the second (reversed, Y) experiment, the procedure is not actually the time reverse of the measurement in the first (original, Z) experiment; for example, the entropy in the apparatus will increase during the measurement, and it will not decrease during the corresponding measurement in the second experiment. And so on.

Some authors (e.g., Aharonov et al. 2017 [4]) hypothesize a deep and fundamental meaning in the ABL symmetry. I remain unconvinced: It is not a real symmetry for the reasons I described when I said it was surprising; it is not an actual symmetry because it only applies to the *conditional* probabilities, i.e., only when we *assume* that $Z_{n+1} = 1$ and $Y_0 = 1$; furthermore, it only applies to the highly idealized concept of a quantum measurement, whereas the general experiments associated with the generalized observables that we will study in Chap. 5 do not have ABL symmetry any more. I would tend to see the reason behind ABL symmetry in the unspectacular fact that the idealized mathematics of quantum measurements in terms of projections is so simple that it will feature symmetries not present in the physical situation.

Aharonov et al. (2017) [4] have hypothesized that the ϕ_0 is as physically real as ψ_0, so there are two wave functions in reality ("two-state vector formalism"), and that ϕ_0 "evolves backwards in time." (Less radical ideas in this direction have been advocated by Vaidman (2010) [50].) I find the terminology of "evolving backwards in time" misleading, as ϕ will be a solution of the Schrödinger equation just as ψ is (only with a final condition instead of an initial condition). It seems clear, though, that if at some time t with $t_0 < t < t_1$ (and, say, $n = 1$), nature knew both $\psi(t) = U_{t-t_0}\psi_0$ and $\phi(t) = U_{t-t_1} P_{1z_1} U_{t_1-t_2}\phi_0$, the outcome z_1 of the A_1-measurement at t_1 would be pre-determined. Moreover, it would also be pre-determined that the observable measured at t_1 must be A_1 and not some other observable because $\phi(t_1-)$ must be an eigenfunction of A_1. But this contradicts the obvious fact that the experimenters might still choose freely, briefly before t_1, between several possible observables to measure at t_1. Another problem with the two-state vector idea is that a fundamental physical theory should apply to the whole universe, but since collapses occur in the quantum formalism when an outside observer intervenes and there are no observers outside the universe, it would seem that the wave function ψ of the universe does not collapse, and then it is unclear how nature should choose ϕ, and for us, there is not much left of the motivation for having ϕ at all.

Another proposal (Bedingham and Maroney 2016 [8]) connects GRW theory with ABL symmetry. Suppose that the whole universe has finite duration; let us say it begins at time 0 and ends at T. We need the following formula for the distribution of the flash process:

Exercise 3.6 (Joint Distribution of GRW Flashes) Verify using (3.62) and (3.64) that the flash process up to time T has distribution density, given the times $0 < t_1 < \ldots < t_n < T$ of the collapses,

$$\rho\Big(\boldsymbol{X}_n = \boldsymbol{x}_n, I_n = i_n, \ldots, \boldsymbol{X}_1 = \boldsymbol{x}_1, I_1 = i_1 \Big| n \text{ flashes}, T_n = t_n, \ldots, T_1 = t_1\Big)$$
$$= \Big\| C_{i_n}(\boldsymbol{x}_n) U_{t_n - t_{n-1}} \cdots C_{i_1}(\boldsymbol{x}_1) U_{t_1 - t_0} \psi_0 \Big\|^2 \tag{3.109}$$

with $t_0 = 0$.

This formula is very similar to (3.105) but with projections replaced by the collapse operators $C_i(\boldsymbol{x})$ that multiply in the i-th variable by a Gaussian centered at $\boldsymbol{x}$. Now (3.109) is not time reversal invariant, but it would become time reversal invariant after changing it to

$$\rho\Big(\boldsymbol{X}_n = \boldsymbol{x}_n, I_n = i_n, \ldots, \boldsymbol{X}_1 = \boldsymbol{x}_1, I_1 = i_1 \Big| n \text{ flashes}, T_n = t_n, \ldots, T_1 = t_1\Big)$$
$$= \mathscr{N}' \Big| \langle \phi_0 | U_{T-t_n} C_{i_n}(\boldsymbol{x}_n) U_{t_n - t_{n-1}} \cdots C_{i_1}(\boldsymbol{x}_1) U_{t_1 - t_0} | \psi_0 \rangle \Big|^2 . \tag{3.110}$$

To make the nature of the change more visible, let us write $\tilde{\psi}$ as an abbreviation for $U_{T-t_n} C_{i_n}(\boldsymbol{x}_n) U_{t_n - t_{n-1}} \cdots C_{i_1}(\boldsymbol{x}_1) U_{t_1 - t_0} \psi_0$; then the right-hand side of (3.109) equals $\langle \tilde{\psi} | \tilde{\psi} \rangle$, whereas that of (3.110) is $\mathscr{N}' \langle \tilde{\psi} | P_{\mathbb{C}\phi_0} | \tilde{\psi} \rangle$. One might hope that the effects of this change are minor because the additional projection operator $P_{\mathbb{C}\phi_0}$ acts at time T in the distant future.

Since the Poisson process is time reversal invariant, the whole flash process is time reversal covariant if we assume (3.110). (That is, the reversed process is the solution of the same laws with the initial wave function ψ_0 replaced by ϕ_0^* and the final wave function ϕ_0 by ψ_0^*.)

Now it is not very clear how ϕ_0 should be chosen, or how that would affect the distribution of the flashes and the empirical predictions of GRW theory, or whether GRW theory would then still be empirically adequate, or in fact whether it would be a relevant advantage to restore time reversal symmetry to GRW theory.

Exercises

Exercise 3.1 can be found in Sect. 3.1.3, Exercise 3.2 in Sect. 3.1.5, Exercise 3.3 in Sect. 3.3.1, Exercise 3.4 in Sect. 3.7.4, and Exercises 3.5 and 3.6 in Sect. 3.7.7.

Exercise 3.7 (Essay Question) Describe the measurement problem of quantum mechanics. Use formulas where appropriate.

Exercise 3.8 (Essay Question) Why does GRW theory make approximately the same predictions as the quantum formalism?

Exercise 3.9 (Essay Question) Explain why Schrödinger's theory Sm has a many-worlds character.

Exercise 3.10 (Robertson–Schrödinger Inequality) Compute both sides of the Robertson–Schrödinger inequality (3.32) for $A = \sigma_1$, $B = \sigma_2$, and $\psi = |z\text{-down}\rangle$. *Hint*: In order to obtain σ_A and σ_B, compute first the probability distribution for A and B according to Born's rule.

Exercise 3.11 (Normal Matrices) Use Theorem 3.1 to show that a $d \times d$ matrix C can be unitarily diagonalized if and only if C commutes with $C^\dagger$. Such a matrix is called "normal." (Hint: write $C = A + iB$.)

Exercise 3.12 (Variance of the Exponential Distribution) Find the variance of the exponential distribution with parameter λ.

Exercise 3.13 (Poisson Process) For the Poisson process with rate $\lambda > 0$, determine for any fixed $t_0 > 0$ the distribution of $X_{t_0} = \#\{k : T_k < t_0\}$, the number of events up to time t_0. Follow two reasonings:

(a) Heuristically, assume that an event occurs in every infinitesimal time interval $[t, t + dt]$ independent of disjoint intervals with probability $\lambda\, dt$.
Hint: Divide $[0, t_0]$ in $n \gg 1$ subintervals of length $dt = t_0/n$.

(b) Rigorously, assume that the random variables $T_1, T_2, \ldots$ are defined to be $T_k = W_1 + \ldots + W_k$ with all waiting times W_k independent and exponentially distributed with parameter λ, i.e., with density $\rho(w) = 1_{w>0}\, \lambda\, e^{-\lambda w}$.
Hint:

$$\mathbb{P}\big(X_{t_0} \geq 2\big) = \mathbb{P}\Big(W_1 + W_2 < t_0\Big) = \int_0^{t_0} dw_1 \int_0^{t_0 - w_1} dw_2\, \rho(w_1)\, \rho(w_2) \quad \text{and}$$

$$\mathbb{P}(X_{t_0} = k) = \mathbb{P}\big(X_{t_0} \geq k\big) - \mathbb{P}\big(X_{t_0} \geq k + 1\big)\,.$$

Exercise 3.14 (GRW Theory) Consider the GRW theory with the constant σ much smaller than the value 10^{-7} m suggested by GRW; say, $\sigma = 10^{-12}$ m. Explain why Heisenberg's uncertainty relation (2.62) implies that a free electron, after being hit by a GRW collapse, could move very fast. Use the uncertainty relation to compute the order of magnitude of how fast it can be (assuming it was more or less at rest before the collapse); the mass of an electron is about 10^{-30} kg and $\hbar \approx 10^{-34}\,\mathrm{kg\,m^2\,s^{-1}}$.

Exercise 3.15 (Empirical Test of GRW Theory: Universal Warming) Assuming that usual wave functions are wider than 10^{-7} m, a GRW collapse will tend to make a wave function narrower in position and, by the Heisenberg uncertainty relation, wider in momentum, thus increasing the average energy. As a consequence, all matter is spontaneously getting warmer all the time. Estimate the rate at which the temperature increases in Kelvin per year, given the parameter values suggested by GRW, $\sigma = 10^{-7}$ m and a collapse rate of $\lambda = 10^{-16}\,\mathrm{s}^{-1}$ per nucleon. Use that an energy increase of ΔE corresponds to a temperature increase ΔT such that $\Delta E = \frac{3}{2} N k \Delta T$ with N the number of molecules and $k \approx 10^{-23}$ J/K the Boltzmann constant. You may assume that a typical molecule contains of order 10 nucleons.

Exercise 3.16 (Primitive Ontology) Argue that GRWf and GRWm are empirically equivalent, i.e., make the same predictions. (A solution to this problem can be found in [31, Sec. 2.4].)

Exercise 3.17 (A Measurement Puzzle) Bob asks you to prepare an electron on which he will perform a single quantum measurement of either σ_x or σ_z without telling you which measurement he did. After his measurement, he will give you the electron back, so you can perform your measurement on it. Your task is to retrodict with certainty the result Bob got if he measured σ_x and the result he got if he measured σ_z. What should you do?

References

1. R.A. Adams, J.J.F. Fournier, *Sobolev Spaces*, 2nd edn. (Academic, Amsterdam, 2003)
2. S.L. Adler, Lower and upper bounds on CSL parameters from latent image formation and IGM heating. J. Phys. A: Math. Theor. **40**, 2935–2957 (2007). http://arxiv.org/abs/quant-ph/0605072
3. Y. Aharonov, P.G. Bergmann, J.L. Lebowitz, Time symmetry in the quantum process of measurement. Phys. Rev. **134**, B1410–B1416 (1964)
4. Y. Aharonov, E. Cohen, T. Landsberger, The two-time interpretation and macroscopic time-reversibility. Entropy **19**(3), 111 (2017)
5. S. Albeverio, F. Gesztesy, R. Høegh-Krohn, H. Holden, *Solvable Models in Quantum Mechanics* (Springer, Heidelberg, 1988)
6. V. Allori, S. Goldstein, R. Tumulka, N. Zanghì, Many-worlds and Schrödinger's first quantum theory. Br. J. Philos. Sci. **62**(1), 1–27 (2011). http://arxiv.org/abs/0903.2211

7. V. Allori, S. Goldstein, R. Tumulka, N. Zanghì, Predictions and primitive ontology in quantum foundations: a study of examples. Br. J. Philos. Sci. **65**, 323–352 (2014). http://arxiv.org/abs/1206.0019
8. D. Bedingham, O. Maroney, Time symmetry in wave function collapse. Phys. Rev. A **95**, 042103 (2017). http://arxiv.org/abs/1607.01940
9. J.S. Bell, The measurement theory of Everett and de Broglie's Pilot wave, in *Quantum Mechanics, Determinism, Causality, and Particles*, ed. by M. Flato, Z. Maric, A. Milojevic, D. Sternheimer, J.P. Vigier (Reidel, Dordrecht, 1976), pp. 11–17. Reprinted as chapter 11 of [13]
10. J.S. Bell, Quantum mechanics for cosmologists, in *Quantum Gravity 2*, ed. by C. Isham, R. Penrose, D. Sciama (Clarendon Press, Oxford, 1981), pp. 611–637. Reprinted as chapter 15 of [13]
11. J.S. Bell, Beables for quantum field theory. Phys. Rep. **137**, 49–54 (1986). Reprinted as chapter 19 of [13]. Also reprinted in: *Quantum Implications: Essays in Honour of David Bohm*, ed. by F.D. Peat, B.J. Hiley (Routledge, London, 1987), p. 227
12. J.S. Bell, Are there quantum jumps?, in *Schrödinger. Centenary Celebration of a Polymath*, ed. by C.W. Kilmister (Cambridge University Press, Cambridge, 1987), pp. 41–52. Reprinted as chapter 22 of [13]
13. J.S. Bell, *Speakable and Unspeakable in Quantum Mechanics* (Cambridge University Press, Cambridge, 1987)
14. N. Bohr, *Atomic Theory and the Description of Nature* (Cambridge University Press, Cambridge, 1934)
15. M. Born, *The Born–Einstein Letters* (Walker and Company, New York, 1971). Translated by I. Born
16. J. Bricmont, *Making Sense of Quantum Mechanics* (Springer, Heidelberg, 2016)
17. H. Brown, D. Wallace, Solving the measurement problem: de Broglie-Bohm loses out to Everett. Foundations Phys. **35**, 517–540 (2005). http://arxiv.org/abs/quant-ph/0403094
18. M. Carlesso, A. Bassi, P. Falferi, A. Vinante, Experimental bounds on collapse models from gravitational wave detectors. Phys. Rev. D **94**, 124036 (2016). http://arxiv.org/abs/1606.04581
19. J.G. Cramer, The transactional interpretation of quantum mechanics. Rev. Modern Phys. **58**, 647–688 (1986)
20. D. Deutsch, Quantum theory of probability and decisions. Proc. R. Soc. Lond. A **455**, 3129–3137 (1999). http://arxiv.org/abs/quant-ph/9906015
21. L. Diósi, Models for universal reduction of macroscopic quantum fluctuations. Phys. Rev. A **40**, 1165–1174 (1989)
22. A. Einstein, Reply to criticisms, in *Albert Einstein, Philosopher-Scientist*, ed. by P.A. Schilpp (Library of Living Philosophers, Evanston, 1949), pp. 665–688
23. H. Everett, The theory of the universal wavefunction. Ph.D. Thesis. Department of Physics, Princeton University, 1955. Reprinted in *The Many-Worlds Interpretation of Quantum Mechanics*, ed. by B. DeWitt, R.N. Graham (Princeton University Press, Princeton, 1973), pp. 3–140
24. H. Everett, Relative state formulation of quantum mechanics. Rev. Modern Phys. **29**, 454–462 (1957)
25. W. Feldmann, R. Tumulka, Parameter diagrams of the GRW and CSL theories of wave function collapse. J. Phys. A: Math. Theor. **45**, 065304 (2012). http://arxiv.org/abs/1109.6579
26. R.P. Feynman, Space-time approach to non-relativistic quantum mechanics. Rev. Modern Phys. **20**, 367–387 (1948)
27. R.P. Feynman, *The Character of Physical Law* (MIT Press, Cambridge, 1967)
28. R.P. Feynman, F.B. Morinigo, W.G. Wagner, *Feynman Lectures on Gravitation*, ed. by B. Hatfield (Addison-Wesley, Boston, 1995). Although printed only in 1995, the lecture was given in 1962
29. G.C. Ghirardi, A. Rimini, T. Weber, Unified dynamics for microscopic and macroscopic systems. Phys. Rev. D **34**, 470–491 (1986)
30. K. Gödel, Über formal unentscheidbare Sätze der Principia Mathematica und verwandter Systeme I. Monatshefte für Mathematik und Physik **38**, 173–198 (1931)

31. S. Goldstein, R. Tumulka, N. Zanghì, The quantum formalism and the GRW formalism. J. Statist. Phys. **149**, 142–201 (2012). http://arxiv.org/abs/0710.0885
32. B.C. Hall, *Quantum Theory for Mathematicians* (Springer, New York, 2013)
33. W. Heisenberg, *Physical Principles of the Quantum Theory* (University of Chicago Press, Chicago, 1930)
34. W. Heisenberg, *Physics and Philosophy* (Harper, New York, 1958)
35. R. Kaltenbaek, Tests in space, in *Do Wave Functions Jump?* ed. by V. Allori, A. Bassi, D. Dürr, N. Zanghì (Springer, Berlin, 2020), pp. 401–411
36. T. Maudlin, Three measurement problems. Topoi **14**(1), 7–15 (1995)
37. T. Maudlin, Can the world be only wave function? in *Many Worlds? Everett, Quantum Theory, and Reality*, ed. by S. Saunders, J. Barrett, A. Kent, D. Wallace (Oxford University Press, Oxford, 2010), pp. 121–143
38. T. Norsen, *Foundations of Quantum Mechanics* (Springer, New York, 2017)
39. J. Park, The concept of transition in quantum mechanics. Foundations Phys. **1**, 23–33 (1970)
40. P. Pearle, Combining stochastic dynamical state-vector reduction with spontaneous localization. Phys. Rev. A **39**, 2277–2289 (1989)
41. H.P. Robertson, The uncertainty principle. Phys. Rev. **34**, 163–164 (1929)
42. S. Ross, *Stochastic Processes* (Wiley, Hoboken, 1996)
43. B. Russell, A.N. Whitehead, *Principia Mathematica* (Cambridge University Press, Cambridge, 1913)
44. E. Schrödinger, Quantisierung als Eigenwertproblem (Vierte Mitteilung). Ann. der Physik **81**, 109–139 (1926). English translation by J.F. Shearer in [45]
45. E. Schrödinger, *Collected Papers on Wave Mechanics* (Chelsea, New York, 1927)
46. E. Schrödinger, Zum Heisenbergschen Unschärfeprinzip. Sitzungsberichte der Preussischen Akademie der Wissenschaften, physikalisch-mathematische Klasse **14**, 296–303 (1930)
47. E. Schrödinger, Die gegenwärtige Situation in der Quantenmechanik. Naturwissenschaften **23**, 807–812, 823–828, 844–849 (1935). English translation by J.D. Trimmer: The present situation in quantum mechanics. Proc. Amer. Philos. Soc. **124**, 323–338 (1980). Reprinted in J.A. Wheeler, W.H. Zurek (eds.), *Quantum Theory and Measurement* (Princeton University Press, Princeton, 1983), pp. 152–167
48. R. Tumulka, Paradoxes and primitive ontology in collapse theories of quantum mechanics, in *Collapse of the Wave Function*, ed. by S. Gao (Cambridge University Press, Cambridge, 2018), pp. 139–159. http://arxiv.org/abs/1102.5767
49. L. Vaidman, On schizophrenic experiences of the neutron or why we should believe in the many-worlds interpretation of quantum theory. Int. Stud. Philos. Sci. **12**, 245–261 (1998). http://arxiv.org/abs/quant-ph/9609006
50. L. Vaidman, Time symmetry and the many-worlds interpretation, in *Many Worlds? Everett, Realism and Quantum Mechanics*, ed. by S. Saunders, J. Barrett, A. Kent, D. Wallace (Oxford University Press, Oxford, 2010), pp. 582–586. http://philsci-archive.pitt.edu/4396/
51. W. Wootters, W. Zurek, A single quantum cannot be cloned. Nature **299**, 802–803 (1982)

4 Nonlocality

4.1 The Einstein–Podolsky–Rosen Argument

In the literature, the "Einstein-Podolsky-Rosen (EPR) paradox" is often mentioned. It is clear from EPR's (1935) [11] article that they did not intend to describe a paradox (as did, e.g., Wheeler when describing the delayed-choice experiment), but rather to describe an argument. They were arguing for the following:

Claim *There are additional variables beyond the wave function.*

I now explain their reasoning in my own words, partly in preparation for Bell's (1964) [3] argument, which builds on EPR's argument.

4.1.1 The EPR Argument

EPR considered two particles in one dimension with entangled wave function

$$\Psi(x_1, x_2) = \delta(x_1 - x_2 + x_0)\,, \tag{4.1}$$

with x_0 a constant; see Fig. 4.1. (This wave function is actually unphysical because it does not lie in Hilbert space; but the same argument can be made with a square-integrable wave function, as I will point out in Sect. 4.1.2.)

An observer, let us call her Alice, measures the position of particle 1. The outcome X_1 is uniformly distributed, and the wave function collapses to

$$\Psi'(x_1, x_2) = \delta(x_1 - X_1)\delta(x_2 - X_1 - x_0)\,, \tag{4.2}$$

R. Tumulka, *Foundations of Quantum Mechanics*, Lecture Notes in Physics 1003,
https://doi.org/10.1007/978-3-031-09548-1_4

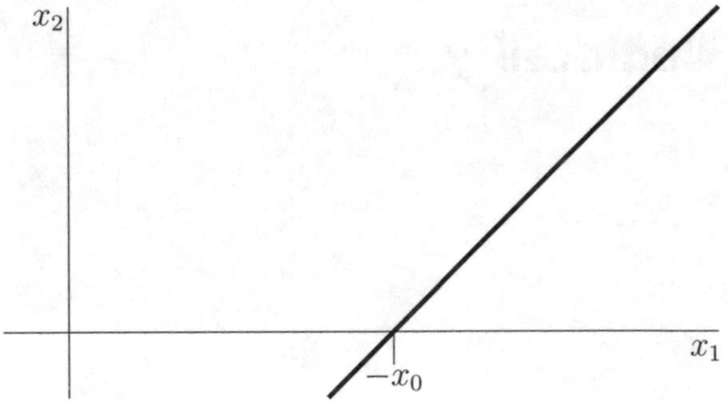

Fig. 4.1 The wave function Ψ considered by EPR is concentrated on a line in the configuration space of two particles, drawn here for a negative value of the constant x_0

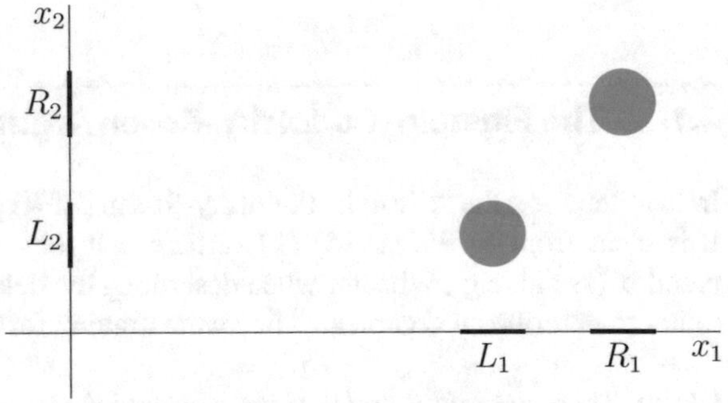

Fig. 4.2 Support of the wave function Ψ in the square-integrable version of the EPR argument

so that another observer, Bob, measuring the position of particle 2, is certain to obtain $X_2 = X_1 + x_0$. It follows that particle 2 had a position even before Bob made his experiment. Now EPR make the assumption that

$$\text{no real change can take place in the second system in consequence of [a measurement on] the first system.} \tag{4.3}$$

This assumption is a special case of *locality*. They took it as obviously true, but it is worthy of a closer examination; we will come back to it in Sect. 4.2. It then follows that particle 2 had a definite position even before Alice made her experiment, despite the fact that Ψ is not an eigenfunction of x_2-position. Quod erat demonstrandum.

4.1.2 Square-Integrable Version

Consider two spatial regions $L_1, R_1 \subset \mathbb{R}^3$ ("the left box" and "the right box") in Alice's lab and two spatial regions $L_2, R_2 \subset \mathbb{R}^3$ in Bob's lab. Consider a square-integrable wave function Ψ on two-particle configuration space that consists of two equally large wave packets, one concentrated in $L_1 \times L_2$ and the other in $R_1 \times R_2$; see Fig. 4.2. First, Alice carries out a quantum measurement of the observable P_{R_1} as in (3.17) on particle 1 and then Bob one of P_{R_2} on particle 2. It follows that Bob's result Z_2 is always equal to Alice's result Z_1; with probability 1/2, they are both 0 and, with probability 1/2, both 1. It follows that the observable P_{R_2} for particle 2 already had a definite value before Bob's measurement. The assumption (4.3) then implies further that it already had a definite value even before Alice's measurement, despite the fact that Ψ is not an eigenfunction. Thus, Ψ is not a complete description of the factual situation.

4.1.3 Further Conclusions

EPR drew further conclusions from their example by considering also momentum. Note that the Fourier transform of Ψ is

$$\widehat{\Psi}(k_1, k_2) = \frac{1}{2\pi} \int dx_1 \int dx_2\, e^{-ik_1x_1} e^{-ik_2x_2}\, \delta(x_1 - x_2 + x_0) \tag{4.4}$$

$$= \frac{1}{2\pi} \int dx_1\, e^{-ik_1x_1} e^{-ik_2(x_1+x_0)} \tag{4.5}$$

$$= e^{-ik_2x_0} \frac{1}{\sqrt{2\pi}} \int dx_1 \frac{e^{-i(k_1+k_2)x_1}}{\sqrt{2\pi}} \tag{4.6}$$

$$= e^{-ik_2x_0}\, \delta(k_1 + k_2) \tag{4.7}$$

$$= e^{ik_1x_0}\, \delta(k_1 + k_2)\,. \tag{4.8}$$

Alice could measure either the position or the momentum of particle 1, and Bob either the position or the momentum of particle 2. If Alice measures position, then, as seen above, the outcome X_1 is uniformly distributed, and Bob, if he chooses to measure position, finds $X_2 = X_1 + x_0$ with certainty. If, alternatively, Alice measures momentum, then the outcome K_1 will be uniformly distributed, and the wave function in momentum representation collapses from $\widehat{\Psi}$ to

$$\widetilde{\Psi}(k_1, k_2) = e^{iK_1x_0}\, \delta(k_1 - K_1)\, \delta(k_2 + K_1) \tag{4.9}$$

so that Bob, if he chooses to measure momentum, is certain to find $K_2 = -K_1$. In the same way as above, it follows that Bob's particle had a position before any of the experiments *and* that it had a momentum!

There even arises a way of simultaneously measuring the position and momentum of particle 2: Alice measures position X_1 and Bob momentum K_2. Since particle 2 has, as just proved, a well-defined position and a well-defined momentum, and since, by (4.3), Alice's measurement did not influence particle 2, K_2 must be the original (pre-measurement) momentum of particle 2. Likewise, $X_1 + x_0$ must have been the original position of particle 2.

4.1.4 Bohm's Version of the EPR Argument Using Spin

Before he discovered Bohmian mechanics, Bohm wrote a textbook (1951) [7] about quantum mechanics in which he followed the orthodox view. In it, he also described the following useful variant of the EPR argument, sometimes called the EPRB experiment (B for Bohm).

Consider two spin-$\frac{1}{2}$ particles with joint spinor $\phi_{s_1 s_2}$ in the two-particle spin space $\mathbb{C}^4 = \mathbb{C}^2 \otimes \mathbb{C}^2$ given by the so-called singlet state[1]

$$\phi = \frac{1}{\sqrt{2}}\Big(|z\text{-up}\rangle_1 |z\text{-down}\rangle_2 - |z\text{-down}\rangle_1 |z\text{-up}\rangle_2\Big) . \tag{4.10}$$

Written as a matrix,

$$\phi = \frac{1}{\sqrt{2}}\begin{pmatrix} 0 & 1 \\ -1 & 0 \end{pmatrix} . \tag{4.11}$$

In the following, we will drop the indices and use the notation $|z\text{-up}\rangle|z\text{-down}\rangle$ instead of $|z\text{-up}\rangle_1|z\text{-down}\rangle_2$ for the tensor product state.

Alice measures σ_3 on particle 1. The outcome Z_1 is ± 1, each with probability $1/2$. If $Z_1 = +1$, then the wave function collapses to

$$\phi'_+ = |z\text{-up}\rangle|z\text{-down}\rangle , \tag{4.12}$$

and Bob, measuring σ_3 on particle 2, is certain to obtain $Z_2 = -1$. If, however, $Z_1 = -1$, then the wave function collapses to

$$\phi'_- = |z\text{-down}\rangle|z\text{-up}\rangle , \tag{4.13}$$

and Bob is certain to obtain $Z_2 = +1$. Thus, always $Z_2 = -Z_1$; one speaks of *perfect anti-correlation*. As a consequence, particle 2 had a definite value of z-spin even before Bob's experiment. Now, from the assumption (4.3), it follows that it had that value even before Alice's experiment. Likewise, particle 1 had a definite value of z-spin before any attempt to measure it.

Moreover, again as in EPR's reasoning, we can consider other observables, say σ_1 and σ_2.

[1] Here is the origin of that name: The rotation group $SO(3)$ (or, rather, its covering group) acts on the Hilbert space (here, on $\mathbb{C}^4$), and the irreducible invariant subspaces of this action are called "multiplets." Here, there are two irreducible invariant subspaces: a three-dimensional one called the "triplet" and a one-dimensional one called the "singlet"; ϕ spans the latter.

Exercise 4.1 (Spin Singlet State)

(a) Verify through direct computation that in the spin space $\mathbb{C}^4 = \mathbb{C}^2 \otimes \mathbb{C}^2$ of two spin-$\frac{1}{2}$ particles,

$$\begin{aligned} &|x\text{-up}\rangle|x\text{-down}\rangle - |x\text{-down}\rangle|x\text{-up}\rangle \\ &= |y\text{-up}\rangle|y\text{-down}\rangle - |y\text{-down}\rangle|y\text{-up}\rangle \\ &= |z\text{-up}\rangle|z\text{-down}\rangle - |z\text{-down}\rangle|z\text{-up}\rangle \end{aligned} \tag{4.14}$$

up to phase factors.

(b) An element of $\mathbb{C}^2 \otimes \mathbb{C}^2$ can be represented by a complex 2×2 matrix $\psi_{s_1 s_2}$. One calls those elements *anti-symmetric* for which $\psi_{s_2 s_1} = -\psi_{s_1 s_2}$. Show that they form a one-dimensional subspace. Explain why it follows from this fact that (4.14) holds up to phase factors and indeed that

$$\begin{aligned} &|\boldsymbol{n}\text{-up}\rangle|\boldsymbol{n}\text{-down}\rangle - |\boldsymbol{n}\text{-down}\rangle|\boldsymbol{n}\text{-up}\rangle \\ &= |z\text{-up}\rangle|z\text{-down}\rangle - |z\text{-down}\rangle|z\text{-up}\rangle \end{aligned} \tag{4.15}$$

up to phase factors for any direction given by $\boldsymbol{n} \in \mathbb{R}^3$ with $|\boldsymbol{n}| = 1$.

It follows from (4.14) that if Alice and Bob both measure x-spin, then their outcomes are also perfectly anti-correlated, and by (4.15), likewise for the spin component in any direction $\boldsymbol{n}$. By the same argument as before, it can be inferred that any spin component, for each particle, has a well-defined value before any experiment.

Finally, Alice and Bob together can measure σ_1 *and* σ_3 of particle 2: Alice measures σ_1 of particle 1 and Bob σ_3 of particle 2. By (4.3) and the perfect anti-correlation, the negative of Alice's outcome is what Bob would have obtained had he measured σ_1; and by (4.3), Bob's outcome is not affected by Alice's experiment.

4.1.5 Einstein's Boxes Argument

We have seen that EPR's argument yields more than just the incompleteness of the wave function. It also yields that particles have well-defined positions *and* momenta. If we only want to establish the incompleteness of the wave function, which seems like a worthwhile goal for a proof, a simpler argument will do. Einstein developed such an argument already in 1927 (before the EPR paper), presented it at a conference, but never published it. It has been reported by, e.g., de Broglie (1964) [10]; a more detailed discussion is given in Norsen (2005) [17].

Consider a single particle whose wave function $\psi(\boldsymbol{x})$ is confined to a box B with impermeable walls and (more or less) uniform in B. Now split B (e.g., by inserting a partition) into two boxes B_1 and B_2; move one box to Tokyo and the other to Paris. There is some nonzero amount of the particle's wave function in Paris and some in Tokyo. Carry out a detection in Paris. Let us assume that

no real change can take place in Tokyo in consequence of a measurement in Paris. (4.16)

(Again a special case of locality.) If we believed that the wave function was a complete description of reality, then there would not be a fact of the matter, before the detection experiment, about whether the particle is in Paris or Tokyo, but afterward, there would be. This contradicts (4.16), so the wave function cannot be complete.

Of course, the assumption (4.16) is understood as "no real change can *immediately* take place in Tokyo in consequence of a measurement in Paris," allowing changes in Tokyo after a while, such as the while it would take a signal to travel from Paris to Tokyo at the speed of light. That is, (4.16) (and similarly (4.3)) is particularly motivated by the theory of relativity, which strongly suggests that signals cannot propagate faster than at the speed of light. On one occasion, Einstein wrote that the faster-than-light effect entailed by insisting on completeness of the wave function was "spukhafte Fernwirkung" (spooky action-at-a-distance).

4.1.6 Too Good to Be True

EPR's argument is, in fact, correct. Nevertheless, it may strike you that its conclusion, the incompleteness of the wave function, is very strong—maybe too strong to be true. After all, it is not true in GRW or many-worlds! How can this be: that EPR proved something that is not true?

This can happen only because the assumption (4.3) is actually not true in these theories. And in Bohmian mechanics, where the wave function is in fact incomplete, it is not true that all spin observables have pre-existing actual values, as would follow from EPR's reasoning. Thus, also in Bohmian mechanics, (4.3) is not true. We will see in the next section that (4.3) is problematical in any version of quantum mechanics. This fact was discovered nearly 30 years after EPR's paper by John Bell.

4.2 Proof of Nonlocality

Two space-time points $x = (s, \boldsymbol{x})$ and $y = (t, \boldsymbol{y})$ are called *spacelike separated* if and only if no signal propagating at the speed of light can reach x from y or y from x. This occurs if and only if

$$|\boldsymbol{x} - \boldsymbol{y}| > c|s - t|, \tag{4.17}$$

with $c = 3 \times 10^8$ m/s the speed of light. Einstein's theory of relativity strongly suggests that signals cannot propagate faster than at the speed of light (*superluminally*). That is, if x and y are spacelike separated, then no signal can be sent from x to y or from y to x. This in turn suggests that

If x and y are spacelike separated then events at x do not influence events at y. (4.18)

This statement is called *locality*. It is true in relativistic versions of classical physics (mechanics, electrodynamics, and also Einstein's relativistic theory of gravity he called the *general theory of relativity*). Bell (1964) [3] proved a result often called *Bell's theorem*:

Locality is sometimes false if certain empirical predictions of the quantum formalism are correct. (4.19)

The relevant predictions have since been experimentally confirmed; the first convincing tests were carried out by Alain Aspect et al. (1982) [2]. Thus, locality is false in our world; this fact is often called *quantum nonlocality*. Our main goal in this chapter is to understand Bell's proof.

Some remarks.

- Einstein believed in locality until his death in 1955. The EPR assumption (4.3) is a special case of locality: If Alice's measurement takes place at x and Bob's at y, and if x and y are spacelike separated, then locality implies that Alice's measurement on particle 1 at x cannot affect particle 2 at y. Conversely, the only situation in which we can be certain that the two particles *cannot* influence each other occurs if Alice's and Bob's experiments are spacelike separated and locality holds true. Ironically, EPR were wrong even though their argument was correct: The premise (4.3) is false. They took locality for granted. Likewise, in Einstein's boxes argument, the assumption (4.16) is a special case of locality: The point of talking about Tokyo and Paris is that these two places are distant, and since there clearly can be influences if we allow more time than distance/c, the assumption is that there cannot be an influence between spacelike separated events.
- Despite nonlocality, it is not possible to send messages faster than light, according to the appropriate relativistic version of the quantum formalism; this fact is often called the *no-signaling theorem*. We will prove it in great generality in Sect. 5.5.9. Put differently, the superluminal influences cannot be used by agents for sending messages.
- Does nonlocality prove relativity wrong? That statement would be too strong. Nonlocality proves a *certain understanding* of relativity wrong. Much of relativity theory, however, remains untouched by nonlocality.
- In relativistic versions of the Schrödinger equation (such as the Dirac equation), wave functions propagate no faster than at the speed c of light. This property of

the Dirac equation is called *propagation locality*. It can be expected to be true in nature, but it is not the same as locality, which we will see is false in nature.

- If x and y are spacelike separated, then relativistic Hamiltonians H contain no interaction term between x and y. This property of H is called *interaction locality*.

  Let me explain this statement. The non-relativistic Schrödinger equation needs to be replaced, in a relativistic theory, by a relativistic equation. The latter is different from the non-relativistic Schrödinger equation in two ways: (i) Already for a free particle, the equation requires a modification such as the Dirac equation. (ii) Instead of interaction potentials, interaction arises from the creation and annihilation of particles. For example, an electron can create a photon, which travels to another electron and is annihilated there. Potentials can only be used as an approximation. Since the creation or absorption of a photon occurs only at the locations of electrons (or other charged particles), and since photon wave functions propagate no faster than at the speed c, there is no interaction term in the Hamiltonian between particles at x and at y.

  So there are two meanings to the word "interaction": first, an interaction term in the Hamiltonian and second, any influence. Bell's proof shows that in the absence of the first type of interaction, the second type can still be present.
- Bell's proof shows for a certain experiment that *either* events at x must have influenced events at y *or* vice versa, but does not tell us who influenced whom.
- Bell's proof applies equally to quantum field theory as to quantum mechanics. For the experiment Bell considered, quantum field theory (including relativistic quantum field theory) makes the same predictions as quantum mechanics, and Bell's proof shows that these predictions (which have been confirmed in experiment) imply nonlocality. Some authors correctly emphasize that certain locality-related properties play a crucial role in relativistic quantum field theory, such as interaction locality or that any two observables referring to spacelike separated space-time regions[2] commute. But this does not conflict with nonlocality, as interaction locality is not the same as locality, and we will see that Bell's proof makes use of commuting observables right from the start.

4.2.1 Bell's Experiment

As in Bohm's version of the EPR example, consider two spin-$\frac{1}{2}$ particles in the singlet state

$$\phi = \frac{1}{\sqrt{2}}\Big(|z\text{-up}\rangle|z\text{-down}\rangle - |z\text{-down}\rangle|z\text{-up}\rangle\Big). \tag{4.20}$$

[2] Two space-time regions A, B are *spacelike separated* if and only if every $a \in A$ is spacelike separated from every $b \in B$.

While keeping their spinor constant, the two particles are brought to distant places. Alice makes an experiment on particle 1 at (or near) space-time point x and Bob one on particle 2 at y; x and y are spacelike separated. Let

$$\mathbb{S}^2 = \{\boldsymbol{n} \in \mathbb{R}^3 : |\boldsymbol{n}| = 1\} \tag{4.21}$$

denote the unit sphere in $\mathbb{R}^3$. Each experimenter chooses a direction in space, corresponding to a $\boldsymbol{n} \in \mathbb{S}^2$, and carries out a Stern–Gerlach experiment in that direction, i.e., a quantum measurement of $\boldsymbol{n} \cdot \boldsymbol{\sigma}$. The difference to Bohm's example is that Alice and Bob can choose different directions. I write $\boldsymbol{\alpha}$ for Alice's unit vector, $\boldsymbol{\beta}$ for Bob's, Z_1 for the random outcome ± 1 of Alice's experiment, and Z_2 for that of Bob's. Let us compute the joint distribution $\mu_{\boldsymbol{\alpha},\boldsymbol{\beta}}$ of Z_1 and Z_2.

Proposition 4.1 *Independent of whether Alice's or Bob's experiment occurs first, the joint distribution of Z_1, Z_2 is*

$$\mu_{\boldsymbol{\alpha},\boldsymbol{\beta}} := \begin{pmatrix} \mathbb{P}(up,up) & \mathbb{P}(up,down) \\ \mathbb{P}(down,up) & \mathbb{P}(down,down) \end{pmatrix} \tag{4.22}$$

$$= \begin{pmatrix} \frac{1}{4} - \frac{1}{4}\boldsymbol{\alpha}\cdot\boldsymbol{\beta} & \frac{1}{4} + \frac{1}{4}\boldsymbol{\alpha}\cdot\boldsymbol{\beta} \\ \frac{1}{4} + \frac{1}{4}\boldsymbol{\alpha}\cdot\boldsymbol{\beta} & \frac{1}{4} - \frac{1}{4}\boldsymbol{\alpha}\cdot\boldsymbol{\beta} \end{pmatrix} \tag{4.23}$$

$$= \begin{pmatrix} \frac{1}{2}\sin^2(\theta/2) & \frac{1}{2}\cos^2(\theta/2) \\ \frac{1}{2}\cos^2(\theta/2) & \frac{1}{2}\sin^2(\theta/2) \end{pmatrix} \tag{4.24}$$

with θ the angle between $\boldsymbol{\alpha}$ and $\boldsymbol{\beta}$.

Proof Assume that Alice's experiment occurs first, and write the initial spinor using (4.15) as

$$\phi = c|\boldsymbol{\alpha}\text{-up}\rangle|\boldsymbol{\alpha}\text{-down}\rangle - c|\boldsymbol{\alpha}\text{-down}\rangle|\boldsymbol{\alpha}\text{-up}\rangle \tag{4.25}$$

with c a complex constant with $|c| = 1/\sqrt{2}$. According to Born's rule, Alice obtains $+1$ or -1, each with probability 1/2. In case $Z_1 = +1$, ϕ collapses to

$$\phi'_+ = |\boldsymbol{\alpha}\text{-up}\rangle|\boldsymbol{\alpha}\text{-down}\rangle\,. \tag{4.26}$$

According to Born's rule, the probability that Bob obtains $Z_2 = +1$ is

$$\mathbb{P}(Z_2 = +1|Z_1 = +1) = \big|\langle\boldsymbol{\beta}\text{-up}|\boldsymbol{\alpha}\text{-down}\rangle\big|^2 = 1 - \big|\langle\boldsymbol{\beta}\text{-up}|\boldsymbol{\alpha}\text{-up}\rangle\big|^2\,. \tag{4.27}$$

Since the angle in Hilbert space between $|\boldsymbol{\beta}\text{-up}\rangle$ and $|\boldsymbol{\alpha}\text{-up}\rangle$ is half the angle between $\boldsymbol{\beta}$ and $\boldsymbol{\alpha}$, and since they are unit vectors in Hilbert space, we have that

$$\big|\langle\boldsymbol{\beta}\text{-up}|\boldsymbol{\alpha}\text{-up}\rangle\big| = \cos(\theta/2) \tag{4.28}$$

and thus

$$\mathbb{P}(Z_2 = +1|Z_1 = +1) = 1 - \cos^2(\theta/2) = \sin^2(\theta/2) \tag{4.29}$$

and the total probability is

$$\mathbb{P}(Z_1 = +1, Z_2 = +1) = \frac{1}{2}\sin^2(\theta/2)\,. \tag{4.30}$$

Since $\cos^2 x = \frac{1}{2} + \frac{1}{2}\cos(2x)$, this value can be rewritten as

$$\mathbb{P}(Z_1 = +1, Z_2 = +1) = \frac{1}{2} - \frac{1}{2}\cos^2(\theta/2) = \frac{1}{2} - \frac{1}{4} - \frac{1}{4}\cos\theta = \frac{1}{4} - \frac{1}{4}\boldsymbol{\alpha}\cdot\boldsymbol{\beta}\,. \tag{4.31}$$

The other three matrix elements can be computed in the same way. Assuming that Bob's experiment occurs first leads to the same matrix. □

Remark 4.1

- Note that the four entries in $\mu_{\boldsymbol{\alpha},\boldsymbol{\beta}}$ are nonnegative and add up to 1, as they should.
- In the case $\boldsymbol{\alpha} = \boldsymbol{\beta}$ corresponding to Bohm's version of the EPR example,

$$\mu_{\boldsymbol{\alpha},\boldsymbol{\alpha}} = \begin{pmatrix} 0 & \frac{1}{2} \\ \frac{1}{2} & 0 \end{pmatrix}, \tag{4.32}$$

  implying the perfect anti-correlation $Z_2 = -Z_1$.
- The *marginal distribution* is the distribution of Z_1 alone, irrespective of Z_2. It is 1/2, 1/2. Likewise for Z_2. Let us assume that Alice's experiment occurs first. Then the fact that the marginal distribution for Z_2 is 1/2, 1/2 amounts to a *no-signaling theorem* for Bell's experiment: Bob cannot infer from Z_2 any information about Alice's choice $\boldsymbol{\alpha}$ because the distribution of Z_2 does not depend on $\boldsymbol{\alpha}$. (The general no-signaling theorem that we will prove in Sect. 5.5.9 covers all possible experiments.)
- The fact that the joint distribution of the outcomes does not depend on the order of experiments means that the observables measured by Alice and Bob can be simultaneously measured. What are these observables, actually? Alice's is the matrix $\sigma_{\boldsymbol{\alpha}} \otimes I$ with components $\sigma_{\boldsymbol{\alpha} s_1 s_1'}\delta_{s_2 s_1'}$, and Bob's is $I \otimes \sigma_{\boldsymbol{\beta}}$ with components

$\delta_{s_1 s_1'} \sigma_{\beta s_2 s_1'}$. (For deeper discussion of the tensor product $A \otimes B$ of two operators A, B, see Sect. 5.5.)

4.2.2 Bell's 1964 Proof of Nonlocality

Let us recapitulate what needs to be shown in Bell's theorem. The claim is that the joint distribution $\mu_{\boldsymbol{\alpha},\boldsymbol{\beta}}$ of Z_1 and Z_2, as a function of $\boldsymbol{\alpha}, \boldsymbol{\beta} \in \mathbb{S}^2$, is such that it cannot be created in a local way (i.e., in the absence of influences) if no information about $\boldsymbol{\alpha}$ and $\boldsymbol{\beta}$ is available beforehand.

Bell's proof has two parts. The first part is the EPR argument (in Bohm's version with spin as described in Sect. 4.1.4), applied to all directions $\boldsymbol{\alpha}$; it shows that if locality is true, then the values of Z_1 and Z_2 must have been determined in advance. In fact, in every run of the experiment, there must exist well-defined values $Z_{1\boldsymbol{\alpha}}$ *for every* $\boldsymbol{\alpha} \in \mathbb{S}^2$ and $Z_{2\boldsymbol{\alpha}} = -Z_{1\boldsymbol{\alpha}}$ even before any measurement, such that Alice's outcome will be $Z_{1\boldsymbol{\alpha}}$ for the $\boldsymbol{\alpha}$ she chooses and Bob's outcome will be $Z_{2\boldsymbol{\beta}} = -Z_{1\boldsymbol{\beta}}$ for the $\boldsymbol{\beta}$ he chooses, *also if* $\boldsymbol{\beta} \neq \boldsymbol{\alpha}$ *and independent of whether Alice's or Bob's experiment occurs first.* In particular, the measurement outcomes must be pre-determined. In sum, if locality is true, quantities $Z_{i\boldsymbol{\alpha}}$ exist in nature for every $i \in \{1, 2\}$ and every $\boldsymbol{\alpha} \in \mathbb{S}^2$ with the properties

$$Z_{i\boldsymbol{\alpha}} = \pm 1 \tag{4.33}$$

$$Z_{1\boldsymbol{\alpha}} = -Z_{2\boldsymbol{\alpha}} \tag{4.34}$$

$$\mathbb{P}(Z_{1\boldsymbol{\alpha}} = -Z_{2\boldsymbol{\beta}}) = \cos^2(\theta_{\boldsymbol{\alpha}\boldsymbol{\beta}}/2) \tag{4.35}$$

where $\theta_{\boldsymbol{\alpha}\boldsymbol{\beta}}$ is the angle between $\boldsymbol{\alpha}$ and $\boldsymbol{\beta}$; the last equation is obtained by adding the off-diagonal entries in (4.24).

Now comes the second part of the proof. It turns out sufficient to consider only three directions $\mathbf{a}, \mathbf{b}, \mathbf{c} \in \mathbb{S}^2$, among which $\boldsymbol{\alpha}$ and $\boldsymbol{\beta}$ can be chosen. Clearly,

$$\mathbb{P}\Big(\{Z_{1\mathbf{a}} = Z_{1\mathbf{b}}\} \cup \{Z_{1\mathbf{b}} = Z_{1\mathbf{c}}\} \cup \{Z_{1\mathbf{c}} = Z_{1\mathbf{a}}\}\Big) = 1\,, \tag{4.36}$$

since at least two of the three (2-valued) variables $Z_{1\boldsymbol{\alpha}}$ must have the same value. Hence, by elementary probability theory,

$$\mathbb{P}\,(Z_{1\mathbf{a}} = Z_{1\mathbf{b}}) + \mathbb{P}\,(Z_{1\mathbf{b}} = Z_{1\mathbf{c}}) + \mathbb{P}\,(Z_{1\mathbf{c}} = Z_{1\mathbf{a}}) \geq 1, \tag{4.37}$$

and using the perfect anti-correlations (4.34), we have that

$$\mathbb{P}\,(Z_{1\mathbf{a}} = -Z_{2\mathbf{b}}) + \mathbb{P}\,(Z_{1\mathbf{b}} = -Z_{2\mathbf{c}}) + \mathbb{P}\,(Z_{1\mathbf{c}} = -Z_{2\mathbf{a}}) \geq 1. \tag{4.38}$$

Equation (4.38) is equivalent to the celebrated *Bell inequality*. It is incompatible with (4.35). For example, when **a**, **b**, and **c** lie in a common plane and the angles between them are all 120°, then, since $\cos 60° = 1/2$, the three relevant quantities $\cos^2(\theta_{\alpha\beta}/2)$ are all $1/4$, implying a value of 3/4 for the left-hand side of (4.38). This completes the proof of Bell's theorem.

4.2.3 Bell's 1976 Proof of Nonlocality

Here is a different proof of nonlocality, first published in Bell (1976) [4]; it is also described in Bell (1981) [5]. It was designed for the purpose of allowing small experimental errors in all probabilities, so that the perfect anti-correlation in the case $\theta = 0$ becomes merely a *near-perfect* anti-correlation, and the conclusion of pre-existing values cannot be drawn.

We want to prove that any theory or hypothesis H about what nature does (i.e., about what happens in reality) in a Bell experiment must either violate locality or predict a joint distribution of the outcomes Z_1, Z_2 that deviates significantly from the quantum mechanical prediction (4.24). To this end, we use the following precise definition of "any theory or hypothesis" H. First, let λ denote a variable (or vector of variables) that collects the values of all physically real variables concerning the particle pair before any measurements; that is, λ describes all physical facts about the pair; it describes the physical state of the pair. For example, λ may be (ϕ, X), where ϕ is the wave function and X some hidden variable (such as Bohm's particle positions or the $Z_{i\alpha}$), or, if H denies the existence of hidden variables, λ may just be ϕ. When we repeat the experiment, λ may have different values in every run, so there will be a probability distribution[3] $\varrho(d\lambda)$. The hypothesis H can provide an arbitrary set Λ and an arbitrary probability distribution ϱ on Λ as the set of possible values and the distribution of λ. Second, after Alice and Bob have chosen their parameters $\boldsymbol{\alpha}$ and $\boldsymbol{\beta}$, nature must choose the outcomes Z_1 and Z_2. The hypothesis H has to state their joint probability distribution

$$P_{\alpha\beta\lambda}(z_1, z_2) \tag{4.39}$$

as a function $P : \mathbb{S}^2 \times \mathbb{S}^2 \times \Lambda \times \{0, 1\} \times \{0, 1\} \to [0, 1]$. If H is a deterministic theory, then these probabilities have to be either 0 or 1, but we also allow the possibility that nature chooses Z_1, Z_2 randomly. The probabilities may depend on $\boldsymbol{\alpha}$, $\boldsymbol{\beta}$, and λ. If, according to some theory, different experiments can be quantum measurements of the same observable $\boldsymbol{\alpha} \cdot \boldsymbol{\sigma} \otimes I$ and the different experiments lead to different outcome distributions given the same λ, then we assume that a particular such experiment gets chosen for each $\boldsymbol{\alpha} \in \mathbb{S}^2$ for specifying the P function. This amounts to saying

[3] We may even avoid the reference to probabilities by considering a large number of runs, say 10^4, and take ϱ to represent the relative frequencies, i.e., to be the sum of 10^4 Dirac delta peaks of amplitude 10^{-4}.

that Alice and Bob each have the choice between a variety of different experiments. In sum, specifying a hypothesis H requires specifying the set[4] Λ, the probability distribution ϱ on Λ, and the function P. Of course, P has to satisfy

$$\sum_{z_1,z_2=\pm 1} P_{\boldsymbol{\alpha}\boldsymbol{\beta}\lambda}(z_1,z_2) = 1. \tag{4.40}$$

Now the probabilities P are not necessarily observable because λ may be neither observable nor controllable. What we observe is the distribution of the pair of outcomes over many runs, and H's prediction for that is the λ-average over P,

$$\int_\Lambda \varrho(d\lambda)\, P_{\boldsymbol{\alpha}\boldsymbol{\beta}\lambda}(z_1,z_2). \tag{4.41}$$

The prediction agrees with quantum mechanics if

$$\int_\Lambda \varrho(d\lambda)\, P_{\boldsymbol{\alpha}\boldsymbol{\beta}\lambda}(z_1,z_2) = \mu_{\boldsymbol{\alpha}\boldsymbol{\beta}}(z_1,z_2). \tag{4.42}$$

When we think of λ, Z_1, Z_2 as three random variables, where λ is drawn with distribution ϱ and then Z_1, Z_2 are drawn with conditional distribution $P_{\boldsymbol{\alpha}\boldsymbol{\beta}\lambda}$, then their joint distribution $\mathbb{P}_{\boldsymbol{\alpha}\boldsymbol{\beta}}$ depends on $\boldsymbol{\alpha}$ and $\boldsymbol{\beta}$. In terms of this distribution, we can also write $\mathbb{P}_{\boldsymbol{\alpha}\boldsymbol{\beta}}(Z_1 = z_1, Z_2 = z_2)$ for (4.41), which is the marginal distribution of the pair (Z_1, Z_2). The conditional distribution $\mathbb{P}_{\boldsymbol{\alpha}\boldsymbol{\beta}}(Z_1 = z_1, Z_2 = z_2|\lambda)$ is the same as $P_{\boldsymbol{\alpha}\boldsymbol{\beta}\lambda}(z_1,z_2)$.

Next, we need a condition on the hypothesis H that characterizes locality or at least follows from locality. Let us assume locality and imagine that Alice's measurement occurs first. Nature must choose Z_1 in Alice's lab; the decision can depend on λ and the experiment chosen by Alice (i.e., on $\boldsymbol{\alpha}$), but by locality, it cannot depend on Bob's choice $\boldsymbol{\beta}$. That is, the marginal distribution of Z_1 given λ,

$$\mathbb{P}_{\boldsymbol{\alpha}\boldsymbol{\beta}}(Z_1 = z_1|\lambda) = \sum_{z_2=\pm 1} P_{\boldsymbol{\alpha}\boldsymbol{\beta}\lambda}(z_1,z_2)\,, \tag{4.43}$$

must be independent of $\boldsymbol{\beta}$. Put differently, there is a function $f_1(\boldsymbol{\alpha}, z_1, \lambda)$ such that

$$\mathbb{P}_{\boldsymbol{\alpha}\boldsymbol{\beta}}(Z_1 = z_1|\lambda) = f_1(\boldsymbol{\alpha}, z_1, \lambda)\,. \tag{4.44}$$

Likewise, after nature in Alice's lab has chosen Z_1, nature in Bob's lab has to choose Z_2, but by locality, nature in Bob's lab knows neither Z_1 nor $\boldsymbol{\alpha}$; the choice of Z_2 can only depend on λ and $\boldsymbol{\beta}$. That is, the conditional distribution of Z_2 given Z_1 and λ

[4] For mathematicians: a measurable space

must be independent of $\boldsymbol{\alpha}$ and Z_1,

$$\mathbb{P}_{\boldsymbol{\alpha\beta}}(Z_2 = z_2 | Z_1 = z_1, \lambda) = \frac{P_{\boldsymbol{\alpha\beta}\lambda}(z_1, z_2)}{\sum_{z_2'=\pm 1} P_{\boldsymbol{\alpha\beta}\lambda}(z_1, z_2')} = f_2(\boldsymbol{\beta}, z_2, \lambda) \tag{4.45}$$

for some function f_2. From the last two equations together, we obtain that

$$P_{\boldsymbol{\alpha\beta}\lambda}(z_1, z_2) = f_1(\boldsymbol{\alpha}, z_1, \lambda)\, f_2(\boldsymbol{\beta}, z_2, \lambda)\,. \tag{4.46}$$

So, this factorization property is a consequence of locality. We will show that there is no H satisfying both (4.42) and (4.46); in fact, we will show that if H satisfies (4.46), its prediction (4.41) cannot even be close to (4.42).

Note that the random quantities Z_1 and Z_2 can very well be *dependent*: their independence would amount to

$$\int_\Lambda \varrho(d\lambda)\, P_{\boldsymbol{\alpha\beta}\lambda}(z_1, z_2) = g_1(\boldsymbol{\alpha}, \boldsymbol{\beta}, z_1)\, g_2(\boldsymbol{\alpha}, \boldsymbol{\beta}, z_2) \tag{4.47}$$

instead of (4.46). The locality condition (4.46) implies that Z_1 and Z_2 are *conditionally independent, given* λ; this is different from being independent. A dependence between Z_1 and Z_2 can arise from the common cause λ. As a famous example, suppose I notice during the day that I left one glove at home; by observing whether the glove in my pocket is right-handed or left-handed, I immediately know the handedness of the glove at home, but no superluminal influence is involved; the analog of (4.46) (without $\boldsymbol{\alpha}$ and $\boldsymbol{\beta}$) is not violated because λ includes the information about which glove is at home and which in my pocket.

Now we draw conclusions about how the locality condition (4.46) restricts which probability distributions can occur as the observed distribution (4.41). To this end, we introduce the *correlation coefficient* defined by

$$\kappa(\boldsymbol{\alpha}, \boldsymbol{\beta}) = \sum_{z_1=\pm 1} \sum_{z_2=\pm 1} z_1 z_2 \mathbb{P}_{\boldsymbol{\alpha\beta}}(Z_1 = z_1, Z_2 = z_2)\,. \tag{4.48}$$

Proposition 4.2 *Locality implies the following version of Bell's inequality known as the CHSH inequality:*[5]

$$\left|\kappa(\boldsymbol{\alpha}, \boldsymbol{\beta}) + \kappa(\boldsymbol{\alpha}, \boldsymbol{\beta}') + \kappa(\boldsymbol{\alpha}', \boldsymbol{\beta}) - \kappa(\boldsymbol{\alpha}', \boldsymbol{\beta}')\right| \leq 2\,. \tag{4.49}$$

[5] The name refers to Clauser, Horne, Shimony, and Holt; in their paper (1969) [9], this inequality was first mentioned, though with a different derivation making the stronger assumption that Z_1, Z_2 are deterministic functions of $\boldsymbol{\alpha}, \boldsymbol{\beta}, \lambda$. By the way, a correlation coefficient can be defined generally for two real-valued random variables and expresses how strongly they tend to deviate from their means together in the same direction; however, I do not know of an intuitive interpretation of the particular combination of four correlation coefficients that appears in this inequality (4.49).

Proof The locality condition (4.46) implies that

$$\kappa(\boldsymbol{\alpha}, \boldsymbol{\beta}) = \int \varrho(d\lambda) \sum_{z_1=\pm 1} \sum_{z_2=\pm 1} z_1 z_2 \mathbb{P}_{\boldsymbol{\alpha}\boldsymbol{\beta}}(Z_1 = z_1, Z_2 = z_2 | \lambda) \tag{4.50a}$$

$$= \int \varrho(d\lambda) \sum_{z_1=\pm 1} \sum_{z_2=\pm 1} z_1 z_2 f_1(\boldsymbol{\alpha}, z_1, \lambda)\, f_2(\boldsymbol{\beta}, z_2, \lambda) \tag{4.50b}$$

$$= \int \varrho(d\lambda)\, E_1(\boldsymbol{\alpha}, \lambda)\, E_2(\boldsymbol{\beta}, \lambda) \tag{4.50c}$$

with

$$E_1(\boldsymbol{\alpha}, \lambda) = \sum_{z_1=\pm 1} z_1\, f_1(\boldsymbol{\alpha}, z_1, \lambda) \quad \text{and} \quad E_2(\boldsymbol{\beta}, \lambda) = \sum_{z_2=\pm 1} z_2\, f_2(\boldsymbol{\beta}, z_2, \lambda). \tag{4.51}$$

Since $Z_i \in \{1, -1\}$, we have that

$$\Big| E_1(\boldsymbol{\alpha}, \lambda) \Big| \leq 1 \quad \text{and} \quad \Big| E_2(\boldsymbol{\beta}, \lambda) \Big| \leq 1\,. \tag{4.52}$$

So,

$$\Big| \kappa(\boldsymbol{\alpha}, \boldsymbol{\beta}) \pm \kappa(\boldsymbol{\alpha}, \boldsymbol{\beta}') \Big| = \left| \int \varrho(d\lambda)\, E_1(\boldsymbol{\alpha}, \lambda) \Big(E_2(\boldsymbol{\beta}, \lambda) \pm E_2(\boldsymbol{\beta}', \lambda) \Big) \right| \tag{4.53a}$$

$$\leq \int \varrho(d\lambda) \left| E_1(\boldsymbol{\alpha}, \lambda) \right| \left| E_2(\boldsymbol{\beta}, \lambda) \pm E_2(\boldsymbol{\beta}', \lambda) \right| \tag{4.53b}$$

$$\leq \int \varrho(d\lambda) \left| E_2(\boldsymbol{\beta}, \lambda) \pm E_2(\boldsymbol{\beta}', \lambda) \right|. \tag{4.53c}$$

Now for any $u, v \in [-1, 1]$,

$$|u + v| + |u - v| \leq 2 \tag{4.54}$$

because

$$(u + v) + (u - v) = 2u \leq 2 \qquad (-u - v) + (u - v) = -2v \leq 2 \tag{4.55}$$

$$(u + v) + (v - u) = 2v \leq 2 \qquad (-u - v) + (v - u) = -2u \leq 2\,. \tag{4.56}$$

That is, regardless of the sign of $u + v$ and that of $u - v$, (4.54) is satisfied. Hence, setting $u = E_2(\boldsymbol{\beta}, \lambda)$ and $v = E_2(\boldsymbol{\beta}', \lambda)$,

$$\Big|\kappa(\boldsymbol{\alpha}, \boldsymbol{\beta}) + \kappa(\boldsymbol{\alpha}, \boldsymbol{\beta}') + \kappa(\boldsymbol{\alpha}', \boldsymbol{\beta}) - \kappa(\boldsymbol{\alpha}', \boldsymbol{\beta}')\Big|$$

$$\leq \Big|\kappa(\boldsymbol{\alpha}, \boldsymbol{\beta}) + \kappa(\boldsymbol{\alpha}, \boldsymbol{\beta}')\Big| + \Big|\kappa(\boldsymbol{\alpha}', \boldsymbol{\beta}) - \kappa(\boldsymbol{\alpha}', \boldsymbol{\beta}')\Big| \tag{4.57a}$$

$$\overset{(4.53c)}{\leq} \int \varrho(d\lambda) \left(\Big|E_2(\boldsymbol{\beta}, \lambda) + E_2(\boldsymbol{\beta}', \lambda)\Big| + \Big|E_2(\boldsymbol{\beta}, \lambda) - E_2(\boldsymbol{\beta}', \lambda)\Big| \right) \tag{4.57b}$$

$$\overset{(4.54)}{\leq} 2\,. \tag{4.57c}$$

□

Since the quantum mechanical prediction $\mu_{\boldsymbol{\alpha},\boldsymbol{\beta}}$ for the Bell experiment has

$$\kappa(\boldsymbol{\alpha}, \boldsymbol{\beta}) = \mu_{\boldsymbol{\alpha},\boldsymbol{\beta}}(++) - \mu_{\boldsymbol{\alpha},\boldsymbol{\beta}}(+-) - \mu_{\boldsymbol{\alpha},\boldsymbol{\beta}}(-+) + \mu_{\boldsymbol{\alpha},\boldsymbol{\beta}}(--) = -\boldsymbol{\alpha} \cdot \boldsymbol{\beta} = -\cos\theta\,, \tag{4.58}$$

setting (in some plane)

$$\boldsymbol{\alpha} = 0^\circ\,, \quad \boldsymbol{\alpha}' = 90^\circ\,, \quad \boldsymbol{\beta} = 45^\circ\,, \quad \boldsymbol{\beta}' = -45^\circ \tag{4.59}$$

leads to

$$\kappa(\boldsymbol{\alpha}, \boldsymbol{\beta}) + \kappa(\boldsymbol{\alpha}, \boldsymbol{\beta}') + \kappa(\boldsymbol{\alpha}', \boldsymbol{\beta}) - \kappa(\boldsymbol{\alpha}', \boldsymbol{\beta}') = -2\sqrt{2}\,, \tag{4.60}$$

violating (4.49). This completes the 1976 proof of Bell's theorem. □

Remark 4.2

- *Inaccuracy.* If the values of $\mathbb{P}_{\boldsymbol{\alpha}\boldsymbol{\beta}}(Z_1 = z_1, Z_2 = z_2)$ are known only with some inaccuracy (because they were obtained experimentally, not from the quantum formalism), then also the $\kappa(\boldsymbol{\alpha}, \boldsymbol{\beta})$ are subject to some inaccuracy. But if (4.49) is violated by more than the inaccuracy, then locality is refuted.
- *Photons.* Experimental tests of Bell's inequality are usually done with photons instead of electrons. For photons, spin is usually called *polarization*, and the Stern–Gerlach magnets are replaced with *polarization analyzers* (also known as *polarizers*), i.e., crystals that are transparent to the $|z\text{-up}\rangle$ part of the wave but reflect (or absorb) the $|z\text{-down}\rangle$ part. Like the Stern–Gerlach magnets, the analyzers can be rotated into any direction. Since photons have spin 1, $\theta/2$ needs to be replaced by θ.

- *Analogy.* The mathematical content of Bell's theorem can also be expressed as follows. Suppose two computers A and B are set up in such a way that, upon input of $\boldsymbol{\alpha}$ into A and $\boldsymbol{\beta}$ into B, A produces a random number $Z_1 = \pm 1$ and B $Z_2 = \pm 1$. Suppose further that A and B cannot communicate, but they have access to some shared information (possibly including the code of some algorithm or some random bits). Then the joint distribution of Z_1, Z_2 cannot agree with $\mu_{\boldsymbol{\alpha},\boldsymbol{\beta}}$. (Note that the communication between the computers corresponds to an influence between Alice's and Bob's experiments, not to communication between Alice and Bob.)

4.3 Discussion of Nonlocality

4.3.1 Nonlocality in Bohmian Mechanics, GRW, Copenhagen, and Many-Worlds

Since we have considered only non-relativistic formulations of these theories, we cannot directly analyze spacelike separated events, but instead we can analyze the case of two systems (e.g., Alice's lab and Bob's lab) without an interaction term between them in the Hamiltonian H.

Bohmian Mechanics Bohmian mechanics is explicitly nonlocal, as the velocity of particle 2 depends on the position of particle 1, *no matter how distant and no matter whether there is an interaction term in the Hamiltonian.* That is where the superluminal influence occurs. (Historically, Bell's nonlocality analysis was inspired by the examination of Bohmian mechanics.)

This influence depends on entanglement: In the absence of entanglement, the velocity of particle 2 is independent of the position of particle 1. The fact that Bohmian mechanics is local for disentangled wave functions shows that it was necessary for proving nonlocality to consider at least two particles and an entangled wave function (such as the singlet state). It can be shown that any entangled wave function violates Bell's inequality for some observables.

Furthermore, the position of particle 1 will depend on the external fields at work near particle 1. That is, for any given initial position of particle 1, its later position will depend on the external fields. An example of an external field is the field of the Stern–Gerlach magnet. To a large extent, we can control external fields at our whim; e.g., we can rotate the Stern–Gerlach magnet. Bohm's equation of motion implies that these fields have an instantaneous influence on the motion of particle 2.

GRW Theory In GRW theories, nonlocality comes in at the point when the wave function collapses, as then it does so *instantaneously over arbitrary distances*.

At least, this trait of the theory *suggests* that GRW is nonlocal, and in fact, that is the ultimate source of the nonlocality. Strictly speaking, however, the definition of nonlocality, i.e., the negation of (4.18), requires that *events* at x and at y influence each other and the value of the wave function $\psi_t(\boldsymbol{x}_1, \boldsymbol{x}_2)$ is linked to *several* space-time points, $(t, \boldsymbol{x}_1)$ and $(t, \boldsymbol{x}_2)$, and thus is not an example of an "event at x." So we need to formulate the proof that GRW theory is nonlocal more carefully; of course, Bell's proof achieves this, but we can give a more direct proof. Since the "events at x" are not given by the wave function itself but by the primitive ontology, we need to consider GRWf and GRWm separately.

In GRWf, consider Einstein's boxes example. The wave function of a particle is half in a box in Paris and half in a box in Tokyo. Let us apply detectors to both boxes at time t and consider the macroscopic superposition of the detectors arising from the Schrödinger equation. It is random whether the first flash (in any detector) after t occurs in Paris or in Tokyo. Suppose it occurs in Tokyo, and suppose it can occur in one of two places in Tokyo, corresponding to the outcomes 0 or 1. If it was 1, then after the collapse, the wave function of the particle is 100% in Tokyo, and later flashes in Paris are certain to occur in a place where they indicate the outcome 0—that is a nonlocal influence of a flash in Tokyo on the flashes in Paris.

Likewise in GRWm: If, after the first collapse, the pointer of the detector in Tokyo, according to the m function, points to 1, then the pointer in Paris immediately points to 0. (You might object that the Tokyo pointer position according to the m function was not the *cause* of the Paris pointer position, but rather both pointer positions were caused by the collapse of the wave function. However, this distinction is not relevant to whether the theory is nonlocal.)

Note that while Bell's proof shows that *any* version of quantum mechanics must be nonlocal, for proving that *GRWf and GRWm* are nonlocal, it is sufficient to consider a simpler situation, that of Einstein's boxes.

Both GRWf and GRWm are already nonlocal when governing a universe containing only one particle; thus, their nonlocality does not depend on the existence of a macroscopic number of particles, and they are even nonlocal in a case (one particle) in which Bohmian mechanics is local. For example, consider a particle with wave function

$$\psi = \frac{1}{\sqrt{2}}\big(|\text{here}\rangle + |\text{there}\rangle\big) \tag{4.61}$$

at time t, as in Einstein's boxes example. Suppose that $|\text{here}\rangle$ and $|\text{there}\rangle$ are two narrow wave packets separated by a distance of 500 million light years. The distance is so large that the first collapse is likely to occur before a light signal can travel between the two places. For GRWf, a flash here precludes a flash there—that is a nonlocal influence. For GRWm, if the wave function collapses to $|\text{here}\rangle$, then m (here) doubles, and m(there) instantaneously goes to zero—that is a nonlocal influence. (As described in Sect. 7.8.4, there is a relativistic version of GRWm in which m(there) goes to zero only after a delay of distance/c or when a collapse centered "there" occurs. Nevertheless, also this theory is nonlocal even for one particle because when a collapse centered "there" occurs, which can happen any time, then m(there) cannot double (as it could in a local theory) but must go to zero.)

Orthodox Quantum Mechanics (OQM) That OQM is nonlocal can also be seen from Einstein's boxes argument: OQM says the outcomes of the detectors are not pre-determined. (That is, there is no fact about where the particle really is before any detectors are applied.) Thus, the outcome of the Tokyo detector must have influenced the Paris detector, or vice versa.

This, of course, was the point of Einstein's boxes argument: He objected to OQM because it is nonlocal.

Many Worlds Many-worlds theories are nonlocal, too. This is not obvious from Bell's argument because the latter is formulated in a single-world framework. Here is why Sm is nonlocal (Allori et al. 2011 [1]). After Alice carries out her Stern–Gerlach experiment, there are two pointers in her lab, one pointing to $+1$ and the other to -1. Then Bob carries out his experiment, and there are two pointers in his lab. Suppose Bob chose the same direction as Alice. Then the world in which Alice's pointer points to $+1$ is the same world as the one in which Bob's pointer points to -1, and this nonlocal fact was created in a nonlocal way by Bob's experiment. The same kind of nonlocality occurs in Sm already in Einstein's boxes experiment: The world in which a particle was detected in Paris is the same as the one in which no particle was detected in Tokyo—a nonlocal fact that arises as soon as both experiments are completed, without the need to wait for the time it takes light to travel from Paris to Tokyo.

How about Bell's many-worlds theories? The second theory, involving a random configuration selected independently at every time, is very clearly nonlocal, for example, in Einstein's boxes experiment: At every time t, nature makes a random decision about whether the particle is in Paris, and if it is, nature ensures immediately that there is no particle in Tokyo. A local theory would require that the particle has a continuous history of traveling, at a speed less than that of light, to either Paris or Tokyo, and this history is missing in Bell's second many-worlds theory. Bell's first many-worlds theory is even more radical, in fact in such a way that the concept of locality is not even applicable. The concept of locality requires that at every point in space, there are local variables whose changes propagate at most at the speed of light. Since in Bell's first many-worlds theory, no association is made between worlds at different times, one cannot even ask how any local variables would change with time. Thus, this theory is nonlocal as well.

4.3.2 Popular Myths About Bell's Theorem

Let P be the statement that, prior to any experiment, there exist values Z_{in} (for all $i = 1, 2$ and $\boldsymbol{n} \in \mathbb{S}^2$) such that Alice and Bob obtain as outcomes $Z_{1\alpha}$ and $Z_{2\beta}$ (the "hidden variables"). Then Bell's 1964 proof, described in Sect. 4.2.2, has the following structure:

$$\text{Part 1 (EPR):} \quad \text{QM} + \text{locality} \Rightarrow \text{P} \tag{4.62}$$

$$\text{Part 2:} \quad \text{QM} \Rightarrow \text{not P} \tag{4.63}$$

$$\text{Conclusion:} \quad \text{QM} \Rightarrow \text{not locality} \tag{4.64}$$

Here, QM means the statement that the statistics of outcomes is given by (4.24).

Certain popular myths about Bell's proof arise from missing part 1 and noticing only part 2 of the argument. (In Bell's (1964) [3] paper, part 1 is formulated in 3 lines and part 2 in 2.5 pages.) Before explaining this point, I mention that already Bell (1981, [5] or [6, p. 143]) had complained about people misunderstanding the EPR argument:

> It is important to note that to the limited degree to which *determinism* plays a role in the EPR argument, it is not assumed but *inferred*. What is held sacred is the principle of 'local causality' – or 'no action at a distance'. [...] It is remarkably difficult to get this point across, that determinism is not a *presupposition* of the analysis.

Here, "determinism" means P, and "local causality" means locality. What Bell wrote about the EPR argument is equally true about his own 1964 nonlocality proof: P plays a "limited role" because it is only an auxiliary statement. While part 2 can be summarized as assuming P (and QM) and deriving a contradiction, the whole argument consisting of part 1 and part 2 together does *not* introduce P as an *assumption* (but infers P); moreover, not-P is not the final conclusion, but not-locality is.

The mistake of missing part 1 leads to the impression that Bell proved that

$$\text{hidden variables are impossible} \tag{4.65}$$

or that

$$\text{hidden variables, while perhaps possible, must be nonlocal.} \tag{4.66}$$

These claims are still widespread and were even more common in the twentieth century.[6] They are convenient for Copenhagenists, who tend to think that coherent theories of the microscopic realm are impossible (see Sect. 3.4.2). Let me explain what is wrong about (4.65) and (4.66).

Statement (4.65) is plainly wrong, since a deterministic hidden-variable theory exists and works, namely, Bohmian mechanics. Bell did not establish the impossi-

[6] For example, recall the title of Clauser et al.'s (1969) [9] paper: "Proposed Experiment to Test Local Hidden-Variable Theories." In the same vein, other authors claimed that Bell's argument excludes "local realism."

bility of a deterministic reformulation of quantum theory, nor did he ever claim to have done so.

Statement (4.66) is true and non-trivial but nonetheless rather misleading. It follows from (4.62) and (4.63) that *any* (single-world) account of quantum phenomena must be nonlocal, not just any hidden-variable account. (Many-world theories are nonlocal as well according to the considerations described in Sect. 4.3.1.) For example, Bell's theorem implies that collapse theories, if their predictions are sufficiently close to the standard quantum formalism (as they are for GRW theory), must be nonlocal, even though in these theories the wave function provides the complete information about a system. In fact, Bell's theorem shows that nonlocality follows from the predictions of the standard quantum formalism. Since experiments have confirmed that nature is indeed governed by these predictions, we must conclude that *nature is nonlocal*.

This kind of misconception about the meaning of Bell's theorem has also arisen from the symbol λ in Bell's 1976 proof: since it is natural to call λ the "hidden variable," it is easy to get the impression that the reasoning concerns hidden-variable theories only. In fact, since the possible values of λ are up to the arbitrary hypothesis H admitted in the proof, λ may or may not involve hidden variables, and the proof equally covers theories with or without hidden variables.

Let us have another look at Bohmian mechanics in this context. Bohmian mechanics has hidden variables, of course, but not the hidden variables $Z_{i\alpha}$ that the EPR argument yields. Then again, once we have chosen, for every $\boldsymbol{\alpha}$, a particular experiment that is a quantum measurement of $\boldsymbol{\alpha} \cdot \boldsymbol{\sigma}$ (let us say, the Stern–Gerlach experiment in direction $\boldsymbol{\alpha}$[7]) and chosen whether Alice's or Bob's experiment takes place first (say, Alice's), the pair of outcomes (Z_1, Z_2) according to Bohmian mechanics is deterministic. It is given by pre-determined hidden variables. And yet, these are still not the hidden variables $Z_{i\alpha}$ of EPR! Rather, they are of the form $Z_{1\alpha}$ and $Z_{2\alpha\beta}$, that is, those on Bob's side depend on Alice's choice $\boldsymbol{\alpha}$. This fact reflects the nonlocality of Bohmian mechanics. When considering $\boldsymbol{\alpha}, \boldsymbol{\beta}$ from among the 3 directions **a**, **b**, **c** violating the Bell inequality (4.38), then instead of the 6 random variables $Z_{i\alpha}$, we have the 12 random variables $Z_{1\alpha}$ and $Z_{2\alpha\beta}$, whose existence is not excluded by Bell's proof.

For further discussion of controversies about Bell's theorem, see Goldstein et al. (2011) [13], Maudlin (2014) [16], or Tumulka (2016) [18].

4.3.3 Simultaneous Quantum Measurements

In Bell's experiment, Alice's and Bob's observables, $\boldsymbol{\alpha} \cdot \boldsymbol{\sigma} \otimes I$ and $I \otimes \boldsymbol{\beta} \cdot \boldsymbol{\sigma}$, commute with each other, so they can be simultaneously measured and possess a joint distribution, which is the distribution $\mu_{\alpha,\beta}$ given in Proposition 4.1. This terminology of "simultaneous measurement" is somewhat misleading as it sounds

[7] Without the inversion considered in Sect. 2.3.9.

as if these two experiments did not influence each other. We now know from Bell's theorem that they are actually not that innocent.

Here is another way in which this terminology is misleading. Suppose quantum measurements of A, B with $AB = BA$ are carried out, A first and B thereafter. We know that the distribution of outcomes would have been the same if we had done the experiments in the other order. But the word "simultaneous" (as it suggests that order is irrelevant) makes it tempting to even think that

$$\text{the outcomes would have been the same if we had done the experiments in the other order.} \tag{4.67}$$

However, this does not follow from the fact that the distribution would have been the same. After all, if Alice rolls a die and Bob rolls a die, then the two outcomes have the same distribution (viz., uniform on $\{1, 2, 3, 4, 5, 6\}$), but they are in general not equal. I call (4.67) the *fallacy of simultaneous measurement*. In Sect. 4.4.2, we will encounter a paradox in which this fallacy plays a role.

Apart from seeing the logical gap in the fallacy, we may find it useful to see a counter-example. Here is how the outcomes of Bell's experiment arise in (non-relativistic) Bohmian mechanics from the initial configuration of the two particles. Suppose Alice first applies a Stern–Gerlach magnetic field in direction $\boldsymbol{\alpha}$ (but no detector yet); then the wave packet of the two particles in six-dimensional configuration splits in two packets of equal size, and the Bohmian trajectories will stay on the same side of the symmetry plane; that is, the outcome is $Z_1 = +1$ if and only if $\boldsymbol{\alpha} \cdot \boldsymbol{X}_1$ is greater than a critical value. Now suppose Bob applies a Stern–Gerlach magnetic field in direction $\boldsymbol{\beta}$; then each of the two wave packets splits again in two packets of sizes given by the appropriate column of $\mu_{\alpha,\beta}$ as in (4.24). (It does not matter whether Alice's detector gets applied before or after Bob's magnet.) When a packet splits, the trajectory follows the "up" packet if and only if $\boldsymbol{\beta} \cdot \boldsymbol{X}_2$ is greater than a critical value. The dependence of the outcomes Z_1, Z_2 on the initial configuration $(\boldsymbol{X}_1, \boldsymbol{X}_2)$ is summarized in Fig. 4.3. The issue will come up again in Sect. 5.7 on no-hidden-variable theorems.

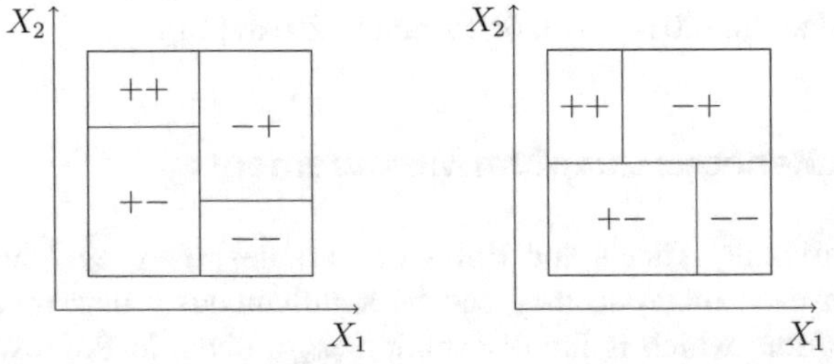

Fig. 4.3 Dependence of the outcome pair $(Z_1, Z_2) \in \{+1, -1\}^2$ of a Bell experiment with $\theta = 70°$ on the initial Bohmian position coordinates $X_1 := \boldsymbol{\alpha} \cdot \boldsymbol{X}_1$ and $X_2 := \boldsymbol{\beta} \cdot \boldsymbol{X}_2$: Left, if Alice's experiment is done first; Right, if Bob's is done first

4.4 Special Topics

4.4.1 Bohr's Reply to EPR

Let us go back once more to EPR. Bohr (1935) [8] wrote a reply to EPR, which was published the same year in the same journal under the same title as EPR's paper. Before we look at it, let us pause for a moment and think about what kind of reply would be possible. EPR argued, assuming locality, that the wave function is incomplete. Given that Bohr insisted that the wave function was complete, he had two options: either argue that there is a mistake in EPR's argument or deny locality, the premise of EPR's argument.

It is hard to make sense of Bohr's reply. It is even hard to say which of the two options he chose. Here is what he wrote. He referred to the following sentence of EPR (1935) [11] that expresses the version of locality that they used in their argument:

> If, without in any way disturbing a system, we can predict with certainty (i.e., with probability equal to unity) the value of a physical quantity, then there exists an element of physical reality corresponding to this physical quantity.

About this, Bohr wrote:

> the wording of the above-mentioned criterion of physical reality proposed by Einstein, Podolsky and Rosen contains an ambiguity as regards the meaning of the expression 'without in any way disturbing a system.' Of course there is in a case like that just considered no question of a mechanical disturbance of the system under investigation during the last critical stage of the measuring procedure. But even at this stage there is essentially the question of *an influence on the very conditions which define the possible types of predictions regarding the future behavior of the system.* [emphasis in the original]

Let us try to understand what he is saying. Clearly, an ambiguity in terminology can easily lead to a mistake in an argument: if you show that A implies B, if you then change the meaning of a word in B so B becomes B', and if you then show that B' implies C, then you have not shown that A implies C. On the other hand, somebody could point to an ambiguity to express that the hypothesis is less plausible than EPR thought, for example, if B is plausible but the argument actually assumes B'. (So talking about an ambiguity can be relevant in both cases, if Bohr sees a mistake in the argument or if he denies the premise.)

In any case, what is the ambiguity? In the quotation above, Bohr offered two readings of the assumption of locality. *Reading 1*, which Bohr agreed with, says that there is no "mechanical disturbance of the system," which sounds very much like saying that there is no change in the physical state of particle 2 as a consequence of Alice's interaction with particle 1. That is, Bohr appears to have agreed with the hypothesis of locality. *Reading 2*, which he disagreed with, is about another kind of influence, on "conditions ... of predictions." I am not sure what that means. Here is a possibility:[8] Elsewhere in his article, Bohr emphasized that measurements of different observables (such as position and momentum) require different experimental setups. Maybe these setups are the "conditions of predictions." Thus, he seems to be saying that EPR's trick for measuring position and momentum simultaneously is *not* a serious simultaneous measurement of position and momentum. However, even if Bohr were right about this, something would not fit together: although EPR talked about position and momentum measurements in the later part of their paper, their basic argument concerns only positions. So no different setups are involved. It seems that Bohr missed that and did not actually answer EPR's basic argument.

After the passage I quoted, Bohr continued,

> Since these conditions constitute an inherent element of the description of any phenomenon to which the term 'physical reality' can be properly attached, we see that the argumentation of the mentioned authors does not justify their conclusion that quantum-mechanical description is essentially incomplete.

This sounds positivistic, as if Bohr was unwilling to consider physical reality itself but kept sliding instead into considering what observers know or how they would describe phenomena. However, one cannot understand EPR's reasoning without thinking about how reality is affected by Alice's actions.

It seems that *reading 1* is what EPR assumed, and I do not see them switch the meaning in the middle of the argument. After all, their argument is very short, and there is only one step that makes use of the locality assumption. I am left wondering whether Bohr understood EPR's argument.

4.4.2 The Frauchiger–Renner Paradox

In this section, we come back to the central topics of Chap. 3: collapse and measurement. The considerations here are not directly related to nonlocality, they rather circle around the measurement process, but they parallel some elements of our discussion in this chapter.

Frauchiger and Renner (2018) [12] gave an argument which, if it were correct, would show that every single-world no-collapse theory of quantum mechanics must be inconsistent. We know already that Bohmian mechanics is a consistent single-world no-collapse theory in agreement with the quantum formalism, so the argument of Frauchiger and Renner must be mistaken. I now describe the argument and point out where it commits Wheeler's fallacy (see Sect. 1.6.6). A similar analysis of the paradox was provided by Lazarovici and Hubert (2019) [14].

[8] Pointed out on pages 129–131 in Maudlin (2011) [15]

We consider two quantum systems C and S, each with a two-dimensional Hilbert space, and four observers $F, \overline{F}, W, \overline{W}$. At time t_0, C gets prepared with the wave function

$$\sqrt{\tfrac{1}{3}}\,|\mathrm{h}\rangle_C + \sqrt{\tfrac{2}{3}}\,|\mathrm{t}\rangle_C\,, \tag{4.68}$$

where $\{|\mathrm{h}\rangle_C, |\mathrm{t}\rangle_C\}$ is an ONB of $\mathscr{H}_C$. At $t_1 > t_0$, $\overline{F}$ carries out a quantum measurement with eigenbasis $\{|\mathrm{h}\rangle_C, |\mathrm{t}\rangle_C\}$. If the outcome Z_C is h, then $\overline{F}$ prepares S with the wave function $\big|\downarrow\big\rangle_S = |z\text{-down}\rangle$; if the outcome Z_C is t, then $\overline{F}$ prepares S with the wave function $\big|\rightarrow\big\rangle_S = |x\text{-up}\rangle$. At $t_2 > t_1$, F carries out a quantum measurement of σ_z on S with outcome $Z_S = \pm$.

At $t_3 > t_2$, $\overline{W}$ carries out a quantum measurement on $C \cup \overline{F}$. Now it is important that we, in our analysis, do *not* collapse wave functions when F and $\overline{F}$ do their measurements but keep the full wave function of the universe, including parts corresponding to outcomes that did not occur; that is analogous to our analysis of the full wave function in our discussion of the measurement problem in Sect. 3.2. (And that would not be correct in a collapse theory.) Explicitly, right after t_1, the wave function of $C \cup \overline{F} \cup S$ is

$$\sqrt{\tfrac{1}{3}}\,|\mathrm{h}\rangle_C|\mathrm{H}\rangle_{\overline{F}}\big|\downarrow\big\rangle_S + \sqrt{\tfrac{2}{3}}\,|\mathrm{t}\rangle_C|\mathrm{T}\rangle_{\overline{F}}\big|\rightarrow\big\rangle_S\,, \tag{4.69}$$

where $|\mathrm{H}\rangle_{\overline{F}}$ is a wave function of $\overline{F}$ (and her lab) according to which she remembers having observed the outcome h and her records of the outcome show h and correspondingly $|\mathrm{T}\rangle_{\overline{F}}$ for outcome t. Now $\overline{W}$ performs a very curious quantum measurement: not (as did the super-observer in the situation of the measurement problem in Sect. 3.2.3) of the macroscopic state of $\overline{F}$'s records, but one with eigenbasis

$$\Big\{|\overline{\mathrm{ok}}\rangle_{C\overline{F}} := \tfrac{1}{\sqrt{2}}|\mathrm{h}\rangle_C|\mathrm{H}\rangle_{\overline{F}} - \tfrac{1}{\sqrt{2}}|\mathrm{t}\rangle_C|\mathrm{T}\rangle_{\overline{F}}\,, \tag{4.70a}$$

$$|\overline{\mathrm{fail}}\rangle_{C\overline{F}} := \tfrac{1}{\sqrt{2}}|\mathrm{h}\rangle_C|\mathrm{H}\rangle_{\overline{F}} + \tfrac{1}{\sqrt{2}}|\mathrm{t}\rangle_C|\mathrm{T}\rangle_{\overline{F}}\Big\}\,. \tag{4.70b}$$

This is a measurement that requires bringing $|\mathrm{h}\rangle|\mathrm{H}\rangle$ and $|\mathrm{t}\rangle|\mathrm{T}\rangle$ back together in configuration space, making them overlap and interfere. I said in Sect. 3.2.3 that this is for all practical purposes impossible for macroscopic systems such as $\overline{F}$; I said macroscopically disjoint wave packets will remain disjoint practically forever—this is decoherence. But this practical difficulty is not my objection to Frauchiger and Renner. Let us assume with Frauchiger and Renner, hypothetically and as a thought experiment, that $\overline{W}$ *can* carry out a measurement corresponding to (4.70), and let us consider what will happen: Right after t_2 (before $\overline{W}$'s measurement), the wave function of $C \cup \overline{F} \cup S \cup F$ is

$$\begin{aligned} &\sqrt{\tfrac{1}{3}}\,|\mathrm{h}\rangle_C|\mathrm{H}\rangle_{\overline{F}}|\downarrow\rangle_S|-\rangle_F \\ &+\sqrt{\tfrac{1}{3}}\,|\mathrm{t}\rangle_C|\mathrm{T}\rangle_{\overline{F}}|\uparrow\rangle_S|+\rangle_F \\ &+\sqrt{\tfrac{1}{3}}\,|\mathrm{t}\rangle_C|\mathrm{T}\rangle_{\overline{F}}|\downarrow\rangle_S|-\rangle_F\,, \end{aligned} \tag{4.71}$$

where $|+\rangle_F$ is a wave function of F (and her lab) according to which she remembers having observed the outcome + and her records of the outcome show + and correspondingly $|-\rangle_F$ for outcome $-$. Right after t_3 (after $\overline{W}$'s measurement), the wave function of $C \cup \overline{F} \cup S \cup F \cup \overline{W}$ is, as a computation shows,

$$\begin{aligned} &\sqrt{\tfrac{1}{3}}\Big(|\mathrm{h}\rangle_C|\mathrm{H}\rangle_{\overline{F}}+|\mathrm{t}\rangle_C|\mathrm{T}\rangle_{\overline{F}}\Big)|\downarrow\rangle_S|-\rangle_F|\overline{\mathrm{FAIL}}\rangle_{\overline{W}} \\ &+\sqrt{\tfrac{1}{12}}\Big(|\mathrm{h}\rangle_C|\mathrm{H}\rangle_{\overline{F}}+|\mathrm{t}\rangle_C|\mathrm{T}\rangle_{\overline{F}}\Big)|\uparrow\rangle_S|+\rangle_F|\overline{\mathrm{FAIL}}\rangle_{\overline{W}} \\ &-\sqrt{\tfrac{1}{12}}\Big(|\mathrm{h}\rangle_C|\mathrm{H}\rangle_{\overline{F}}-|\mathrm{t}\rangle_C|\mathrm{T}\rangle_{\overline{F}}\Big)|\uparrow\rangle_S|+\rangle_F|\overline{\mathrm{OK}}\rangle_{\overline{W}} \end{aligned} \tag{4.72}$$

with a self-explanatory notation for the states of $\overline{W}$ corresponding to the possible outcomes obtained by $\overline{W}$.

Finally, at $t_4 > t_3$, W carries out a quantum measurement on $S \cup F$ with eigenbasis

$$\Big\{|\mathrm{ok}\rangle_{SF} := \tfrac{1}{\sqrt{2}}|\downarrow\rangle_S|-\rangle_F - \tfrac{1}{\sqrt{2}}|\uparrow\rangle_S|+\rangle_F\,, \tag{4.73a}$$

$$|\mathrm{fail}\rangle_{SF} := \tfrac{1}{\sqrt{2}}|\downarrow\rangle_S|-\rangle_F + \tfrac{1}{\sqrt{2}}|\uparrow\rangle_S|+\rangle_F\Big\}\,. \tag{4.73b}$$

Right after t_4, the wave function of $C \cup \overline{F} \cup S \cup F \cup \overline{W} \cup W$ is, as a computation shows,

$$\begin{aligned} &\sqrt{\tfrac{3}{16}}\Big(|\mathrm{h}\rangle_C|\mathrm{H}\rangle_{\overline{F}}+|\mathrm{t}\rangle_C|\mathrm{T}\rangle_{\overline{F}}\Big)\Big(|\downarrow\rangle_S|-\rangle_F+|\uparrow\rangle_S|+\rangle_F\Big)|\overline{\mathrm{FAIL}}\rangle_{\overline{W}}|\mathrm{FAIL}\rangle_W \\ &+\sqrt{\tfrac{1}{48}}\Big(|\mathrm{h}\rangle_C|\mathrm{H}\rangle_{\overline{F}}+|\mathrm{t}\rangle_C|\mathrm{T}\rangle_{\overline{F}}\Big)\Big(|\downarrow\rangle_S|-\rangle_F-|\uparrow\rangle_S|+\rangle_F\Big)|\overline{\mathrm{FAIL}}\rangle_{\overline{W}}|\mathrm{OK}\rangle_W \\ &-\sqrt{\tfrac{1}{48}}\Big(|\mathrm{h}\rangle_C|\mathrm{H}\rangle_{\overline{F}}-|\mathrm{t}\rangle_C|\mathrm{T}\rangle_{\overline{F}}\Big)\Big(|\downarrow\rangle_S|-\rangle_F+|\uparrow\rangle_S|+\rangle_F\Big)|\overline{\mathrm{OK}}\rangle_{\overline{W}}|\mathrm{FAIL}\rangle_W \\ &+\sqrt{\tfrac{1}{48}}\Big(|\mathrm{h}\rangle_C|\mathrm{H}\rangle_{\overline{F}}-|\mathrm{t}\rangle_C|\mathrm{T}\rangle_{\overline{F}}\Big)\Big(|\downarrow\rangle_S|-\rangle_F-|\uparrow\rangle_S|+\rangle_F\Big)|\overline{\mathrm{OK}}\rangle_{\overline{W}}|\mathrm{OK}\rangle_W\,. \end{aligned} \tag{4.74}$$

In particular, the last line shows that

$$\text{there is positive probability that } \overline{W} \text{ obtains } \overline{\text{OK}} \text{ and } W \text{ obtains OK.} \tag{4.75}$$

So far everybody agrees. Now Frauchiger and Renner reason as follows:

1. According to standard quantum rules, F can obtain the result $Z_S = +$ only if the outcome of $\overline{F}$'s measurement was $Z_C = \text{t}$; see (4.71).
2. $\overline{W}$ can obtain $\overline{\text{OK}}$ only if $Z_S = +$; see (4.72).
3. W can obtain OK only if $Z_C = \text{h}$: after all, $\overline{W}$'s measurement on $C \cup \overline{F}$ and W's measurement on $S \cup F$ commute, and the terms with $|\text{t}\rangle_C$ in (4.71) combine to $\sqrt{2/3}\,|\text{t}\rangle_C|\text{T}\rangle_{\overline{F}}|\text{fail}\rangle_{SF}$.

Thus, it is impossible that $\overline{W}$ obtains $\overline{\text{OK}}$ and W obtains OK, in contradiction to (4.75).

Here is how the argument commits Wheeler's fallacy: First, if there are no hidden variables, then there never was a fact about which values $\overline{F}$ or F obtained as outcome; that is, Z_C and Z_S do not exist; as a consequence, it is not valid to reason that Z_C must have been either h or t, and if it was h then etc.

Second, let us consider the possibility of hidden variables, concretely in Bohmian mechanics: there was a fact which way F's pointer was pointing right after her measurement, so there was a fact about the value of Z_S; likewise with $\overline{F}$ and Z_C. However, the trajectory can switch packets when two packets overlap, so it is not certain that the trajectory moves in the way a single wave packet would move in the absence of the other wave packet. In Wheeler's setup, $|\text{upper slit}\rangle$ evolves to $|\text{near lower detector}\rangle$, and $|\text{lower slit}\rangle$ evolves to $|\text{near upper detector}\rangle$, but it is a fallacy to believe that a particle detected by the lower detector at the later time must have gone through the upper slit at the earlier time (go back to Sect. 1.6.6 to understand this point). In the present situation, if you follow the trajectories in configuration space, you find that indeed $\overline{W}$ can obtain $\overline{\text{OK}}$ only if $Z_S = +$ and $Z_C = \text{t}$ (because the corresponding packets do not overlap any more in configuration space), but not that W can obtain OK only if Z_C was h. Step 3 above is wrong. To be sure, the packet with $|\text{t}\rangle_C$, if alone, would end up completely with $|\text{FAIL}\rangle_W$ (that was the basis of step 3), but the configuration can switch packets when they overlap. That is the mistake.

A couple of more remarks. Ordinarily, records are made to be permanent, for example, written notes do not spontaneously change, but $\overline{W}$'s action on $\overline{F}$ and W's on F are out of the ordinary, for example, because they make wave packets overlap that ordinarily do not. Ordinarily, you can collapse a wave function after a quantum measurement, and if you collapsed (4.69) to the term containing $|\text{t}\rangle_C$, then $\mathbb{P}(\text{OK for } W) = 0$. But here, collapse is incorrect and thus cannot serve as an argument for step 3.

Another remark concerns another angle at the mistake: You can also say it is an instance of the fallacy of simultaneous measurement that I explained in Sect. 4.3.3 and that consists of assuming that whenever two observables can be simultaneously measured, then the outcomes do not depend on the order. In the present situation, the two commuting measurements are those of $\overline{W}$ and W. If W had carried out his measurement first, then the trajectories would have been such that W obtains OK only if Z_C = h—just as claimed in step 3. But we consider the other order of experiments ($\overline{W}$ first), and then the trajectories are different, and step 3 is wrong.

Exercises

Exercise 4.1 can be found in Sect. 4.1.4.

Exercise 4.2 (Essay Question) Describe the Einstein–Podolsky–Rosen argument (either in terms of position and momentum or in terms of spin matrices).

Exercise 4.3 (Essay Question) Describe Einstein's boxes argument.

Exercise 4.4 (Marginal and Conditional Distribution) Consider two random variables X, Y that assume only values ± 1. Their joint distribution can be described by a 2×2 table of probabilities. **(a)** Give a generic example of such a table (i.e., one without symmetries). For your table, compute **(b)** the marginal distribution of X and **(c)** that of Y, as well as **(d)** the conditional distribution of X, given that $Y = +1$, **(e)** the expectation value $\mathbb{E}(X)$, and **(f)** $\mathbb{E}(XY)$.

Exercise 4.5 (Distinguish Ensembles) In this variant of Bell's experiment, a source generates

(a) either 10,000 particle pairs in the spin singlet state

$$\tfrac{1}{\sqrt{2}}\big(|z\text{-up}\rangle|z\text{-down}\rangle - |z\text{-down}\rangle|z\text{-up}\rangle\big) = \tfrac{1}{\sqrt{2}}\big(|x\text{-up}\rangle|x\text{-down}\rangle - |x\text{-down}\rangle|x\text{-up}\rangle\big)$$

(b) or, in a random order, 5000 pairs in $|z\text{-up}\rangle|z\text{-down}\rangle$ and 5000 in $|z\text{-down}\rangle|z\text{-up}\rangle$

(c) or, in a random order, 5000 pairs in $|x\text{-up}\rangle|x\text{-down}\rangle$ and 5000 in $|x\text{-down}\rangle|x\text{-up}\rangle$.

Alice and Bob are far from each other, and each receives one particle of every pair. By carrying out (local) Stern–Gerlach experiments on their particles and comparing their results afterward, how can they decide whether the source was of type (a), (b), or (c)? (Is there another way if they only need to distinguish (a) and (b)?)

Exercise 4.6 (Nonlocality and Entanglement) Any proof of nonlocality must involve an *entangled* wave function of *at least two* particles. Explain.

References

1. V. Allori, S. Goldstein, R. Tumulka, N. Zanghì, Many-worlds and Schrödinger's first quantum theory. British J. Philosophy Sci. **62**(1), 1–27 (2011). https://arxiv.org/abs/0903.2211
2. A. Aspect, J. Dalibard, G. Roger, Experimental test of Bell's inequalities using time-varying analyzers. Phys. Rev. Lett. **49**, 1804–1807 (1982)
3. J.S. Bell, On the Einstein-Podolsky-Rosen paradox. Physics **1**, 195–200 (1964). Reprinted as chapter 2 of [6]
4. J.S. Bell, The theory of local beables. Epistemological Letters **9**, 11 (1976). Reprinted in: Dialectica **39**, 85 (1985) and as chapter 7 of [6]
5. J.S. Bell, Bertlmann's socks and the nature of reality. J. de Phys. **42**(C2), 41–61 (1981). Reprinted as chapter 16 of [6]
6. J.S. Bell, *Speakable and Unspeakable in Quantum Mechanics* (Cambridge University Press, 1987)
7. D. Bohm, *Quantum Theory* (Prentice-Hall, New York, 1951)
8. N. Bohr, Can quantum-mechanical description of physical reality be considered complete? Phys. Rev. **48**, 696–702 (1935)
9. J.F. Clauser, M.A. Horne, A. Shimony, R.A. Holt, Proposed experiment to test local hidden-variable theories. Phys. Rev. Lett. **23**, 880–884 (1969)
10. L. de Broglie, *The Current Interpretation of Wave Mechanics. A Critical Study* (Elsevier, 1964)
11. A. Einstein, B. Podolsky, N. Rosen, Can quantum-mechanical description of physical reality be considered complete? Physical Review **47**, 777–780 (1935)
12. D. Frauchiger, R. Renner, Quantum theory cannot consistently describe the use of itself. Nature Communications **9**(1), 3711 (2018). https://arxiv.org/abs/1604.07422
13. S. Goldstein, T. Norsen, D.V. Tausk, N.N. Zanghì, Bell's Theorem. Scholarpedia **6**(10), 8378 (2011). https://www.scholarpedia.org/article/Bell%27s_theorem
14. D. Lazarovici, M. Hubert, How quantum mechanics can consistently describe the use of itself. Scientific Reports **9**, 470 (2019). https://arxiv.org/abs/1809.08070
15. T. Maudlin, *Quantum Non-Locality and Relativity*, 3rd edn. (Wiley-Blackwell, Oxford, 2011)
16. T. Maudlin, What Bell did. J. Phys. A Math. Theoret. **47**, 424010 (2014)
17. T. Norsen, Einstein's Boxes. Am. J. Phys. **73**(2), 164–176 (2005). https://arxiv.org/abs/quant-ph/0404016
18. R. Tumulka, The assumptions of Bell's proof, in *Quantum Nonlocality and Reality – 50 Years of Bell's Theorem* ed. by M. Bell, S. Gao (Cambridge University Press, 2016), pp. 79–90. https://arxiv.org/abs/1501.04168

General Observables 5

5.1 POVMs: General Observables

Let

$$\mathbb{S}(\mathscr{H}) = \{\psi \in \mathscr{H} : \|\psi\| = 1\} \tag{5.1}$$

denote the unit sphere in Hilbert space.

5.1.1 Definition

An observable as we considered it so far is mathematically represented by a self-adjoint operator. A *generalized observable* is mathematically represented by a *positive-operator-valued measure (POVM)*, as we will explain now.

Definition 5.1 An operator is called *positive* if and only if it is self-adjoint and all (generalized) eigenvalues are greater than or equal to zero. (In linear algebra, a positive operator is commonly called "positive semi-definite.") Equivalently, a bounded operator $A : \mathscr{H} \to \mathscr{H}$ is positive if and only if

$$\langle\psi|A|\psi\rangle \geq 0 \quad \text{for every } \psi \in \mathscr{H}\,. \tag{5.2}$$

The sum of two positive operators is again a positive operator, whereas the product of two positive operators is in general not even self-adjoint. Note that every projection is a positive operator. We also write $A \geq 0$ to express that A is positive; $A \geq B$ means that $A - B \geq 0$.

R. Tumulka, *Foundations of Quantum Mechanics*, Lecture Notes in Physics 1003,
https://doi.org/10.1007/978-3-031-09548-1_5

As a first, rough definition, we can say the following: A *POVM* is a family of positive operators E_z such that

$$\sum_z E_z = I\,. \tag{5.3}$$

(Refined definition later.)

Example 5.1

1. $E_1 = \begin{pmatrix} 1/2 & \\ & 1/3 \end{pmatrix}$, $E_2 = \begin{pmatrix} 1/2 & \\ & 2/3 \end{pmatrix}$.

   In fact, all (generalized) eigenvalues of E_z must lie in [0, 1] because if $E_\zeta \psi = \eta\psi$, then

$$\langle\psi|\psi\rangle = \langle\psi|I|\psi\rangle \overset{(5.3)}{=} \langle\psi|E_\zeta|\psi\rangle + \langle\psi|\sum_{z\neq\zeta} E_z|\psi\rangle \geq \langle\psi|E_\zeta|\psi\rangle = \eta\langle\psi|\psi\rangle\,, \tag{5.4}$$

   so $\eta \leq 1$.
2. $E_1 = \begin{pmatrix} 1 & \\ & 0 \end{pmatrix}$, $E_2 = \begin{pmatrix} 0 & \\ & 1 \end{pmatrix}$. In the special case in which all operators E_z are projection operators, E is called a *projection-valued measure (PVM)*. In this case, the subspaces to which E_z and $E_{z'}$ ($z \neq z'$) project must be mutually orthogonal (see Exercise 5.12 at the end of Chap. 5).
3. Every *self-adjoint matrix* defines a PVM: Let $z = \alpha$ run through the eigenvalues of A and let E_α be the projection to the eigenspace of A with eigenvalue α,

$$E_\alpha = \sum_\lambda |\phi_{\alpha,\lambda}\rangle\langle\phi_{\alpha,\lambda}|\,. \tag{5.5}$$

   Then their sum is I, as easily seen from the point of view of an orthonormal basis of eigenvectors of A. So E is a PVM, the *spectral PVM* of A. Example 2 above is of this form for $A = \sigma_3$.
4. A POVM E and a vector $\psi \in \mathbb{S}(\mathscr{H})$ together define a probability distribution over z as follows:

$$\mathbb{P}_\psi(z) = \langle\psi|E_z|\psi\rangle\,. \tag{5.6}$$

   To see this, note that $\langle\psi|E_z|\psi\rangle$ is a nonnegative real number since E_z is a positive operator, and

$$\sum_z \mathbb{P}_\psi(z) = \sum_z \langle\psi|E_z|\psi\rangle = \langle\psi|I|\psi\rangle = \|\psi\|^2 = 1. \tag{5.7}$$

5. *Fuzzy position observable*:

$$E_z\psi(x) = \frac{1}{\sqrt{2\pi\sigma^2}} e^{-\frac{(x-z)^2}{2\sigma^2}} \psi(x) . \tag{5.8}$$

Each E_z is a positive operator (but not a projection) because

$$\langle\psi|E_z|\psi\rangle = \int_{\mathbb{R}} dx\, \psi^*(x) \frac{1}{\sqrt{2\pi\sigma^2}} e^{-\frac{(x-z)^2}{2\sigma^2}} \psi(x) \geq 0 . \tag{5.9}$$

The E_z add to unity in the continuous sense:

$$\int_{\mathbb{R}} E_z\, dz = I . \tag{5.10}$$

Indeed,

$$\int_{\mathbb{R}} dz\, E_z\psi(x) = \frac{1}{\sqrt{2\pi\sigma^2}} \psi(x) \int_{\mathbb{R}} dz\, e^{-\frac{(x-z)^2}{2\sigma^2}} = \psi(x) . \tag{5.11}$$

The case of a continuous variable z brings us to the general definition of a POVM, which I will formulate rigorously although we do not aim at rigor in general. The definition is, in fact, quite analogous to the rigorous definition of a probability distribution in measure theory: A *measure* associates a value (i.e., a number or an operator) not with a point but with a set: $E(B)$ instead of E_z, where $B \subseteq \mathscr{Z}$ and $\mathscr{Z}$ is the set of all z's. More precisely, let $\mathscr{Z}$ be a set and $\mathscr{B}$ a σ-algebra of subsets of $\mathscr{Z}$,[1] the family of the "measurable sets." A *probability measure* is a mapping $\mu : \mathscr{B} \to [0, 1]$ such that for any $B_1, B_2, \ldots \in \mathscr{B}$ with $B_i \cap B_j = \emptyset$ for $i \neq j$,

$$\mu\Big(\bigcup_{n=1}^{\infty} B_n\Big) = \sum_{n=1}^{\infty} \mu(B_n) . \tag{5.12}$$

Definition 5.2 A *POVM* on the measurable space $(\mathscr{Z}, \mathscr{B})$ acting on the Hilbert space $\mathscr{H}$ is a mapping E from $\mathscr{B}$ to the set of bounded operators on $\mathscr{H}$ such that each $E(B)$ is positive, $E(\mathscr{Z}) = I$, and for any $B_1, B_2, \ldots \in \mathscr{B}$ with $B_i \cap B_j = \emptyset$ for $i \neq j$,

$$E\Big(\bigcup_{n=1}^{\infty} B_n\Big) = \sum_{n=1}^{\infty} E(B_n) , \tag{5.13}$$

[1] A *σ-algebra* is a family $\mathscr{B}$ of subsets of $\mathscr{Z}$ such that $\emptyset \in \mathscr{B}$ and, for every $B_1, B_2, B_3, \ldots$ in $\mathscr{A}$ also $B_1^c := \mathscr{Z} \setminus B_1 \in \mathscr{B}$ and $B_1 \cup B_2 \cup \ldots \in \mathscr{B}$. It follows that $\mathscr{Z} \in \mathscr{B}$ and $B_1 \cap B_2 \cap \ldots \in \mathscr{B}$. A set $\mathscr{Z}$ equipped with a σ-algebra is also called a *measurable space*. The σ-algebra usually considered on $\mathbb{R}^n$ consists of the "Borel sets" and is called the "Borel σ-algebra."

where the series on the right-hand side converges strongly, i.e., $\sum_n E(B_n)\psi$ converges in $\mathscr{H}$ for every $\psi \in \mathscr{H}$.[2]

It follows that a POVM E and a vector $\psi \in \mathbb{S}(\mathscr{H})$ together define a probability measure on $\mathscr{Z}$ as follows:

$$\mu_\psi(B) = \langle\psi|E(B)|\psi\rangle\,. \tag{5.14}$$

Exercise 5.1 Verify the definition of a probability measure.

Again, one defines a PVM to be a POVM such that every $E(B)$ is a projection. In the special case in which $\mathscr{Z}$ is a countable set and $\mathscr{B}$ consists of all subsets, any POVM satisfies

$$E(B) = \sum_{z\in B} E_z \tag{5.15}$$

with $E_z = E(\{z\})$, so in that case Definition 5.2 boils down to the earlier definition around (5.3). The fuzzy position observable of Example 5 corresponds to $\mathscr{Z} = \mathbb{R}$, $\mathscr{B}$ the Borel sets, and $E(B)$ the multiplication operator

$$E(B)\psi(x) = \int_B dz \frac{1}{\sqrt{2\pi\sigma^2}} e^{-\frac{(x-z)^2}{2\sigma^2}}\,\psi(x)\,, \tag{5.16}$$

which multiplies by the function $1_B * g$, where 1_B is the characteristic function of B, g is the Gaussian density function, and $*$ means convolution.

It turns out that every observable is a generalized observable; that is, every self-adjoint operator A defines a PVM E with $E(B)$ the projection to the so-called spectral subspace of B. If there is an ONB of eigenvectors of A, then the spectral subspace of B is the closed span of all eigenspaces with eigenvalues in B; that is, in that case $E(\{z\})$ is the projection to the eigenspace of eigenvalue z (and 0 if z is not an eigenvalue). In the case of a general self-adjoint operator A, the following is a reformulation of the spectral theorem:

Theorem 5.1 *For every self-adjoint operator A there is a uniquely defined PVM E on the real line with the Borel σ-algebra (the "spectral PVM" of A) such that*

$$A = \int_{\mathbb{R}} \alpha\, E(d\alpha)\,. \tag{5.17}$$

[2] It is equivalent to merely demand that the series on the right-hand side converges weakly, i.e., that $\sum_n \langle\psi|E(B_n)|\psi\rangle$ converges for every $\psi \in \mathscr{H}$. Indeed, the weak convergence implies $E(\emptyset) = 0$ and thus finite additivity, whence $E(\cup_{n=1}^\infty B_n) - \sum_{n=1}^N E(B_n) = E(\cup_{n=N+1}^\infty B_n)$. Call this operator T_N. We have that $0 \le T_N$ and $0 \le E(\mathscr{Z}\setminus \cup_{n=N+1}^\infty B_n) = I - T_N$, so $0 \le T_N \le I$. Thus, $T_N^2 \le T_N$ and $\|T_N\psi\|^2 = \langle\psi|T_N^2|\psi\rangle \le \langle\psi|T_N|\psi\rangle \to 0$ as $N \to \infty$.

To explain the last equation: In the same way as one can define the integral $\int_{\mathscr{Z}} f(z)\,\mu(dz)$ of a measurable function $f : \mathscr{Z} \to \mathbb{R}$ relative to a measure μ, one can define an operator-valued integral $\int_{\mathscr{Z}} f(z)\,E(dz)$ relative to a POVM E. Equation (5.17) is a generalization of the relation

$$A = \sum_{\alpha} \alpha\, E_\alpha \tag{5.18}$$

for self-adjoint matrices A. In the literature, the spectral PVM is also sometimes called the *spectral measure*. Another way of characterizing the spectral PVM is through the equation

$$E(B) = 1_B(A) \tag{5.19}$$

for any set $B \subseteq \mathbb{R}$; here, 1_B is the characteristic function of the set B, and we apply this function to the operator A. (For example, in the representation in which A is a multiplication operator multiplying by the function f, $1_B(A)$ is the multiplication operator multiplying by $1_B \circ f$.)

If several self-adjoint operators $A_1, \ldots, A_n$ commute pairwise, then they can be diagonalized simultaneously, i.e., there is a PVM E on $\mathbb{R}^n$ such that for every $k \in \{1, \ldots, n\}$,

$$A_k = \int_{\mathbb{R}^n} \alpha_k\, E(d\boldsymbol{\alpha})\,. \tag{5.20}$$

Example 5.2 The PVM diagonalizing the three position operators X_1, X_2, X_3 on $L^2(\mathbb{R}^3)$ is

$$E(B)\psi(\boldsymbol{x}) = \begin{cases} \psi(\boldsymbol{x}) & \text{if } \boldsymbol{x} \in B \\ 0 & \text{if } \boldsymbol{x} \notin B\,, \end{cases} \tag{5.21}$$

mentioned before in (3.17). Equivalently, $E(B)$ is the multiplication by the characteristic function of B.

Example 5.3 It was shown in Exercise 3.5 in Sect. 3.7.7 that if we make consecutive ideal quantum measurements of observables corresponding to self-adjoint operators $A_1, \ldots, A_n$ (which need not commute with each other) with purely discrete spectra at times $0 < t_1 < \ldots < t_n$ respectively on a system with initial wave function $\psi_0 \in \mathbb{S}(\mathscr{H})$, then the joint distribution of the outcomes $Z_1, \ldots, Z_n$ is given by (3.105). Equivalently, the distribution is given by

$$\mathbb{P}\Big((Z_1, \ldots, Z_n) \in B\Big) = \langle\psi_0|E(B)|\psi_0\rangle \tag{5.22}$$

for all (Borel) subsets $B \subseteq \mathbb{R}^n$, where

$$E(B) = \sum_{(z_1 \dots z_n) \in B \cap S} L^\dagger L \tag{5.23}$$

with $S = \text{spectrum}(A_1) \times \cdots \times \text{spectrum}(A_n)$,

$$L = P_{nz_n} U_{t_n - t_{n-1}} \cdots P_{1z_1} U_{t_1 - t_0}\,, \tag{5.24}$$

and P_{jz_j} the projection to the eigenspace of A_j with eigenvalue z_j.

Exercise 5.2 (E is a POVM) Show that E is a POVM on $\mathbb{R}^n$. (In fact, it is concentrated on S, a countable subset of $\mathbb{R}^n$.)

Example 5.4 In GRWf, the joint distribution of all flashes is of the form

$$\mathbb{P}(F \in B) = \langle \Psi_0 | G(B) | \Psi_0 \rangle \tag{5.25}$$

for all sets $B \subseteq \mathscr{Z}$, with Ψ_0 the initial wave function and G a POVM on the history space $\mathscr{Z}$ of flashes,

$$\mathscr{Z} = \Big\{ \big((t_1, \boldsymbol{x}_1, i_1), (t_2, \boldsymbol{x}_2, i_2), \dots \big) \in (\mathbb{R}^4 \times \{1 \dots N\})^\infty : 0 < t_1 < t_2 < \dots \Big\}. \tag{5.26}$$

Derivation Consider first the joint distribution of the first two flashes for $N = 1$ particle: The probability of $T_1 \in [t_1, t_1 + dt_1]$ is $1_{t_1 > 0}\, e^{-\lambda t_1}\, \lambda\, dt_1$; given T_1, the probability of $\boldsymbol{X}_1 \in d^3\boldsymbol{x}_1$ is, according to (3.59), $\|C(\boldsymbol{x}_1)\Psi_{T_1-}\|^2$ with $\Psi_{T_1-} = e^{-iHT_1}\Psi_0$ and $C(\boldsymbol{x}_1)$ the collapse operator defined in (3.57). Given T_1 and $\boldsymbol{X}_1$, the probability of $T_2 \in [t_2, t_2 + dt_2]$ is $1_{t_2 > t_1}\, e^{-\lambda(t_2 - t_1)}\, \lambda\, dt_2$; given T_1, $\boldsymbol{X}_1$, and T_2, the probability of $\boldsymbol{X}_2 \in d^3\boldsymbol{x}_2$ is $\|C(\boldsymbol{x}_2) e^{-iH(T_2 - T_1)} \Psi_{T_1+}\|^2$ with $\Psi_{T_1+} = C(\boldsymbol{X}_1)\Psi_{T_1-}$. Putting these formulas together, the joint distribution of T_1, $\boldsymbol{x}_1$, T_2, and $\boldsymbol{X}_2$ is given by

$$\mathbb{P}\Big(T_1 \in [t_1, t_1 + dt_1], \boldsymbol{X}_1 \in d^3\boldsymbol{x}_1, T_2 \in [t_2, t_2 + dt_2], \boldsymbol{X}_2 \in d^3\boldsymbol{x}_2\Big)$$

$$= 1_{0 < t_1 < t_2}\, e^{-\lambda t_2}\, \lambda^2 \Big\| C(\boldsymbol{x}_2) e^{-iH(t_2 - t_1)} C(\boldsymbol{x}_1) e^{-iHt_1} \Psi_0 \Big\|^2\, dt_1\, d^3\boldsymbol{x}_1\, dt_2\, d^3\boldsymbol{x}_2 \tag{5.27}$$

$$= \langle \Psi_0 | G(dt_1 \times d^3\boldsymbol{x}_1 \times dt_2 \times d^3\boldsymbol{x}_2) | \Psi_0 \rangle \tag{5.28}$$

with

$$G(dt_1 \times d^3\boldsymbol{x}_1 \times dt_2 \times d^3\boldsymbol{x}_2) = 1_{0<t_1<t_2}\, e^{-\lambda t_2}\,\lambda^2\, \times$$
$$\times\; e^{iHt_1}C(\boldsymbol{x}_1)e^{iH(t_2-t_1)}C(\boldsymbol{x}_2)^2e^{-iH(t_2-t_1)}C(\boldsymbol{x}_1)e^{-iHt_1}\,dt_1\,d^3\boldsymbol{x}_1\,dt_2\,d^3\boldsymbol{x}_2\,, \tag{5.29}$$

which is self-adjoint and positive because (5.28) is always real and ≥ 0. It follows that also $G(B)$, obtained by summing (i.e., integrating) over all infinitesimal volume elements in B, is self-adjoint and positive. Additivity holds by construction, and $G(\mathscr{Z}) = I$ because (5.28) is a probability distribution (so $\langle\Psi_0|G(\mathscr{Z})|\Psi_0\rangle = 1$ for every Ψ_0 with $\|\Psi_0\| = 1$). Thus, G is a POVM. For the joint distribution of more than two flashes or more than one particle, the reasoning proceeds in a similar way. For the joint distribution of *all* (infinitely many) flashes, the rigorous proof requires some more technical steps (carried out in my paper [42]) but bears no surprises. □

The fact that the distribution of the flashes is governed by a POVM is relevant to the no-signaling theorem in Sect. 5.5.9 and to the setup of a relativistic GRW theory in Sect. 7.8.

5.1.2 The Main Theorem About POVMs

Main theorem about POVMs *For every quantum physical experiment $\mathscr{E}$ on a quantum system S whose possible outcomes lie in a space $\mathscr{Z}$, there exists a POVM E on $\mathscr{Z}$ such that, whenever S has wave function ψ at the beginning of $\mathscr{E}$, the random outcome Z has probability distribution given by*

$$\mathbb{P}(Z \in B) = \langle\psi|E(B)|\psi\rangle\,. \tag{5.30}$$

We will prove this statement in Bohmian mechanics and GRWf. It plays the role of Born's rule for POVMs. The experiment $\mathscr{E}$ consists of coupling S to an apparatus A at some initial time t_{i}, letting $S \cup A$ evolve up to some final time t_{f}, and then reading off the result Z from A. It is assumed that S and A are not entangled at the beginning of $\mathscr{E}$:

$$\Psi_{S\cup A}(t_{\mathrm{i}}) = \psi \otimes \phi_A \tag{5.31}$$

with ϕ_A the ready state of A. (The main theorem of POVMs can also be proven for the case in which t_{f} is itself chosen by the experiment; e.g., the experiment might wait for a detector to click, and the outcome Z may be the time of the click. I give the proof only for the simpler case in which t_{f} is fixed in advance.) I will further assume that $\mathscr{E}$ has only finitely many possible outcomes Z; actually, this assumption is not needed for the proof, but it simplifies the consideration a bit and is satisfied in every realistic scenario.

Let us first consider an experiment in the framework of Bohmian mechanics. Since the outcome is read off from the pointer position, it can be written as

$$Z = \zeta\big(Q(t_f)\big)\,, \tag{5.32}$$

where Q is the Bohmian configuration and ζ is called the *calibration function*. (In practice, the function ζ depends only on the configuration *of the apparatus*, in fact only on its macroscopic features, not on microscopic details. However, the arguments that follow apply to arbitrary calibration functions.) We can even obtain a formula for E:

Law of Operators For the experiment $\mathscr{E}$ defined by ϕ_A, t_i, t_f, and $H_{S\cup A}$,

$$E_z\psi = \langle\phi_A|U^\dagger P_{B_z} U(\psi\otimes\phi_A)\rangle_y \tag{5.33}$$

with

$$U = e^{-iH_{S\cup A}(t_f - t_i)}\,, \tag{5.34}$$

$$B_z = \{q \in \mathbb{R}^{3N} : \zeta(q) = z\}\,, \tag{5.35}$$

and P_B the position projection defined in (3.17).

In words, (5.33) can be expressed as follows: For given ψ, form $\psi\otimes\phi$, then apply the operator $U^\dagger P_{B_z} U$, and finally take the *partial inner product* with ϕ. The partial inner product of a function $\Psi(x, y)$ with the function $\phi(y)$ is a function of x defined as

$$\langle\phi|\Psi\rangle_y(x) = \int dy\,\phi^*(y)\,\Psi(x, y)\,. \tag{5.36}$$

A special case of the law of operators was already included in Definition 3.2 of quantum measurements in Sect. 3.2.6.

Proof of the Main Theorem from Bohmian Mechanics Let U and B_z be given by (5.34) and (5.35). Then,

$$\mathbb{P}(Z = z) = \mathbb{P}\big(Q(t_f) \in B_z\big) \tag{5.37a}$$

$$= \int_{B_z} |\Psi(q, t_f)|^2\, dq \tag{5.37b}$$

$$= \langle\Psi(t_f)|P_{B_z}|\Psi(t_f)\rangle \tag{5.37c}$$

$$= \langle\psi\otimes\phi_A|U^\dagger P_{B_z} U|\psi\otimes\phi_A\rangle \tag{5.37d}$$

$$= \langle\psi|E_z|\psi\rangle_S\,, \tag{5.37e}$$

where $\langle\cdot|\cdot\rangle_S$ denotes the inner product in the Hilbert space of the system S alone (as opposed to the Hilbert space of $S \cup A$) and E_z is defined through (5.33). We now verify that E is a POVM. First, E_z is a positive operator because

$$\langle\psi|E_z|\psi\rangle = \langle\Psi(t_f)|P_{B_z}|\Psi(t_f)\rangle \geq 0 \tag{5.38}$$

for every ψ. Second, $\sum_z E_z = I$ because

$$\sum_z E_z\psi = \sum_z \langle\phi_A|U^\dagger P_{B_z} U(\psi \otimes \phi_A)\rangle_y \tag{5.39a}$$

$$= \langle\phi_A|U^\dagger \sum_z P_{B_z} U(\psi \otimes \phi_A)\rangle_y \tag{5.39b}$$

$$= \langle\phi_A|U^\dagger I U(\psi \otimes \phi_A)\rangle_y \tag{5.39c}$$

$$= \langle\phi_A|I(\psi \otimes \phi_A)\rangle_y = \psi\,. \tag{5.39d}$$

Here, we have used that

$$\sum_z P_{B_z} = I\,, \tag{5.40}$$

that $U^\dagger U = I$, and that the partial inner product of $\psi \otimes \phi$ with ϕ returns ψ. Equation (5.40) follows from the fact that the sets B_z form a partition of configuration space $\mathbb{R}^{3N}$ (i.e., they are mutually disjoint and together cover the entire configuration space, $\cup_z B_z = \mathbb{R}^{3N}$). This, in turn, follows from the assumption that the calibration function ζ is defined everywhere in $\mathbb{R}^{3N}$.[3] Thus, the proof is complete. □

Proof of the Main Theorem from GRWf In GRWf, the main theorem is also valid, but with a slightly different POVM E^{GRW}, one that is not given by the law of operators (5.33) but by a different law of operators. To this end, let $F = \{(T_1, \boldsymbol{X}_1, I_1), (T_2, \boldsymbol{X}_2, I_2), \ldots\}$ be the set of flashes (of both S and A) from t_i onward. We know from Example 5.4 that the distribution of F (i.e., the joint distribution of all flashes after t_i) is given by $\Psi(t_i)$ and some POVM G:

$$\mathbb{P}(F \in B) = \langle\Psi(t_i)|G(B)|\Psi(t_i)\rangle\,. \tag{5.41}$$

---

[3] The physical meaning of this assumption is that the experiment always has *some* outcome. You may worry about the possibility that the experiment could not be completed as planned due to power outage, asteroid impact, or whatever. This possibility can be taken into account by introducing a further element f for "failed" into the set $\mathscr{Z}$ of possible outcomes.

Since the outcome Z of the experiment is read off from A after t_{i}, it is a function of F,

$$Z = \zeta(F) . \tag{5.42}$$

(Z is a function of F because the flashes define where the pointers point and what the shape of the ink on a sheet of paper is. It would even be realistic to assume that Z depends only on the flashes of the apparatus, but this restriction is not needed for further argument.)

Let $B_z = \{f : \zeta(f) = z\}$, the set of flash patterns having outcome z. Then,

$$\mathbb{P}(Z = z) = \mathbb{P}\big(F \in B_z\big) \tag{5.43a}$$

$$= \langle\Psi(t_{\mathrm{i}})|G(B_z)|\Psi(t_{\mathrm{i}})\rangle \tag{5.43b}$$

$$= \langle\psi|E_z^{\mathrm{GRW}}|\psi\rangle \tag{5.43c}$$

with

$$E_z^{\mathrm{GRW}}\psi = \langle\phi_A|G(B_z)|\psi \otimes \phi_A\rangle_y . \tag{5.44}$$

The fact that E_z^{GRW} may be different from the E_z obtained from Bohmian mechanics agrees with the fact that the same experiment (using the same initial wave function of the apparatus, etc.) may yield different outcomes in GRW than in Bohmian mechanics. (However, since we know the two theories make very very similar predictions, E_z^{GRW} will usually be very very close to E_z.) To see that E_z^{GRW} is a POVM, we note that

$$\langle\psi|E_z^{\mathrm{GRW}}|\psi\rangle = \langle\Psi(t_{\mathrm{i}})|G(B_z)|\Psi(t_{\mathrm{i}})\rangle \geq 0 \tag{5.45}$$

and

$$\sum_z E_z^{\mathrm{GRW}}\psi = \langle\phi_A|\sum_z G(B_z)|\psi \otimes \phi_A\rangle_y \tag{5.46a}$$

$$= \langle\phi_A|G(\cup_z B_z)|\psi \otimes \phi_A\rangle_y \tag{5.46b}$$

$$= \langle\phi_A|I|\psi \otimes \phi_A\rangle_y = \psi \tag{5.46c}$$

using $\cup_z B_z = \mathscr{Z}$. This completes the proof. □

The main theorem about POVMs is equally valid in orthodox quantum mechanics (OQM). However, since OQM does not permit a coherent analysis of measurement processes (as it suffers from the measurement problem), we cannot give a complete proof of the main theorem from OQM, although the same reasoning as given in the proof from Bohmian mechanics could be regarded as compelling in OQM. At the same time, the main theorem undercuts the spirit of OQM, which is to leave the measurement process unanalyzed and to introduce observables by postulate. Put differently, the main theorem about POVMs makes it harder to ignore the measurement problem.

By the way, as far as I know, it is an open question how to decide for a given POVM E whether there is an experiment $\mathscr{E}$ having POVM E, i.e., whether E can be written in the form (5.33) for some ϕ_A, t_i, t_f, and an $H_{S\cup A}$ that is a Schrödinger operator $-\nabla^2 + V$ for some potential $V : \mathbb{R}^{3N_S+3N_A} \to \mathbb{R}$, or likewise for GRW theory.

5.1.3 Limitations to Knowledge

The first and most important thing we should want to measure about a quantum system would be the most central quantity of quantum mechanics: the wave function. That would actually be a *genuine measurement* (i.e., the empirical determination of the value of a quantity that has a value), as opposed to most quantum measurements. However, to measure ψ is impossible!

Corollary 5.1 *There is no experiment with $Z = \psi$ or $Z = \mathbb{C}\psi$. That is, one cannot measure the wave function of a given system, not even up to a global phase, if we allow the system to come with an arbitrary wave function.*

Proof Suppose there was an experiment with $Z = \psi$. Then, for any given ψ, Z is deterministic, i.e., its probability distribution is concentrated on a single point, $\mathbb{P}(Z = \phi) = \delta(\phi - \psi)$. The dependence of this distribution on ψ is not quadratic, and thus not of the form $\langle\psi|E_\phi|\psi\rangle$ for any POVM E. The argument remains valid when we replace ψ by $\mathbb{C}\psi$. □

This fact amounts to a limitation to knowledge in any version of quantum mechanics in which wave functions are part of the ontology, which includes all interpretations of quantum mechanics that we have talked about: Suppose Alice chooses a direction in space $\boldsymbol{n}$, prepares a spin-$\frac{1}{2}$ particle in the state $|\boldsymbol{n}\text{-up}\rangle$, and hands that particle over to Bob. Then, by Corollary 5.1, Bob has no way of discovering $\boldsymbol{n}$ if Alice does not give the information away. The best thing Bob can do is, in fact, a Stern–Gerlach experiment in any direction he likes, say in the z-direction; then he obtains one bit of information, up or down; if the result was "up" then it is more likely that $\boldsymbol{n}$ lies on the upper hemisphere than on the lower.

The situation about measuring ψ is different if we are given *many* systems and we know that all of them have the same ψ. If we can carry out quantum measurements of all self-adjoint operators (as we can, e.g., in spin space $\mathbb{C}^2$), then we can determine ψ up to a phase. In fact, then we can do different experiments on different systems and observe the distribution of outcomes for each of them, which will allow us to determine $\mathbb{C}\psi$ to arbitrary precision and with arbitrarily high degree of certainty if the given ensemble is large enough. (As a simple consideration in this direction, if we carry out quantum measurements of all 1d projections and can determine the distribution of outcomes of each, then $\mathbb{C}\psi$ is the subspace whose projection yields outcome "1" with probability 1.)

Here is another limitation to knowledge.

Corollary 5.2 *There is no experiment in Bohmian mechanics that can measure the instantaneous velocity of a particle with unknown wave function.*

Proof Again, the distribution of the velocity $\mathrm{Im}\nabla\psi/\psi(Q)$ with $Q \sim |\psi|^2$ is not quadratic in ψ. □

In contrast, the *asymptotic velocity* can be measured, and its probability distribution is in fact quadratic in ψ: Recall from (2.41) that it is given by $(m/\hbar)^3\,|\widehat{\psi}(m\boldsymbol{u}/\hbar)|^2$. Another statement to contrast Corollary 5.2 to is that the particle's instantaneous velocity *can* be measured if the wave function is known. For example, a mechanism that measures the particle's position and computes the velocity from Bohm's equation of motion would fulfill the task.

The impossibility of measuring instantaneous velocity if ψ is unknown goes along with the impossibility to measure the entire trajectory without disturbing it. If we wanted to measure the trajectory, for example, by repeatedly measuring the positions every Δt with inaccuracy Δx, then the measurements will collapse the wave function, with the consequence that the observed trajectory is very different from what the trajectory would have been had we not intervened. Some authors regard this as an argument against Bohmian mechanics. Bell (1987) [6, Sec. 1] disagreed:

> To admit things not visible to the gross creatures that we are is, in my opinion, to show a decent humility, and not just a lamentable addiction to metaphysics.

That is, Bell criticized the positivistic idea that anything real can always be measured. Indeed, this idea seems rather dubious, given that wave functions cannot be measured. In the next section, we will make an even stronger case against this idea.

5.1.4 Limitations to Knowledge as a General Fact

So there are limitations to knowledge in any version of quantum mechanics in which wave functions are part of the ontology. Are there limitations to knowledge in *every* version of quantum mechanics? The answer is *yes*, and we can prove it!

Consider a quantum system with Hilbert space $\mathscr{H}$ of dimension > 1, and let $\mathcal{POVM}$ denote the set of all POVMs acting on $\mathscr{H}$. We want to allow an arbitrary theory of reality. The theory can tell us what the ontology is, and what the possible states of the ontology are. These states λ are often called the "ontic states," as opposed to the "quantum state" ψ. For example, in Bohmian mechanics, the ontic state of a system is the pair (Q, ψ). The set of all ontic states will be denoted by Λ. For our purposes, we assume the following about the given theory:

- There is an index set $\mathcal{EXP}$ representing the possible experiments.
- In agreement with the main theorem about POVMs, with every experiment $\mathscr{E} \in \mathcal{EXP}$ there is associated a POVM $E_{\mathscr{E}} \in \mathcal{POVM}$.
- Any experiment $\mathscr{E} \in \mathcal{EXP}$ can be carried out on a system in any ontic state λ, and the probability distribution of the outcome Z is specified by the theory and denoted by $P_{\lambda,\mathscr{E}}(dz)$.
- For every $\psi \in \mathbb{S}(\mathscr{H})$, there is an experimental procedure to "prepare the quantum state ψ," which leads to a random λ whose probability distribution is specified by the theory and denoted by $\varrho^{\psi}(d\lambda)$.

Correspondingly, we can give a mathematical formalization of what we require of an *ontological model* of the quantum system with Hilbert space $\mathscr{H}$:

Definition 5.3 An *ontological model of* $\mathscr{H}$ consists of a set[4] Λ, a probability distribution ϱ^{ψ} over Λ for every $\psi \in \mathbb{S}(\mathscr{H})$, a set $\mathcal{EXP}$, a mapping $E : \mathcal{EXP} \to \mathcal{POVM}$, and for every $\lambda \in \Lambda$ and $\mathscr{E} \in \mathcal{EXP}$ a probability measure $P_{\lambda,\mathscr{E}}$ on the value space $\mathscr{Z}$ of $E_{\mathscr{E}}$.

Definition 5.4 We say that the ontological model *reproduces the quantum predictions* if and only if for every $\psi \in \mathbb{S}(\mathscr{H})$ and every $\mathscr{E} \in \mathcal{EXP}$,

$$\int_{\Lambda} \varrho^{\psi}(d\lambda)\, P_{\lambda,\mathscr{E}}(B) = \langle\psi|E_{\mathscr{E}}(B)|\psi\rangle \quad \forall B \subseteq \mathscr{Z} \tag{5.47}$$

with $\mathscr{Z}$ the value space.

Theorem 5.2 (Limitations to Knowledge Are Inevitable [47]) *Consider an ontological model that reproduces the quantum predictions. Suppose that all POVMs consisting of a projection P and $I - P$ lie in the range of E.*[5] *Then there is no experiment that measures λ, i.e., no experiment that would always, when applied to a system in the state λ, yield the outcome λ.*

Proof We assume that such an experiment $\mathscr{G}$ exists and will derive a contradiction. Let $G := E_{\mathscr{G}}$. If the outcome equals λ, then, in particular, the distribution of the outcome, $\langle\psi|G|\psi\rangle$, agrees with the distribution ϱ^{ψ} that λ had already before the

---

[4] Technically, all sets mentioned are assumed to be measurable spaces, all subsets measurable sets, and all functions measurable functions.

[5] For example, for a single spin with $\mathscr{H} = \mathbb{C}^2$, every 1d projection P occurs for a Stern–Gerlach experiment in some direction $\boldsymbol{n} \in \mathbb{S}(\mathbb{R}^3)$ (in fact, $\boldsymbol{n} = \boldsymbol{\omega}(\phi)$ if $P = |\phi\rangle\langle\phi|$).

experiment, so

$$\varrho^{\psi}(d\lambda) = \langle\psi|G(d\lambda)|\psi\rangle\,. \tag{5.48}$$

Thus,

$$\left\langle\psi\right|\int_{\Lambda} G(d\lambda)\, P_{\lambda,\mathscr{E}}(B)\left|\psi\right\rangle = \int_{\Lambda} \langle\psi|G(d\lambda)|\psi\rangle\, P_{\lambda,\mathscr{E}}(B) \tag{5.49a}$$

$$= \int_{\Lambda} \varrho^{\psi}(d\lambda)\, P_{\lambda,\mathscr{E}}(B) \tag{5.49b}$$

$$= \langle\psi|E_{\mathscr{E}}(B)|\psi\rangle\,. \tag{5.49c}$$

Now for any two bounded operators R and S, if $\langle\psi|R|\psi\rangle = \langle\psi|S|\psi\rangle$ for all $\psi \in \mathbb{S}(\mathscr{H})$, then $R = S$. Hence,

$$\int_{\Lambda} G(d\lambda)\, P_{\lambda,\mathscr{E}}(B) = E_{\mathscr{E}}(B)\,. \tag{5.50}$$

Now choose, for every 1d subspace g of $\mathscr{H}$, an experiment $\mathscr{E}(g) \in \mathcal{EXP}$ so that $E_{\mathscr{E}(g)}$ is the POVM with $\mathscr{Z} = \{0, 1\}$, $E_{\mathscr{E}(g)}(\{1\}) = P_g$ (the projection to g), and $E_{\mathscr{E}(g)}(\{0\}) = I - P_g = P_{g^\perp}$. For $B = \{1\}$, we obtain that

$$\int_{\Lambda} G(d\lambda)\, P_{\lambda,\mathscr{E}(g)}(\{1\}) = P_g\,. \tag{5.51}$$

To this equation, apply $\langle\chi|$ and $|\chi\rangle$ for $\chi \in g^\perp$:

$$\int_{\Lambda} \langle\chi|G(d\lambda)|\chi\rangle\, P_{\lambda,\mathscr{E}(g)}(\{1\}) = 0\,. \tag{5.52}$$

Since $\langle\chi|G(d\lambda)|\chi\rangle \geq 0$ and $P_{\lambda,\mathscr{E}(g)}(\{1\})$ is a function of λ with values in $[0, 1]$, the integral can vanish only if everywhere either $\langle\chi|G(d\lambda)|\chi\rangle$ or $P_{\lambda,\mathscr{E}(g)}(\{1\})$ vanishes. Thus, for every subset A of the set

$$\Lambda_g := \left\{\lambda \in \Lambda : P_{\lambda,\mathscr{E}(g)}(\{1\}) > 0\right\} \tag{5.53}$$

where $P_{\lambda,\mathscr{E}(g)}(\{1\})$ is nonzero, $\langle\chi|G(A)|\chi\rangle = 0$, so $G(A) = \alpha\, P_g$ for some $\alpha \in [0, 1]$. In fact,

$$G(\Lambda_g) = P_g\,, \tag{5.54}$$

because if it were strictly less than P_g, then also the left-hand side of (5.51) would be strictly less than P_g. Moreover, for 1d subspaces $g \neq h$,

$$G(\Lambda_g \cap \Lambda_h) = 0 \tag{5.55}$$

because it equals αP_g and βP_h, which is only possible if $\alpha = \beta = 0$.

Now let ψ_1 and ψ_2 be mutually orthogonal unit vectors, set $\psi_3 = \frac{1}{\sqrt{2}}(\psi_1 + \psi_2)$ and $g_i = \mathbb{C}\psi_i$ for $i = 1, 2, 3$. Then

$$G\big(\Lambda_{g_1} \cup \Lambda_{g_2} \cup \Lambda_{g_3}\big) = G(\Lambda_{g_1}) + G(\Lambda_{g_2}) + G(\Lambda_{g_3}) = P_{g_1} + P_{g_2} + P_{g_3}\,, \tag{5.56}$$

which has eigenvalue 2 in the direction of ψ_3, in contradiction to $G(S) \leq G(\Lambda) = I$ for every $S \subseteq \Lambda$. □

Let us pause a bit to consider the significance of this conclusion. Limitations to knowledge are inevitable. Regardless of which view of quantum mechanics we prefer, limitations to knowledge are a fact. There are facts in nature that we cannot find out empirically. Nature can keep a secret.

Limitations to knowledge are perhaps rather shocking. They may seem to go against the principles of science. Upon reflection, we should rather say that they go against the principles of positivism. Actually, we can say that positivism is provably wrong.

We will see other proofs that limitations to knowledge arise in every theory of quantum mechanics in Sect. 5.3 in the context of the PBR theorem and in Sect. 5.4.4 in the context of statistical density matrices.

In classical mechanics, there do not appear to be any limitations to knowledge. They are a feature of quantum physics. Maybe the shocking feel of limitations to knowledge is a relict of classical thinking. There is also a parallel between the limitations to knowledge and Sadi Carnot's (1796–1832) 1824 theory of heat engines: Not all of the energy contained in a system can be extracted in a useful form (i.e., as work); not all of the information contained in a system can be extracted in a useful form (i.e., as human knowledge).

5.1.5 Limitations to Knowledge in Theories We Know

Now that we know that limitations to knowledge are a fact, it is of interest to take another look at some instantiations in Bohmian mechanics, GRW theory, and many-worlds.

Bohmian Mechanics In Bohmian mechanics, there are limitations to knowing the exact positions of particles, mentioned briefly already in Sect. 2.2.2. Let $\varphi(\boldsymbol{x})$ be a particle's conditional wave function as defined in (3.45). It has been shown (Dürr, Goldstein, and Zanghì 1992 [13]) in great generality that inhabitants of a Bohmian universe (made out of Bohmian particles) cannot know the particle's position $\boldsymbol{X}$ more precisely than that it has $|\varphi|^2$ distribution. The basic reason is easy to understand: Suppose information about $\boldsymbol{X}$ (say, whether $\boldsymbol{X} \in B$ or $\boldsymbol{X} \in B^c = \mathbb{R}^3 \setminus B$ for some region $B \subset \mathbb{R}^3$) is contained in the configuration Y of Alice's lab (in the shape of the ink in her notebook, or in the configuration of her computer, or in that of her brain); say, Y lies in some region Ω_B in configuration space if $\boldsymbol{X} \in B$ but in some disjoint region Ω_{B^c} if $\boldsymbol{X} \in B^c$. Then the total configuration $Q = (\boldsymbol{X}, Y)$ can only lie in $(B \times \Omega_B) \cup (B^c \times \Omega_{B^c})$. But Q is $|\Psi|^2$-distributed, so the total wave function Ψ must be supported in $(B \times \Omega_B) \cup (B^c \times \Omega_{B^c})$. And then the conditional wave function $\varphi(\boldsymbol{x})$ is given by $\Psi(\boldsymbol{x}, Y)$ up to a normalizing factor. Thus, if $Y \in \Omega_B$, then φ is supported in B, and if $Y \in \Omega_{B^c}$, then φ is supported in B^c. In each case, the $|\varphi|^2$ distribution is compatible with Alice's knowledge about $\boldsymbol{X}$. Conversely, if $|\varphi|^2$ has a certain width σ then Alice cannot know $\boldsymbol{X}$ more precisely than with inaccuracy σ. This fact is called *absolute uncertainty*.

More generally, a basic property of the conditional wave function is the *fundamental conditional probability formula*

$$\mathbb{P}\big(\boldsymbol{X} \in d\boldsymbol{x} \big| Y\big) = |\varphi(\boldsymbol{x})|^2 \, d\boldsymbol{x} \,, \tag{5.57}$$

which follows directly from the fact that the pair $(\boldsymbol{X}, Y)$ has joint distribution density $|\Psi|^2$. This relation implies that the full information contained in Y provides no more probabilistic knowledge about $\boldsymbol{X}$ than that it is $|\varphi|^2$ distributed.

As another consequence of limitations to knowledge, there are also *limitations to control*: Inhabitants of a Bohmian universe can prepare a system or particle to have a particular wave function φ, but they *cannot* prepare the particle's position $\boldsymbol{X}$ to be a particular point $\boldsymbol{x}_0$, unless $\varphi(\boldsymbol{x}) = \delta^3(\boldsymbol{x} - \boldsymbol{x}_0)$. So, the "hidden variable" $\boldsymbol{X}_k(t)$ could better be called an "uncontrollable variable." In fact, they cannot prepare $\boldsymbol{X}$ any more accurately, or any differently, than being random with distribution $|\varphi|^2$. After all, if they could *prepare* $\boldsymbol{X}$ more accurately, then they would afterward also *know* $\boldsymbol{X}$ more accurately than according to $|\varphi|^2$, or else the particle's conditional wave function would not be the desired function φ.

GRW Theory In GRWf theory, an obvious thing to want to measure is the times and locations of the flashes, or how many flashes occur in a certain space-time region. This is another example of a *genuine measurement*, one that finds out values that exist in nature, as opposed to quantum measurements. Also in GRWm, even though there are no flashes, we may want to measure the times and centers of the collapses, which is when–where the flashes would be in GRWf. Again, there are strong limitations to the possibility of detecting collapses. The question has been analyzed by Cowan et al. (2016) [11]; here are some of the findings.

Macroscopically and with macroscopic inaccuracy, the number of flashes in a space-time region can be determined. However, on the microscopic level, this is not fully possible, but only with limited probability of success. For example, consider a wave function as in the Einstein box example, consisting of two widely separated packets,

$$\psi = \frac{1}{\sqrt{2}}|\text{here}\rangle + \frac{1}{\sqrt{2}}|\text{there}\rangle\,. \tag{5.58}$$

Let us suppose that each packet is narrower than σ, so it will not be changed significantly by a collapse with center in its support. That is, to a good degree of approximation, ψ collapses, if it collapses, to either $|\text{here}\rangle$ or $|\text{there}\rangle$. Suppose we let the system sit during $[t_1, t_2]$ and want to determine at t_2 whether it has collapsed or not, i.e., we want to determine the number $C \in \{0, 1\}$ of collapses. Suppose the probability of a collapse during $[t_1, t_2]$ is $0 < p < 1$, so

$$\mathbb{P}\big(\psi(t_2) = \psi\big) = 1 - p\,, \quad \mathbb{P}\big(\psi(t_2) = |\text{here}\rangle\big) = p/2\,, \quad \mathbb{P}\big(\psi(t_2) = |\text{there}\rangle\big) = p/2\,. \tag{5.59}$$

Then no experiment can determine C with certainty. Any experiment can only yield probabilistic information about C, and the probability of correctly retrodicting C on the basis of this information is bounded away from 1:

Proposition 5.1 (Cowan et al. 2014 [10], 2016 [11]) *Define the reliability $R(\mathscr{E})$ of an experiment $\mathscr{E}$ with outcome Z in the value space $\mathscr{Z} = \{0, 1\}$ as the probability of correctly retrodicting C, $R(\mathscr{E}) = \mathbb{P}(Z = C)$. For every $\mathscr{E}$,*

$$R(\mathscr{E}) \leq \begin{cases} 1 - \frac{p}{2} & \textit{if } p \in (0, \frac{2}{3}] \\ p & \textit{if } p \in [\frac{2}{3}, 1)\,. \end{cases} \tag{5.60}$$

These bounds are sharp, i.e., attained by some $\mathscr{E}$.

The graph of the right-hand side of (5.60) is shown in Fig. 5.1. We might expect to obtain more information by interacting with the system already before t_2; but it turns out that this cannot be done without changing the probability distribution of the flashes. We might also consider situations in which we do not know the initial wave function; as readers might have guessed, the possibilities of detecting collapses then get even worse.

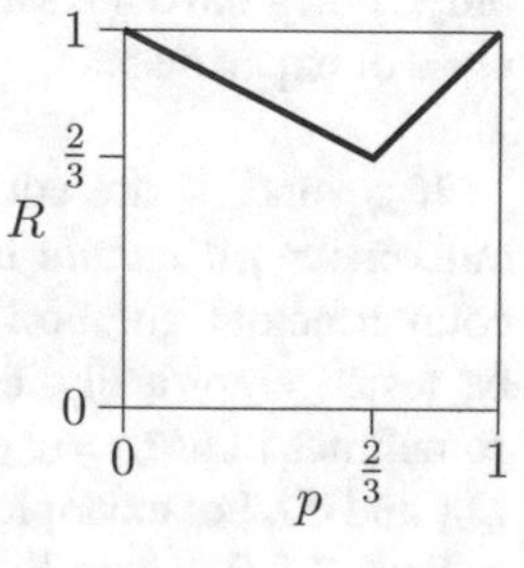

Fig. 5.1 The upper bound on the reliability of any experiment detecting collapses given in (5.60) as a function of p

The upshot is that in a GRW universe, nature knows the exact number of collapses in a system during $[t_1, t_2]$, and observers do not.

Many-Worlds In many-worlds theories, limitations to knowledge are in place, too, and in fact play a central role. Let me briefly mention two important ones, apart from the impossibility to measure wave functions: In the Sm theory, it is impossible to measure the matter density $m(t, \boldsymbol{x})$. And in any many-worlds theory, it is impossible for inhabitants of one world to observe the other worlds.

In retrospect, the positivistic idea that "what cannot be observed is not real" sounds rather absurd. For example, it could have been applied to conclude that no more events can happen inside a spaceship once it has crossed the horizon (surface) of a black hole because no signals can leave a black hole. As Mihalis Dafermos once put it (2015),

> People in black holes have rights, too!

5.1.6 The Concept of Observable

I can now say something about *what an observable ultimately is*, as the concept of POVM gives us a new perspective on the concept of an observable. Before POVMs, by an "observable" we meant a self-adjoint operator. In the literature, people often talk about self-adjoint operators as if they were quantities. Of course, strange and mysterious kinds of quantities they must be, as they may have indeterminate values and may fail to commute. Enter POVMs. They are not only more generally applicable, they also play a different role in two ways: First, they are associated with experiments, and second, they translate unit vectors in Hilbert space into probability distributions—distributions that govern the outcome of the experiment. Now the concept of observable is no longer about quantities, it is about probability distributions. Here is the natural general definition:

Definition 5.5 Two experiments (that can be carried out on arbitrary wave functions $\psi \in \mathscr{H}$ with norm 1) are *equivalent in law* if and only if for every $\psi \in \mathbb{S}(\mathscr{H})$, they have the same distribution of the outcome. (Thus, they are equivalent in law if and only if they have the same POVM.) An *observable* is a corresponding equivalence class of experiments.

If $\mathscr{E}_1$ and $\mathscr{E}_2$ are equivalent in law and a particular run of $\mathscr{E}_1$ has yielded the outcome z_1, it *cannot* be concluded that $\mathscr{E}_2$ would have yielded z_1 as well. The counterfactual question, "what would z_2 have been if we had run $\mathscr{E}_2$?" cannot be tested empirically, but it can be analyzed in Bohmian mechanics; there, one sometimes finds $z_2 \neq z_1$ (for the same Q_S and ψ in both experiments, but different Q_A and ϕ). For example, we have encountered this phenomenon as "contextuality" in Sect. 2.3.9, where $\mathscr{E}_1$ was a Stern–Gerlach experiment in the z direction and $\mathscr{E}_2$

the Stern–Gerlach experiment with inverted polarity as depicted in Fig. 2.5. Then $\mathscr{E}_1$ and $\mathscr{E}_2$ are equivalent in law, although in Bohmian mechanics, the two experiments will often yield different results when applied to the same 1-particle wave function and position.

This situation illustrates why the term “observable” can be rather misleading: It is intended to suggest “observable quantity,” but an observable is not even a well-defined quantity to begin with (as the outcome Z depends on Q_A and ϕ), it is *a class of experiments with equal probability distributions*, just like “rolling a die” is an equivalence class of experiments (the distribution is uniform on $\{1, 2, 3, 4, 5, 6\}$) that have different outcomes. Dice are “contextual”!

This point is also connected to Wheeler’s fallacy. Recall the delayed-choice experiment, but now consider detecting the particle either directly at the slits or far away, ignoring the interference region. As $\mathscr{E}_1$, we put detectors directly at the slits and say that the outcome is $Z_1 = +1$ if the particle was detected in the upper slit and $Z_1 = -1$ if in the lower one. This is a kind of position measurement that can be represented in the 2d Hilbert space formed by wave functions of the form

$$\psi = c_1|\text{upper slit}\rangle + c_2|\text{lower slit}\rangle\,, \tag{5.61}$$

so $\mathbb{P}(Z_1 = +1) = |c_1|^2$. Relative to the basis $\{|\text{upper slit}\rangle, |\text{lower slit}\rangle\}$, the POVM is the spectral PVM of σ_3. As $\mathscr{E}_2$, we put the detectors far away and say that $Z_2 = +1$ if the particle was detected in the lower cluster and $Z_2 = -1$ if in the upper cluster. ψ evolves to

$$\psi' = c_1|\text{lower cluster}\rangle + c_2|\text{upper cluster}\rangle\,, \tag{5.62}$$

so $\mathbb{P}(Z_2 = +1) = |c_1|^2$. So, Z_1 and Z_2 have the same distribution for any choice of $(c_1, c_2) \in \mathbb{S}(\mathbb{C}^2)$, $\mathscr{E}_1$ and $\mathscr{E}_2$ have the same POVM, and the two experiments are equivalent in law, although we know that the Bohmian particle often passes through the lower slit and still ends up in the lower cluster.

Now comes the point that has confused a number of authors:[6] Since $\mathscr{E}_1$ measures the “position observable,” and since $\mathscr{E}_1$ and $\mathscr{E}_2$ “measure” the same observable, it is clear that $\mathscr{E}_2$ also measures the position observable. People concluded that $\mathscr{E}_2$ “measures through which slit the particle went”—Wheeler’s fallacy! People concluded further that since the Bohmian trajectory may pass through the upper slit while $Z_2 = -1$, Bohmian mechanics must somehow disagree with measured facts about which slit the particle went through. Bad, bad Bohm, they concluded. (Some authors called Bohm’s trajectories “surrealistic,” perhaps alluding to the dreamlike, absurd content of Salvador Dalí’s paintings, to brand the realist view as absurd.) Of course, it is the other way around: The “measurement” did not at all measure which slit the particle went through.

---

[6] For example (using a different but similar setup), Englert et al. (1992) [16] (this is Berthold-Georg Englert, not Nobel laureate François Englert).

Here is a variant of the example, due to Englert et al. (1992) [16]. In a double-slit experiment with a spin-$\frac{1}{2}$ particle, arrange that a pure spin-up wave emanates from the upper slit, and a pure spin-down wave from the lower slit,

$$\psi = c_1|\uparrow\rangle|\text{upper slit}\rangle + c_2|\downarrow\rangle|\text{lower slit}\rangle\,. \tag{5.63}$$

At the screen, after measuring the position, also measure the z-spin. (Practically, we could make a hole in the screen, so that only the part of the wave function arriving at a certain position can move on, and place a Stern–Gerlach experiment behind the hole.) There will be no interference pattern in the position results because the partial waves from each slit do not interfere with each other. Now let Z be the result of the spin measurement. In the 2d Hilbert space formed by the wave functions of the form (5.63), the POVM is again the spectral PVM of σ_3, and this motivated people to say that the spin measurement was really a measurement of the position observable at the moment of passing the slits, and that if $Z = +1$ then the particle went through the upper slit—which is not true according to Bohm's equation of motion (2.120) for a spin-$\frac{1}{2}$ particle. Wheeler's fallacy once more!

5.2 Time of Detection

5.2.1 The Problem

Suppose we set up a detector, wait for the arrival of the particle at the detector, and measure the time T at which the detector clicks. What is the probability distribution of T? This is a natural question not covered by the usual quantum formalism because there is no self-adjoint operator for time. But from the main theorem about POVMs, it is clear that there must be a POVM E such that

$$\mathbb{P}(T \in B) = \langle\psi_0|E(B)|\psi_0\rangle\,. \tag{5.64}$$

That is, time of detection is a generalized observable. In this section we take a look at this POVM E.

Suppose that we form a surface $\Sigma \subset \mathbb{R}^3$ out of little detectors so we can measure the time and the location at which the quantum particle first crosses Σ. Suppose further that, as depicted in Fig. 5.2, Σ divides physical space $\mathbb{R}^3$ into two regions, Ω and its complement, and the particle's initial wave function ψ_0 is concentrated in Ω. The outcome of the experiment is the pair $Z = (T, \boldsymbol{X})$ of the time $T \in [0, \infty)$ of detection and the location $\boldsymbol{X} \in \Sigma$ of detection; should no detection ever occur, then we write $Z = \infty$. So the value space of E is $\mathscr{Z} = [0, \infty) \times \Sigma \cup \{\infty\}$, and E acts on $L^2(\Omega)$. We want to compute the distribution of Z from ψ_0.

Let us compare the problem to Born's rule. In Born's rule, we choose a time t_0 and measure the three position coordinates at time t_0; here, if we take Ω to be the half space $\{(x, y, z) : x > x_0\}$ and Σ its boundary plane $\{(x, y, z) : x = x_0\}$, then we choose the value of one position coordinate (x_0) and measure the time as well as the

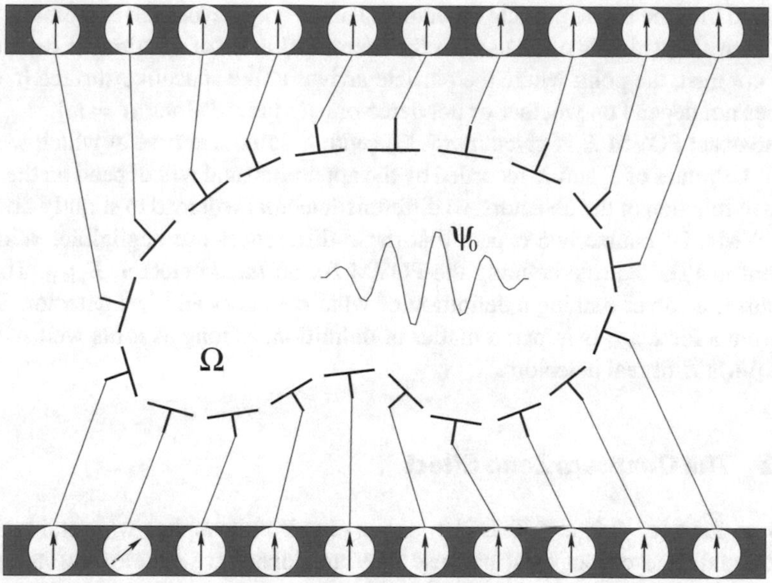

Fig. 5.2 A quantum particle in a region Ω surrounded by a surface $\Sigma = \partial\Omega$ made out of detectors (symbolized by $\perp$'s), each of which is connected to a pointer. In part, the figure depicts the situation before the experiment, as the initial wave function ψ_0 is symbolized by a wave, and in part the situation after the experiment, as the location of detection is indicated by one pointer in the triggered position (Adapted from Dürr and Teufel (2009) [12, Fig. 16.2] with permission by Springer)

other two position coordinates when the particle reaches that value. Put differently in terms of space-time $\mathbb{R}^4 = \{(t, x, y, z)\}$, Born's rule concerns measuring where the particle intersects the spacelike 3-surface $\{t = t_0\}$, and our problem concerns measuring where the particle intersects the timelike 3-surface $\{x = x_0\}$. We could say that we need a Born rule for timelike 3-surfaces.

I should make three caveats, though.

- I have used language such as "particle arriving at a surface" that presupposes the existence of trajectories although we know that some theories of quantum mechanics (GRWm and GRWf) claim that there are no trajectories, and still these theories are approximately empirically equivalent to Bohmian mechanics, so the time and location of the detector click would have approximately the same distribution as in Bohmian mechanics. Our problem really concerns the distribution of the detection events, and we should keep in mind that in some theories the trajectory language cannot be taken seriously.
- Even in Bohmian mechanics, there is a crucial difference between the case with the spacelike surface and the one with the timelike surface: The point where the particle arrives on the timelike surface $\{x = x_0\}$ may depend on whether or not detectors are present on that surface. A detector that does not click may still affect

ψ and thus the future particle trajectory. That is why I avoid the expression "time of arrival" (which is often used in the literature) in favor of "time of detection." In contrast, the point where the particle arrives at the spacelike surface $\{t = t_0\}$ does not depend on whether or not detectors are placed along $\{t = t_0\}$.

- The exact POVM E is given by (5.33) (with t_f some late time at which we read off the values of T and X recorded by the apparatus) and will depend on the exact wave function of the detectors, so different detectors will lead to slightly different POVMs. Of course, we expect that these differences are negligible. What we want is a simple rule defining the POVM for an *ideal* detector, E_{ideal}. That, of course, involves making a definition of what counts as an ideal detector. So the formula for E_{ideal} is in part a matter of definition, as long as it fits well with the POVMs E of real detectors.

5.2.2 The Quantum Zeno Effect

Zeno of Elea (c. 490–c. 430 BCE) was a Greek philosopher who claimed that motion and time cannot exist because they are inherently paradoxical notions, a claim which he tried to support by formulating various paradoxes, including one involving Achilles and a turtle. In modern times, Alan Turing (of computer science fame, lived 1912–1954) reportedly first discovered the following effect, which was later named after Zeno because of its paradoxical flavor: Suppose a quantum particle moves in 1d, and its initial wave function $\psi_0(x)$ is concentrated in the negative half axis $(-\infty, 0)$. We want to model, as a kind of time measurement, a detector, located at the origin, that clicks when the particle arrives. To this end, we choose a small time resolution $\tau > 0$ and perform, at times $\tau, 2\tau, 3\tau, \ldots$, a quantum measurement of $1_{x \geq 0}$, i.e., of whether the particle is in the right half axis. The ideal detector would seem to correspond to the limit $\tau \to 0$; however, in this limit, the probability that the detector *ever* clicks is 0. Misra and Sudarshan (1977) [31] wrote:

> A watched pot never boils.

Exercise 5.3 (Quantum Zeno Effect) Prove the following simplified version: In a 2d Hilbert space $\mathbb{C}^2$, let $\psi_0 = (1, 0)$ evolve with Hamiltonian $H = \sigma_1$, interrupted by a quantum measurement of σ_3 at times $n\tau$ for all $n \in \mathbb{N}$. For any fixed $T > 0$, the probability that any of the $\approx T/\tau$ measurements in the time interval $[0, T]$ yields the result -1 tends to 0 as $\tau \to 0$. (Hint: Show that $\psi_{\tau-} = (\cos\tau, -i\sin\tau)$, that the probability of always obtaining $+1$ in m consecutive trials is $\cos^{2m}\tau$, and that this quantity tends to 1 as $\tau \to 0$, $m \to \infty$ so that $m\tau \to T$.)

5.2.3 Allcock's Paradox

Allcock (1969) [2] considered a "soft detector," i.e., one for which the particle may fly through the detector volume for a while before being detected. An as effective description, Allcock proposed an imaginary potential. For example, for a single particle in 1d and a detector in the right half axis, he considered the Schrödinger equation

$$i\hbar\frac{\partial\psi}{\partial t} = -\frac{\hbar^2}{2m}\frac{\partial^2\psi}{\partial x^2} - iv1_{x\geq 0}\psi \tag{5.65}$$

with $v > 0$ a constant. The time evolution then is not unitary, and the Hamiltonian $H = -\frac{\hbar^2}{2m}\partial_x^2 - iv1_{x>0}$ is not self-adjoint.

Exercise 5.4 (Imaginary Potential) Derive from (5.65) the continuity equation

$$\frac{\partial\rho}{\partial t} = -\mathrm{div}\, j - \tfrac{2v}{\hbar}1_{x\geq 0}\rho \tag{5.66}$$

for $\rho = |\psi|^2$ and $j = \frac{\hbar}{m}\mathrm{Im}[\psi^*\partial\psi/\partial x]$.

Equation (5.66) is the evolution of the probability density of a particle that moves along Bohmian trajectories and disappears spontaneously (stochastically) at rate $2v/\hbar$ whenever it stays in the region $x \geq 0$. $\|\psi_t\|^2$ is a decreasing function of t and represents the probability that the particle has not been detected (and disappeared from the model) yet.

To obtain an effective description of a hard detector (i.e., one that will detect the particle as soon as it reaches the region $x \geq 0$), Allcock assumed that ψ_0 is concentrated in $\{x < 0\}$ and took the limit $v \to \infty$, but found that the particle never gets detected ($\|\psi_t\|^2 = \text{const.}$)! That is parallel to the quantum Zeno effect.

Exercise 5.5 (Simplified Version of Allcock's Paradox) In a 2d Hilbert space $\mathbb{C}^2$, let $\psi_0 = (1, 0)$ evolve with the (non-self-adjoint) Hamiltonian

$$H_v = \begin{pmatrix} 0 & 1 \\ 1 & -iv \end{pmatrix}. \tag{5.67}$$

Prove that for every $t > 0$, $\psi_t = e^{-iH_v t/\hbar}\psi_0 \to \psi_0$ as $v \to \infty$. (Hint: Express ψ_t in terms of the eigenvectors and eigenvalues of H_v and consider their limits as $v \to \infty$.)

5.2.4 The Absorbing Boundary Rule

The question of what E_{ideal} is is not fully settled; I will describe the most plausible proposal, the *absorbing boundary rule* (Werner 1987 [51], Tumulka 2016 [44]). Such a rule was for a long time believed to be impossible because of the quantum

Fig. 5.3 The outward unit normal vector $\boldsymbol{n}(\boldsymbol{x})$ to the surface

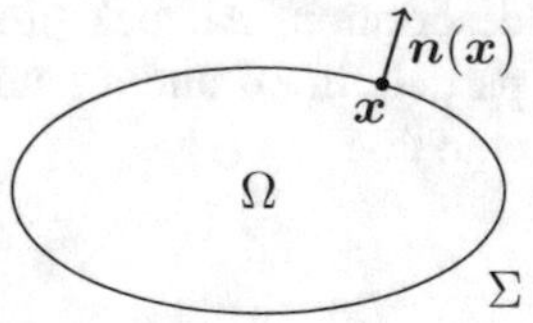

Zeno effect and Allcock's paradox. Henceforth I will write E instead of E_{ideal}. Let $\Sigma = \partial\Omega$, ψ_0 be concentrated in Ω, $\|\psi_0\| = 1$, and let $\kappa > 0$ be a constant of dimension 1/length (it will be a parameter of the detector). Here is the rule:

Absorbing Boundary Rule Solve the Schrödinger equation

$$i\hbar\frac{\partial\psi}{\partial t} = -\frac{\hbar^2}{2m}\nabla^2\psi + V\psi \tag{5.68}$$

in Ω with potential $V : \Omega \to \mathbb{R}$ and boundary condition

$$\frac{\partial\psi}{\partial n}(\boldsymbol{x}) = i\kappa\psi(\boldsymbol{x}) \tag{5.69}$$

at every $\boldsymbol{x} \in \Sigma = \partial\Omega$, with $\partial/\partial n$ the outward normal derivative on the surface, $\partial\psi/\partial n := \boldsymbol{n}(\boldsymbol{x}) \cdot \nabla\psi(\boldsymbol{x})$ with $\boldsymbol{n}(\boldsymbol{x})$ the outward unit normal vector to Σ at $\boldsymbol{x} \in \Sigma$ (see Fig. 5.3).

Then, the rule asserts:

$$\mathbb{P}_{\psi_0}\Big(t_1 \le T < t_2, \boldsymbol{X} \in B\Big) = \int_{t_1}^{t_2} dt \int_B d^2\boldsymbol{x}\; \boldsymbol{n}(\boldsymbol{x}) \cdot \boldsymbol{j}^{\psi_t}(\boldsymbol{x}) \tag{5.70}$$

for any $0 \le t_1 < t_2$ and any set $B \subseteq \Sigma$, with $d^2\boldsymbol{x}$ the surface area element and $\boldsymbol{j}^{\psi}$ the probability current vector field (1.22). In other words, the joint probability density of T and $\boldsymbol{X}$ relative to $dt\, d^2\boldsymbol{x}$ is the normal component of the current across the boundary, $j_n^{\psi_t}(\boldsymbol{x}) = \boldsymbol{n}(\boldsymbol{x}) \cdot \boldsymbol{j}^{\psi_t}(\boldsymbol{x})$. Furthermore,

$$\mathbb{P}_{\psi_0}(Z = \infty) = 1 - \int_0^\infty dt \int_\Sigma d^2\boldsymbol{x}\; \boldsymbol{n}(\boldsymbol{x}) \cdot \boldsymbol{j}^{\psi_t}(\boldsymbol{x})\,. \tag{5.71}$$

This completes the statement of the rule. □

Let us study the properties of the rule. To begin with, the boundary condition (5.69) implies that the current vector $\boldsymbol{j}$ at the boundary is always outward-pointing: Indeed, for every $\boldsymbol{x} \in \Sigma$,

$$\boldsymbol{n}(\boldsymbol{x}) \cdot \boldsymbol{j}(\boldsymbol{x}) = \frac{\hbar}{m}\mathrm{Im}\Big(\psi(\boldsymbol{x})^* \frac{\partial \psi}{\partial n}(\boldsymbol{x})\Big) = \frac{\hbar}{m}\mathrm{Im}\Big(\psi(\boldsymbol{x})^* i\kappa\psi(\boldsymbol{x})\Big) = \frac{\hbar\kappa}{m}|\psi(\boldsymbol{x})|^2 \geq 0\,. \tag{5.72}$$

For this reason, (5.69) is called an *absorbing boundary condition*: It implies that there is never any current coming out of the boundary. In particular, the right-hand side of (5.70) is nonnegative.

So the rule invokes a new kind of time evolution for a 1-particle wave function as an effective treatment of the whole system formed by one particle and the detectors together. It is useful to picture the Bohmian trajectories for this time evolution. Equation (5.72) implies that the Bohmian velocity field $\boldsymbol{v}(\boldsymbol{x})$ is always outward-pointing at the boundary, $\boldsymbol{n}(\boldsymbol{x}) \cdot \boldsymbol{v}(\boldsymbol{x}) > 0$ for all $\boldsymbol{x} \in \Sigma$; in fact, the normal velocity is prescribed, $\boldsymbol{n}(\boldsymbol{x}) \cdot \boldsymbol{v}(\boldsymbol{x}) = \hbar\kappa/m$. In particular, Bohmian trajectories can cross Σ only in the outward direction; when they do, they end on Σ, as ψ is not defined behind Σ. Put differently, no Bohmian trajectories begin on Σ; they all begin at $t = 0$ in Ω with $|\psi_0|^2$ distribution. In fact, the right-hand side of (5.70) is exactly the probability distribution of the space-time point at which the Bohmian trajectory reaches the boundary. That is not surprising, as in a Bohmian world we would expect the detector to click when and where the particle reaches the detecting surface. As a further consequence, the right-hand side of (5.71) is exactly the probability that the Bohmian trajectory never reaches Σ. In particular, (5.70) and (5.71) together define a probability distribution on $\mathscr{Z}$. If we had evolved ψ_0 with the Schrödinger equation on $\mathbb{R}^3$ without boundary condition on Σ, then some Bohmian trajectories may cross Σ several times in both directions; this illustrates that the trajectory in the presence of detectors can be different from what it would have been in the absence of detectors.

Since probability can only be lost at the boundary, never gained,

$$\|\psi_t\|^2 = \int_\Omega d^2\boldsymbol{x}\, |\psi_t(\boldsymbol{x})|^2 \tag{5.73}$$

can only decrease with t, never increase. So here we are dealing with a new kind of Schrödinger equation whose time evolution is not unitary as the norm of ψ is not conserved. Correspondingly, the Hamiltonian H, defined by $-\frac{\hbar^2}{2m}\nabla^2 + V$ with the boundary condition (5.69), is not self-adjoint. The time evolution operators W_t, defined by the property $W_t\psi_0 = \psi_t$, have the following properties: First, they are not unitary but satisfy $\|W_t\psi\| \leq \|\psi\|$; such operators are called *contractions*. Second, $W_s W_t = W_{s+t}$ and $W_0 = I$; a family $(W_t)_{t\geq 0}$ with this property is called a *semigroup*. Thus, the W_t form a *contraction semigroup*. Using the Hille–Yosida theorem from functional analysis, one can prove:

Theorem 5.3 (Teufel et al. 2019 [40]) *Suppose the potential function* V *is bounded. For every* $\kappa > 0$*, the Schrödinger equation* (5.68) *with the boundary condition* (5.69) *defines a contraction semigroup* $(W_t)_{t\geq 0}$, $W_t : L^2(\Omega) \to L^2(\Omega)$.

In particular, (5.68) *with* (5.69) *possesses a unique solution* $\psi(t, \boldsymbol{x})$ *for every initial datum* $\psi_0 \in L^2(\Omega)$.

In fact, $\|\psi_t\|^2$ is the probability that the Bohmian particle is still somewhere in Ω at time t, that is, has not reached the boundary yet. In particular, as an alternative to (5.71) we can write

$$\mathbb{P}(Z = \infty) = \lim_{t\to\infty} \|\psi_t\|^2 . \tag{5.74}$$

The conclusions from our considerations about Bohmian trajectories can also be obtained from the Ostrogradsky–Gauss integral theorem (divergence theorem) in four dimensions: The four-vector field $j = (\rho, \boldsymbol{j})$ has vanishing four-divergence, as that is what the continuity equation (1.21) expresses. Integrating the divergence over $[0, t] \times \Omega$ yields

$$0 = \int_0^t dt' \int_\Omega d^3\boldsymbol{x}\, \mathrm{div}\, j(t', \boldsymbol{x}) \tag{5.75}$$

$$= \int_\Omega d^3\boldsymbol{x}\, \rho(t, \boldsymbol{x}) - \int_\Omega d^3\boldsymbol{x}\, \rho(0, \boldsymbol{x}) + \int_0^t dt' \int_\Sigma d^2\boldsymbol{x}\, \boldsymbol{n}(\boldsymbol{x}) \cdot \boldsymbol{j}(t', \boldsymbol{x}) \tag{5.76}$$

$$= \|\psi_t\|^2 - 1 + \int_0^t dt' \int_\Sigma d^2\boldsymbol{x}\, \boldsymbol{n}(\boldsymbol{x}) \cdot \boldsymbol{j}(t', \boldsymbol{x}) . \tag{5.77}$$

Since the last integrand is nonnegative, $\|\psi_t\|^2$ is decreasing with time and equals 1—the flux of $\boldsymbol{j}$ into the boundary during $[0, t]$. In particular,

$$\lim_{t\to\infty} \|\psi_t\|^2 = 1 - \int_0^\infty dt' \int_\Sigma d^2\boldsymbol{x}\, \boldsymbol{n}(\boldsymbol{x}) \cdot \boldsymbol{j}(t', \boldsymbol{x}) , \tag{5.78}$$

so (5.71) is nonnegative, and (5.70) and (5.71) together define a probability distribution.

So what is the POVM E? It is given by

$$E\big(dt \times d^2\boldsymbol{x}\big) = \frac{\hbar\kappa}{m} W_t^\dagger\, |\boldsymbol{x}\rangle\langle\boldsymbol{x}|\, W_t\, dt\, d^2\boldsymbol{x} \tag{5.79}$$

$$E(\{\infty\}) = \lim_{t\to\infty} W_t^\dagger W_t . \tag{5.80}$$

Since the $E(dt)$ are not projections, there are in general no eigenstates of detection time.

Variants of the absorbing boundary rule have been developed for moving surfaces, systems of several detectable particles, and particles with spin [43].

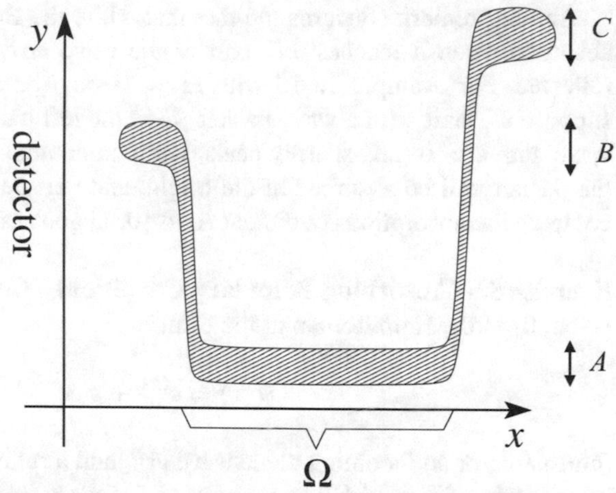

Fig. 5.4 Region in configuration space where one should expect the wave function to propagate in, as explained in the text

Here is the reasoning leading to the absorbing boundary rule. For simplicity, let Ω be an interval in 1d, let x be the coordinate of the particle P, and let y be the configuration of the detectors D. The whole system $S = P \cup D$ evolves unitarily with initial wave function $\Psi_0 = \psi_0 \otimes \varphi_0$. Let A be the region of y-configurations in which the detectors have not clicked (where the "ready state" φ_0 is concentrated), B where the left detector has fired, and C the right one. So, Ψ_0 is concentrated in $\Omega \times A$; see Fig. 5.4.

The interaction between P and D occurs, not in the interior of $\Omega \times A$, but only near the boundary $\partial\Omega \times A$: Any probability current in $\Omega \times A$ that reaches $\partial\Omega \times A$ will be transported quickly to $\partial\Omega \times B$ or $\partial\Omega \times C$ and then remain in $\mathbb{R} \times B$ or $\mathbb{R} \times C$, regions of configuration space that are macroscopically separated from $\Omega \times A$. Due to this separation, parts of Ψ that have reached B or C will not be able to propagate back to A and interfere there with parts of Ψ that have not yet left A; that is, the detection is practically irreversible, resulting in decoherence between the parts of the wave function in A, B, and C, and the motion of the Bohmian configuration from A to B or C is one-way. As a consequence, the x-component of the current at $\partial\Omega \times A$ should point outward. We are thus led to the following picture: (i) The Schrödinger equation (1.1) holds for ψ inside Ω. (ii) Something happens at $\partial\Omega$, which should not depend sensitively on the details of the initial detector state φ_0. (iii) The evolution of ψ_t in Ω is still linear, but no longer unitary because ψ_t corresponds to only a part of the full wave function Ψ_t, i.e., the part in A. (iv) The current $\boldsymbol{j}^{\psi_t}(\boldsymbol{x})$ at $\boldsymbol{x} \in \partial\Omega$ always points outward. (v) The evolution of Ψ_t in A is autonomous, i.e., not affected by whatever Ψ_t looks like in $\mathbb{R} \times B$ or $\mathbb{R} \times C$, as those parts cannot propagate back to $\mathbb{R} \times A$. (vi) Thus, the evolution of ψ_t in Ω should be autonomous, depending only on few parameters ("κ") encoding properties of the detectors. These features suggest an absorbing boundary condition at $\partial\Omega$ for ψ_t.

Another remark concerns the fact that while the Bohmian particle is sure to be absorbed when it reaches $\partial\Omega$, part of the wave arriving at the boundary will be reflected. For example, in 1d with $\Omega = (-\infty, 0]$ and the detector at the origin, suppose we start with a wave packet ψ_0 in the left half axis that is close to a plane wave with $k > 0$ (i.e., sharply peaked in momentum space around k). Then part of the packet will be absorbed at the origin, and part be reflected. A quick recipe to compute the absorption coefficient $A_k \in [0, 1]$ goes as follows.

Exercise 5.6 (Absorbing Boundary Condition) Consider an eigenfunction ψ : $(-\infty, 0]$ of the Hamiltonian of the form

$$\psi(x) = e^{ikx} + c_k\, e^{-ikx} \tag{5.81}$$

(consisting of an incoming plane wave e^{ikx} and a reflected wave $c_k\, e^{-ikx}$). Use the boundary condition (5.69) to compute c_k for every $k > 0$.

Since the strength of the reflected wave is $|c_k|^2$, that is the fraction of the wave that gets reflected, while the fraction $A_k := 1 - |c_k|^2$ gets absorbed; Fig. 5.5 shows a plot of A_k. The maximum occurs at $k = \kappa$, so the value κ characterizes the energy $\hbar^2\kappa^2/2m$ at which the detector is maximally efficient.

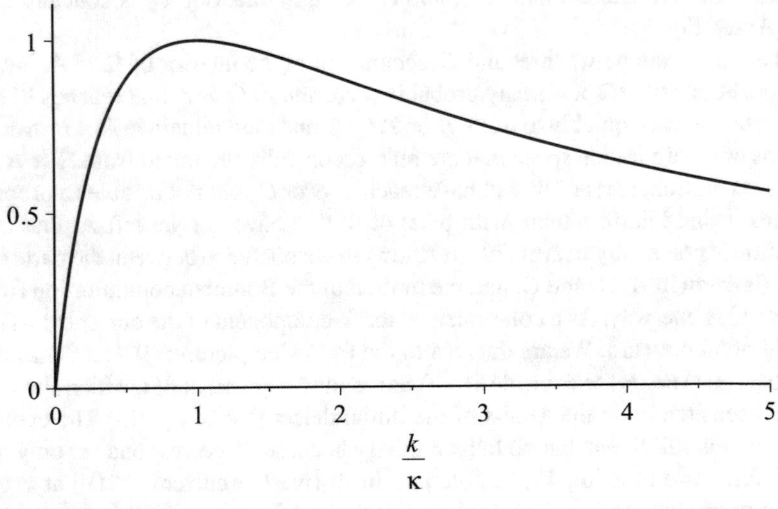

Fig. 5.5 Graph of the absorption strength A_k of the ideal detecting surface as a function of wave number k in units of κ. The maximum attained at $k = \kappa$ is equal to 1, corresponding to complete absorption

5.2.5 Time–Energy Uncertainty Relation

Already in the early days of quantum mechanics, various authors expected an uncertainty relation between time and energy,

$$\sigma_T \, \sigma_E \geq \frac{\hbar}{2} \,. \tag{5.82}$$

However, it is not obvious what should be meant by "the time observable." Now, if we fix a region $\Omega \subset \mathbb{R}^3$, then the time T of detection on $\partial\Omega$ is a natural observable to consider, and we can ask whether (5.82) holds. For the distribution given by the absorbing boundary rule with $V = 0$, this is actually the case, but this fact requires a bit of explanation. Consider $\psi \in \mathbb{S}(L^2(\Omega))$, and let us first suppose that for this ψ, $\mathbb{P}(Z = \infty) = 0$. Then ψ defines a probability distribution for T on $[0, \infty)$, and σ_T is defined to be the standard deviation of this distribution. And what is meant here by "energy"? The Hamiltonian corresponding to the Schrödinger equation (5.68) with absorbing boundary condition (5.69) is not even self-adjoint! It can have complex eigenvalues. What is worse, it may not even be diagonalizable, as it may fail to be a normal operator (i.e., one that commutes with its adjoint), and thus it may not possess a spectral PVM even on $\mathbb{C}$. What is meant instead by the "energy distribution" is that we regard ψ as a function on $\mathbb{R}^3$ by setting it to 0 outside of Ω and then consider its Born distribution for $H_{\text{free}} = -\frac{\hbar^2}{2m}\nabla^2$, the self-adjoint Hamiltonian on $\mathbb{R}^3$ without boundary. Presumably, one would anyway be inclined to understand the "energy distribution" of a wave function this way, regardless of whether it is concentrated in Ω or not. Of this distribution, σ_E is the standard deviation. Then one can prove (5.82) under suitable conditions (but the proof is rather lengthy and tricky and therefore shall not be reproduced here):

Theorem 5.4 (Kiukas et al. 2012 [25, App. A1–3] and Tumulka 2016 [45]) *Suppose that $\Omega \subset \mathbb{R}^3$ is open, has a sufficiently regular boundary $\partial\Omega$ (locally Lipschitz and piecewise C^1), and is such that*

$$\lim_{t\to\infty} W_t^\dagger W_t = 0 \tag{5.83}$$

(equivalently, $\mathbb{P}(Z = \infty) = 0$ for every $\psi \in L^2(\Omega)$). Then, for every C^∞ wave function ψ with compact support in Ω and $\|\psi\| = 1$, (5.82) is true.

Now what if $\mathbb{P}(Z = \infty) > 0$? Then $\sigma_T = \infty$, and (5.82) is trivially true but vacuous as $\sigma_E > 0$. It is then of interest what can be said about the conditional standard deviation of T, given that the particle gets detected at all:

$$\tilde{\sigma}_T := \sqrt{\text{Var}(T|Z \neq \infty)} \,. \tag{5.84}$$

Theorem 5.5 ([25, App. A1–3] and [45]) *Suppose that $\Omega \subset \mathbb{R}^3$ is open and has sufficiently regular boundary $\partial\Omega$. Then, for every C^∞ wave function ψ with compact support in Ω and $\|\psi\| = 1$,*

$$\tilde{\sigma}_T\, \sigma_E \geq \sqrt{\mathbb{P}(Z \neq \infty)}\, \frac{\hbar}{2}\,. \tag{5.85}$$

5.2.6 Historical Notes

Although the question about the distribution of time T and place X of detection seems natural and obvious, it was not discussed for a long time after it had become clear that there is no self-adjoint operator that satisfies the same commutation relation with the Hamiltonian H as position with momentum. The earliest paper I know to have made a guess about the distribution of T is Aharonov and Bohm (1961) [1]. The authors considered a particle in 1d in $\Omega = (-\infty, 0]$ and a detector at the origin. They noted that classically, the time of arrival at the origin is

$$t = -\frac{mx}{p} \tag{5.86}$$

if m is the mass, $x < 0$ the initial position and $p > 0$ the initial momentum. They suggested to "quantize" this relation and obtain a self-adjoint operator $\hat{T}$ for arrival time by replacing the position variable x by the position operator $\hat{X}$ and the momentum variable p by the momentum operator $\hat{P} = -i\hbar\partial_x$. Different orderings of the factors would lead to different operators, but a Hermitian choice would be

$$\hat{T} = -m\hat{P}^{-1/2}\hat{X}\hat{P}^{-1/2} \tag{5.87}$$

if only $\hat{P}$ were positive. Well, the idea is strongly influenced by Copenhagen ways of thinking. The idea is creative, but we see here some disadvantages of the Copenhagen attitude of introducing operators for observables, not through analysis of the measurement process, but through analogy or quantization: The proposal is a blind guess that has nothing to do with how a detector works. Note also that no generalization to arbitrary surfaces $\partial\Omega$ in higher dimension offers itself.

Over the years, there have been numerous proposals for how to compute the distribution of T and X, none of them anywhere near the absorbing boundary rule; I will not discuss them here. The one exception is Werner (1987) [51], where the boundary condition (5.69) and the probability distribution (5.70) were considered as an example of a detection time distribution; however, the spirit of Werner's paper was to regard this distribution as just a particular example while "any contraction semigroup determines a natural arrival time observable" [51], whereas we ask here, which observable or POVM (or which contraction semigroup, for that matter) we should use if we want to represent ideal detectors on $\partial\Omega$. In any case, the boundary condition (5.69) did not receive much attention in the literature; for example, in an 86-page review paper (Muga and Leavens 2000 [32]) on arrival times, it was mentioned in passing but not even written down; the authors did not realize how reasonable it is as a model of an ideal detector.

5.3 Ontic Ψ Versus Epistemic Ψ

In this section, we are concerned with the question whether the wave function is real—whether ψ has to be a beable in all possible theories of quantum mechanics. It is a beable in all the theories we have encountered, but we ask here whether alternatives are possible. The Pusey–Barrett–Rudolph (PBR) theorem, which I describe below, asserts that, under reasonable assumptions on the admissible theories, the answer is *no*.

Some authors thought that a wave function is analogous to a probability distribution. In classical mechanics, the factual, actual, physical situation of a system corresponds to its phase point; a probability distribution over phase space may represent an observer's knowledge or may characterize a preparation mechanism that will produce a different phase point every time it is used. But for a single system, nature only knows its phase point and need not know its probability distribution. That is, there need not be a fact in the world about what "the" probability distribution is. Likewise, in quantum mechanics, we may wonder whether for every wave function ψ there is a probability distribution ϱ^ψ over elements $\lambda \in \Lambda$ that represent the factual, actual, physical situation. In this case, the λ are called the *ontic* states (from the Greek for "being"), while probability distributions are called *epistemic* states (from the Greek for "knowledge"). If a theory says that ψ corresponds to a probability distribution over ontic states, then one says that ψ is epistemic according to that theory. In contrast, if a theory says that nature knows what ψ is (at least up to a global phase factor), then one says that ψ is ontic according to that theory.

Now one may well imagine that the ontic state λ contains, not a wave function ψ, but some other variables that are mathematically equivalent to ψ, such as the Taylor or Fourier coefficients of ψ. In this case, we still say that ψ is ontic. The question we are asking is whether every λ determines a ψ (always up to a phase). That is, we say that the theory is *ψ-ontic* if for $\mathbb{C}\psi \neq \mathbb{C}\phi$, ϱ^ψ and ϱ^ϕ are always concentrated on disjoint sets $\Lambda^\psi, \Lambda^\phi \subseteq \Lambda$ (i.e., they do not overlap). Otherwise we say it is *ψ-epistemic* (i.e., ϱ^ψ and ϱ^ϕ sometimes overlap; see Fig. 5.6).

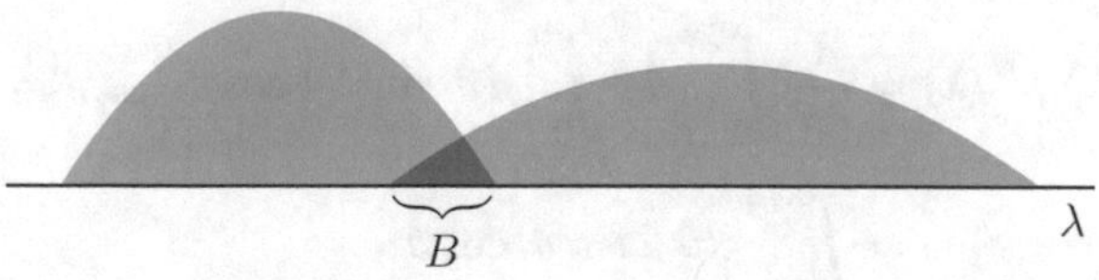

Fig. 5.6 In a ψ-epistemic model, the probability distributions associated with different wave functions, $\mathbb{C}\psi \neq \mathbb{C}\phi$, can overlap, here in the set $B \subset \Lambda$

We use again the formalized notion of an ontological model as in Definition 5.3 in Sect. 5.1.4.

Example 5.5 An example of a ψ-ontic ontological model for $\mathscr{H} = L^2(\mathbb{R}^{3N})$ is provided by Bohmian mechanics: $\Lambda = \mathbb{S}(\mathscr{H}) \times \mathbb{R}^{3N}$, $\lambda = (\psi, Q)$, $\varrho^\psi = \delta_\psi \times |\psi|^2$ (i.e., in an ensemble prepared for ψ, the wave function is always ψ and the configuration is $|\psi|^2$-distributed); $\mathcal{EXP}$ contain all experiments as defined by a ready state $\phi_A \in \mathbb{S}(L^2(\mathbb{R}^{3N_A}))$ of the apparatus, a duration $t_f - t_i$, a Schrödinger-type Hamiltonian on $L^2(\mathbb{R}^{3N+3N_A})$, and a calibration function $\zeta : \mathbb{R}^{3N_A} \to \mathscr{Z}$; the mapping $E : \mathcal{EXP} \to \mathcal{POVM}$ is given by the main theorem about POVMs; finally, choose $Q_A \sim |\phi_A|^2$ randomly, and let $P_{\lambda,\mathscr{E}}(B)$ be the probability that the Bohmian motion starting at t_i from (Q, Q_A) leads to a configuration at t_f with $\zeta \in B$.

Example 5.6 Here is an example of a ψ-epistemic ontological model (Kochen and Specker 1967 [26]): Let $\mathscr{H} = \mathbb{C}^2$ (thought of as spin space) and $\Lambda = \mathbb{S}(\mathbb{R}^3) = \mathbb{S}^2$ (thought of as directions in physical space); we write its elements in three-vector notation as $\boldsymbol{\lambda}$. The distribution ϱ^ψ can be characterized by its density $p^\psi(\boldsymbol{\lambda})$ relative to the normalized uniform measure u on $\mathbb{S}^2$; $u(B)$ is $(4\pi)^{-1}$ times the surface area of $B \subseteq \mathbb{S}^2$, or, in spherical coordinates, $u(d\vartheta \times d\varphi) = (4\pi)^{-1} \sin\vartheta \, d\vartheta \, d\varphi$. The density of ϱ^ψ is defined to be

$$p^\psi(\boldsymbol{\lambda}) = 4\,\boldsymbol{\omega}(\psi) \cdot \boldsymbol{\lambda}\, 1_{\boldsymbol{\omega}(\psi)\cdot\boldsymbol{\lambda}>0} \tag{5.88}$$

with $\boldsymbol{\omega}(\psi) = \psi^* \boldsymbol{\sigma} \psi$ as in (2.107) the direction in $\mathbb{R}^3$ associated with ψ.

For every POVM with a finite value space, we introduce a single experiment $\mathscr{E}$ and define

$$P_{\boldsymbol{\lambda},\mathscr{E}}(z) := e_{z,1} 1_{\boldsymbol{\omega}(|z,1\rangle)\cdot\boldsymbol{\lambda}>0} + e_{z,2} 1_{\boldsymbol{\omega}(|z,1\rangle)\cdot\boldsymbol{\lambda}\leq 0}\,, \tag{5.89}$$

where $|z, 1\rangle, |z, 2\rangle$ is an ONB diagonalizing $E_{\mathscr{E},z}$ with eigenvalues $e_{z,1}$ and $e_{z,2}$,

$$E_{\mathscr{E},z} = e_{z,1}|z, 1\rangle\langle z, 1| + e_{z,2}|z, 2\rangle\langle z, 2|\,. \tag{5.90}$$

In particular, a quantum measurement of $\boldsymbol{n} \cdot \boldsymbol{\sigma}$ yields $+1$ if $\boldsymbol{\lambda}$ lies in the hemisphere centered at $\boldsymbol{n}$ and -1 otherwise.

We verify that p^ψ is a probability density: Clearly $p^\psi(\boldsymbol{\lambda}) \geq 0$, and it is normalized because

$$\int_{\mathbb{S}^2} u(d^2\boldsymbol{\lambda})\, p^\psi(\boldsymbol{\lambda}) = \frac{1}{4\pi} \int_0^{2\pi} d\varphi \int_0^{\pi} d\vartheta \ \sin\vartheta\, 4\cos\vartheta\, 1_{\cos\vartheta>0} \tag{5.91a}$$

$$= \int_0^{\pi/2} d\vartheta\, 2\sin\vartheta\ \cos\vartheta \tag{5.91b}$$

$$= \int_0^{\pi/2} d\vartheta \; \sin 2\vartheta \tag{5.91c}$$

$$= -\tfrac{1}{2}\cos\pi + \tfrac{1}{2}\cos 0 = 1 \tag{5.91d}$$

using spherical coordinates with the z axis in the direction of $\boldsymbol{\omega}(\psi)$.

This ontological model reproduces the quantum predictions: It is shown in, e.g., Leifer (2014) [29, App. B] that for every $\phi \in \mathbb{S}(\mathbb{C}^2)$,

$$\int_{\mathbb{S}^2} u(d^2\boldsymbol{\lambda})\, p^{\psi}(\boldsymbol{\lambda})\, 1_{\boldsymbol{\omega}(\phi)\cdot\boldsymbol{\lambda}>0} = |\langle\psi|\phi\rangle|^2 . \tag{5.92}$$

(About one page of calculation.) From this we obtain, using $\boldsymbol{\omega}(|z,1\rangle) = -\boldsymbol{\omega}(|z,2\rangle)$, that

$$\int_{\Lambda} \varrho^{\psi}(d\lambda)\, P_{\lambda,\mathscr{E}}(z) = \int_{\mathbb{S}^2} u(d^2\boldsymbol{\lambda})\, p^{\psi}(\boldsymbol{\lambda}) \Big(e_{z,1} 1_{\boldsymbol{\omega}(|z,1\rangle)\cdot\boldsymbol{\lambda}>0} + e_{z,2} 1_{\boldsymbol{\omega}(|z,1\rangle)\cdot\boldsymbol{\lambda}\le 0}\Big) \tag{5.93a}$$

$$= \int_{\mathbb{S}^2} u(d^2\boldsymbol{\lambda})\, p^{\psi}(\boldsymbol{\lambda}) \Big(e_{z,1} 1_{\boldsymbol{\omega}(|z,1\rangle)\cdot\boldsymbol{\lambda}>0} + e_{z,2} 1_{\boldsymbol{\omega}(|z,2\rangle)\cdot\boldsymbol{\lambda}>0}\Big) \tag{5.93b}$$

$$\overset{(5.92)}{=} e_{z,1}|\langle\psi|z,1\rangle|^2 + e_{z,2}|\langle\psi|z,2\rangle|^2 = \langle\psi|E_{\mathscr{E},z}|\psi\rangle . \tag{5.93c}$$

What this example shows is that it is possible, at least for a single spinor, that nature does not know ψ but only a random direction $\boldsymbol{\lambda}$ that is continuously distributed on the hemisphere around $\boldsymbol{\omega}(\psi)$. This is enough for nature to reproduce the quantum predictions. Note that for two different spinors $\tilde{\psi} \neq \psi$ (except when $\boldsymbol{\omega}(\tilde{\psi}) = -\boldsymbol{\omega}(\psi)$), ϱ^{ψ} and $\varrho^{\tilde{\psi}}$ overlap nontrivially. The fact that a quantum measurement of $|\psi\rangle\langle\psi|$ in a ψ-prepared (i.e., ϱ^{ψ}-distributed) system yields 1 with certainty might have suggested that nature knows ψ, but we see here that this need not be the case. The PBR theorem addresses the question whether ψ-epistemic ontological models are possible for more general systems.

5.3.1 The Pusey–Barrett–Rudolph Theorem

Now we consider the system $S_1 \cup \ldots \cup S_n$ consisting of n copies of S and imagine preparing each S_j independently with some wave function $\psi_j \in \mathbb{S}(\mathscr{H})$. The joint system is thus prepared in the joint quantum state $\psi_1 \otimes \cdots \otimes \psi_n \in \mathbb{S}(\mathscr{H}^{\otimes n})$, while each S_j has an ontic state $\lambda_j \in \Lambda$ distributed according to ϱ^{ψ_j}, so that the joint ontic state $(\lambda_1, \ldots, \lambda_n) \in \Lambda^n$ is distributed according to $\varrho^{\psi_1} \times \cdots \times \varrho^{\psi_n}$. Now we ask what happens if we measure on this n-fold system an arbitrary POVM $(E_z^{(n)})_z$ acting on $\mathscr{H}^{\otimes n}$. It seems natural to demand of a theory that it provides a probability $P(\lambda_1, \ldots, \lambda_n, z)$ also for this situation.[7]

Definition 5.6 Let $n \in \mathbb{N}$. An ontological model that reproduces the quantum predictions is *n-extendible* if and only if for every POVM $E^{(n)}$ (with finite value space) acting on $\mathscr{H}^{\otimes n}$, there is a function $P : \Lambda^n \times \mathscr{Z} \to [0, 1]$ such that

$$\sum_{z \in \mathscr{Z}} P(\lambda_1, \ldots, \lambda_n, z) = 1 \tag{5.94}$$

for all $(\lambda_1, \ldots, \lambda_n) \in \Lambda^n$, and the quantum predictions for $\mathscr{H}^{\otimes n}$ are reproduced:

$$\int_\Lambda \varrho^{\psi_1}(d\lambda_1) \cdots \int_\Lambda \varrho^{\psi_n}(d\lambda_n)\, P(\lambda_1, \ldots, \lambda_n, z) = \langle \psi_1 \otimes \cdots \otimes \psi_n | E_z | \psi_1 \otimes \cdots \otimes \psi_n \rangle \quad \forall z \in \mathscr{Z}. \tag{5.95}$$

The requirement that an ontological model be n-extendible for every $n \in \mathbb{N}$ is also called the *preparation independence postulate* in the literature.

Theorem 5.6 (PBR theorem; Pusey, Barrett, and Rudolph (2012) [36]) *Any ontological model that reproduces the quantum predictions and is n-extendible for every $n \in \mathbb{N}$ is ψ-ontic.*

Proof We need to show for $\psi, \phi \in \mathbb{S}(\mathscr{H})$ with $\psi \notin \mathbb{C}\phi$ that ϱ^ψ and ϱ^ϕ do not overlap. The full proof can be found in [36] or [29]. Here, we limit ourselves to the simple case $|\langle\psi|\phi\rangle|^2 = 1/2$. In the 2d subspace of $\mathscr{H}$ spanned by ψ and ϕ, let $\{|0\rangle, |1\rangle\}$ be an ONB with $|0\rangle = \psi$, and let $|\pm\rangle = \frac{1}{\sqrt{2}}(|0\rangle \pm |1\rangle)$, possibly times a phase to make $|+\rangle = \phi$. If ϱ^ψ and ϱ^ϕ overlapped, then there would be a set $B \subset \Lambda$ and a number $q > 0$ such that $\varrho^\psi(B) \geq q$ and $\varrho^\phi(B) \geq q$ (see Fig. 5.6). Now we use that the ontological model is two-extendible and consider preparing any of the four quantum states:

$$|0\rangle \otimes |0\rangle, \quad |0\rangle \otimes |+\rangle, \quad |+\rangle \otimes |0\rangle, \quad |+\rangle \otimes |+\rangle. \tag{5.96}$$

In each case, with probability $\geq q^2 > 0$, $(\lambda_1, \lambda_2) \in B^2 \subset \Lambda^2$; see Fig. 5.7.

[7] It is perhaps not realistic to assume that every POVM can be realized through some experiment. Actually, the PBR theorem can still be proved under the weaker assumption of a much smaller set of POVMs, which plausibly can be realized experimentally.

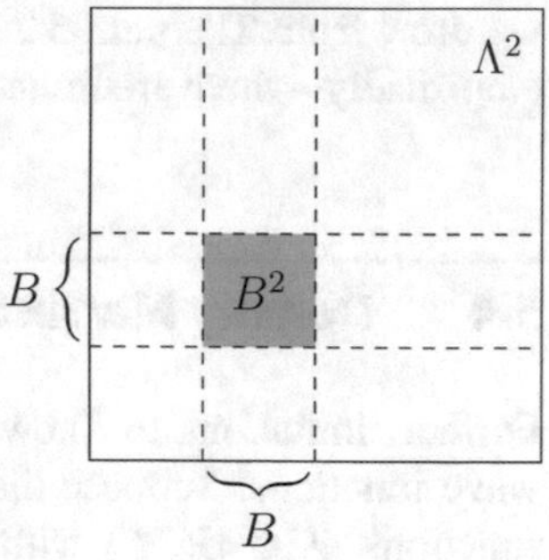

Fig. 5.7 Regardless of which of the states in (5.96) is prepared, the actual (λ_1, λ_2) will lie in the little square B^2 with probability at least $q^2 > 0$

Now we measure the following PVM E on $\mathscr{Z} = \{1, 2, 3, 4, 5\}$ acting on $\mathscr{H}^{\otimes 2}$:

$$E_1 = |\xi_1\rangle\langle\xi_1| \text{ with } |\xi_1\rangle = \tfrac{1}{\sqrt{2}}\big(|0\rangle \otimes |1\rangle + |1\rangle \otimes |0\rangle\big) \tag{5.97a}$$

$$E_2 = |\xi_2\rangle\langle\xi_2| \text{ with } |\xi_2\rangle = \tfrac{1}{\sqrt{2}}\big(|0\rangle \otimes |-\rangle + |1\rangle \otimes |+\rangle\big) \tag{5.97b}$$

$$E_3 = |\xi_3\rangle\langle\xi_3| \text{ with } |\xi_3\rangle = \tfrac{1}{\sqrt{2}}\big(|+\rangle \otimes |1\rangle + |-\rangle \otimes |0\rangle\big) \tag{5.97c}$$

$$E_4 = |\xi_4\rangle\langle\xi_4| \text{ with } |\xi_4\rangle = \tfrac{1}{\sqrt{2}}\big(|+\rangle \otimes |-\rangle + |-\rangle \otimes |+\rangle\big) \tag{5.97d}$$

$$E_5 = P_{((\mathbb{C}|0\rangle \oplus \mathbb{C}|1\rangle)^{\otimes 2})^\perp} \,. \tag{5.97e}$$

One easily verifies that $\xi_1, \xi_2, \xi_3, \xi_4$ form an orthonormal system spanning $(\mathbb{C}|0\rangle \oplus \mathbb{C}|1\rangle)^{\otimes 2}$. By (5.95),

$$\int \varrho^{\psi_1}(d\lambda_1) \int \varrho^{\psi_2}(d\lambda_2)\, P(\lambda_1, \lambda_2, 5) = \langle \psi_1 \otimes \psi_2 | E_5 | \psi_1 \otimes \psi_2 \rangle = 0 \tag{5.98}$$

for every $\psi_1, \psi_2 \in \mathbb{C}|0\rangle \oplus \mathbb{C}|1\rangle$, so $P(\lambda_1, \lambda_2, 5) = 0$, and so the other four $P(\lambda_1, \lambda_2, z)$ add up to 1. As the next step, suppose $|0\rangle \otimes |0\rangle$ gets prepared; since that is orthogonal to ξ_1, the outcome $Z = 1$ has probability 0. Thus, $P(\lambda_1, \lambda_2, z = 1) = 0$ for (λ_1, λ_2) in support of $\varrho^\psi \times \varrho^\psi$, in particular in B^2. Now we repeat the argument with $z = 2, 3, 4$: Prepare the z-th of the four states in (5.96), which is orthogonal to ξ_z, so the probability of outcome z vanishes, so $P(\lambda_1, \lambda_2, z)$ must vanish in support of the appropriate measure, and in particular in B^2. But this leads to a contradiction to $\sum_{z=1}^4 P(\lambda_1, \lambda_2, z) = 1$ in B^2. So, ϱ^ψ and ϱ^ϕ cannot overlap. □

Remark 5.1 We have seen in Sect. 5.1.3 that the wave function of a single system cannot be measured, not even up to phase. Now the PBR theorem shows that in theories of reality under reasonable assumption, the wave function up to phase is a real thing—for each individual system, there is a fact in nature about what $\mathbb{C}\psi$ is. Thus, the PBR theorem also provides an alternative proof of the conclusion we drew from Theorem 5.2: That there are facts in nature that we cannot find out empirically—there are limitations to knowledge.

5.4 Density Matrix and Mixed State

Further limitations to knowledge are connected to probability distributions over wave functions. Suppose that we have a mechanism that generates random wave functions $\Psi \in \mathbb{S}(\mathscr{H})$ with probability distribution μ on the unit sphere $\mathbb{S}(\mathscr{H})$ in $\mathscr{H}$. Then *it is impossible to determine* μ *empirically.* In fact, there exist different distributions $\mu_1 \neq \mu_2$ that are *empirically indistinguishable*, i.e., they lead to the same distribution of outcomes Z for any experiment. We call such distributions *empirically equivalent* (which is an equivalence relation) and show that the equivalence classes are in one-to-one correspondence with certain operators known as *density matrices* or *density operators*.

To describe these matters, we need the mathematical concept of *trace*.

5.4.1 Trace

Definition 5.7 The *trace of a matrix* $A = (A_{mn})$ is the sum of its diagonal elements. The *trace of an operator* T is defined to be the sum of the diagonal elements of its matrix representation $T_{nm} = \langle n|T|m\rangle$ relative to an arbitrary ONB $\{|n\rangle\}$,

$$\operatorname{tr} T = \sum_{n=1}^{\infty} \langle n|T|n\rangle \,. \tag{5.99}$$

Every positive operator either has finite trace or has trace $+\infty$, and the value of the trace does not depend on the choice of ONB. The *trace class* is the set of those operators T for which the positive operator $\sqrt{T^\dagger T}$ has finite trace. For every operator from the trace class, the trace is finite and does not depend on the ONB.

The trace has the following properties for all operators $A, B, \ldots$ from the trace class (e.g., Reed and Simon 1980 [37, Sec. VI.6]):

(i) The trace is linear:

$$\operatorname{tr}(A + B) = \operatorname{tr} A + \operatorname{tr} B\,, \quad \operatorname{tr}(\lambda A) = \lambda \operatorname{tr} A \tag{5.100}$$

for all $\lambda \in \mathbb{C}$.

(ii) The trace is invariant under cyclic permutation of factors:

$$\operatorname{tr}(AB \cdots YZ) = \operatorname{tr}(ZAB \cdots Y)\,. \tag{5.101}$$

In particular $\operatorname{tr}(AB) = \operatorname{tr}(BA)$ and $\operatorname{tr}(ABC) = \operatorname{tr}(CAB)$, which is, however, not always the same as $\operatorname{tr}(CBA)$.

(iii) If an operator T can be diagonalized, i.e., if there exists an orthonormal basis of eigenvectors, then $\operatorname{tr}(T)$ is the sum of the eigenvalues, counted with multiplicity (= degree of degeneracy).

(iv) The trace of the adjoint operator $T^\dagger$ is the complex conjugate of the trace of T: $\operatorname{tr}(T^\dagger) = \operatorname{tr}(T)^*$.

(v) The trace of a self-adjoint operator T is real.

(vi) If T is a positive operator then $\operatorname{tr}(T) \geq 0$.

5.4.2 The Trace Formula in Quantum Mechanics

Suppose that (by whatever mechanism) we have generated a random wave function $\Psi \in \mathbb{S}(\mathscr{H})$ with probability distribution μ on $\mathbb{S}(\mathscr{H})$. Then for any experiment $\mathscr{E}$ with POVM E, the probability distribution of the outcome Z is

$$\mathbb{P}(Z \in B) = \mathbb{E}\langle\Psi|E(B)|\Psi\rangle = \int_{\mathbb{S}(\mathscr{H})} \mu(d\psi)\,\langle\psi|E(B)|\psi\rangle = \operatorname{tr}(\rho_\mu\, E(B))\,, \tag{5.102}$$

where $\mathbb{E}$ means expectation, and

$$\rho_\mu = \mathbb{E}|\Psi\rangle\langle\Psi| = \int_{\mathbb{S}(\mathscr{H})} \mu(d\psi)\,|\psi\rangle\langle\psi| \tag{5.103}$$

is called the *density operator* or *density matrix* (rarely: *statistical operator*) of the distribution μ. Equation (5.102) (i.e., the equality of the leftmost and the rightmost side) is called the *trace formula*. It was discovered by John von Neumann (1927) [49], except that von Neumann did not know POVMs and considered only PVMs. In case the distribution μ is concentrated on discrete points on $\mathbb{S}(\mathscr{H})$, (5.103) becomes

$$\rho_\mu = \mathbb{E}|\Psi\rangle\langle\Psi| = \sum_\psi \mu(\psi)\,|\psi\rangle\langle\psi|\,. \tag{5.104}$$

In order to verify (5.102), note first that

$$\operatorname{tr}\Big(|\psi\rangle\langle\psi|\,E\Big) = \langle\psi|E|\psi\rangle \tag{5.105}$$

because, if we choose the basis $\{|n\rangle\}$ in (5.99) such that $|1\rangle = \psi$, then the summands in (5.99) are $\langle n|\psi\rangle\langle\psi|E|n\rangle$, which for $n = 1$ is $\langle\psi|E|\psi\rangle$ and for $n > 1$ is zero because $\langle n|1\rangle = 0$. By linearity, we also have that

$$\mathrm{tr}\Big(\sum_j \mu(\psi_j)|\psi_j\rangle\langle\psi_j|\,E\Big) = \sum_j \mu(\psi_j)\,\langle\psi_j|E|\psi_j\rangle\,, \tag{5.106}$$

which yields (5.102) for any μ that is concentrated on finitely many points ψ_j on $\mathbb{S}(\mathscr{H})$. One can prove (5.102) for arbitrary probability distribution μ by considering limits.

5.4.3 Pure and Mixed States

A density matrix is also often called a *quantum state*. If $\rho = |\psi\rangle\langle\psi|$ with $\|\psi\| = 1$, then ρ is usually called a *pure quantum state*, otherwise a *mixed quantum state*.

Proposition 5.2 *A density matrix ρ is always a positive operator with* $\mathrm{tr}\,\rho = 1$. *Conversely, every positive operator ρ with* $\mathrm{tr}\,\rho = 1$ *is a density matrix, i.e.,* $\rho = \rho_\mu$ *for some probability distribution μ on* $\mathbb{S}(\mathscr{H})$.

Proof Indeed, a sum (or integral) of positive operators is positive, and $\mu(\psi_j)|\psi_j\rangle\langle\psi_j|$ is positive. Furthermore,

$$\mathrm{tr}(\rho_\mu) = \mathrm{tr}(\rho_\mu\, E(\mathscr{Z})) = \int_{\mathbb{S}(\mathscr{H})} \mu(d\psi)\,\langle\psi|E(\mathscr{Z})|\psi\rangle = \int_{\mathbb{S}(\mathscr{H})} \mu(d\psi) = 1\,. \tag{5.107}$$

Concerning the converse, here is one such μ: Find an orthonormal basis $\{|\phi_n\rangle : n \in \mathbb{N}\}$ of eigenvectors of ρ with eigenvalues $p_n \in [0,\infty)$. Then

$$\sum_n p_n = \mathrm{tr}\,\rho = 1\,. \tag{5.108}$$

Now let μ be the distribution that gives probability p_n to ϕ_n; its density matrix is just the ρ we started with. □

5.4.4 Empirically Equivalent Distributions

Let us draw conclusions from the trace formula (5.102). It implies that the distribution of the outcome Z depends on μ only through ρ_μ. Different distributions μ_a, μ_b can have the same $\rho = \rho_{\mu_a} = \rho_{\mu_b}$; that is, the mapping $\mu \mapsto \rho_\mu$ is *many-to-one* (i.e., not injective). For example, if $\mathscr{H} = \mathbb{C}^2$ then the uniform distribution over $\mathbb{S}(\mathscr{H})$ has $\rho = \frac{1}{2}I$, and for every orthonormal basis $|\phi_1\rangle, |\phi_2\rangle$ of $\mathbb{C}^2$ the probability distribution

$$\tfrac{1}{2}\delta_{\phi_1} + \tfrac{1}{2}\delta_{\phi_2} \tag{5.109}$$

also has $\rho = \frac{1}{2}I$. Such two distributions μ_a, μ_b will lead to the same distribution of outcomes for any experiment and are therefore *empirically equivalent*.

Example 5.7 Suppose Alice chooses between two options; she chooses μ to be either μ_a or μ_b. Suppose that each μ is of the form (5.109) but for different ONBs. Then she chooses $n = 10{,}000$ points ψ_i on $\mathbb{S}(\mathscr{H})$ at random independently with μ, then she prepares n systems with wave functions ψ_i, and then she hands these systems over to Bob with the challenge to determine whether $\mu = \mu_a$ or $\mu = \mu_b$. As a consequence of (5.102), Bob cannot determine that by means of experiments on the n systems. However, by the PBR theorem (Theorem 5.6), there is a fact in nature about what the wave function ψ_i of each system is. Thus, there is a fact in nature about whether the distribution is actually μ_a or μ_b. The fact that Bob cannot determine μ empirically is another limitation to knowledge.[8]

We now turn to another proof, not based on the PBR theorem, showing that in any version of quantum mechanics there must be distinct, empirically equivalent ensembles; we formulate the result as Theorem 5.7 below.

To be sure, we know already that there are distinct, empirically equivalent ensembles of wave functions ψ, but before the PBR theorem, it was not clear whether the beables λ of the theory (i.e., the ontic variables, those representing the ontology of the theory) would have to determine ψ (I mean, up to phase). In fact, Example 5.6 (which is not a satisfactory theory but a toy example) illustrates how λ could fail to determine ψ, with the consequence that nature knows λ but not ψ, and that there is no fact in reality about what ψ is. And in this situation, it seems conceivable that two different distributions over ψ might correspond to the *same* distribution over λ. If that were so, there would be two consequences: First, it would be unsurprising that different distributions over ψ can be empirically equivalent (because they would correspond to the same ensemble in reality), and second, it would seem possible that there is no limitation to knowledge of the kind that two distinct ensembles could be empirically equivalent. Theorem 5.7 excludes this.

As a consequence, the same reasoning as in Example 5.7 (Alice challenges Bob to determine which ensemble she chose) then shows that in any version of quantum mechanics there must be limitations to knowledge—in fact, the specific limitation

---

[8] In a prior publication [11], I argued, without referring to the PBR theorem or a similar one, that already the fact alone that Alice knows for each system what its wave function ψ_i is and can make the prediction that a quantum measurement of $|\psi_i\rangle\langle\psi_i|$ will yield 1 with certainty shows that nature knows that system i has wave function ψ_i. This argument is mistaken, as Example 5.6 shows. I am grateful to Matthew Leifer for pointing out this mistake to me. The same mistake was made in [33, p. 314].

that two physically different situations (two physical states of an n-partite system, corresponding to two different statistical distributions of the physical states of the parts) cannot be distinguished empirically.[9]

We consider ontological models but defined in a slightly different way than in Definition 5.3, as appropriate for dealing with density matrices. So consider a quantum system with Hilbert space $\mathscr{H}$ with $\dim \mathscr{H} \geq 2$. We assume first that a space Λ of ontic states λ is given. Next, that not every probability distribution over Λ is preparable, only those belonging to a certain set $\mathcal{C} \subseteq \mathcal{PM}(\Lambda)$, where $\mathcal{PM}(\Omega)$ denotes the set of all probability measures on Ω. (This, of course, is a limitation to control.) We take the set $\mathcal{C}$ to be convex[10] because, if we have two preparation procedures, we might choose one of them randomly. Let $\mathcal{POVM}$ denote again the set of all POVMs acting on $\mathscr{H}$ and $\mathcal{DM}$ the set of all density matrices in $\mathscr{H}$. We assume as given an index set $\mathcal{EXP}$ representing the set of experiments on this system and a mapping $E : \mathcal{EXP} \to \mathcal{POVM}$ associating to every experiment its POVM. For any ontic state $\lambda \in \Lambda$ and experiment $\mathscr{E} \in \mathcal{EXP}$, let $P_{\lambda,\mathscr{E}}$ denote the probability distribution of the outcome, given that the system is in λ. We assume further that the model *reproduces the quantum predictions* in the sense that for every $\varrho \in \mathcal{C}$, there is a density matrix D_ϱ on $\mathscr{H}$ such that, for every experiment $\mathscr{E} \in \mathcal{EXP}$,

$$\int_\Lambda \varrho(d\lambda)\, P_{\lambda,\mathscr{E}}(B) = \mathrm{tr}\Big(D_\varrho\, E_{\mathscr{E}}(B)\Big) \tag{5.110}$$

for every $B \subseteq \mathbb{R}$. We assume further that $D : \mathcal{C} \to \mathcal{DM}$ is surjective (so every pure or mixed quantum state can be accounted for by the model).

Theorem 5.7 (Empirically Equivalent Distributions [41,46]) *Consider an ontological model as just described that reproduces the quantum predictions, and suppose that all POVMs consisting of a projection P and $I - P$ lie in the range of E. Then the mapping $D : \mathcal{C} \to \mathcal{DM}$ cannot be injective.*

This means that certain distributions $\varrho_a \neq \varrho_b$ over Λ that can occur according to the theory are empirically equivalent, $D_{\varrho_a} = D_{\varrho_b}$.

Outline of Proof (The Full Proof Is Given in [41, 46]) It suffices to consider the case $\dim \mathscr{H} = 2$; for every other Hilbert space, the result follows by restriction to a 2d subspace. We assume that D is injective and will derive a contradiction.

---

[9] The proof of Theorem 5.7, contained in an unpublished thesis I wrote in 1998 [41], was perhaps the earliest proof showing that limitations to knowledge are inevitable, though I did not put the result this way in [41].

[10] A subset Ω of a vector space is called *convex* if and only if $sa + (1 - s)b \in \Omega$ for all $s \in [0, 1]$ and all $a, b \in \Omega$.

First, we show that D is convex linear.[11] If $\varrho_a, \varrho_b \in \mathcal{C}$ and $s \in [0, 1]$, then $\varrho := s\varrho_a + (1-s)\varrho_b$ lies in $\mathcal{C}$ by assumption, and

$$\operatorname{tr}\Big(D_\varrho\, E_{\mathscr{E}}(B)\Big) \overset{(5.110)}{=} \int_\Lambda \varrho(d\lambda)\, P_{\lambda\mathscr{E}}(B) \tag{5.111a}$$

$$= s\int_\Lambda \varrho_a(d\lambda)\, P_{\lambda\mathscr{E}}(B) + (1-s)\int_\Lambda \varrho_b(d\lambda)\, P_{\lambda\mathscr{E}}(B) \tag{5.111b}$$

$$\overset{(5.110)}{=} s\operatorname{tr}\Big(D_{\varrho_a}\, E_{\mathscr{E}}(B)\Big) + (1-s)\operatorname{tr}\Big(D_{\varrho_b}\, E_{\mathscr{E}}(B)\Big) \tag{5.111c}$$

$$= \operatorname{tr}\Big((sD_{\varrho_a} + (1-s)D_{\varrho_b})E_{\mathscr{E}}(B)\Big). \tag{5.111d}$$

By assumption, every projection P can occur as $E_{\mathscr{E}}(B)$, and the relation $\operatorname{tr}(D_1\, P) = \operatorname{tr}(D_2\, P)$ can hold for all projections only if $D_1 = D_2$; thus, $D_\varrho = sD_{\varrho_a} + (1-s)D_{\varrho_b}$.

Second, since D is bijective, it has an inverse $D^{-1} : \mathcal{DM} \to \mathcal{C}$. The inverse of any bijective convex linear mapping D between convex sets is convex linear because if $W_a, W_b \in \mathcal{DM}$ and $W := sW_a + (1-s)W_b$ with $s \in [0, 1]$, set $\varrho_a := D^{-1}(W_a)$, $\varrho_b := D^{-1}(W_b)$, $\varrho := s\varrho_a + (1-s)\varrho_b$, and observe that $D(\varrho) = sD(\varrho_a) + (1-s)D(\varrho_b) = sW_a + (1-s)W_b = W$, so $\varrho = D^{-1}(W)$.

Next, define the measure

$$\varrho_1 := 2\, D^{-1}(\tfrac{1}{2}I) \tag{5.112}$$

(the subscript 1 indicates that it is the "whole" measure, or the one corresponding to the identity operator). For every 1d subspace g of $\mathscr{H}$, define

$$\varrho_g := D^{-1}(P_g) \tag{5.113}$$

with P_g the projection to g. By assumption, we can choose for every 1d subspace h an $\mathscr{E}_h \in \mathcal{EXP}$ so that $E_{\mathscr{E}_h}(0) = P_{h^\perp}$ and $E_{\mathscr{E}_h}(1) = P_h$. By (5.110),

$$\int_\Lambda \varrho_g(d\lambda)\, P_{\lambda,\mathscr{E}_h} = \operatorname{tr}(P_g P_h)\delta_1 + \operatorname{tr}(P_g P_{h^\perp})\delta_0\,, \tag{5.114}$$

where δ_x means the measure on $\mathbb{R}$ with weight 1 in the point x and 0 everywhere else (i.e., with "density" $\delta(\cdot - x)$). Define the subset $\Lambda_h \subseteq \Lambda$ by

$$\Lambda_h := \big\{\lambda \in \Lambda : P_{\lambda,\mathscr{E}_h} = \delta_1\big\}\,. \tag{5.115}$$

[11] A mapping f on a convex domain is called *convex linear* if and only if $f(sa + (1-s)b) = sf(a) + (1-s)f(b)$ for all $s \in [0, 1]$ and all a, b in the domain of f.

Through several steps (about 1.5 pages of proof), one can derive that

$$\varrho_1(\Lambda_g \cap \Lambda_h) = \mathrm{tr}(P_g P_h) . \tag{5.116}$$

The remainder of the proof will show that there are no sets Λ_g in any measure space $(\Lambda, \mathscr{F}, \varrho_1)$ with $\varrho_1(\Lambda) = 2$ that make the relation (5.116) true. If there were, let λ be a random point in Λ with distribution $\frac{1}{2}\varrho_1$ and define the $\{0, 1\}$-valued random variables $X_g = 1_{\Lambda_g}(\lambda)$. Since $\varrho_1(\Lambda_g) = \varrho_1(\Lambda_g \cap \Lambda_g) = \mathrm{tr}(P_g P_g) = \mathrm{tr}(P_g) = 1$, we have for any g, h that

$$\mathbb{P}(X_g = 1, X_h = 1) = \tfrac{1}{2}\mathrm{tr}(P_g P_h) \tag{5.117a}$$

$$\mathbb{P}(X_g = 0, X_h = 1) = \tfrac{1}{2} - \tfrac{1}{2}\mathrm{tr}(P_g P_h) \tag{5.117b}$$

$$\mathbb{P}(X_g = 1, X_h = 0) = \tfrac{1}{2} - \tfrac{1}{2}\mathrm{tr}(P_g P_h) \tag{5.117c}$$

$$\mathbb{P}(X_g = 0, X_h = 0) = \tfrac{1}{2}\mathrm{tr}(P_g P_h) . \tag{5.117d}$$

Now consider for g the three subspaces:

$$g_1 = \mathbb{C}\begin{pmatrix}1\\0\end{pmatrix}, \quad g_2 = \mathbb{C}\begin{pmatrix}1/2\\ \sqrt{3}/2\end{pmatrix}, \quad g_3 = \mathbb{C}\begin{pmatrix}1/2\\ -\sqrt{3}/2\end{pmatrix}. \tag{5.118}$$

Since for any two of these, $\mathrm{tr}(P_g P_h) = 1/4$, any two of the three random variables X_{g_i} would have to have joint distribution:

	0	1
0	1/8	3/8
1	3/8	1/8

. (5.119)

If p_{ijk} denotes the probability that $X_{g_1} = i, X_{g_2} = j, X_{g_3} = k$, then

$$p_{000} + p_{001} = 1/8 \tag{5.120a}$$

$$p_{001} + p_{011} = 3/8 \tag{5.120b}$$

$$p_{011} + p_{111} = 1/8 , \tag{5.120c}$$

and (5.120a)–(5.120b)+(5.120c) yields $p_{000} + p_{111} = -1/8$, which is impossible. □

5.4.5 Density Matrix and Dynamics

If the random vector Ψ evolves according to the Schrödinger equation, $\Psi_t = e^{-iHt/\hbar}\Psi$, the distribution changes into μ_t and the density matrix into

$$\rho_t = e^{-iHt/\hbar}\rho e^{iHt/\hbar} . \tag{5.121}$$

Indeed,

$$\rho_t = \mathbb{E}|\Psi_t\rangle\langle\Psi_t| = \int\limits_{\mathbb{S}(\mathscr{H})} \mu(d\psi)\, e^{-iHt/\hbar}|\psi\rangle\langle\psi|e^{iHt/\hbar} \tag{5.122}$$

$$= e^{-iHt/\hbar}\Bigg(\int\limits_{\mathbb{S}(\mathscr{H})} \mu(d\psi)\,|\psi\rangle\langle\psi|\Bigg)e^{iHt/\hbar} = e^{-iHt/\hbar}\rho e^{iHt/\hbar}. \tag{5.123}$$

In analogy to the Schrödinger equation, (5.121) can be written as a differential equation,

$$\frac{d\rho_t}{dt} = -\tfrac{i}{\hbar}[H, \rho_t], \tag{5.124}$$

known as the *von Neumann equation*. The step from (5.121) to (5.124) is based on the fact that

$$\frac{d}{dt}e^{At} = Ae^{At} = e^{At}A. \tag{5.125}$$

Note that a pure quantum state evolves to a pure quantum state: If $\rho = |\psi\rangle\langle\psi|$, then $\rho_t = |\psi_t\rangle\langle\psi_t|$.

This is different for the GRW evolution. It leads, for fixed initial wave function Ψ_0, to a random wave function Ψ_t at $t > 0$, and the density matrix ρ_t associated with the distribution μ_t of Ψ_t obeys the differential equation:

$$\frac{d\rho_t}{dt} = -\tfrac{i}{\hbar}[H, \rho_t] + \lambda\sum_{i=1}^{N}\int\limits_{\mathbb{R}^3} d^3\boldsymbol{x}\; C_i(\boldsymbol{x})\,\rho_t\, C_i(\boldsymbol{x}) - N\lambda\rho_t\,. \tag{5.126}$$

An equation for the time evolution of a density matrix is called a *master equation*.

Derivation of (5.126) Suppose Ψ_t has distribution μ_t; let us determine μ_{t+dt}. With probability $1 - N\lambda dt$, no collapse occurs, and Ψ_t follows the unitary evolution. With probability $\lambda\, dt$, a collapse hits particle i, the collapse center $\boldsymbol{X}$ gets chosen randomly with probability density $\|C_i(\boldsymbol{x})\Psi_t\|^2$, and Ψ_t gets replaced by $C_i(\boldsymbol{X})\Psi_t/\|C_i(\boldsymbol{X})\Psi_t\|$. Thus,

$$\rho_{t+dt} = (1 - N\lambda dt)e^{-iHdt/\hbar}\rho_t e^{iHdt/\hbar}$$
$$+ \lambda dt\sum_{i=1}^{N}\int\limits_{\mathbb{S}(\mathscr{H})} \mu_t(d\psi)\int\limits_{\mathbb{R}^3} d^3\boldsymbol{x}\,\|C_i(\boldsymbol{x})\psi\|^2\frac{|C_i(\boldsymbol{x})\psi\rangle\langle C_i(\boldsymbol{x})\psi|}{\|C_i(\boldsymbol{x})\psi\|^2}. \tag{5.127}$$

To first order in dt,

$$(1 - N\lambda dt)e^{-iHdt/\hbar}\rho_t e^{iHdt/\hbar} = (1 - N\lambda dt)\Big(\rho_t - \tfrac{i}{\hbar}[H, \rho_t]dt + O(dt^2)\Big) \tag{5.128}$$

$$= \rho_t - \tfrac{i}{\hbar}[H, \rho_t]dt - N\lambda\, \rho_t\, dt + O(dt^2)\,. \tag{5.129}$$

Putting these expressions together and simplifying yields (5.126). □

Since a fixed initial wave function evolves to a random wave function, which of course differs from the unitarily evolved one by more than just a phase, it is clear that the master equation (5.126) will evolve pure states ρ_0 into mixed ones. (This can also be checked directly from (5.126).) On the other hand, (5.126) is equally valid if we start with an arbitrary distribution μ_0 of Ψ_0.

A master equation of the form:

$$\frac{d\rho_t}{dt} = -\tfrac{i}{\hbar}[H, \rho_t] + \sum_\alpha \Big(L_\alpha \rho_t L_\alpha^\dagger - \tfrac{1}{2}\Big\{L_\alpha^\dagger L_\alpha, \rho_t\Big\}\Big)\,, \tag{5.130}$$

where $\{A, B\} = AB + BA$ means the anti-commutator, the sum over α runs through some index set or can be replaced by an integral, and L_α are some operators, is called a *Gorini–Kossakowski–Sudarshan–Lindblad equation* or just a *Lindblad equation*. They have applications not just for collapse theories but also for the calculation of the effects of decoherence (more explanation in Sect. 5.5.5).

5.5 Reduced Density Matrix and Partial Trace

There is another way in which density matrices arise, leading to what is called the *reduced density matrix*, as opposed to the statistical density matrix of the previous section. Suppose that the system under consideration is *bipartite*, i.e., consists of two parts, system a and system b, so that its Hilbert space is $\mathscr{H} = \mathscr{H}_a \otimes \mathscr{H}_b$. At this point, we need to talk more about the tensor product.

5.5.1 Tensor Product

Let $\{|n_a\rangle_a\}_{n_a}$ be an ONB of $\mathscr{H}_a$ and $\{|n_b\rangle_b\}_{n_b}$ one of $\mathscr{H}_b$. Then the products $|n_a\rangle_a \otimes |n_b\rangle_b = |n_a n_b\rangle$ form an ONB of $\mathscr{H}$. Since an operator T can be expressed relative to an ONB by its matrix elements $T_{mn} = \langle m|T|n\rangle$, an operator on $\mathscr{H}_a \otimes \mathscr{H}_b$ can be represented by the numbers $T_{m_a m_b n_a n_b}$. Given operators $T_a : \mathscr{H}_a \to \mathscr{H}_a$ and $T_b : \mathscr{H}_b \to \mathscr{H}_b$, the products of their coefficients $T_{a,m_a n_a} T_{b,m_b n_b}$ are the coefficients of an operator on $\mathscr{H}_a \otimes \mathscr{H}_b$ called $T_a \otimes T_b$, the tensor product of T_a and T_b. For comparison, if a vector $\psi_a \in \mathscr{H}_a$ has coefficients ψ_{a,n_a} and $\psi_b \in \mathscr{H}_b$ coefficients

ψ_{b,n_b}, then the tensor product of vectors, $\psi_a \otimes \psi_b$, has coefficients $\psi_{a,n_a}\,\psi_{b,n_b}$ relative to the $|n_a n_b\rangle$.

Exercise 5.7 (Tensor Product) Show that $T_a \otimes T_b$ satisfies and is uniquely characterized by the following condition:

$$(T_a \otimes T_b)(\psi_a \otimes \psi_b) = (T_a\psi_a) \otimes (T_b\psi_b) \tag{5.131}$$

for all $\psi_a \in \mathscr{H}_a$ and $\psi_b \in \mathscr{H}_b$. Conclude that the definition of $T_a \otimes T_b$ does not depend on the choice of ONBs, and show further that it has the following properties:

(i) $(T_a \otimes T_b)^\dagger = T_a^\dagger \otimes T_b^\dagger$
(ii) $(T_a \otimes T_b)(S_a \otimes S_b) = (T_a S_a) \otimes (T_b S_b)$
(iii) $\operatorname{tr}(T_a \otimes T_b) = (\operatorname{tr} T_a)(\operatorname{tr} T_b)$.

Note that for two non-interacting systems, the Hamiltonian is of the form

$$H = H_a \otimes I_b + I_a \otimes H_b\,, \tag{5.132}$$

and the time evolution is

$$e^{-iHt} = e^{-iH_at} \otimes e^{-iH_bt}\,, \tag{5.133}$$

i.e., of the form $U_t = U_{a,t} \otimes U_{b,t}$.

Exercise 5.8 (Main Theorem About POVMs and Tensor Product) Show that in Bohmian mechanics, an experiment on $S = a \cup b$ in which the apparatus interacts only with system a but not with system b has a POVM of the form:

$$E(B) = E_a(B) \otimes I_b \tag{5.134}$$

for all $B \subseteq \mathscr{Z}$.[12]

The statement about (5.134) holds as a theorem also in GRW theory. The statement is regarded as true also in orthodox quantum mechanics.

[12] To this end, adapt the proof of the main theorem of POVMs. Suppose the experiment $\mathscr{E}$ begins at time t_i and ends at time t_f, and suppose the wave function of system a, system b, and the apparatus A together at time t_i is $\Psi(t_\mathrm{i}) = \psi \otimes \phi_A$ with $\psi \in \mathscr{H}_a \otimes \mathscr{H}_b$, so $\Psi(t_\mathrm{i}) \in \mathscr{H}_a \otimes \mathscr{H}_b \otimes \mathscr{H}_A$. Use (5.132) and (5.133). Assume further that the outcome Z is a function ζ of the configuration Q_A of the apparatus at time t_f.

5.5.2 Definition of the Reduced Density Matrix

Now we consider a fixed, non-random wave function, but one for $a \cup b$ together, $\psi \in \mathbb{S}(\mathscr{H}_a \otimes \mathscr{H}_b)$. We carry out an experiment on a alone, so (5.134) applies, and ask what the distribution of the outcome is. In this case, another density matrix plays a role, as I will explain in the following. The relevant equation is

$$\mathbb{P}(Z \in B) = \langle\psi|E(B)|\psi\rangle = \mathrm{tr}\big(\rho_\psi\, E_a(B)\big) \tag{5.135}$$

with the *reduced density matrix of system a*

$$\rho_\psi = \mathrm{tr}_b|\psi\rangle\langle\psi|\,, \tag{5.136}$$

where tr_b means the *partial trace* over $\mathscr{H}_b$, a concept I will discuss in the next section. The reduced density matrix and the trace formula for it were discovered by Lev Landau (1908–1968) in 1927 [28].

5.5.3 Partial Trace

Consider again ONBs $\{|n_a\rangle_a\}_{n_a}$ and $\{|n_b\rangle_b\}_{n_b}$.

Definition 5.8 If the operator T on $\mathscr{H} = \mathscr{H}_a \otimes \mathscr{H}_b$ has matrix elements $T_{m_a m_b n_a n_b}$ relative to the $|n_a n_b\rangle$, then its *partial trace* is the operator $S = \mathrm{tr}_b\, T$ on $\mathscr{H}_a$ with matrix elements

$$S_{m_a n_a} = \sum_{n_b} T_{m_a n_b n_a n_b}\,. \tag{5.137}$$

It turns out that the partial trace does not depend on the choice of ONBs. We will also write

$$S = \sum_{n_b} {}_b\langle n_b|T|n_b\rangle_b\,, \tag{5.138}$$

where the inner products are partial inner products. The partial trace has the following properties:

(i) It is linear:

$$\mathrm{tr}_b(R+T) = \mathrm{tr}_b(R) + \mathrm{tr}_b(T)\,, \quad \mathrm{tr}_b(\lambda T) = \lambda\,\mathrm{tr}_b(T). \tag{5.139}$$

(ii) $\mathrm{tr}(\mathrm{tr}_b(T)) = \mathrm{tr}(T)$. Here, the first tr symbol means the trace in $\mathscr{H}_a$, the second one the partial trace, and the last one the trace in $\mathscr{H}_a \otimes \mathscr{H}_b$. This property follows from (5.137) by setting $m_a = n_a$ and summing over n_a. In particular, tr_b maps the trace class of $\mathscr{H}_a \otimes \mathscr{H}_b$ to that of $\mathscr{H}_a$.

(iii) $\mathrm{tr}_b(T^\dagger) = (\mathrm{tr}_b\, T)^\dagger$. The adjoint of the partial trace is the partial trace of the adjoint. In particular, if T is self-adjoint then so is $\mathrm{tr}_b\, T$.

(iv) $\mathrm{tr}_b(T_a \otimes T_b) = (\mathrm{tr}\, T_b)T_a$.

(v) If T is a positive operator then so is $\mathrm{tr}_b\, T$. (This is best seen from (5.138).)

(vi) $\mathrm{tr}_b\big[R(T_a \otimes I_b)\big] = (\mathrm{tr}_b\, R)T_a$.

(vii) $\mathrm{tr}_b\big[R(I_a \otimes T_b)\big] = \mathrm{tr}_b\big[(I_a \otimes T_b)R\big]$.

The operation tr_b of taking the partial trace is also called *tracing out the b system*, in analogy to the operation of integrating out the y variable to obtain the marginal $\int dy\, \rho(x, y)$ of a probability distribution $\rho(x, y)$.

5.5.4 The Trace Formula Again

We are now ready to prove the trace formula (5.135). From properties (vi) and (ii) we obtain that

$$\mathrm{tr}\big[R(T_a \otimes I_b)\big] = \mathrm{tr}\big[(\mathrm{tr}_b\, R)T_a\big]. \tag{5.140}$$

Setting $R = |\psi\rangle\langle\psi|$ and $T_a = E_a(B)$, we find that $\mathrm{tr}_b\, R = \rho_\psi$ and

$$\langle\psi|E_a(B) \otimes I_b|\psi\rangle = \mathrm{tr}\big[|\psi\rangle\langle\psi|(E_a(B) \otimes I_b)\big] = \mathrm{tr}\big[\rho_\psi E_a(B)\big], \tag{5.141}$$

which proves (5.135).

From properties (ii) and (v) it follows also that ρ_ψ is a positive operator with trace 1. Conversely, every positive operator ρ on $\mathscr{H}_a$ with $\mathrm{tr}\,\rho = 1$ arises as a reduced density matrix. Indeed, if $\rho = \sum_n p_n |\phi_n\rangle\langle\phi_n|$ with $p_n \geq 0, \sum_n p_n = 1$ and orthonormal ϕ_n, then choose any ONB $\{\chi_m\}$ of $\mathscr{H}_b$ and set $\psi = \sum_n \sqrt{p_n}\phi_n \otimes \chi_n$. Then $\psi \in \mathscr{H}_a \otimes \mathscr{H}_b$, $\|\psi\| = 1$, and $\mathrm{tr}_b|\psi\rangle\langle\psi| = \rho$.

Finally, statistical density matrices as in (5.103) and reduced density matrices can be combined into a *statistical reduced density matrix*: If $\Psi \in \mathscr{H}_a \otimes \mathscr{H}_b$ is random then set

$$\rho = \mathbb{E}\,\mathrm{tr}_b\, |\Psi\rangle\langle\Psi| = \mathrm{tr}_b\, \mathbb{E}\, |\Psi\rangle\langle\Psi|. \tag{5.142}$$

Again, we find that the probability distribution of the outcome Z of an experiment on the a-system is given by the usual trace formula:

$$\mathbb{P}(Z \in B) = \mathrm{tr}\big(\rho\, E_a(B)\big)\,. \tag{5.143}$$

5.5.5 The Measurement Problem and Density Matrices

Statistical and reduced density matrices sometimes get confused; here is an example. Consider again the wave function of the measurement problem,

$$\Psi = \sum_\alpha \Psi_\alpha\,, \tag{5.144}$$

the wave function of an object and an apparatus after a quantum measurement of the observable $A = \sum \alpha P_\alpha$. Suppose that Ψ_α, the contribution corresponding to the outcome α, is of the form

$$\Psi_\alpha = c_\alpha\, \psi_\alpha \otimes \phi_\alpha\,, \tag{5.145}$$

where $c_\alpha = \|P_\alpha \psi\|$, ψ is the initial object wave function ψ, $\psi_\alpha = P_\alpha\psi/\|P_\alpha\psi\|$, and ϕ_α with $\|\phi_\alpha\| = 1$ is a wave function of the apparatus after having measured α. Since the ϕ_α have disjoint supports in configuration space, they are mutually orthogonal; thus, they are a subset of some orthonormal basis $\{\phi_n\}_n$. The reduced density matrix of the object is

$$\rho_\Psi = \mathrm{tr}_b\, |\Psi\rangle\langle\Psi| = \sum_n \langle\phi_n|\Psi\rangle\langle\Psi|\phi_n\rangle = \sum_\alpha |c_\alpha|^2\, |\psi_\alpha\rangle\langle\psi_\alpha|\,. \tag{5.146}$$

This is the same density matrix as the statistical density matrix associated with the probability distribution μ of the collapsed wave function ψ',

$$\mu = \sum_\alpha |c_\alpha|^2\, \delta_{\psi_\alpha}\,, \tag{5.147}$$

since

$$\rho_\mu = \sum_\alpha |c_\alpha|^2 |\psi_\alpha\rangle\langle\psi_\alpha|\,. \tag{5.148}$$

It is sometimes claimed that this fact solves the measurement problem. The argument is this: From (5.144) we obtain (5.146), which is the same as (5.148), which means that the system's wave function has distribution (5.147), so we have a random outcome α. This argument is incorrect, as the mere fact that two situations—one with Ψ as in (5.144), the other with random ψ'—define the same density matrix for the object does not mean the two situations are physically equivalent. And obviously from (5.144), the situation after a quantum measurement involves neither a random outcome nor a random wave function. As John Bell once put it,

> *and* is not *or*.

It is sometimes taken as the definition of *decoherence* that the reduced density matrix is (approximately) diagonal in the eigenbasis of the relevant operator A. In Sect. 3.2.2 I had defined decoherence as the situation that two or more wave packets Ψ_α are macroscopically disjoint in configuration space (and thus remain disjoint for the relevant future). The connection between the two definitions is that the latter implies the former if Ψ_α is of the form (5.145).

It is common to call a density matrix that is a one-dimensional projection a *pure state* and otherwise a *mixed state*, even if it is a reduced density matrix and thus does not arise from a mixture (i.e., from a probability distribution μ). Beware of the potential confusion!

5.5.6 POVM and Collapse

For the quantum measurement of a self-adjoint operator, we had the Born rule (that governed the distribution of outcome) and the projection postulate (that tells us how to collapse the wave function). We have seen that for a general experiment, the distribution of the outcome is governed by a POVM. But how, in this situation, must the wave function be collapsed?

To answer this question, we proceed as for proving the main theorem of POVMs. We form a tensor product of the object S and the apparatus A, evolve both together from the initial time t_i until the final time t_f of the experiment, conditionalize on the actual outcome $Z = z$ by projecting to the region of configuration space compatible with z, and then ask about the quantum state of the object. However, the object might still be entangled with the apparatus, even after conditionalizing on a particular outcome. This suggests tracing out the apparatus and using the reduced density matrix ρ_S of S as its post-experiment quantum state.

(As an alternative, it appears we might use its conditional wave function. However, this would mean to conditionalize on the exact configuration of the apparatus, whereas we want to conditionalize on the more limited information that $Z = z$.)

In fact, the reduced density matrix ρ_S is the mathematically correct description in the sense that for a subsequent experiment using a second, separate apparatus that interacts only with the object but not with the first apparatus (of the first experiment), the distribution takes the form

$$\mathbb{P}(Z_2 \in B) = \mathrm{tr}\big(\rho_S \, E_2(B)\big) \tag{5.149}$$

(all operators act in $\mathscr{H}_S$). Now, if we allow density matrices as mathematical descriptions of the collapsed state, then we need to also allow density matrices as initial states. But that causes no trouble at all. In fact, it fits exactly with our use of POVMs, which has always led to the same trace formula regardless of whether the density matrix is a statistical, a reduced, or a statistical reduced density matrix. More than that, in order to compute the object's reduced density matrix at t_f, it is not necessary to know whether the object's initial density matrix at t_i is a statistical, reduced, or a statistical reduced density matrix, provided that the apparatus does not interact with any previous apparatus. As a consequence of this situation, the appropriate collapse rule maps $\rho_S(t_i)$ to $\rho_S(t_f)$. We will formulate this rule in Sect. 5.5.8.

A linear mapping that maps operators to operators (rather than vectors to vectors) is called a *superoperator*. Let us first spend a bit of time with the math of superoperators.

5.5.7 Completely Positive Superoperators

Let $TRCL(\mathscr{H})$ denote the trace class of $\mathscr{H}$. (In finite dimension, this is just the set of all operators on $\mathscr{H}$.) A *superoperator* means a $\mathbb{C}$-linear mapping that acts on operators in $\mathscr{H}$, particularly on density matrices; we will here consider superoperators of the form

$$\mathscr{C} : TRCL(\mathscr{H}_1) \to TRCL(\mathscr{H}_2) \tag{5.150}$$

(including the possibility $\mathscr{H}_2 = \mathscr{H}_1$). In terms of ONBs $\{|n_1\rangle_1\}_{n_1}$ of $\mathscr{H}_1$ and $\{|n_2\rangle_2\}_{n_2}$ of $\mathscr{H}_2$, $\mathscr{C}$ can be expressed through its matrix elements $\mathscr{C}_{m_2n_2m_1n_1}$ as follows: $\mathscr{C}$ will map an operator $\rho \in TRCL(\mathscr{H}_1)$ with matrix elements $\rho_{m_1n_1}$ to the operator σ whose matrix elements are

$$\sigma_{m_2n_2} = \sum_{m_1,n_1} \mathscr{C}_{m_2n_2m_1n_1}\,\rho_{m_1n_1}\,. \tag{5.151}$$

The fact that $\mathscr{C}$ has four indices allows us to re-interpret $\mathscr{C}$ as an operator $\tilde{\mathscr{C}}$ on $\mathscr{H}_1 \otimes \mathscr{H}_2$ that maps the vector ψ with coefficients $\psi_{n_1n_2}$ to the vector ϕ with coefficients

$$\phi_{m_1m_2} = \sum_{n_1,n_2} \mathscr{C}_{m_2n_2m_1n_1}\,\psi_{n_1n_2}\,. \tag{5.152}$$

Definition 5.9 A superoperator $\mathscr{C}$ is called *completely positive* if for every integer $k \geq 0$ and every positive operator $\rho \in \mathbb{C}^{k\times k} \otimes TRCL(\mathscr{H}_1)$, $(I_k \otimes \mathscr{C})(\rho)$ is positive, where I_k denotes the identity operator on $\mathbb{C}^{k\times k}$.

Here, $\mathbb{C}^{k\times k}$ means the space of complex $k \times k$ matrices. We note that $\mathbb{C}^{k\times k} \otimes TRCL(\mathscr{H}_1) = TRCL(\mathbb{C}^k \otimes \mathscr{H}_1)$, so $I_k \otimes \mathscr{C}$ maps operators on $\mathbb{C}^k \otimes \mathscr{H}_1$ to operators on $\mathbb{C}^k \otimes \mathscr{H}_2$. For $k = 0$ the condition says that $\mathscr{C}$ maps positive operators (from the trace class) on $\mathscr{H}_1$ to positive operators on $\mathscr{H}_2$. (One might have thought that if $\mathscr{C}$ maps positive operators on $\mathscr{H}_1$ to positive operators on $\mathscr{H}_2$, then $I_k \otimes \mathscr{C}$ maps positive operators on $\mathbb{C}^k \otimes \mathscr{H}_1$ to positive operators on $\mathbb{C}^k \otimes \mathscr{H}_2$. However, this is not the case, which is why we demand it explicitly.)

Proposition 5.3 *Suppose $\mathscr{H}_1$ and $\mathscr{H}_2$ have finite dimension. Then $\mathscr{C} : TRCL(\mathscr{H}_1) \to TRCL(\mathscr{H}_2)$ is completely positive if and only if the operator $\tilde{\mathscr{C}}$ given by* (5.152) *is positive.*

Proof Let $\{|n_1\rangle_1\}_{n_1}$ and $\{|n_2\rangle_2\}_{n_2}$ be ONBs again. Fix a $k \geq 0$ and let ρ be a positive operator on $\mathbb{C}^k \otimes \mathscr{H}_1$ with matrix elements $\rho_{\beta m_1 \alpha n_1}$, where α, β run from 1 to k. Let $\{\omega_i\}_i$ be an ONB of eigenvectors of ρ in $\mathbb{C}^k \otimes \mathscr{H}_1$ with eigenvalues $\lambda_i \geq 0$. Then

$$\rho_{\beta m_1 \alpha n_1} = \sum_i \lambda_i \omega_{i\beta m_1} \omega^*_{i\alpha n_1} \tag{5.153}$$

and

$$[(I_k \otimes \mathscr{C})(\rho)]_{\beta m_2 \alpha n_2} = \sum_{m_1, n_1} \mathscr{C}_{m_2 n_2 m_1 n_1} \rho_{\beta m_1 \alpha n_1} \tag{5.154}$$

$$= \sum_{m_1, n_1, i} \mathscr{C}_{m_2 n_2 m_1 n_1} \lambda_i \omega_{i\beta m_1} \omega^*_{i\alpha n_1} . \tag{5.155}$$

Thus, for any $\chi \in \mathbb{C}^k \otimes \mathscr{H}_2$,

$$\langle \chi | (I_k \otimes \mathscr{C})(\rho) | \chi \rangle_{\mathbb{C}^k \otimes \mathscr{H}_2} = \sum_{\alpha, \beta, m_2, n_2} \chi^*_{\beta m_2} [(I_k \otimes \mathscr{C})(\rho)]_{\beta m_2 \alpha n_2} \chi_{\alpha n_2} \tag{5.156a}$$

$$= \sum_{\alpha \beta m_2 n_2 m_1 n_1 i} \lambda_i \chi^*_{\beta m_2} \omega_{i \beta m_1} \mathscr{C}_{m_2 n_2 m_1 n_1} \omega^*_{i \alpha n_1} \chi_{\alpha n_2} \tag{5.156b}$$

$$= \sum_i \lambda_i \langle \phi_i | \tilde{\mathscr{C}} | \phi_i \rangle \tag{5.156c}$$

with $\phi_{i n_1 n_2} = \sum_\alpha \omega^*_{i \alpha n_1} \chi_{\alpha n_2}$.

We first show that if $\tilde{\mathscr{C}}$ is positive, then $(I_k \otimes \mathscr{C})(\rho)$ is positive: If $\tilde{\mathscr{C}}$ is positive then $\langle \phi_i | \tilde{\mathscr{C}} | \phi_i \rangle \geq 0$; since $\lambda_i \geq 0$, (5.156c) is ≥ 0.

We next show that if $\mathscr{C}$ is completely positive, then $\tilde{\mathscr{C}}$ is positive: For $j = 1, 2$ let $d_j = \dim \mathscr{H}_j$. For any $\phi \in \mathscr{H}_1 \otimes \mathscr{H}_2$, let $k = d_1 d_2$, write $\alpha = (m_1, m_2)$, and set $\omega_{m_1 m_2 n_1} = d_2^{-1/2} \delta_{m_1 n_1}$ and $\chi_{m_1 m_2 n_2} = d_2^{1/2} \delta_{m_2 n_2} \phi_{m_1 m_2}$ and $\rho = |\omega\rangle\langle\omega|$. Then

$$\sum_\alpha \omega^*_{\alpha n_1} \chi_{\alpha n_2} = \sum_{m_1 m_2} \delta_{m_1 n_1} \delta_{m_2 n_2} \phi_{m_1 m_2} = \phi_{n_1 n_2} \tag{5.157}$$

and

$$0 \le \langle \chi | (I_k \otimes \mathscr{C})(\rho) | \chi \rangle_{\mathbb{C}^k \otimes \mathscr{H}_2} \overset{(5.156c)}{=} \langle \phi | \tilde{\mathscr{C}} | \phi \rangle \,, \tag{5.158}$$

as claimed. □

Completely positive superoperators are also often called *completely positive maps* (CPMs). They arise as a description of how a density matrix changes under the collapse caused by an experiment: If ρ is the density matrix before the collapse, then $\mathscr{C}(\rho)/\mathrm{tr}\,\mathscr{C}(\rho)$ is the density matrix afterward. The simplest example of a completely positive superoperator is

$$\mathscr{C}(\rho) = P\rho P \,, \tag{5.159}$$

where P is a projection. Note that for a density matrix ρ, $\mathscr{C}(\rho)$ is not, in general, a density matrix because completely positive superoperators do not, in general, preserve the trace.

In order to establish the complete positivity of a given superoperator, the following facts are useful: If ρ_2 is a density matrix on $\mathscr{H}_2$ then the mapping $\mathscr{C} : TRCL(\mathscr{H}_1) \to TRCL(\mathscr{H}_1 \otimes \mathscr{H}_2)$ given by $\mathscr{C}(\rho) = \rho \otimes \rho_2$ is completely positive. Conversely, the partial trace $\rho \mapsto \mathrm{tr}_2\, \rho$ is a completely positive superoperator $TRCL(\mathscr{H}_1 \otimes \mathscr{H}_2) \to TRCL(\mathscr{H}_1)$. For any bounded operator $R : \mathscr{H}_1 \to \mathscr{H}_2$, $\rho \mapsto R\rho R^\dagger$ is a completely positive superoperator $TRCL(\mathscr{H}_1) \to TRCL(\mathscr{H}_2)$. The composition of completely positive superoperators is completely positive. Positive multiples of a completely positive superoperator are completely positive. Finally, when a family of completely positive superoperators is summed or integrated over, the result is completely positive.

A canonical form of completely positive superoperators is provided by the theorem of Choi and Kraus:

Theorem 5.8 (Choi (1975) [9] and Kraus (1983) [27]) *For every bounded completely positive superoperator $\mathscr{C} : TRCL(\mathscr{H}_1) \to TRCL(\mathscr{H}_2)$ there exist bounded operators $R_i : \mathscr{H}_1 \to \mathscr{H}_2$ so that*

$$\mathscr{C}(\rho) = \sum_{i \in \mathscr{I}} R_i \, \rho \, R_i^\dagger \,, \tag{5.160}$$

where $\mathscr{I}$ is a finite or countable index set.

5.5.8 The Main Theorem About Superoperators

We can now formulate the analog of the main theorem about POVMs concerning the post-experiment quantum state.

Main theorem about superoperators *With every experiment $\mathscr{E}$ on the system S during $[t_i, t_f]$ with finite value space $\mathscr{Z}$ is associated a family $(\mathscr{C}_z)_{z\in\mathscr{Z}}$ of completely positive superoperators acting on $TRCL(\mathscr{H}_S)$ such that, whenever $Z = z$, the density matrix of the object after the experiment is*

$$\rho_{t_f} = \frac{\mathscr{C}_z(\rho_{t_i})}{\operatorname{tr}\mathscr{C}_z(\rho_{t_i})} . \tag{5.161}$$

$\mathscr{C}_z$ is related to the POVM E_z by

$$\operatorname{tr}\big(\rho\, E_z\big) = \operatorname{tr}\mathscr{C}_z(\rho) . \tag{5.162}$$

In particular, $\sum_{z\in\mathscr{Z}}\mathscr{C}_z$ is trace-preserving. Explicitly, $\mathscr{C}_z$ is given by

$$\mathscr{C}_z(\rho) = \operatorname{tr}_A\Big([I_S\otimes P_{A,z}]U[\rho\otimes\rho_A]U^\dagger[I_S\otimes P_{A,z}]\Big) , \tag{5.163}$$

where A means the apparatus, $P_{A,z}$ is the projection to the subspace of apparatus states in which the pointer is pointing to the value z, U the unitary time evolution of object and apparatus together from t_i to t_f, and ρ_A the density matrix of the ready state of the apparatus.

Proof The proof proceeds along similar lines as the proof of the main theorem about POVMs. Here is an outline; for more detail and discussion see [23].

Starting from $\rho_{S\cup A}(t_i) = \rho\otimes\rho_A$, solve the Schrödinger equation for the object and the apparatus together up to t_f to obtain $\rho_{S\cup A}(t_f)$. The probability of pointer position z is the trace of

$$[I_S\otimes P_{A,z}]\rho_{S\cup A}(t_f)[I_S\otimes P_{A,z}] . \tag{5.164}$$

(By virtue of property (vii) of the partial trace in Sect. 5.5.3, we could as well omit one of the factors $[I_S\otimes P_{A,z}]$ inside the (partial) trace, also in (5.163).) This, together with the definition (5.163), proves (5.162). Now we determine the resulting density matrix. To conditionalize on the pointer position z means to replace $\rho_{S\cup A}(t_f)$ by (5.164), divided by its trace as a normalizing factor. But note that its trace (over $S\cup A$) is just the trace (over S) of $\mathscr{C}_z(\rho)$. Now taking (5.164) and tracing out the apparatus yields (5.161). □

Readers may feel that this theorem is a rather complicated statement. I have two comments on this. First, this is the price we pay for an honest statement that is not limited to the idealized situation of a quantum measurement of a self-adjoint operator. Second, the complexity is largely owed to the fact that it is an *operational* statement, one describing what will come out of an experiment. Note the contrast with the simplicity of the *ontological* description of Bohmian mechanics in Sect. 1.6.1.

By the way, as far as I know, it is an open question how to decide for a given finite family $(\mathscr{C}_z)_z$ of completely positive superoperators whether there is an experiment $\mathscr{E}$ associated with it, i.e., whether $\mathscr{C}_z(\rho)$ can be written in the form (5.163) for some ρ_A, t_i, t_f, and an $H_{S\cup A}$ that is a Schrödinger operator $-\nabla^2 + V$ for some potential $V : \mathbb{R}^{3N_S+3N_A} \to \mathbb{R}$, or likewise for GRW theory.

5.5.9 The No-Signaling Theorem

This is a consequence of the quantum formalism:

No-Signaling Theorem *If system a is located in Alice's lab and system b in Bob's, and if the total Hamiltonian contains no interaction term between the two labs, then the statistical reduced density matrix ρ_a of system a is not affected by anything Bob does. As a consequence, the marginal distribution of the outcome of any experiment that Alice conducts on a is not affected by anything Bob does. In particular, the nonlocal influences between a and b implied by Bell's theorem cannot be used for sending signals.*

To prove the no-signaling theorem, we will verify that:

(i) ρ_a does not depend on the Hamiltonian of system b (and thus not on any external fields that Bob may apply to b).
(ii) ρ_a is not affected by any experiment Bob performs on b.

Moreover, a no-signaling theorem holds also for GRW theory, and this conclusion is based on the further fact that if there is no interaction term between a and b, then:

(iii) ρ_a is not affected by any GRW collapse on b.

To verify (i), note that in the absence of interaction the unitary time evolution operator is $U_t = U_{a,t} \otimes U_{b,t}$. Thus, the reduced density matrix evolves according to

$$\rho_{a,t} = \mathrm{tr}_b |U_t\psi\rangle\langle U_t\psi| \tag{5.165a}$$

$$= \mathrm{tr}_b\Big[U_t|\psi\rangle\langle\psi|U_t^\dagger\Big] \tag{5.165b}$$

$$= \mathrm{tr}_b\Big[(U_{a,t} \otimes U_{b,t})|\psi\rangle\langle\psi|(U_{a,t}^\dagger \otimes U_{b,t}^\dagger)\Big] \tag{5.165c}$$

$$= \mathrm{tr}_b\Big[(U_{a,t} \otimes I_b)|\psi\rangle\langle\psi|(U_{a,t}^\dagger \otimes (U_{b,t}^\dagger U_{b,t}))\Big] \tag{5.165d}$$

$$= \mathrm{tr}_b\Big[(U_{a,t} \otimes I_b)|\psi\rangle\langle\psi|(U_{a,t}^\dagger \otimes I_b)\Big] \tag{5.165e}$$

$$= U_{a,t}[\mathrm{tr}_b\,|\psi\rangle\langle\psi|]U_{a,t}^\dagger = U_{a,t}\rho_\psi U_{a,t}^\dagger\,, \tag{5.165f}$$

which does not depend on $U_{b,t}$. The argument extends without difficulty to statistical reduced density matrices.

For (ii), we need to consider how an experiment that interacts only with system b of a composite system $a \cup b$ (as in Exercise 5.8 but with the roles of a and b interchanged) will collapse the state of $a \cup b$:

Theorem 5.9 *In Bohmian mechanics, an experiment in which the apparatus interacts only with system b but not with system a has completely positive superoperators of the form*

$$\mathscr{C}_z = \mathscr{U}_a \otimes \mathscr{C}_{b,z}\,, \tag{5.166}$$

where $\mathscr{U}_a(\rho_a) = U_a \rho_a U_a^\dagger$ *is the unitary evolution from* t_i *to* t_f *on* $TRCL(\mathscr{H}_a)$.

Proof This follows from (5.163) by noting that $\mathscr{H}_S = \mathscr{H}_a \otimes \mathscr{H}_b$, as well as $I_S = I_a \otimes I_b$ and $U = U_a \otimes U_{b\cup A}$, and exploiting the rules of the partial trace. □

We now verify (ii). Bob conducts an experiment $\mathscr{E}$ and obtains outcome z with probability $\mathrm{tr}\, \mathscr{C}_z(\rho_{a\cup b})$; the density matrix $\rho_{a\cup b}$ collapses to $\mathscr{C}_z(\rho_{a\cup b})/\mathrm{tr}\, \mathscr{C}_z(\rho_{a\cup b})$. Thus, the statistical density matrix of $a \cup b$ after $\mathscr{E}$ is $\sum_z \mathscr{C}_z(\rho_{a\cup b})$. Using (5.166), we obtain, as the reduced statistical density matrix after $\mathscr{E}$,

$$\rho_a(t_\mathrm{f}) = \mathrm{tr}_b \sum_z \mathscr{C}_z(\rho_{a\cup b}) \tag{5.167a}$$

$$= \mathrm{tr}_b \sum_z (\mathscr{U}_a \otimes \mathscr{C}_{b,z})(\rho_{a\cup b}) \tag{5.167b}$$

$$= \mathscr{U}_a \mathrm{tr}_b \sum_z \mathscr{C}_{b,z}(\rho_{a\cup b}) \tag{5.167c}$$

$$= \mathscr{U}_a \mathrm{tr}_b \rho_{a\cup b} \tag{5.167d}$$

$$= \mathscr{U}_a \rho_a(t_\mathrm{i}) \tag{5.167e}$$

because $\sum_z \mathscr{C}_{b,z}$ is trace-preserving. This is the same as how ρ_a would have evolved without Bob's experiment.

To verify (iii), suppose that ψ is a function of $N = N_a + N_b$ variables in $\mathbb{R}^3$, and that, at a particular time t, a GRW collapse hits particle i which belongs to b, so

$$\psi_{t+} = \frac{C_i(X)\psi_{t-}}{\|C_i(X)\psi_{t-}\|} \tag{5.168}$$

with random collapse center $\boldsymbol{X}$ chosen with probability density $\|C_i(\boldsymbol{x})\psi_{t-}\|^2$ at $\boldsymbol{x}$. Then the statistical reduced density matrix of a after the collapse is given by

$$\rho_{t+} = \mathrm{tr}_b \int_{\mathbb{R}^3} d^3\boldsymbol{x}\, \|C_i(\boldsymbol{x})\psi_{t-}\|^2 \, \frac{C_i(\boldsymbol{x})|\psi_{t-}\rangle\langle\psi_{t-}|C_i(\boldsymbol{x})^\dagger}{\|C_i(\boldsymbol{x})\psi_{t-}\|^2} \tag{5.169a}$$

$$= \int_{\mathbb{R}^3} d^3\boldsymbol{x}\, \mathrm{tr}_b\Big[C_i(\boldsymbol{x})|\psi_{t-}\rangle\langle\psi_{t-}|C_i(\boldsymbol{x})^\dagger\Big] \tag{5.169b}$$

$$\overset{\text{(vii)}}{=} \int_{\mathbb{R}^3} d^3\boldsymbol{x}\, \mathrm{tr}_b\Big[|\psi_{t-}\rangle\langle\psi_{t-}|C_i(\boldsymbol{x})^\dagger C_i(\boldsymbol{x})\Big] \tag{5.169c}$$

$$= \mathrm{tr}_b\Big[|\psi_{t-}\rangle\langle\psi_{t-}| \int_{\mathbb{R}^3} d^3\boldsymbol{x}\, C_i(\boldsymbol{x})^\dagger C_i(\boldsymbol{x})\Big] \tag{5.169d}$$

$$= \mathrm{tr}_b|\psi_{t-}\rangle\langle\psi_{t-}| \tag{5.169e}$$

because $\int d^3\boldsymbol{x}\, C_i(\boldsymbol{x})^\dagger C_i(\boldsymbol{x}) = I$ as in (3.60)–(3.61). This completes the proof of the no-signaling theorem.

One more remark: The cancelation of $\|C_i(\boldsymbol{x})\psi_{t-}\|^2$ in (5.169a) is possible because this factor is the probability density of the random collapse center $\boldsymbol{X}$. This density can be written as $|\psi_{t-}|^2 * g_\sigma$ (the Born density convolved with a Gaussian of width σ), and when we first introduced it in (3.59) and (3.64), this convolution may have seemed surprising: Why not simply take $\boldsymbol{X} \sim |\psi_{t-}|^2$? We now see that the simpler choice would not have led to a cancelation and thus would have spoiled the no-signaling property. That is the reason for $|\psi_{t-}|^2 * g_\sigma$.

5.5.10 Canonical Typicality

This is an application of reduced density matrices in quantum statistical mechanics. The main goal of quantum statistical mechanics is to derive facts of thermodynamics from a quantum mechanical analysis of systems with a macroscopic number of particles (say, $N > 10^{20}$). One of the rules of quantum statistical mechanics asserts that if a quantum system S is in thermal equilibrium at absolute temperature $T \geq 0$, then it has density matrix

$$\rho_{\mathrm{can}} = \frac{1}{Z} e^{-\beta H_S}\,, \tag{5.170}$$

where H_S is the system's Hamiltonian, $\beta = 1/kT$ with $k = 1.38 \cdot 10^{-23}$ J/K the Boltzmann constant, and $Z = \mathrm{tr}\, e^{-\beta H}$ the normalizing factor; ρ_{can} is called the *canonical density matrix* with inverse temperature β.

While this rule has long been used, its justification is rather recent[13] and goes as follows. Suppose that S is coupled to another system B (the "heat bath"), and suppose that S and B together have wave function $\psi \in \mathscr{H}_S \otimes \mathscr{H}_B$ and Hamiltonian H with pure point spectrum (this comes out for systems confined to finite volume). Let $I_{\text{mc}} = [E - \Delta E, E]$ be an energy interval whose length ΔE is small on the macroscopic scale but large enough for I_{mc} to contain very many eigenvalues of H; I_{mc} is called a *micro-canonical energy shell*. Let $\mathscr{H}_{\text{mc}}$ be the corresponding spectral subspace, i.e., the range of $1_{I_{\text{mc}}}(H)$, and u_{mc} the uniform probability distribution over $\mathbb{S}(\mathscr{H}_{\text{mc}})$.

Theorem 5.10 (Canonical Typicality, Informal Statement) *If B is sufficiently "large," and if the interaction between S and B is negligible,*

$$H \approx H_S \otimes I_B + I_S \otimes H_B \,, \tag{5.171}$$

then for most ψ relative to u_{mc}, the reduced density matrix of S is approximately canonical for some value of β, i.e.,

$$\text{tr}_B|\psi\rangle\langle\psi| \approx \rho_{\text{can}} \,. \tag{5.172}$$

In order to prepare a typical state $\psi \in \mathbb{S}(\mathscr{H}_{\text{mc}})$ (and thus at thermal equilibrium between S and B), it will be relevant to have some interaction between S and B. Large interaction terms in H, however, will lead to deviations from the form (5.170). It is relevant for (5.172) that S and B are entangled: If they were not, then the reduced density matrix of S would be pure, whereas ρ_{can} is usually highly mixed (i.e., has many eigenvalues that are significantly nonzero).

Canonical typicality explains why we see canonical density matrices: Because "most" wave functions of $S \cup B$ lead to a canonical density matrix for S.

5.5.11 The Possibility of a Fundamental Density Matrix

In Bohmian mechanics, a particle is guided by a wave function ψ. Here is a variant of Bohmian mechanics in which a particle is guided by a density matrix ρ (Dürr et al. 2005 [14]). This is an example of a theory in which a density matrix appears that is neither a statistical nor a reduced density matrix; it is a fundamental density matrix, part of the fundamental ontology. In this theory, the physical state of the universe is given by the pair (Q, ρ) with Q the particle configuration and ρ the density matrix of the universe.

Here is the definition of the theory. The density matrix ρ evolves unitarily; that is,

$$\rho_t = e^{-iHt/\hbar}\rho_0 e^{iHt/\hbar} \quad \text{or} \quad \frac{d\rho_t}{dt} = -\tfrac{i}{\hbar}[H, \rho_t] \,. \tag{5.173}$$

[13] This was discovered by several groups independently (Gemmer, Mahler, and Michel 2004 [18]; Popescu, Short, and Winter 2006 [35]; Goldstein et al. 2006 [22]). Preliminary considerations in this direction can already be found in Schrödinger (1952) [38].

A density matrix ρ can also be expressed as a function $\rho(q,q') = \langle q|\rho|q'\rangle$ with $|q\rangle$ the position (or configuration) basis. For the Schrödinger Hamiltonian $H = -\frac{\hbar^2}{2m}\nabla^2 + V$, the unitary evolution can be expressed as

$$\frac{\partial \rho_t(q,q')}{\partial t} = \frac{i}{\hbar}\frac{\hbar^2}{2m}\big(\nabla_q^2\,\rho_t(q,q') - \nabla_{q'}^2\,\rho_t(q,q')\big) - \frac{i}{\hbar}\big(V(q) - V(q')\big)\rho_t(q,q') \tag{5.174}$$

(For simplicity, we are taking all masses to be equal, $m_k = m$; but everything described here still works if the masses differ, it just needs to be written down differently.)

We replace Bohm's equation of motion,

$$\frac{dQ}{dt} = \frac{\hbar}{m}\mathrm{Im}\frac{\psi^*(q)\nabla_q\psi(q)}{\psi^*(q)\psi(q)}\bigg|_{q=Q_t}, \tag{5.175}$$

by

$$\frac{dQ}{dt} = \frac{\hbar}{m}\mathrm{Im}\frac{\nabla_q\,\rho(q,q')}{\rho(q,q)}\bigg|_{q=q'=Q_t}. \tag{5.176}$$

Finally, the distribution density $|\psi_t(q)|^2$ gets replaced with the formula

$$\rho_t(q,q)\,. \tag{5.177}$$

This completes the definition.

We make two mathematical observations. First, if ρ is a pure density matrix (i.e., a 1d projection) $|\psi\rangle\langle\psi|$, so $\rho(q,q') = \psi(q)\,\psi^*(q')$, then it stays pure for all times, the equation of motion (5.176) reduces to Bohm's (5.175), and the density (5.177) reduces to $|\psi_t(q)|^2$.

Second, an equivariance theorem holds in the sense that the density $\rho(q,q)$ is equivariant relative to the equation of motion (5.176): If $Q_0 \sim \rho_0(q,q)$, then $Q_t \sim \rho_t(q,q)$ at all times t. This is a consequence of the fact that the unitary evolution (5.174) implies the continuity equation

$$\frac{\partial\rho(q,q,t)}{\partial t} = -\nabla_q\cdot\Big(\rho(q,q,t)\,v^{\rho_t}(q)\Big) \tag{5.178}$$

with $v^{\rho_t}(Q_t)$ the right-hand side of (5.176).

Indeed, by setting $q' = q$ in (5.174),

$$\frac{\partial\rho(q,q,t)}{\partial t} = \frac{i\hbar}{2m}\Big[\nabla_q^2\,\rho_t(q,q') - \nabla_{q'}^2\,\rho_t(q,q')\Big]_{q'=q}. \tag{5.179}$$

Since $\rho_t(q, q') = \rho_t(q', q)^*$, it follows that

$$\nabla^2_{q'}\rho_t(q, q') = \nabla^2_{q'}\rho_t(q', q)^* = (\nabla^2_{q'}\rho_t(q', q))^* = \left[\nabla^2_q\rho_t(q, q')\right]^*_{q\leftrightarrow q'}, \tag{5.180}$$

so

$$\left[\nabla^2_{q'}\rho_t(q, q')\right]_{q'=q} = \left[\nabla^2_q\rho_t(q, q')\right]^*_{q'=q} \tag{5.181}$$

and

$$\frac{\partial\rho(q, q, t)}{\partial t} = \frac{i\hbar}{2m}\left[2i\,\mathrm{Im}\nabla^2_q\,\rho_t(q, q')\right]_{q'=q}. \tag{5.182}$$

On the other hand,

$$-\nabla_q\cdot\Big(\rho(q, q, t)\, v^{\rho_t}(q)\Big) = -\nabla_q\cdot\left[\frac{\hbar}{m}\mathrm{Im}\nabla_q\,\rho(q, q')\right]_{q'=q} \tag{5.183a}$$

$$= -\left[(\nabla_q + \nabla_{q'})\cdot\frac{\hbar}{m}\mathrm{Im}\nabla_q\,\rho(q, q')\right]_{q'=q} \tag{5.183b}$$

$$= -\left[\frac{\hbar}{m}\mathrm{Im}\nabla^2_q\,\rho(q, q') + \frac{\hbar}{m}\mathrm{Im}\nabla_{q'}\cdot\nabla_q\,\rho(q, q')\right]_{q'=q} \tag{5.183c}$$

$$= -\left[\frac{\hbar}{m}\mathrm{Im}\nabla^2_q\,\rho(q, q')\right]_{q'=q} \tag{5.183d}$$

because $[\nabla_{q'}\cdot\nabla_q\,\rho(q, q')]_{q'=q}$ is real-valued, as

$$\nabla_{q'}\cdot\nabla_q\,\rho(q, q') = \nabla_{q'}\cdot\nabla_q\,\rho(q', q)^* = \nabla_q\cdot\nabla_{q'}\,\rho(q', q)^* = [\nabla_{q'}\cdot\nabla_q\,\rho(q, q')]^*_{q\leftrightarrow q'}. \tag{5.184}$$

□

Exercise 5.9 In the presence of spin, the density matrix gets spin indices as in $\rho_{ss'}(q, q')$ with $s, s' \in \{1, 2\}^N$ for N particles. Find the appropriate generalization of (5.176) and (5.177) for this situation. If $V(q)$ is assumed to be a self-adjoint $2^N \times 2^N$ matrix, what changes in the equivariance proof?

5.6 Quantum Logic

The expression "quantum logic" is used in the literature for (at least) three different things:

- A certain piece of mathematics that is rather pretty

- A certain analogy between two formalisms that is rather limited
- A certain philosophical idea that is rather wrongheaded

Logic is the collection of those statements and rules that are valid in every conceivable universe and every conceivable situation. Some people have suggested that logic simply consists of the rules for the connectives "and," "or," and "not," with "$\forall x \in M$" an extension of "and" and "$\exists x \in M$" an extension of "or" to (possibly infinite) ranges M. I would say that viewpoint is not completely right (because of Gödel's incompleteness theorem[14]) and not completely wrong. Be that as it may, let us focus for a moment on the operations "and" (conjunction $A \wedge B$), "or" (disjunction $A \vee B$), and "not" (negation $\neg A$), and let us ignore infinite conjunctions or disjunctions.

5.6.1 Boolean Algebras

A *Boolean algebra* is a set $\mathscr{A}$ of elements $A, B, C, \ldots$ of which operations $A \wedge B$, $A \vee B$, and $\neg A$ are defined such that the following rules hold:

- $\wedge$ and $\vee$ are associative, commutative, and idempotent ($A \wedge A = A$ and $A \vee A = A$).
- Absorption laws: $A \wedge (A \vee B) = A$ and $A \vee (A \wedge B) = A$.
- There are elements $0 \in \mathscr{A}$ ("false") and $1 \in \mathscr{A}$ ("true") such that for all $A \in \mathscr{A}$, $A \wedge 0 = 0$, $A \wedge 1 = A$, $A \vee 0 = A$, $A \vee 1 = 1$.
- Complementation laws: $A \wedge \neg A = 0$, $A \vee \neg A = 1$.
- Distributive laws: $A \wedge (B \vee C) = (A \wedge B) \vee (A \wedge C)$ and $A \vee (B \wedge C) = (A \vee B) \wedge (A \vee C)$.

It follows from these axioms that $\neg(\neg A) = A$, and that de Morgan's laws hold, $\neg A \vee \neg B = \neg(A \wedge B)$ and $\neg A \wedge \neg B = \neg(A \vee B)$.

The laws of logic for "and," "or," and "not" are exactly the laws that hold in every Boolean algebra, with $A, B, C, \ldots$ playing the role of statements or propositions or conditions. Another case in which these axioms are satisfied is that $A, B, C, \ldots$ are *sets*, more precisely subsets of some set Ω, $A \wedge B$ means the intersection $A \cap B$, $A \vee B$ means the union $A \cup B$, $\neg A$ means the complement $A^c = \Omega \setminus A$, 0 means the empty set $\emptyset$, and 1 means the full set Ω. That is, every family $\mathscr{A}$ of subsets of Ω that contains Ω and is closed under complement and intersection (in particular, every σ-algebra) is a Boolean algebra. (It turns out that also, conversely, every Boolean algebra can be realized as a family of subsets of some set Ω.)

[14] Gödel (1931) [20] provided an example of a statement that is true about the natural numbers, so it follows from the Peano axioms, but cannot be derived from the Peano axioms using the standard rules of logic, thus showing that these rules are incomplete.

Now let $A, B, C, \ldots$ be *subspaces* of a Hilbert space $\mathscr{H}$ (more precisely, closed subspaces, which makes no difference in finite dimension where every subspace is closed); let $A \wedge B := A \cap B$, $A \vee B := \overline{\text{span}}(A \cup B)$ (the smallest closed subspace containing both A and B), and let $\neg A := A^\perp = \{\psi \in \mathscr{H} : \langle\psi|\phi\rangle = 0 \,\forall \phi \in A\}$ be the orthogonal complement of A; let $0 = \{0\}$ be the zero-dimensional subspace and $1 = \mathscr{H}$ the full subspace. Then all axioms except distributivity are satisfied. So this structure is no longer a Boolean algebra; it is called an *orthomodular lattice* or simply *lattice*. Hence, a distributive lattice is a Boolean algebra, and the closed subspaces form a non-distributive lattice $\mathbb{L}(\mathscr{H})$.

That is nice mathematics, and we will see more of that in a moment. The analogy I mentioned holds between $\mathbb{L}(\mathscr{H})$ and Boolean algebras, often understood as representing the rules of logic. The analogy is that both are lattices. In order to emphasize the analogy, some authors call the elements of $\mathbb{L}(\mathscr{H})$ "propositions" and the operations $\wedge$, $\vee$, and $\neg$ "and," "or," and "not." They call $\mathbb{L}(\mathscr{H})$ the "quantum logic" and say things like, $A \in \mathbb{L}(\mathscr{H})$ is a yes–no question that you can ask about a quantum system, as you can carry out a quantum measurement of the projection to A and get result 0 (no) or 1 (yes).

Here is why the analogy is rather limited. Let me give two examples.

- First, consider a spin-$\frac{1}{2}$ particle with spinor $\psi \in \mathbb{C}^2$, and consider the words "*ψ lies in* $\mathbb{C}|\text{up}\rangle$." These words sound very much like a proposition, let me call it $\mathscr{P}$, and indeed they naturally correspond to a subspace of $\mathscr{H} = \mathbb{C}^2$, viz., $\mathbb{C}|\text{up}\rangle$. Now the negation of $\mathscr{P}$ is, of course, "*ψ lies in* $\mathscr{H} \setminus \mathbb{C}|\text{up}\rangle$," whereas the orthogonal complement of $\mathbb{C}|\text{up}\rangle$ is $\mathbb{C}|\text{down}\rangle$. Let me say that again in different words: The negation of "spin is up" is not "spin is down," but "spin is in any direction but up."
- Second, consider the delayed-choice experiment in the form discussed at the end of Sect. 5.1.6: Forget about the interference region and consider just the two options of either putting detectors in the two slits or putting detectors far away. The first option has the PVM $P_{\text{upper slit}} + P_{\text{lower slit}} = I$, the second to the PVM $U^\dagger P_{\text{lower cluster}} U + U^\dagger P_{\text{upper cluster}} U = I$, where U is the unitary time evolution from the slits to the far regions where the detectors are placed. The two PVMs are identical, as $U^\dagger P_{\text{lower cluster}} U = P_{\text{upper slit}}$ (and likewise for the other projection); that is, we have two experiments associated with the same observable. If we think of subspaces as propositions, then it is natural to think of *the particle passes through the upper slit* as a proposition and identify it with the subspace A that is the range of $P_{\text{upper slit}}$. But if we carry out the second option, detect the particle in the lower cluster, and say that we have confirmed the proposition A and thus that the particle passed through the upper slit, then we have committed Wheeler's fallacy.

The philosophical idea that I mentioned is that logic as we know it is false, that it applies in classical physics but not in quantum physics, and that a different kind of logic with different rules applies in quantum physics—a *quantum logic*. Why did I call that a rather wrongheaded idea? Because logic is, by definition, what is true in every conceivable situation. So logic cannot depend on physical laws and cannot be revised by empirical science. As Tim Maudlin once nicely said:

> There is no point in arguing with somebody who does not believe in logic.

Bell (1990) [8] wrote:

> When one forgets the role of the apparatus, as the word "measurement" makes all too likely, one despairs of ordinary logic—hence "quantum logic." When one remembers the role of the apparatus, ordinary logic is just fine.

5.6.2 Quantum Measures

Nevertheless, there is more mathematics relevant to $\mathbb{L}(\mathscr{H})$, something analogous to probability theory. Recall that a probability distribution on a set Ω is a normalized measure, that is, a mapping μ from subsets of Ω to $[0, 1]$ that is σ-additive and satisfies $\mu(1) = \mu(\Omega) = 1$. The domain of definition of μ is a σ-algebra, which is a Boolean algebra with slightly stronger technical requirements. By analogy, we define that a *normalized quantum measure* is a mapping $\hat{\mu} : \mathbb{L}(\mathscr{H}) \to [0, 1]$ that satisfies $\hat{\mu}(1) = \hat{\mu}(\mathscr{H}) = 1$ and is σ-additive, i.e.,

$$\hat{\mu}\Big(\bigvee_{n=1}^{\infty} A_n\Big) = \sum_{n=1}^{\infty} \hat{\mu}(A_n) \tag{5.185}$$

whenever $A_n \perp A_m$ for all $n \neq m$. (The relation $A \perp B$ can be expressed through lattice operations as $A \leq (\neg B)$, with $A \leq C$ defined to mean $A \vee C = C$ or, equivalently, $A \wedge C = A$. In $\mathbb{L}(\mathscr{H})$, $A \leq B \Leftrightarrow A \subseteq B$.) The following result is due to Andrew Gleason (1921–2008) in 1957 [19]:

Theorem 5.11 (Gleason's Theorem) *Suppose the dimension of $\mathscr{H}$ is at least 3 and at most countably infinite. Then the normalized quantum measures are exactly the mappings $\hat{\mu}$ of the form*

$$\hat{\mu}(A) = \mathrm{tr}(\rho P_A) \quad \forall A \in \mathbb{L}(\mathscr{H}), \tag{5.186}$$

where P_A denotes the projection to A and ρ is a density matrix (i.e., a positive operator with trace 1).

This amazing parallel between probability measures and density matrices has led some authors to call elements of $\mathbb{L}(\mathscr{H})$ "events" (as one would call subsets of Ω). Again, this is a rather limited analogy, for the same reasons as above.

5.7 No-Hidden-Variables Theorems

This name refers to a collection of theorems that aim at proving the impossibility of hidden variables. This aim may seem strange in view of the fact that Bohmian mechanics is a hidden-variable theory, is consistent, and makes predictions in agreement with quantum mechanics. So how could hidden variables be impossible? A first observation concerns what is meant by "hidden variables." Most no-hidden-variable theorems (NHVTs) address the idea that every observable A (a self-adjoint operator) has a "true value" v_A in nature (the "hidden variable"), and that a quantum measurement of A yields v_A as its outcome. This idea should sound dubious to you because we have discussed already that observables are really equivalence classes of experiments, not all of which yield the same value. Moreover, we know that in Bohmian mechanics, a true value is associated with position but not with every observable, in particular not with spin observables. Hence, in this sense of "hidden variables," Bohmian mechanics is really a *no-hidden-variables theory*.

But this is not the central reason why the NHVTs do not exclude Bohmian mechanics. Suppose we choose, in Bohmian mechanics, one experiment from every equivalence class. (The experiment could be specified by specifying the wave function and configuration of the apparatus together with the joint Hamiltonian of object and apparatus as well as the calibration function.) For example, for every spin observable $\boldsymbol{n} \cdot \boldsymbol{\sigma}$ we could say we will measure it by a Stern–Gerlach experiment in the direction $\boldsymbol{n}$ and subsequent detection of the object particle. Then the outcome $Z_{\boldsymbol{n}}$ of the experiment is a function of the object wave function ψ and the object configuration Q, so we have associated with every observable $A = \boldsymbol{n} \cdot \boldsymbol{\sigma}$ a value v_A which comes out if we choose to carry out the experiment associated with $\boldsymbol{n} \cdot \boldsymbol{\sigma}$ (although it does not mean anything like the "true value of $\boldsymbol{n} \cdot \boldsymbol{\sigma}$"). And it is this situation that NHVTs claim to exclude! So we are back at an apparent conflict between Bohmian mechanics and NHVTs.

It may occur to you that even a much simpler example than Bohmian mechanics will prove the possibility of hidden-variable theories. Suppose we choose, as a trivial model, for every self-adjoint operator A a random value v_A independently of all other $v_{A'}$ with the Born distribution,

$$\mathbb{P}(v_A = \alpha) = \|P_\alpha \psi\|^2 . \tag{5.187}$$

Then we have not provided a real theory of quantum mechanics such as Bohmian mechanics provides, but we have provided a clearly consistent possibility for which values the variables v_A could have that agrees with the probabilities seen in experiment. Therefore, all NHVTs must make some further assumptions about the hidden variables v_A that are violated in the trivial model, as well as in Bohmian mechanics. We now take a look at several NHVTs and their assumptions.

5.7.1 Bell's NHVT

Bell's theorem implies a NHVT, or rather, the second half of Bell's (1964) [4] proof *is* a NHVT. Let me explain. In the trivial model introduced around (5.187), we have not specified how the v_A change with time. They may change according to some law under the unitary time evolution; more importantly for us now, they may change whenever ψ collapses. That is, when a quantum measurement of A is carried out, we should expect the $v_{A'}$ $(A' \neq A)$ to change. However, there is an exception if we believe in locality. Then we should expect that Alice's measurement of $\boldsymbol{\alpha} \cdot \boldsymbol{\sigma}_a$ (on her particle a) will not alter the value of any spin observable $\boldsymbol{\beta} \cdot \boldsymbol{\sigma}_b$ acting on Bob's particle. But Bell's analysis shows that this is impossible.

We can sum up this conclusion and formulate it mathematically as the following theorem. According to the hidden-variable hypothesis, every observable A from a certain collection of observables has an actual value v_A. This value will be different in every run of the experiment, it will be random; so v_A is a random variable. Since in each run *each* v_A has a definite value, the random variables v_A possess a *joint* probability distribution. (Mathematicians also sometimes express this situation by saying that "the random variables v_A are defined on the same probability space.")

Theorem 5.12 (Bell's (1964) [4] NHVT) *Consider a joint distribution of random variables v_A, where A runs through the collection of observables*

$$\mathscr{A} \cup \mathscr{B} = \{\boldsymbol{\alpha} \cdot \boldsymbol{\sigma}_a : \boldsymbol{\alpha} \in \mathbb{S}^2\} \cup \{\boldsymbol{\beta} \cdot \boldsymbol{\sigma}_b : \boldsymbol{\beta} \in \mathbb{S}^2\}. \tag{5.188}$$

Suppose that a quantum measurement of $A \in \mathscr{A}$ yields v_A and does not alter the value of v_B for any $B \in \mathscr{B}$, and that a subsequent quantum measurement of $B \in \mathscr{B}$ yields v_B. Then the joint distribution of the outcomes satisfies Bell's inequality (4.49). *In particular, it disagrees with the distribution of outcomes predicted by the quantum formalism.*

In short, local hidden variables are impossible. (Here, "local" means that v_A cannot be affected at spacelike separation. As we have discussed in Sect. 4.3.2, it should be kept in mind that Bell's full proof, of which Bell's NHVT is just a part, shows that all local theories are impossible.)

5.7.2 Von Neumann's NHVT

John von Neumann presented a NHVT in his book (1932) [50]. It is clear that for a hidden-variable model to agree with the predictions of quantum mechanics, every v_A can only have values that are eigenvalues of A, and its marginal distribution must be the Born distribution. Von Neumann made an additional assumption that, as we will discuss afterward, is not very reasonable: He assumed that whenever an observable C is a linear combination of observables A and B,

$$C = \alpha A + \beta B\,, \quad \alpha, \beta \in \mathbb{R}\,, \tag{5.189}$$

then v_C is the same linear combination of v_A and v_B,

$$v_C = \alpha v_A + \beta v_B\,. \tag{5.190}$$

Theorem 5.13 (von Neumann's (1932) [50] NHVT) *Suppose* $2 \le \dim \mathscr{H} \le \infty$ *and* $\psi \in \mathbb{S}(\mathscr{H})$*, let* $\mathscr{A}$ *be the set of all self-adjoint operators on* $\mathscr{H}$*, and consider a joint distribution of random variables* v_A *for all* $A \in \mathscr{A}$*. Suppose that* (5.190) *holds whenever* (5.189) *does. Then for some* A *the marginal distribution of* v_A *disagrees with the Born distribution associated with* A *and* ψ*.*

Proof Suppose first that $\dim \mathscr{H} = 2$ and choose an ONB of $\mathscr{H}$, so we can identify $\mathscr{H} = \mathbb{C}^2$. Let $A = \sigma_1$, $B = \sigma_3$, and C be the spin observable in the direction at 45° between the x- and the z-direction; then

$$C = \frac{1}{\sqrt{2}} A + \frac{1}{\sqrt{2}} B\,. \tag{5.191}$$

Each of A, B, C has eigenvalues ± 1; thus, if v_A is an eigenvalue of A, v_B one of B, and (5.190) holds, then

$$v_C = \frac{1}{\sqrt{2}} v_A + \frac{1}{\sqrt{2}} v_B \tag{5.192}$$

can only assume the values 0, $\frac{1}{\sqrt{2}}$, or $-\frac{1}{\sqrt{2}}$, and thus cannot be an eigenvalue of C. Thus, the distribution of v_C cannot be the Born distribution, as claimed.

If the dimension of $\mathscr{H}$ is > 2, we can apply the same argument as follows. Choose an ONB and let the matrix of A (B) be block diagonal with upper-left 2×2 block given by σ_1 (respectively, σ_3) and lower-right $(\dim \mathscr{H} - 2) \times (\dim \mathscr{H} - 2)$ block given by (say) $5I$, C be given by (5.191); then each of A, B, C has eigenvalues $1, -1$, and 5, and (5.192) cannot be an eigenvalue if v_A and v_B are. □

As emphasized by Bell (1966) [5], the assumption (5.190) is not reasonable—there is no physical reason to expect (5.190) to hold. For example, let again $\mathscr{H} = \mathbb{C}^2$, $A = \sigma_1$, $B = \sigma_3$, and C the spin observable in the direction at 45° between the x- and the z-direction, so $C = \frac{1}{\sqrt{2}} A + \frac{1}{\sqrt{2}} B$. The obvious experiment for C is the Stern–Gerlach experiment in direction $\boldsymbol{n} = (\frac{1}{\sqrt{2}}, 0, \frac{1}{\sqrt{2}})$, whereas those for A and B would have the axis of the magnetic field point in the x- and the z-direction. Of course, the experiment for C is not based on measuring A and B and then combining their results, but is a completely different experiment. Thus, there is no reason to expect that its outcome was a linear combination of what we would have obtained, had we applied a magnetic field in the x- or the z-direction. That is why von Neumann's assumption is not reasonable.

5.7.3 Gleason's NHVT

From Theorem 5.11, one can obtain a different NHVT:

Theorem 5.14 (Gleason's (1957) [19] NHVT) *Suppose* $3 \leq \dim \mathscr{H} \leq \infty$ *and* $\psi \in \mathbb{S}(\mathscr{H})$, *let* $\mathscr{A}$ *be the set of all self-adjoint operators on* $\mathscr{H}$, *and consider a joint distribution of random variables* v_A *for all* $A \in \mathscr{A}$. *Suppose that whenever* $A, B \in \mathscr{A}$ *commute, then a quantum measurement of* A *yields* v_A *and does not alter the value of* v_B, *and that a subsequent quantum measurement of* B *yields* v_B. *Then the distribution of the outcomes disagrees for some experiment with the prediction of the quantum formalism using* ψ.

The idea here is that while in general a quantum measurement of A may change the values of v_B for $B \neq A$, this should not happen if A and B can be "simultaneously measured." A key step in the proof is this:

Lemma 5.1 *Suppose there existed a family of random variables* v_A *as described whose distribution agrees with the quantum prediction based on* ψ. *If* $A, B \in \mathscr{A}$ *commute, then* $A + B$ *and* AB *are self-adjoint as well, and*

$$v_{A+B} = v_A + v_B \quad \textit{and} \quad v_{AB} = v_A v_B\,. \tag{5.193}$$

Proof We give the proof for AB, the one for $A + B$ is analogous. First, since the Born distribution for ψ and A is always concentrated on the spectrum of A, v_A must lie in the spectrum of A. Second, we note that whenever $AB = BA$, also $C := AB$ is self-adjoint and commutes with both A and B.

As the next step, we show that $v_C = v_A v_B$. To this end, consider three subsequent quantum measurements: first of A, then of B, and then of C. By Theorem 3.1 (the spectral theorem for commuting self-adjoint operators), there is an ONB of joint eigenvectors ϕ_n of A and B. But if $A\phi_n = \alpha_n \phi_n$ and $B\phi_n = \beta_n \phi_n$, then $C\phi_n = AB\phi_n = A(\beta_n \phi_n) = \beta_n(A\phi_n) = \beta_n \alpha_n \phi_n$; so, the ϕ_n are also eigenvectors of C with eigenvalues $\alpha_n \beta_n$. Since C commutes with both A and B, the A measurement (yielding v_A) does not change v_B or v_C, the subsequent B measurement (yielding v_B) does not change v_A or v_C, and a final C measurement yields v_C. On the other hand, the outcome of the A measurement must have been one of the α_n, the outcome of the subsequent B measurement a β with a joint eigenvector with α_n (say, β_m with $\alpha_m = \alpha_n$), and the outcome of the C measurement then $\alpha_m \beta_m$. So, $v_C = v_A v_B$. □

Proof of Theorem 5.14 from Theorem 5.11 We can assume $\dim \mathscr{H} = 3$ because otherwise we consider a 3d subpsace $\mathscr{K}$ containing ψ and extend the self-adjoint operators $A_{\mathscr{K}}$ on $\mathscr{K}$ by setting $A = A_{\mathscr{K}} P_{\mathscr{K}}$.

Keep $\psi \in \mathbb{S}(\mathscr{H})$ fixed and suppose there existed a family of random variables v_A as described whose distribution agrees with the quantum prediction based on ψ. Define, for every subspace S of $\mathscr{H}$,

$$\hat{\mu}(S) = v_{P_S}\,. \tag{5.194}$$

Then $\hat{\mu}$ is a quantum measure: First, v_{P_S} must be one of the eigenvalues of P_S, that is, either 0 or 1. Second, for $S = \mathscr{H}$, $P_S = I$ and $v_I = 1$ because the spectrum of I is $\{1\}$, so $\hat{\mu}(\mathscr{H}) = 1$. Third, we verify σ-additivity: In finite dimension this follows from additivity for two summands, $\hat{\mu}(S_1 \vee S_2) = \hat{\mu}(S_1) + \hat{\mu}(S_2)$ whenever $S_1 \perp S_2$, which we check now. Note that $A := P_{S_1}$ and $B := P_{S_2}$ commute and set $C := P_{S_1 \vee S_2} = P_{S_1} + P_{S_2}$. By Lemma 5.1, $v_C = v_A + v_B$.

Now we can apply Theorem 5.11 to obtain the existence of a density matrix ρ with $\hat{\mu}(S) = \operatorname{tr}(\rho P_S)$ for every $S \in \mathbb{L}(\mathscr{H})$. But while $\hat{\mu}(S) \in \{0, 1\}$, $\operatorname{tr}(\rho P_S)$ will also assume values strictly between 0 and 1 for suitable subspaces S, contradiction. □

Theorem 5.14 is also often called the Kochen–Specker theorem because Simon Kochen and Ernst Specker (1967) [26] gave a proof of it that was not based on Theorem 5.11. (Kochen and Specker originally stated stronger assumptions than Theorem 5.14, but their proof could be so formulated that it yields Theorem 5.14.) Further proofs of Theorem 5.14 were given by Specker (1960) [39], Bell (1966) [5], Mermin (1990) [30], and Peres (1991) [34].

Here is a proof following Peres (1991) [34] that does not use Theorem 5.11; it covers only the case $\dim \mathscr{H} \geq 4$ but is particularly simple: Suppose there existed a family of random variables v_A as described whose distribution agrees with the quantum prediction based on ψ. By Lemma 5.1, with probability 1, the random function $v : \mathscr{A} \to \mathbb{R}$ has the properties $v_{AB} = v_A v_B$ whenever $AB = BA$ and $v_{-I} = -1$. The contradiction is that according to the following lemma, such a function does not exist.

Lemma 5.2 *For $\dim \mathscr{H} \geq 4$, there is no function $v : \mathscr{A} \to \mathbb{R}$ such that $v_{-I} = -1$ and $v_{AB} = v_A v_B$ whenever $AB = BA$.*

Proof Regard a 4d subspace of $\mathscr{H}$ containing ψ as $\mathbb{C}^2 \otimes \mathbb{C}^2$, and let σ_k^i be the Pauli matrix σ_k as in (2.105) acting on the i-th factor. We use the following properties of the Pauli matrices:

$$(\sigma_k^i)^2 = I, \quad \sigma_x^i \sigma_y^i = -\sigma_y^i \sigma_x^i, \quad \sigma_j^1 \sigma_k^2 = \sigma_k^2 \sigma_j^1\,. \tag{5.195}$$

Verify that

$$\sigma_x^1\sigma_y^2\sigma_y^1\sigma_x^2\sigma_x^1\sigma_x^2\sigma_y^1\sigma_y^2 = -I \tag{5.196}$$

by moving σ_x^1 from the first to fourth place, collecting a minus sign, and then simplifying. Define the self-adjoint matrices

$$A = \sigma_x^1\sigma_y^2, \quad B = \sigma_y^1\sigma_x^2, \quad C = \sigma_x^1\sigma_x^2, \quad D = \sigma_y^1\sigma_y^2, \quad E = AB, \quad F = CD\,, \tag{5.197}$$

so (5.196) can be written as $EF = -I$. Verify that $[A, B] = 0 = [C, D] = [E, F]$. By assumption,

$$-1 = v_{-I} = v_{EF} = v_E v_F = v_{AB} v_{CD} = v_A v_B v_C v_D = v_{\sigma_x^1} v_{\sigma_y^2} v_{\sigma_y^1} v_{\sigma_x^2} v_{\sigma_x^1} v_{\sigma_x^2} v_{\sigma_y^1} v_{\sigma_y^2}\,. \tag{5.198}$$

However, the last expression is ≥ 0 since every factor appears twice. □

Instead of Lemma 5.2, we can also use Theorem 5.12 (Bell's NHVT): Indeed, any $\boldsymbol{\alpha}\cdot\boldsymbol{\sigma}_a$ commutes with any $\boldsymbol{\beta}\cdot\boldsymbol{\sigma}_b$, so the assumption of Theorem 5.12 is satisfied under the assumption of Theorem 5.14. In particular, the assumption of Theorem 5.14 is violated in any nonlocal hidden-variable theory.

The fact that the outcome of a quantum measurement of A may depend on which other observable B commuting with A is "measured" simultaneously with A is sometimes called "contextuality" in the literature (cf. Sect. 2.3.9). Correspondingly, a theory satisfying the assumption of Theorem 5.14 (and whose empirical predictions therefore deviate from the quantum formalism) is called a theory of non-contextual hidden variables.

Upshot So what is the upshot of all these no-hidden-variable theorems? Some authors such as von Neumann thought they prove the Copenhagen idea that a coherent theory of quantum mechanics is impossible, that reality in itself is paradoxical. But we know that Bohmian mechanics is a counterexample. What the NHVTs show is that "measurement" is a misnomer: that we should not think of a quantum measurement of an operator A as discovering the "true value" associated with A. Allow me to repeat this Bell (1990) [8] quote:

> On this list of bad words from good books, the worst of all is *measurement*.

On top of that, the NHVTs bring to light that another terminology is misleading as well: that of "simultaneous measurement" of commuting operators A, B. The terminology suggests that the A measurement will not change the value v_B of B and vice versa, and thus that the two quantum measurements involved are somehow innocent. We saw already in Sect. 4.3.3 that this is in general wrong in Bohmian mechanics ("fallacy of simultaneous measurement"); now we see that this is in general wrong in every hidden-variable theory.

5.7.4 Hidden Variables and Ontology

Besides Bohmian mechanics, other hidden-variable theories have been proposed, also with hidden variables that are not particle positions. Various authors proposing hidden variables think of them as the actual values of certain observables; for example, this is so for the so-called modal interpretations (van Fraassen 1972 [48], Bacciagaluppi and Dickson 1999 [3]). I have discussed already two problems that can arise with this idea: first, how the no-hidden-variable theorems put restrictions on this possibility, and second, that "observables" are really equivalence classes of experiments, not quantities that have a value. Here I want to mention another problem: the ontological meaning of the "actual values." This problem is absent in Bohmian mechanics.

The problem has to do with ontology and with what are conceivable ontologies for physical theories. To illustrate the problem, suppose that a particular theory postulates that velocities, but not positions, have actual values. Then what would be the ontology? Would the particles have trajectories? Presumably not, as that would mean they have actual positions. But then it is unclear what it would mean for them to have velocities.

Let me put this point in terms of a creation myth. In some ancient mythologies, God has helper angels called demiurgs (Greek for "craftsmen") who build the creation according to God's plan. If God wrote in his plan that the particles have velocities but not positions, then I think the demiurgs sent out to build the world would come back to ask God what exactly they are supposed to make, given that the particles are expected to not have physical trajectories.

We can make up further variants of this story. If A and B are self-adjoint operators, and if we can leave aside questions of domains, then $A + B$ is a self-adjoint operator, too. If we mean, when saying "the position observables have actual values," that the particles have physical trajectories, and if A is position and B momentum, then what is the physical meaning of saying "$A + B$ has actual values, but not A or B alone"? What should the demiurgs make? In sum, I have trouble making sense of the proposal of "actual values of the observable A" for most A, while there is no such problem with Bohm's choice of preferring positions. In Bohmian mechanics, actually, position is not thought of as an "observable," it is thought of as a "beable." The "position observable" is simply the equivalence class of those experiments that measure (in the literal sense) the actual (Bohmian) position, and this equivalence class contains also experiments that do not literally measure the actual position.

Let me turn to another hidden-variables theory, one that has been proposed by Jürg Fröhlich [17] under the name "ETH approach," where ETH stands for "events, trees, histories" (but is also the abbreviation of the Eidgenössische Technische Hochschule in Zurich, where Fröhlich used to teach). If I understand correctly what

I heard Fröhlich say, the ETH approach cannot be formulated for non-relativistic quantum mechanics in $L^2(\mathbb{R}^{3N}, \mathbb{C}^d)$ but only for certain types of quantum field theories. Be that as it may, a key idea of the approach is, first, that not all self-adjoint operators are actually observables; that is, not for every self-adjoint operator A is there an experiment that is a quantum measurement of A. (This claim is supported by considerations for relativistic space-time, especially for curved space-time, where the information visible to an observer moving along the world line γ is limited to the past of γ, which may be a proper subset of space-time even if γ extends infinitely far to the future.) Second, given that not all A's are observables, it is impossible for observers to distinguish certain superpositions from certain mixtures. (For example, if $\mathscr{H} = \mathscr{H}_1 \oplus \mathscr{H}_2$ and only block diagonal operators are observables, then observers cannot distinguish between superpositions $c_1\psi_1 + c_2\psi_2$ of $\psi_1 \in \mathscr{H}_1$, $\psi_2 \in \mathscr{H}_2$ and mixtures of ψ_1, ψ_2 with weights $|c_1|^2, |c_2|^2$.) Third, Fröhlich argues that if a certain superposition and a certain mixture are indistinguishable at time t, then they will also be indistinguishable at any $t' > t$. He then, fourth, postulates that whenever a certain mathematical condition, expressing that a superposition and a mixture have newly become indistinguishable, holds, then nature randomly picks the value of a new hidden variable selecting an element of the mixture (in the example, selecting the value 1 or 2 with probability $|c_1|^2$, respectively $|c_2|^2$); alternatively, one could equivalently postulate that the wave function spontaneously jumps to a member of the mixture (in the example, from $c_1\psi_1 + c_2\psi_2$ to ψ_1 or ψ_2 with probability $|c_1|^2$, respectively $|c_2|^2$). Leaving aside that this postulate seems strongly inspired by the dubious positivistic idea that what cannot be measured is not real, it will be crucial for the viability of the theory that the hidden variables are sufficient to uniquely select one of the terms in essentially *every* macroscopic superposition and that the hidden variables are consistent for all observers. But even if all that comes out as desired, the problem of the demiurgs remains: What kind of ontology should the demiurgs build when the plan demands that the operator A has value α? That is where Fröhlich's approach fails to make sense to me.

5.8 Special Topics

5.8.1 The Decoherent Histories Interpretation

Another view of quantum mechanics was proposed by Murray Gell-Mann, James Hartle, Bob Griffiths, and Roland Omnès under the name *decoherent histories* or *consistent histories*; see Griffiths (2002) [24] and references therein.

Even before I describe this view, I need to say that it fails to provide a possible way the world may be, or a realist picture of quantum mechanics. The ultimate reason is, in my humble opinion, that the proponents of this view do not think in terms of reality. Rather, they think in terms of words and phrases, and a central element of their interpretation is to set up rules for which phrases are legitimate or justified. In realist theories, words refer to objects in reality (like "particle" in Bohmian mechanics), and phrases or statements can be true and justified because they express a fact about reality. Now the spirit of decoherent histories is more that statements are just sequences of words, and since you have rules for which statements to regard as justified, you do not think about the situation in reality. The proponents of this view have no consistent picture of reality in mind, and no such picture is in sight. So if you are looking for such a picture, you will be disappointed.

The motivation for this view comes from the fact that quantum mechanics provides via the Born rule a probability distribution over configurations (or the index set of another ONB in Hilbert space $\mathscr{H}$) *at a fixed time t* but not over *histories* (such as paths in configuration space). It may seem that if quantum mechanics provided a probability distribution over histories, then the interpretation of quantum mechanics would be straightforward: One of these histories occurs, and it occurs with the probability dictated by quantum mechanics. Now "histories" is taken to mean not just paths in configuration space, but the following broader concept: Consider, for simplicity, just a finite set of times $\{t_1, t_2, \ldots, t_r\}$, and for each t_i an ONB $\{\phi_{in} : n \in \mathbb{N}\}$ of $\mathscr{H}$ (such as the eigenbasis of an observable); now one talks of the ray $\mathbb{C}\phi_{in}$ as an "event" at time t_i, and a "history" now means a list $(\mathbb{C}\phi_{1n_1}, \ldots, \mathbb{C}\phi_{rn_r})$ of such events, or briefly just the indices $(n_1, \ldots, n_r)$. Then, for some choices of ONBs, the Born rule *does* provide a probability distribution over the set of histories $(n_1, \ldots, n_r)$: Suppose that the unitary time evolution is given by U_t $(t \in \mathbb{R})$, and that $\phi_{i+1,n} = U_{t_{i+1}-t_i}\phi_{in}$ for all $i \in \{1, \ldots, r-1\}$ and $n \in \mathbb{N}$. Then, trivially, an initial wave function agreeing with some ϕ_{1n} at t_1 would agree with some basis vector at each t_i. Furthermore, an arbitrary initial wave function $\psi \in \mathbb{S}(\mathscr{H})$ at t_1—defines a probability distribution over n and thus a probability distribution over histories, i.e.,

$$\mathbb{P}(n_1, \ldots, n_r) = \delta_{n_1 n_2}\delta_{n_2 n_3} \cdots \delta_{n_{r-1} n_r} \left|\langle\phi_{1n_1}|\psi\rangle\right|^2 . \tag{5.199}$$

One can be more general by allowing for coarse graining. Suppose we allow a sequence of subspaces $Y := (\mathscr{K}_1, \ldots, \mathscr{K}_n)$ of arbitrary dimension as a description of a history; let P_i denote the projection to $\mathscr{K}_i$ and $U_i := U_{t_i - t_{i-1}}$. If we made quantum measurements of P_i at all t_i, then the probability of obtaining 1111...1 would be

$$\|K(Y)\psi\|^2 \text{ with } K(Y) = P_n U_n \cdots P_1 U_1 . \tag{5.200}$$

The decoherent histories approach uses the same formula (5.200) to assign probabilities to histories Y, but only to histories belonging to special families $\mathscr{F}$ of histories, so-called decoherent families, which are closed under coarse graining and for which (5.200) is additive in Y. A calculation shows that that is the case whenever

$$\operatorname{Re}\langle\psi|K(Y)^\dagger K(Y')|\psi\rangle = 0 \qquad \forall Y, Y' \in \mathscr{F}, \tag{5.201}$$

a condition called the *decoherence condition*.

Now the decoherent histories interpretation postulates that for decoherent families, the statement "The history Y has probability (5.200)" is justified; for families that are not decoherent, in contrast, it is postulated that there simply do not exist probabilities. For example, in Wheeler's delayed-choice experiment (Sect. 1.6.6) with the screen in the far position, consider as t_1 the time when the electron passes through the double slit, t_2 the time when the electron arrives at the screen, ϕ_{11} the wave packet in the upper slit and ϕ_{12} in the lower, ϕ_{21} the wave packet at the upper cluster on the screen and ϕ_{22} at the lower, and $\psi = \psi_{t_1} = \frac{1}{\sqrt{2}}\phi_{11} + \frac{1}{\sqrt{2}}\phi_{12}$. Since the unitary evolution is $\phi_{11} \to \phi_{22}$ and $\phi_{12} \to \phi_{21}$, the decoherent histories view attributes

$$\text{prob. } \tfrac{1}{2} \text{ to "passed the upper slit and arrived at the lower cluster"} \tag{5.202a}$$

$$\text{prob. } 0 \text{ to "passed the lower slit and arrived at the lower cluster"} \tag{5.202b}$$

$$\text{prob. } 0 \text{ to "passed the upper slit and arrived at the upper cluster"} \tag{5.202c}$$

$$\text{prob. } \tfrac{1}{2} \text{ to "passed the lower slit and arrived at the upper cluster."} \tag{5.202d}$$

You can see that the decoherent histories view commits Wheeler's fallacy (see Sect. 1.6.6). But this is not the main problem.

The main problem is that the decoherent histories view does not commit itself to any particular decoherent family. If there was only one decoherent family, we could assume that nature chooses one history from that family with the probabilities given above, and that that history represents the reality. But since there are many decoherent families, it remains unclear what the reality is supposed to be. Does nature choose one history from each family? If the electron went through the upper slit in one family, does it have to go through the upper slit in all other families containing this event? As Goldstein (1998) [21] pointed out, the no-hidden-variables theorems imply that this is impossible. Then which family is the one connected to our reality? (And, by the way, why bother about other families?) The proponents of decoherent histories have no answer to this, and that is, I think, because they do not think in terms of reality.

I also note that the motivation of the decoherent histories approach is problematical as it takes for granted that *events* should correspond mathematically to *eigenspaces* of observables. I have pointed out in Sect. 5.6 on quantum logic why that analogy is a bad one. By relying on it, the decoherent histories approach takes the words "observable" and "measurement" too literally, as if observables really were observable quantities and as if quantum measurements were procedures to find their values.

5.8.2 The Hilbert–Schmidt Inner Product

This section is about another application of the concept of trace providing a different perspective on operators that is rather natural and sometimes useful.

The complex $n \times m$ matrices form a vector space, and addition and scalar multiplication in this vector space are just what we would get if we wrote the

nm entries in a long list. Writing the entries this way also suggests the following definition of an inner product between such matrices A, B:

$$\langle A|B\rangle = \sum_{i=1}^{n}\sum_{j=1}^{m} A_{ij}^{*} B_{ij} = \operatorname{tr}(A^{\dagger}B)\,. \tag{5.203}$$

This is called the *Hilbert–Schmidt inner product*. The last expression can be equally applied to linear mappings $A, B : \mathscr{H}_1 \to \mathscr{H}_2$ between finite-dimensional Hilbert spaces without even choosing orthonormal bases.

The idea can be extended to infinite-dimensional Hilbert spaces $\mathscr{H}_1, \mathscr{H}_2$ as follows (including the possibility $\mathscr{H}_1 = \mathscr{H}_2$). Not for operators A, B will $A^{\dagger}B$ belong to the trace class. But if A is a *Hilbert-Schmidt operator*, i.e., if

$$\operatorname{tr}(A^{\dagger}A) < \infty\,, \tag{5.204}$$

and likewise for B, then it can be shown that $A^{\dagger}B$ is a trace class operator, so $\operatorname{tr}(A^{\dagger}B)$ is well defined and finite. In fact, the set of all Hilbert-Schmidt operators from $\mathscr{H}_1$ to $\mathscr{H}_2$, equipped with the Hilbert-Schmidt inner product, is itself a Hilbert space. In terms of any ONBs $\{|n_1\rangle_1\}$ of $\mathscr{H}_1$ and $\{|n_2\rangle_2\}$ of $\mathscr{H}_2$, the condition (5.204) amounts to

$$\sum_{n_1,n_2} \left|{}_2\langle n_2|A|n_1\rangle_1\right|^2 < \infty\,. \tag{5.205}$$

If $\mathscr{H}_1 = \mathscr{H}_2$, then the Hilbert-Schmidt operators form a subset of the trace class of $\mathscr{H}_1$ (a proper subset if $\dim \mathscr{H}_1 = \infty$). If $A : \mathscr{H}_1 \to \mathscr{H}_1$ can be unitarily diagonalized with eigenvalues λ_n, then $\operatorname{tr}(A^{\dagger}A) = \sum_n |\lambda_n|^2$, whereas belonging to the trace class is equivalent to $\sum_n |\lambda_n| < \infty$. So, the Hilbert–Schmidt operators $\mathscr{H}_1 \to \mathscr{H}_1$ compare to the trace class operators like the square-summable sequences to the summable sequences, or like L^2 functions to L^1 functions.

Exercises

Exercises 5.1 and 5.2 can be found in Sect. 5.1.1, Exercise 5.3 in Sect. 5.2.2, Exercises 5.4 and 5.5 in Sect. 5.2.3, Exercise 5.6 in Sect. 5.2.4, Exercises 5.7 and 5.8 in Sect. 5.5.1, and Exercise 5.9 in Sect. 5.5.11.

Exercise 5.10 (Positive Operators) Are the following statements about operators on $\mathbb{C}^d$ true or false? Justify your answers.

1. $R^{\dagger}R$ is always a positive operator.
2. If E is a positive operator, then so is $R^{\dagger}ER$.
3. The positive operators form a subspace of the space of self-adjoint operators.

4. The sum of two projections is positive only if they commute.
5. e^{At} is a positive operator for every self-adjoint A and $t \in \mathbb{R}$.

Exercise 5.11 (Tensor Product of Positive Operators) Show that if R and S are positive operators, then so is $R \otimes S$. *Suggestion*: diagonalize R and S to obtain a diagonalization of $R \otimes S$.

Exercise 5.12 (Sum of Projections) Let $\mathscr{H}$ be a Hilbert space of finite dimension, let P_1 and P_2 be projections in $\mathscr{H}$, $P_i = P_i^\dagger$ and $P_i^2 = P_i$, and let $\mathscr{H}_i$ be the range of P_i. Show that if $Q := P_1 + P_2$ is also a projection ($Q = Q^\dagger$ and $Q^2 = Q$), then **(a)** $\mathscr{H}_1 \perp \mathscr{H}_2$, and **(b)** the range $\mathscr{K}$ of Q is the span of $\mathscr{H}_1 \cup \mathscr{H}_2$.

Exercise 5.13 (POVMs)

(a) Suppose E_1 and E_2 are POVMs on $\mathscr{Z}_1$ and $\mathscr{Z}_2$, respectively, both acting on $\mathscr{H}$; let $q_1, q_2 \in [0, 1]$ with $q_1 + q_2 = 1$. Show that $E(B) := q_1 E_1(B \cap \mathscr{Z}_1) + q_2 E_2(B \cap \mathscr{Z}_2)$ defines a POVM on $\mathscr{Z}_1 \cup \mathscr{Z}_2$.

(b) Suppose experiment $\mathscr{E}_1$ has distribution of outcomes $\langle\psi|E_1(\cdot)|\psi\rangle$, and $\mathscr{E}_2$ has distribution of outcomes $\langle\psi|E_2(\cdot)|\psi\rangle$. Describe an experiment with distribution of outcomes $\langle\psi|E(\cdot)|\psi\rangle$.

(c) Give an example of a POVM for which the E_z do not pairwise commute. *Suggestion*: Choose $E_1(z)$ that does not commute with $E_2(z')$ for $\mathscr{Z}_1 \cap \mathscr{Z}_2 = \emptyset$.

Exercise 5.14 (Main Theorem About POVMs) The proof of the main theorem from Bohmian mechanics assumes that at the initial time t_i of the experiment, the joint wave function factorizes, $\Psi_{t_\mathrm{i}} = \psi \otimes \phi$. What if factorization is not exactly satisfied, but only approximately? Then the probability distribution of the outcome Z is still approximately given by $\langle\psi|E(\cdot)|\psi\rangle$. To make this statement precise, suppose that

$$\Psi_{t_\mathrm{i}} = c\psi \otimes \phi + \Delta\Psi\,, \tag{5.206}$$

where $\|\Delta\Psi\| \ll 1$, $\|\psi\| = \|\phi\| = 1$, and $c = \sqrt{1 - \|\Delta\Psi\|^2}$ (which is close to 1). Use the Cauchy–Schwarz inequality,

$$\big|\langle f|g\rangle\big| \leq \|f\|\,\|g\|\,, \tag{5.207}$$

to show that, for any $B \subseteq \mathscr{Z}$,

$$\Big|\mathbb{P}(Z \in B) - \langle\psi|E(B)|\psi\rangle\Big| < 3\|\Delta\Psi\|\,. \tag{5.208}$$

Exercise 5.15 (Spectral PVM) According to the spectral theorem, every self-adjoint operator A in $\mathscr{H}$ is unitarily equivalent to a multiplication operator M_f in $L^2(\Omega)$ that multiplies by $f : \Omega \to \mathbb{R}$. Let $U : \mathscr{H} \to L^2(\Omega)$ denote that unitary operator, so $A = U^{-1}M_fU$. Give an explicit expression for the spectral PVM of A in terms of U and M.

Exercise 5.16 (Commuting POVM) Let E be a POVM on the finite set $\mathscr{Z}$ acting on $\mathscr{H}$. Show that if the E_z commute pairwise, then there are a unitary $U : \mathscr{H} \to L^2(\Omega)$ and functions $f_z : \Omega \to [0, 1]$ such that

$$\sum_z f_z(\omega) = 1 \quad \forall \omega \in \Omega$$

and $E_z = U^{-1} f_z U$.

Exercise 5.17 (Cannot Distinguish Non-orthogonal State Vectors with POVMs) In Exercise 2.9(b), it was shown that Bob, when allowed to use a quantum measurement of *any self-adjoint operator* on a given particle, is unable to decide with certainty whether the quantum state was $(1, 0)$ or $\frac{1}{\sqrt{2}}(1, 1)$. What if Bob is allowed to use *any experiment whatsoever*? Use the main theorem about POVMs.

Exercise 5.18 (Statistical Density Matrix) Show that a probability distribution μ has $\rho_\mu = |\psi\rangle\langle\psi|$ if and only if μ is concentrated on $\mathbb{C}\psi$, i.e., $\Psi = e^{i\Theta}\psi$ with a random global phase factor.

Exercise 5.19 (Reduced Density Matrix) Let $\psi \in \mathbb{S}(\mathscr{H}_a \otimes \mathscr{H}_b)$. Show that the reduced density matrix $\rho_\psi = \mathrm{tr}_b|\psi\rangle\langle\psi|$ is pure (i.e., a 1d projection) if and only if ψ factorizes, $\psi = \psi_a \otimes \psi_b$.

References

1. Y. Aharonov, D. Bohm, Time in the quantum theory and the uncertainty relation for time and energy. Phys. Rev. **122**, 1649 (1961)
2. G.R. Allcock, The time of arrival in quantum mechanics II. The individual measurement. Ann. Phys. **53**, 286–310 (1969)
3. G. Bacciagaluppi, M. Dickson, Dynamics for modal interpretations. Found. Phys. **29**, 1165–1201 (1999)
4. J.S. Bell, On the Einstein-Podolsky-Rosen paradox. Physics **1**, 195–200 (1964). Reprinted as chapter 2 of [7]
5. J.S. Bell, On the problem of hidden variables in quantum mechanics. Rev. Modern Phys. **38**, 447–452 (1966). Reprinted as chapter 1 of [7]
6. J.S. Bell, Are there quantum jumps? in *Schrödinger. Centenary Celebration of a Polymath,* ed. by C.W. Kilmister (Cambridge University Press, 1987), pp. 41–52. Reprinted as chapter 22 of [7]
7. J.S. Bell, *Speakable and Unspeakable in Quantum Mechanics* (Cambridge University Press, 1987)
8. J.S. Bell, Against "measurement", in *Sixty-Two Years of Uncertainty. Historical, Philosophical, and Physical Inquiries into the Foundations of Quantum Physics,* ed. by A.I. Miller *NATO ASI*

Series B, vol. 226 (Plenum Press, New York, 1990). Reprinted in: Physics World **3**(8), 33–40 (1990)
9. M. Choi, Completely positive linear maps on complex matrices. Linear Algebra Appl. **10**, 285–290 (1975)
10. C.W. Cowan, R. Tumulka, Can one detect whether a wave function has collapsed? J. Phys. A Math. Theoret. **47**, 195303 (2014). https://arxiv.org/abs/1307.0810
11. C.W. Cowan, R. Tumulka, Epistemology of wave function collapse in quantum physics. British J. Philos. Sci. **67**(2), 405–434 (2016). https://arxiv.org/abs/1307.0827
12. D. Dürr, S. Teufel, *Bohmian Mechanics* (Springer, Heidelberg, 2009)
13. D. Dürr, S. Goldstein, N. Zanghì, Quantum equilibrium and the origin of absolute uncertainty. J. Stat. Phys. **67**, 843–907 (1992). https://arxiv.org/abs/quant-ph/0308039. Reprinted in [15]
14. D. Dürr, S. Goldstein, R. Tumulka, N. Zanghì, On the role of density matrices in Bohmian mechanics. Found. Phys. **35**, 449–467 (2005). https://arxiv.org/abs/quant-ph/0311127
15. D. Dürr, S. Goldstein, N. Zanghì, *Quantum Physics Without Quantum Philosophy* (Springer, Heidelberg, 2013)
16. B.-G. Englert, M.O. Scully, G. Süssmann, H. Walther, Surrealistic Bohm Trajectories. Zeitschrift für Naturforschung A **47**, 1175–1186 (1992)
17. J. Fröhlich. A brief review of the ETH-approach to quantum mechanics. In *Frontiers in analysis and probability*, ed. by N. Anantharaman, A. Nikeghbali, M.T. Rassias (Cham, Springer, 2020), pp. 21–45. http://arxiv.org/abs/1905.06603
18. J. Gemmer, G. Mahler, M. Michel, *Quantum Thermodynamics. Emergence of Thermodynamic Behavior within Composite Quantum Systems.* Lecture Notes in Physics 657 (Springer, Heidelberg, 2004)
19. A.M. Gleason, Measures on the closed subspaces of a Hilbert space. Indiana Univ. Math. J. **6**, 885–893 (1957)
20. K. Gödel, Über formal unentscheidbare Sätze der Principia Mathematica und verwandter Systeme I. Monatshefte für Mathematik und Physik **38**, 173–198 (1931)
21. S. Goldstein, Quantum theory without observers. Physics Today **51**(3), 42–46 and (4) 38–42 (1998)
22. S. Goldstein, J.L. Lebowitz, R. Tumulka, N. Zanghì, Canonical typicality. Phys. Rev. Lett. **96**, 050403 (2006). https://arxiv.org/abs/cond-mat/0511091
23. S. Goldstein, R. Tumulka, N. Zanghì, The quantum formalism and the GRW formalism. J. Stat. Phys. **149**, 142–201 (2012). https://arxiv.org/abs/0710.0885
24. R.B. Griffiths, *Consistent Quantum Theory* (Cambridge University Press, 2002)
25. J. Kiukas, A. Ruschhaupt, P.O. Schmidt, R.F. Werner. Exact energy–time uncertainty relation for arrival time by absorption. J. Phys. A Math. Theor. **45** 185301 (2012). http://arxiv.org/abs/1109.5087
26. S. Kochen, E.P Specker, The problem of hidden variables in quantum mechanics. J. Math. Mech. **17**, 59–87 (1967)
27. K. Kraus, *States, Effects, and Operations* (Springer, Heidelberg, 1983)
28. L. Landau, Das Dämpfungsproblem in der Wellenmechanik. Zeitschrift für Physik **45**, 430–441 (1927). English translation: The damping problem in wave mechanics. In *Collected Papers of L.D. Landau,* ed. by D. ter Haar (Gordon and Breach, New York, 1965), pp. 8–18
29. M.S. Leifer, Is the quantum state real? An extended review of ψ-ontology theorems. Quanta **3**, 67–155 (2014). https://arxiv.org/abs/1409.1570
30. N.D. Mermin, Simple unified form for the major no-hidden-variables theorems. Phys. Rev. Lett. **65**, 3373–3376 (1990)
31. B. Misra, E.C.G. Sudarshan, The Zeno's paradox in quantum theory. J. Math. Phys. **18**, 756–763 (1977)
32. J.G. Muga, R. Leavens, Arrival time in quantum mechanics. Physics Reports **338**, 353 (2000)
33. R. Penrose, *Shadows of the Mind* (Oxford University Press, Oxford, 1994)
34. A. Peres, Two simple proofs of the Kochen-Specker theorem. J. Phys. A Math. General **24**, L175–L178 (1991)

35. S. Popescu, A.J. Short, A. Winter, Entanglement and the foundation of statistical mechanics. Nature Physics **21**(11), 754–758 (2006)
36. M.F. Pusey, J. Barrett, T. Rudolph, On the reality of the quantum state. Nature Physics **8**, 475–478 (2012). https://arxiv.org/abs/1111.3328
37. M. Reed, B. Simon, *Methods of Modern Mathematical Physics, Vol. 1 (revised edition)* (Academic Press, 1980)
38. E. Schrödinger, *Statistical Thermodynamics, Second Edition* (Cambridge University Press, 1952)
39. E. Specker, Die Logik nicht gleichzeitig entscheidbarer Aussagen. Dialectica **14**, 239–246 (1960)
40. S. Teufel, R. Tumulka, Existence of Schrödinger Evolution with Absorbing Boundary Condition (2019). https://arxiv.org/abs/1912.12057
41. R. Tumulka, Es gibt kein treues stochastisches Modell der Quantenmechanik (in German). Diplom thesis. Department of Mathematics, Johann-Wolfgang-Goethe-Universität, Frankfurt am Main, Germany, 1998
42. R. Tumulka, A Kolmogorov extension theorem for POVMs. Lett. Math. Phys. **84**, 41–46 (2008). https://arxiv.org/abs/0710.3605
43. R. Tumulka, Detection Time Distribution for Several Quantum Particles (2016). https://arxiv.org/abs/1601.03871
44. R. Tumulka, Distribution of the Time at Which an Ideal Detector Clicks (2016). https://arxiv.org/abs/1601.03715
45. R. Tumulka, Energy-Time Uncertainty Relation for Absorbing Boundaries (2020). https://arxiv.org/abs/2005.14514
46. R. Tumulka, Empirically Equivalent Distributions in Ontological Models of Quantum Mechanics (2022). https://arxiv.org/abs/2205.04331
47. R. Tumulka, Limitations to Genuine Measurements in Ontological Models of Quantum Mechanics (2022). https://arxiv.org/abs/2205.05520
48. B.C. van Fraassen, A formal approach to the philosophy of science, in *Paradigms and Paradoxes: The Philosophical Challenge of the Quantum Domain,* ed. by R. Colony (University of Pittsburgh Press, 1972), pp. 303–366
49. J. von Neumann, Wahrscheinlichkeitstheoretischer Aufbau der Quantenmechanik. Göttinger Nachrichten **1**(10), 245–272 (1927). Reprinted in *John von Neumann: Collected Works Vol. I,* ed. by A.H. Taub (Pergamon Press, Oxford, 1961)
50. J. von Neumann, *Mathematische Grundlagen der Quantenmechanik* (Springer, Heidelberg, 1932). English translation by R.T. Beyer published as J. von Neumann: *Mathematical Foundation of Quantum Mechanics* (Princeton University Press, 1955)
51. R. Werner, Arrival time observables in quantum mechanics. Annales de l'Institut Henri Poincaré, section A **47**, 429–449 (1987)

Particle Creation

6

6.1 Identical Particles

There are two more rules of the quantum formalism that we have not covered yet: the symmetrization postulate for identical particles, also known as the boson–fermion alternative, and the spin–statistics rule. They are the subject of this section. We begin by stating them.

6.1.1 Symmetrization Postulate

There are several species of particles: electrons, photons, quarks, neutrinos, muons, and more. Particles belonging to the same species are said to be *identical*.

Symmetrization Postulate *If particle i and j are identical, then*

$$\psi_{\ldots s_i \ldots s_j \ldots}(\ldots \boldsymbol{x}_i \ldots \boldsymbol{x}_j \ldots) = \pm\psi_{\ldots s_j \ldots s_i \ldots}(\ldots \boldsymbol{x}_j \ldots \boldsymbol{x}_i \ldots)\,, \tag{6.1}$$

where the right-hand side has indices s_i and s_j interchanged, variables $\boldsymbol{x}_i$ and $\boldsymbol{x}_j$ interchanged, and all else are kept equal. Some species, called bosonic, always have the plus sign; the others, called fermionic, always minus.

Particles belonging to a bosonic species are called *bosons*, those belonging to a fermionic species *fermions*.

Spin-Statistics Rule *Species with integer spin are bosonic, those with half-odd spin are fermionic.*

R. Tumulka, *Foundations of Quantum Mechanics*, Lecture Notes in Physics 1003,
https://doi.org/10.1007/978-3-031-09548-1_6

It follows that for a system of N identical fermions and any permutation σ of $\{1,\ldots,N\}$ (i.e., any bijective mapping $\{1,\ldots,N\}\to\{1,\ldots,N\}$),

$$\psi_{s_{\sigma(1)}\ldots s_{\sigma(N)}}(\boldsymbol{x}_{\sigma(1)}\ldots\boldsymbol{x}_{\sigma(N)}) = (-)^\sigma\,\psi_{s_1\ldots s_N}(\boldsymbol{x}_1\ldots\boldsymbol{x}_N)\,, \tag{6.2}$$

where $(-)^\sigma$ denotes the *sign* of the permutation σ, i.e., $+1$ for an even permutation and -1 for an odd one. (In the following, the word "permutation" will always refer to a permutation of the particles.) Since for any two permutations σ,ρ, $(-)^{\sigma\circ\rho} = (-)^\sigma(-)^\rho$, and any *transposition* (i.e., exchange of two elements of $\{1,\ldots,N\}$) is odd, a permutation is even if and only if it can be obtained as the composition of an even number of transpositions. A function on $\mathbb{R}^{3N}=(\mathbb{R}^3)^N$ satisfying (6.2) is also said to be *anti-symmetric* under permutations, while a function satisfying (6.2) without the factor $(-)^\sigma$, as appropriate for a system of N identical bosons, is said to be *symmetric*.

Exercise 6.1 (Subspaces of Symmetric and Anti-Symmetric Functions) Show that in $L^2((\mathbb{R}^3)^N,(\mathbb{C}^d)^{\otimes N})$, the anti-symmetric functions form a subspace $\mathscr{H}_{\text{anti}}$, and the symmetric ones form a subspace $\mathscr{H}_{\text{sym}}$. Show further that, for $N>1$, $\mathscr{H}_{\text{anti}}\perp\mathscr{H}_{\text{sym}}$, so in particular $\mathscr{H}_{\text{anti}}\cap\mathscr{H}_{\text{sym}}=\{0\}$.

Exercise 6.2 (Symmetrizer and Anti-Symmetrizer) Show that the projection P_{anti} to $\mathscr{H}_{\text{anti}}$ (also known as the "anti-symmetrizer") can be expressed as

$$P_{\text{anti}} = \frac{1}{N!}\sum_{\sigma\in S_N}(-)^\sigma\,\Pi_\sigma\,, \tag{6.3}$$

where S_N denotes the group of all permutations of $\{1,\ldots,N\}$ (which has $N!$ elements) and Π_σ is the unitary operator on $L^2((\mathbb{R}^3)^N,(\mathbb{C}^d)^{\otimes N})$ that carries out the permutation σ,

$$(\Pi_\sigma\psi)_{s_1\ldots s_N}(\boldsymbol{x}_1\ldots\boldsymbol{x}_N) = \psi_{s_{\sigma(1)}\ldots s_{\sigma(N)}}(\boldsymbol{x}_{\sigma(1)}\ldots\boldsymbol{x}_{\sigma(N)})\,. \tag{6.4}$$

Likewise, the projection P_{sym} to $\mathscr{H}_{\text{sym}}$ (the "symmetrizer") is

$$P_{\text{sym}} = \frac{1}{N!}\sum_{\sigma\in S_N}\Pi_\sigma\,. \tag{6.5}$$

The *Pauli principle* is another name for the statement for any fermionic species such as electrons that the wave function of N identical particles has to be anti-symmetric. Sometimes people express it by saying that "two fermions cannot occupy the same state"; this is a very loose way of speaking that would not convey the situation to anyone who does not understand it already, as a particle belonging to an N-particle system does not have a state (i.e., a wave function) of its own, only the system has a wave function.

It may seem surprising that not every wave function on $(\mathbb{R}^3)^N$ is physically possible. On the other hand, it may seem natural that wave functions of identical particles have to be symmetric, and thus surprising that they can also be anti-symmetric. In fact, it seems surprising that there can be two different kinds of identical particles!

To some extent, explanations of these facts are known; this is what we will talk about in this section. The core of the reasoning concerns topology and can be generalized to arbitrary connected Riemannian manifolds; this will be described in Appendix A.1. Another question that arises and will be discussed in this section is whether and how theories such as Bohmian mechanics and GRW are compatible with identical particles.

6.1.2 Schrödinger Equation and Symmetry

If the initial wave function satisfies the symmetrization postulate, and if the Hamiltonian is invariant under permutations of particles of the same species, then the wave function automatically satisfies the symmetrization postulate at every other time. Specifically, for a system of N identical particles, a Hamiltonian H on $L^2((\mathbb{R}^3)^N)$ is permutation invariant if and only if it commutes with every Π_σ. For example, $-\sum_{i=1}^N(\hbar^2/2m_i)\nabla_i^2$ is permutation invariant if all masses are equal; a multiplication operator V is permutation invariant if the function $V:(\mathbb{R}^3)^N\to\mathbb{R}$ is, i.e.,

$$V(\boldsymbol{x}_{\sigma(1)},\ldots,\boldsymbol{x}_{\sigma(N)})=V(\boldsymbol{x}_1,\ldots,\boldsymbol{x}_N). \tag{6.6}$$

The sum and exponentials of permutation-invariant operators are permutation invariant; permutation-invariant operators commute with P_{anti} and P_{sym}; therefore, they map $\mathscr{H}_{\text{anti}}$ to $\mathscr{H}_{\text{anti}}$ and $\mathscr{H}_{\text{sym}}$ to $\mathscr{H}_{\text{sym}}$, as claimed.

6.1.3 The Space of Unordered Configurations

Another basic observation in this context is that the elements of the space $\mathbb{R}^{3N}=(\mathbb{R}^3)^N$ that we usually take as configuration space are *ordered* configurations $(\boldsymbol{x}_1,\ldots,\boldsymbol{x}_N)$, i.e., N-tuples of points in $\mathbb{R}^3$. In nature, of course, electrons are not ordered; that is, they are not numbered from 1 to N, and there is no fact about which electron is electron number 1. So in reality, there is an *unordered* configuration $\{\boldsymbol{x}_1,\ldots,\boldsymbol{x}_N\}$, a set of N points in $\mathbb{R}^3$. The set of all unordered configurations of N particles will henceforth be denoted

$${}^N\mathbb{R}^3:=\big\{q\subset\mathbb{R}^3:\#q=N\big\} \tag{6.7}$$

(with the number N thought of as a left index to the symbol $\mathbb{R}^3$). The use of this set has been introduced particularly by Leinaas and Myrheim (1977) [17].

Another way of representing an unordered configuration mathematically is to consider ordered configurations but declare that two configurations that are permutations of each other are equivalent. Then an unordered configuration, and thus a physical configuration, corresponds to an equivalence class of ordered configurations. Such an equivalence class has $N!$ elements unless the ordered configuration contains two particles at the same point in 3-space ("collision configuration"). Since the collision configurations are exceptions (they form a set of measure zero in $\mathbb{R}^{3N}$) and will not play a role in the following, we will remove them from the ordered configuration space and consider the set of collision-free configurations,

$$\mathbb{R}^{3,N}_{\neq} := \left\{(\boldsymbol{x}_1, \ldots, \boldsymbol{x}_N) \in (\mathbb{R}^3)^N : \boldsymbol{x}_i \neq \boldsymbol{x}_j \forall i \neq j\right\} \tag{6.8}$$

$$= (\mathbb{R}^3)^N \setminus \bigcup_{1\leq i<j\leq N} \Delta_{ij} \tag{6.9}$$

with $\Delta_{ij} \subset (\mathbb{R}^3)^N$ the set where $\boldsymbol{x}_i = \boldsymbol{x}_j$, a codimension-3 subspace (the ij-"diagonal"). So, the "forgetful mapping" or "unordering mapping"

$$\pi : (\boldsymbol{x}_1, \ldots, \boldsymbol{x}_N) \mapsto \{\boldsymbol{x}_1, \ldots, \boldsymbol{x}_N\}, \tag{6.10}$$

which forgets the ordering, maps $\mathbb{R}^{3,N}_{\neq}$ to ${}^N\mathbb{R}^3$; it is many-to-one, in fact always $N!$-to-one.

The unordered configuration space ${}^N\mathbb{R}^3$ inherits a topology via the mapping π. For readers familiar with manifolds, I mention that π carries the manifold structure from $\mathbb{R}^{3,N}_{\neq}$ to ${}^N\mathbb{R}^3$, as well as the metric; as a consequence, ${}^N\mathbb{R}^3$ is a Riemannian manifold.[1] It has curvature zero but is topologically nontrivial. We will investigate its topology more closely in Appendix A.1.

6.1.4 Identical Particles in Bohmian Mechanics

In view of the fact that in reality, particle configurations are unordered, the Bohmian configuration $Q(t)$ should be an element of ${}^N\mathbb{R}^3$, but Bohmian mechanics as

[1] Indeed, it is known that for a discrete group G acting on a manifold M by diffeomorphisms, the quotient space M/G is again a manifold if the action is "properly discontinuous," a property equivalent to that any two points $x \neq y$ of M have open neighborhoods U_x and U_y such that there are only a finite number of group elements g with $g(U_x)$ meeting U_y. This is always satisfied if G is finite. When, in addition, M is a Riemannian manifold and G acts by isometries, then M/G inherits a Riemannian metric. This is the case with the action of the permutation group on $M = \mathbb{R}^{3,N}_{\neq}$.

we defined it in Sect. 1.6 leads to curves $t \mapsto \widehat{Q}(t)$ in $\mathbb{R}^{3N}$. But that is not a problem, for the following reason (Goldstein 1987 [13]). Given any initial unordered configuration $Q(0) \in {}^N\mathbb{R}^3$, there are $N!$ possible orderings $\widehat{Q}(0) \in \mathbb{R}^{3N}$ of it. They lie in the set $\pi^{-1}(Q(0))$, where

$$\pi^{-1}(q) := \left\{\hat{q} \in \mathbb{R}^{3,N}_{\neq} : \pi(\hat{q}) = q\right\}. \tag{6.11}$$

If $\widehat{Q}_1(0)$ and $\widehat{Q}_2(0)$ are two orderings, then they are related through some permutation σ,

$$\widehat{Q}_2(0) = \sigma \widehat{Q}_1(0), \tag{6.12}$$

where we used the notation

$$\sigma(\boldsymbol{x}_1, \ldots, \boldsymbol{x}_N) = (\boldsymbol{x}_{\sigma(1)}, \ldots, \boldsymbol{x}_{\sigma(N)}). \tag{6.13}$$

Suppose that H is permutation invariant and the wave function is either symmetric or anti-symmetric at $t = 0$ and thus at any t. If we solve Bohm's equation of motion on the ordered configuration space $\mathbb{R}^{3N}$,

$$\frac{d\widehat{Q}}{dt} = \hat{v}^{\psi}(\widehat{Q}) \quad \text{with } \hat{v}^{\psi} = \frac{\hbar}{m}\mathrm{Im}\frac{\psi^*\nabla\psi}{\psi^*\psi}, \tag{6.14}$$

we obtain curves $t \mapsto \widehat{Q}_1(t)$ and $t \mapsto \widehat{Q}_2(t)$ with the property that $\widehat{Q}_1(t)$ and $\widehat{Q}_2(t)$ are still, for any $t \in \mathbb{R}$, related through the same permutation σ,

$$\widehat{Q}_2(t) = \sigma \widehat{Q}_1(t). \tag{6.15}$$

Indeed, that follows from

$$\hat{v}(\sigma \widehat{Q}) = \sigma \hat{v}(\widehat{Q}), \tag{6.16}$$

a consequence of the (anti-)symmetry of ψ.

As a consequence of (6.15), $\pi(\widehat{Q}_2(t)) = \pi(\widehat{Q}_1(t))$. That is, Bohmian mechanics defines for every $Q(0) \in {}^N\mathbb{R}^3$ a unique trajectory $t \mapsto Q(t)$ in ${}^N\mathbb{R}^3$. Put differently, the arbitrary choice of ordering does not affect the motion of the particles.

A different perspective on this fact is that Bohm's law of motion can also be formulated directly on the unordered space ${}^N\mathbb{R}^3$ in the form

$$\frac{dQ}{dt} = v^{\psi}(Q(t)) \tag{6.17}$$

where v^{ψ} is a vector field on the manifold ${}^N\mathbb{R}^3$ obtained from the vector field $\hat{v}$ on $\mathbb{R}^{3,N}_{\neq}$ by "projecting down" using the projection mapping π. Technically speaking, it is the *tangent mapping* $D\pi$ (also known as the *differential* or *total derivative* of π), applied to $\hat{v}(\hat{q})$, that yields $v(q)$,

$$v(q) := D\pi|_{\hat{q}}(\hat{v}(\hat{q})) . \tag{6.18}$$

It is crucial that different orderings $\hat{q} \in \pi^{-1}(q)$ of q yield the same vector $v(q)$; this fact is expressed by formula (6.16), and in words it means that although different orderings assign different numbers to each particle, they agree, if one of the particles in located at $\boldsymbol{x} \in \mathbb{R}^3$, about the 3-velocity of the particle at $\boldsymbol{x}$.

Conversely, this fact can be regarded as an explanation of the boson–fermion alternative: Since for a general ψ on $\mathbb{R}^{3N}$ that is neither symmetric nor anti-symmetric, the vector field $\hat{v}^\psi$ defined by (6.14) violates (6.16), it fails to define a vector field v on ${}^N\mathbb{R}^3$ (or, for that matter, trajectories $t \mapsto Q(t)$ in ${}^N\mathbb{R}^3$). Thus, a wave function ψ of N identical particles should be either symmetric or anti-symmetric.

It may seem surprising that Bohmian mechanics can get along at all with identical particles, for the following reason. Some authors have proposed that the reason why general, asymmetric wave functions on $\mathbb{R}^{3N}$ are unphysical is the impossibility to decide which of the electrons at time t_1 is which of the electrons at time t_2; if electrons had trajectories, then that would define which electron at t_1 is which electron at t_2; since in orthodox quantum mechanics, electrons do not have trajectories, there is a symmetrization postulate in quantum mechanics but not in classical mechanics. We have seen why this reasoning is questionable.

In fact, there is a sense in which Bohmian mechanics gets along better with identical particles than orthodox quantum mechanics: While the space $\mathcal{Q}$ of physically possible configurations is the unordered one ${}^N\mathbb{R}^3$, the space $\widehat{\mathcal{Q}}$ on which the wave function is defined is the ordered one $\mathbb{R}^{3N}$. In Bohmian mechanics, it is not necessary that $\mathcal{Q}$ be the same as $\widehat{\mathcal{Q}}$, as long as Q belongs to $\mathcal{Q}$ and ψ on $\widehat{\mathcal{Q}}$ still defines a vector field v^ψ on $\mathcal{Q}$. In orthodox quantum mechanics, in contrast, there is no element of the ontology that could bridge between $\mathcal{Q}$ and $\widehat{\mathcal{Q}}$, so it remains unintelligible how ψ could be defined on any other space than $\mathcal{Q}$.

6.1.5 Identical Particles in GRW Theory

GRW theory as formulated in Sect. 3.3 has the following problem with the symmetrization postulate: A collapse of the wave function, corresponding to multiplication by a Gaussian function in one of the $\boldsymbol{x}_j$ as in (3.62) and (3.63), usually leads to a wave function Ψ_{T+} that no longer obeys (6.1) (in particular, for identical particles, no longer is symmetric or anti-symmetric). Thus, to accommodate the symmetrization postulate, the equations of GRW theory need to be adjusted. Here is how (Dove and Squires 1995 [5], Tumulka 2006 [27]).

For a universe with N particles, collapses occur with rate $N\lambda$. If the number r of different species of particles is greater than 1, then the species I of a collapse is chosen randomly with

$$\mathbb{P}(I = i) = N_i/N\,, \tag{6.19}$$

where N_i is the number of particles of species i. Equivalently, for each species i, collapses occur with rate $N_i\lambda$. Define the collapse operator for species I and center $\boldsymbol{X}$ by

$$C_I(\boldsymbol{X})\Psi(\boldsymbol{x}_1,\ldots,\boldsymbol{x}_N) = \Big(\sum_{j\in\mathscr{I}_I} g_{\boldsymbol{X},\sigma}(\boldsymbol{x}_j)\Big)^{1/2}\Psi(\boldsymbol{x}_1,\ldots,\boldsymbol{x}_N)\,, \tag{6.20}$$

where $\mathscr{I}_I$ is the set of the labels of all particles belonging to species I. For a collapse at time T, choose the location $\boldsymbol{X}$ of the flash randomly with density

$$\rho(\boldsymbol{X} = \boldsymbol{x}) = \|C_I(\boldsymbol{x})\Psi_{T-}\|^2 \tag{6.21}$$

and collapse the wave function according to

$$\Psi_{T+} = \frac{C_I(\boldsymbol{X})\Psi_{T-}}{\|C_I(\boldsymbol{X})\Psi_{T-}\|}\,. \tag{6.22}$$

For example, for N identical particles, instead of multiplying by a Gaussian function in one $\boldsymbol{x}_j$, we multiply by the square root of the sum of N Gaussians, all with the same center and width but applied to different variables. Since the collapse operator is a multiplication operator by a permutation-invariant function $(\sum_j g(\boldsymbol{x}_j))^{1/2}$, it is a permutation-invariant operator and thus maps $\mathscr{H}_{\text{sym}} \to \mathscr{H}_{\text{sym}}$ and $\mathscr{H}_{\text{anti}} \to \mathscr{H}_{\text{anti}}$. Likewise, Ψ_{T+} given by (6.22) still satisfies (6.1). The empirical predictions of this symmetrized version of the GRW theory are not exactly the same as those of the version described in Sect. 3.3, but both are close to those of the quantum formalism.

6.2 Hamiltonians of Particle Creation

In nature, particles can be created and annihilated: Photons can be emitted and absorbed; an electron and a positron can annihilate each other radiating off a photon; in radioactive beta decay, a down quark mutates into an up quark while emitting a negatively charged W boson, and the latter decays into an electron and an anti-neutrino. The obvious conclusion (although other explanations have been suggested) is that the number of particles can change with time. So far, we assumed a constant number N of particles, so we now want to extend our framework. Often, and in particular for photons, particle creation is considered in relativistic space–time. In this chapter, we use non-relativistic models for simplicity.

6.2.1 Configuration Space of a Variable Number of Particles

Let us consider the issue for a moment from a Bohmian point of view. If particle trajectories can begin and end, then the configuration space $\mathcal{Q}$, the set of *all* possible configurations, should contain configurations with any number of particles, that is,

$$\mathcal{Q} = \Gamma(\mathbb{R}^3) := \{q \subset \mathbb{R}^3 : \#q < \infty\}, \tag{6.23}$$

where $\#q$ means the number of elements of the set q. In words, $\mathcal{Q}$ is the set of all finite subsets of $\mathbb{R}^3$. Here, we have set up $\mathcal{Q}$ as the space of *unordered* configurations of a variable number of particles of a single species. For ℓ species, $\mathcal{Q}$ should be $\Gamma(\mathbb{R}^3)^\ell$, the Cartesian product of ℓ copies of $\Gamma(\mathbb{R}^3)$. The set $\Gamma(\mathbb{R}^3)$ is the union of the unordered N-particle configuration spaces,

$$\Gamma(\mathbb{R}^3) = \bigcup_{N=0}^{\infty} {}^N\mathbb{R}^3 = \bigcup_{N=0}^{\infty} \{q \subset \mathbb{R}^3 : \#q = N\}. \tag{6.24}$$

The notation ${}^N\mathbb{R}^3$ with the left index was introduced in (6.7). As a subset of $\Gamma(\mathbb{R}^3)$, ${}^N\mathbb{R}^3$ is called the *N-particle sector* or just *N-sector* of $\Gamma(\mathbb{R}^3)$; that is, the sectors are the level sets of the particle number function $q \mapsto \#q$; in particular, different sectors are disjoint. It is sometimes convenient to use a space of *ordered* configurations of a variable number of particles,

$$\Gamma_o(\mathbb{R}^3) := \bigcup_{N=0}^{\infty} \mathbb{R}^{3N}, \tag{6.25}$$

with the sectors $\mathbb{R}^{3N}$ understood as disjoint. This is almost the same as

$$\widehat{\mathcal{Q}} = \bigcup_{N=0}^{\infty} \mathbb{R}^{3,N}_{\neq}, \tag{6.26}$$

a space that arises as the universal covering space of $\mathcal{Q}$ (as explained in Appendix A.1), the only difference being whether collision configurations are included or not.

The moving particles are represented by a moving point $Q(t)$ in configuration space; whenever particles are created and annihilated, $t \mapsto Q(t)$ jumps between sectors.

For later use, we describe the natural notion of volume (measure μ) on $\Gamma(\mathbb{R}^3)$: We simply use $3N$-dimensional volume on each sector. In formulas, and taking into account the difference between ordered configurations (in $\mathbb{R}^{3N}$) and unordered ones, this means that any (Borel) set $S \subseteq \Gamma(\mathbb{R}^3)$ has measure

$$\mu(S) = \sum_{N=0}^{\infty} \frac{1}{N!} \mathrm{vol}_{3N}\big(\pi^{-1}(S \cap {}^N\mathbb{R}^3)\big), \tag{6.27}$$

where vol_{3N} is the volume (Lebesgue measure) in $\mathbb{R}^{3N}$ and $\pi : \mathbb{R}^{3,N}_{\neq} \to {}^N\mathbb{R}^3$ is the forgetful mapping (6.10). The factor $1/N!$ compensates the fact that every point in ${}^N\mathbb{R}^3$ has $N!$ pre-images with respect to π. On the ordered configuration space $\Gamma_\mathrm{o}(\mathbb{R}^3)$, we use the measure

$$\mu_\mathrm{o}(S_\mathrm{o}) = \sum_{N=0}^{\infty} \mathrm{vol}_{3N}(S_\mathrm{o} \cap \mathbb{R}^{3N}). \tag{6.28}$$

6.2.2 Fock Space

The Fock Space of Spinless Bosons

The wave function should be expected to be a function on configuration space and to obey the symmetrization postulate (6.1). Let us focus first on the simplest case of a single species of spinless bosons. Then the wave function is a complex-valued function on $\Gamma(\mathbb{R}^3)$ that is square-integrable relative to the measure μ,

$$\psi \in L^2\big(\Gamma(\mathbb{R}^3), \mathbb{C}, \mu\big) =: L^2(\Gamma(\mathbb{R}^3)). \tag{6.29}$$

In this space, the inner product is of course defined as

$$\langle\phi|\psi\rangle = \int_{\Gamma(\mathbb{R}^3)} \mu(dq)\, \phi(q)^*\, \psi(q) \tag{6.30a}$$

$$= \sum_{N=0}^{\infty} \frac{1}{N!} \int_{\mathbb{R}^{3N}} d^{3N}\hat{q}\, \phi(\pi(\hat{q}))^*\, \psi(\pi(\hat{q})). \tag{6.30b}$$

A general ψ that is nonzero on several sectors must be regarded as a superposition of different particle numbers. Any function on a disjoint union of sectors can also be written as a list of functions, one on each sector:

$$\psi = (\psi^{(0)}, \psi^{(1)}, \psi^{(2)}, \ldots) \tag{6.31}$$

with $\psi^{(N)} = \psi|_{{}^N\mathbb{R}^3}$ the restriction to the N-sector, $\psi^{(N)} : {}^N\mathbb{R}^3 \to \mathbb{C}$; we call $\psi^{(N)}$ the N-particle sector of ψ.

The $|\psi|^2$ distribution is then a probability distribution over the configuration space $\Gamma(\mathbb{R}^3)$, provided ψ is normalized, i.e.,

$$1 = \|\psi\|^2 = \langle\psi|\psi\rangle = \sum_{N=0}^{\infty} \frac{1}{N!} \int_{\mathbb{R}^{3N}} d^{3N}\hat{q}\, \left|\psi(\pi(\hat{q}))\right|^2$$

$$= \sum_{N=0}^{\infty} \int_{{}^N\mathbb{R}^3} \mu(d^{3N}q)\, \left|\psi(q)\right|^2 . \qquad (6.32)$$

The $|\psi|^2$ distribution, that is, the probability distribution on $\mathcal{Q} = \Gamma(\mathbb{R}^3)$ with density $|\psi|^2$ relative to the volume measure μ, contains in particular a probability distribution over the particle number,

$$\mathbb{P}(N = n) = \int_{{}^n\mathbb{R}^3} \mu(d^{3n}q)\, |\psi(q)|^2 = \|\psi^{(n)}\|^2_{L^2({}^n\mathbb{R}^3)} . \qquad (6.33)$$

As a consequence of (6.30), if ϕ is concentrated on the sector N_1 and ψ on $N_2 \neq N_1$, their inner product vanishes. Put differently, the subspace $\mathscr{H}^{(N)}$ in $\mathscr{H} := L^2(\Gamma(\mathbb{R}^3))$ consisting of the wave functions concentrated in the N-sector of configuration space is orthogonal in $\mathscr{H}$ to $\mathscr{H}^{(N')}$ with $N' \neq N$. In fact, using that $\mathscr{H}^{(N)}$ can be identified with $L^2({}^N\mathbb{R}^3)$ (where ${}^N\mathbb{R}^3$ is understood as equipped with the measure $N!^{-1}\mathrm{vol}_{3N} \circ \pi^{-1}$),

$$L^2(\Gamma(\mathbb{R}^3)) = \bigoplus_{N=0}^{\infty} L^2({}^N\mathbb{R}^3) , \qquad (6.34)$$

where $\oplus$ means orthogonal sum. This space is called the *Fock space of spinless bosons*, and $\mathscr{H}^{(N)}$ the N-particle sector of $\mathscr{H}$.

In general, the orthogonal sum $\mathscr{H}_1 \oplus \mathscr{H}_2$ is defined as the set of pairs $\psi = (\psi_1, \psi_2)$ with inner product $\langle\phi|\psi\rangle = \langle\phi_1|\psi_1\rangle_{\mathscr{H}_1} + \langle\phi_2|\psi_2\rangle_{\mathscr{H}_2}$. The *infinite* orthogonal sum $\oplus_{N=0}^{\infty}\mathscr{H}_N$ is defined as the set of sequences $(\psi_0, \psi_1, \psi_2, \ldots)$ with $\psi_N \in \mathscr{H}_N$ such that $\sum_{N=0}^{\infty} \|\psi_N\|^2_{\mathscr{H}_N} < \infty$. One can show that the inner product, defined by

$$\langle\phi|\psi\rangle = \sum_{N=0}^{\infty} \langle\phi_N|\psi_N\rangle , \qquad (6.35)$$

is always finite for such sequences (i.e., the series is absolutely convergent).

The Fock Space of Spinless Fermions

Let us turn to another example, a single species of spinless fermions (even though that is against the spin–statistics rule). The N-particle wave function should be a function $\psi^{(N)} : \mathbb{R}^{3N} \to \mathbb{C}$ that is anti-symmetric against permutation of particles. Correspondingly, the full wave function is a function $\psi : \Gamma_{\circ}(\mathbb{R}^3) \to \mathbb{C}$ that is anti-symmetric on each sector; the inner product is

$$\langle\phi|\psi\rangle = \sum_{N=0}^{\infty} \frac{1}{N!} \int_{\mathbb{R}^{3N}} d^{3N}\hat{q}\, \phi(\hat{q})^* \, \psi(\hat{q})\,. \tag{6.36}$$

Again, the Hilbert space is of the form

$$\mathscr{H} = \bigoplus_{N=0}^{\infty} \mathscr{H}^{(N)}\,, \tag{6.37}$$

where the N-particle Hilbert space $\mathscr{H}^{(N)}$ is the set of those square-integrable functions on $\mathbb{R}^{3N}$ that are anti-symmetric against permutations of the particle labels. (We will often simply call them "anti-symmetric functions," although they are not anti-symmetric against permutation of the three space directions.) $\mathscr{H}^{(N)}$ is equipped with the inner product

$$\langle\phi_N|\psi_N\rangle = \frac{1}{N!} \int_{\mathbb{R}^{3N}} d^{3N}\hat{q}\, \phi(\hat{q})^* \, \psi(\hat{q})\,. \tag{6.38}$$

Most authors, however, prefer to use a different definition and leave out the factor $1/N!$:

$$\langle\phi_N|\psi_N\rangle = \int_{\mathbb{R}^{3N}} d^{3N}\hat{q}\, \phi(\hat{q})^* \, \psi(\hat{q})\,. \tag{6.39}$$

Note that one can switch between (6.38) and (6.39) by changing ϕ and ψ by a factor of $\sqrt{N!}$. The definition (6.39) is less physically natural but more convenient and so widespread that we will adopt it as well. It has the consequence that $\mathscr{H}^{(N)}$ is a subspace of $L^2(\mathbb{R}^{3N})$, that is, that the scalar product in $\mathscr{H}^{(N)}$ is that of $L^2(\mathbb{R}^{3N})$. One therefore writes

$$\mathscr{H}^{(N)} = S_- L^2(\mathbb{R}^{3N})\,, \tag{6.40}$$

where S_- means the anti-symmetrization operator for N particles and $S_- L^2(\mathbb{R}^{3N})$ means the image of $L^2(\mathbb{R}^{3N})$ under S_-, that is, the subspace of anti-symmetric functions. The space (6.37) is called the *Fock space of spinless fermions*.

As a consequence of the convention (6.39) instead of (6.38), the probability associated with a subset B of the unordered configuration space $\Gamma(\mathbb{R}^3)$ according to Born's rule reads

$$\mathbb{P}(B) = \int_B \mu(dq)\, \rho(q) \tag{6.41a}$$

$$= \int_B \mu(dq)\, (\#q)!\, \big|\psi(\hat{q})\big|^2 \tag{6.41b}$$

$$= \sum_{N=0}^{\infty} \int_{\pi^{-1}(B\cap {}^N\mathbb{R}^3)} d^{3N}\hat{q}\, \big|\psi(\hat{q})\big|^2 , \tag{6.41c}$$

where $\hat{q}$ in (6.41b) means any ordering of q and $\pi^{-1}(B \cap {}^N\mathbb{R}^3)$ the pre-image of the N-particle sector $B \cap {}^N\mathbb{R}^3$ of B under the forgetful mapping (6.10), that is, the set of all orderings of configurations in B. This probability distribution corresponds to a PVM P on $\Gamma(\mathbb{R}^3)$ acting on the Fock space of spinless fermions ("position PVM"), with $P(B)$ given by the multiplication operator $1_{\pi^{-1}(B)}$ multiplying by the characteristic function of the set of all orderings of configurations in B; here, we have applied the forgetful mapping π to all N, so

$$\pi : \Gamma_o(\mathbb{R}^3) \to \Gamma(\mathbb{R}^3)\,, \quad \pi(\boldsymbol{x}_1, \ldots, \boldsymbol{x}_N) = \{\boldsymbol{x}_1, \ldots, \boldsymbol{x}_N\}\,. \tag{6.42}$$

(Configurations with collisions, $\boldsymbol{x}_i = \boldsymbol{x}_j$ for some $i \neq j$, will be mapped to a different sector, but this does not matter because such configurations form a set of measure 0.)

Analogously, it is common also for the Fock space (6.29) of spinless bosons to express its elements as functions on the ordered configuration space $\Gamma_o(\mathbb{R}^3)$ and to change the definition of the inner product (6.30b) by a factor of $N!$ to

$$\langle\phi|\psi\rangle = \sum_{N=0}^{\infty} \int_{\mathbb{R}^{3N}} d^{3N}\hat{q}\, \phi(\hat{q})^*\, \psi(\hat{q})\,. \tag{6.43}$$

Then, the Fock space can be written as (6.37) with

$$\mathscr{H}^{(N)} = S_+ L^2(\mathbb{R}^{3N}) \tag{6.44}$$

with S_+ the symmetrization operator.

General Fock Space

From (6.40) and (6.44), it is not hard to guess the general pattern of how to build the Hilbert space of a single species: It is

$$\mathscr{H} = \mathscr{F}_\pm := \Gamma_\pm(\mathscr{H}_1) := \bigoplus_{N=0}^{\infty} S_\pm \mathscr{H}_1^{\otimes N} \tag{6.45}$$

for a given 1-particle Hilbert space $\mathscr{H}_1$. Here,

$$\mathscr{H}_1^{\otimes N} := \underbrace{\mathscr{H}_1 \otimes \cdots \otimes \mathscr{H}_1}_{N \text{ factors}} \tag{6.46}$$

means the N-th *tensor power* (i.e., N-fold tensor product with itself), and $S_\pm$ means the (anti-)symmetrization operator with respect to permutations of the factors in the N-fold tensor product. Equation (6.45) is the general definition of *Fock space* $\mathscr{F}_\pm$, called the bosonic (+) or fermionic (−) Fock space over $\mathscr{H}_1$. For spinless bosons or fermions, $\mathscr{H}_1 = L^2(\mathbb{R}^3)$; for spin-$\frac{1}{2}$ particles, $\mathscr{H}_1 = L^2(\mathbb{R}^3, \mathbb{C}^2)$; for spin-$s$ particles, $\mathscr{H}_1 = L^2(\mathbb{R}^3, \mathbb{C}^{2s+1})$.

Given a 1-particle position PVM P_1 on $\mathbb{R}^3$ acting on $\mathscr{H}_1$, we obtain a position PVM $P_\pm$ on $\Gamma(\mathbb{R}^3)$ acting on $\mathscr{F}_\pm$ as follows.

Proposition 6.1 (Corollary 7 in Section 4.4 of [9]) *Given two POVMs, E_a on $\mathcal{Q}_a$ acting on $\mathscr{H}_a$ and E_b on $\mathcal{Q}_b$ acting on $\mathscr{H}_b$. If $\mathcal{Q}_a$ and $\mathcal{Q}_b$ satisfy the technical condition of being standard Borel spaces,*[2] *then there is a unique POVM E on $\mathcal{Q}_a \times \mathcal{Q}_b$ acting on $\mathscr{H}_a \otimes \mathscr{H}_b$ satisfying*

$$E(B_a \times B_b) = E_a(B_a) \otimes E_b(B_b) \tag{6.47}$$

for all $B_a \subseteq \mathcal{Q}_a$ and $B_b \subseteq \mathcal{Q}_b$. We call E the tensor product POVM *and use the symbol $E_a \otimes E_b$ for it. If E_a and E_b are PVMs, then E is a PVM.*

Note that $E(B)$ will *not* be a product operator if B is *not* a product set. We now form $P_1 \otimes P_1$ and by iteration $P_1^{\otimes N}$, a PVM on $(\mathbb{R}^3)^N$ acting on $\mathscr{H}_1^{\otimes N}$. For any set B_o that is invariant under permutations or, equivalently, that is of the form $B_o = \pi^{-1}(B)$ for some $B \subseteq {}^N\mathbb{R}^3$, the operator $P_1^{\otimes N}(B_o)$ is permutation invariant and thus will map (anti-)symmetric vectors in $\mathscr{H}_1^{\otimes N}$ to (anti-)symmetric ones; let $P_{N,\pm}(B)$ be the restriction of $P_1^{\otimes N}(\pi^{-1}(B))$ to the (anti-)symmetric subspace of $\mathscr{H}_1^{\otimes N}$, and set

$$(P_\pm(B)\psi)^{(N)} = P_{N,\pm}(B \cap {}^N\mathbb{R}^3)\psi^{(N)}\,. \tag{6.48}$$

[2] A *standard Borel space* is a measurable space isomorphic to a complete separable metric space with its Borel σ-algebra. Spaces that arise in practice are usually standard Borel spaces, including $\mathbb{R}^d$ and Hilbert spaces of countable dimension ("separable"), so this condition is not much of a restriction.

Two Species

Now consider two species, and let us call them x-particles and y-particles. A configuration consists of a number M of x-particles and a number N of y-particles; thus, the (unordered) configuration space is

$$\mathcal{Q} = \bigcup_{M,N=0}^{\infty} \underbrace{\mathcal{Q}^{(M,N)}}_{{}^M\mathbb{R}^3_x \times {}^N\mathbb{R}^3_y} = \Gamma(\mathbb{R}^3) \times \Gamma(\mathbb{R}^3)\,, \tag{6.49}$$

in agreement with the general principle that the configuration space of two systems together is the Cartesian product of the configuration spaces of each system. (Here, the first system is the x-particles, the second the ys.) If the xs are fermions and the ys bosons (say, both spinless), then $\psi^{(M,N)}$, the (M, N)-sector of the wave function, should be a function $\mathbb{R}^{3M} \times \mathbb{R}^{3N} \to \mathbb{C}$ that is anti-symmetric against permutation of $\boldsymbol{x}_1, \ldots, \boldsymbol{x}_M$ and symmetric against permutation of $\boldsymbol{y}_1, \ldots, \boldsymbol{y}_N$, in agreement with the formulation of the symmetrization principle in (6.1). The corresponding Hilbert space is

$$\mathscr{H} = \bigoplus_{M=0}^{\infty} \bigoplus_{N=0}^{\infty} S_{x-} S_{y+} L^2(\mathbb{R}^{3M} \times \mathbb{R}^{3N}) \tag{6.50a}$$

$$= \bigoplus_{M=0}^{\infty} \bigoplus_{N=0}^{\infty} \Big(S_- L^2(\mathbb{R}^{3M})\Big) \otimes \Big(S_+ L^2(\mathbb{R}^{3N})\Big) \tag{6.50b}$$

$$= \left(\bigoplus_{M=0}^{\infty} S_- L^2(\mathbb{R}^{3M})\right) \otimes \left(\bigoplus_{N=0}^{\infty} S_+ L^2(\mathbb{R}^{3N})\right) \tag{6.50c}$$

$$= \mathscr{F}_- \otimes \mathscr{F}_+\,, \tag{6.50d}$$

the tensor product of the Fock spaces associated with the species x and y, in agreement with the general principle that the Hilbert space of two systems together is the tensor product of the Hilbert spaces of each system. Here, we have used:

Exercise 6.3 (Distributive Laws for $\oplus$ and $\otimes$) Show that

$$\mathscr{H} \otimes (\oplus_i \mathscr{K}_i) = \oplus_i (\mathscr{H} \otimes \mathscr{K}_i) \quad \text{and} \quad (\oplus_j \mathscr{H}_j) \otimes \mathscr{K} = \oplus_j (\mathscr{H}_j \otimes \mathscr{K})\,, \tag{6.51}$$

(a) for finite and (b) for countably infinite orthogonal sums.

For more than two species, the Hilbert space is the tensor product of one Fock space for each species, $\mathscr{H} = \mathscr{F}_1 \otimes \cdots \otimes \mathscr{F}_\ell$ with each $\mathscr{F}_i$ either a bosonic or a fermionic Fock space.

6.2.3 Example: Emission–Absorption Model

As a simple example, let us consider a toy model with two species, x-particles that are fermions and y-particles that are bosons, where the x's can emit and absorb y's ("emission–absorption model," sometimes also called the "Lee model"); that is, we consider particle reactions of the form[3]

$$x \leftrightarrows x + y\,. \tag{6.52}$$

We write

$$x^M = (\boldsymbol{x}_1, \ldots, \boldsymbol{x}_M)\,, \quad y^N = (\boldsymbol{y}_1, \ldots, \boldsymbol{y}_N) \tag{6.53}$$

for an x-configuration and a y-configuration, where the superscript reminds us of the number of particles in the configuration. We sometimes write $\psi(x^M, y^N)$ and sometimes $\psi^{(M,N)}(x^M, y^N)$ although specifying the sector is redundant as it is uniquely determined already by the configuration.

The Hilbert space $\mathscr{H}$ is given by (6.50), and the wave function (or vector in $\mathscr{H}$) evolves with time according to the usual relation

$$i\hbar \frac{\partial \psi}{\partial t} = H\psi \tag{6.54}$$

with Hamiltonian H defined by

$$\begin{aligned}
(H\psi)^{(M,N)}(x^M, y^N) = &-\sum_{j=1}^{M} \frac{\hbar^2}{2m_x} \nabla^2_{\boldsymbol{x}_j} \psi^{(M,N)}(x^M, y^N) \\
&- \sum_{k=1}^{N} \frac{\hbar^2}{2m_y} \nabla^2_{\boldsymbol{y}_k} \psi^{(M,N)}(x^M, y^N) + N\,E_0\,\psi^{(M,N)}(x^M, y^N) \\
&+ g\sqrt{N+1} \sum_{j=1}^{M} \psi^{(M,N+1)}(x^M, (y^N, \boldsymbol{x}_j)) \\
&+ \frac{g}{\sqrt{N}} \sum_{j=1}^{M} \sum_{k=1}^{N} \delta^3(\boldsymbol{x}_j - \boldsymbol{y}_k)\, \psi^{(M,N-1)}(x^M, y^N \setminus \boldsymbol{y}_k)\,.
\end{aligned} \tag{6.55}$$

[3] In this model, every creation and annihilation event changes the number of particles by 1. The standard model of particle physics also knows events changing the number of particles by 2, and everything said in this chapter should carry over to this case. On the other hand, these events might also be a close succession of events changing the number of particles by 1.

Here, $m_x, m_y > 0$ are the masses of an x- and a y-particle; $E_0 \geq 0$ is the energy that must be expended for creating a y-particle (realistically, $E_0 = m_y c^2$); g is another real constant that determines the strength of particle creation and annihilation (called the "coupling constant" or the "charge" of an x-particle); $(y^N, \boldsymbol{x}_j)$ means a configuration of $N+1$ y-particles with $\boldsymbol{y}_{N+1} = \boldsymbol{x}_j$; and

$$y^N \setminus \boldsymbol{y}_k = (\boldsymbol{y}_1, \ldots, \boldsymbol{y}_{k-1}, \boldsymbol{y}_{k+1}, \ldots, \boldsymbol{y}_N)\,. \tag{6.56}$$

For $N = 0$, the last line of (6.55) is absent, as $\sum_{k=1}^{N}$ is understood as comprising N terms.

The factors $\sqrt{N+1}$ and $\sqrt{N}$ are owed to the convention (6.39) for the inner product and would be absent in the more natural convention (6.38). Indeed, the switch from (6.38) to (6.39) requires replacing ψ by $N!^{-1/2}\psi$, which leads to the square root factors in (6.55).

It is visible from (6.55) that the sector $\psi^{(M,N)}$ couples to the neighboring sectors $\psi^{(M,N+1)}$ and $\psi^{(M,N-1)}$. Thus, if ψ at time 0 is concentrated on a single sector, then it will not be so concentrated at other times. Put differently, the time evolution involves the creation and annihilation of y-particles. One can say that the x-particles interact with each other through the exchange of y-particles.

Non-Rigorous Theorem 6.1 *While H is not well defined (see Sect. 6.2.5), it looks self-adjoint in the non-rigorous calculation.*

Derivation

$$\langle\phi|H\psi\rangle = \sum_{M=0}^{\infty}\sum_{N=0}^{\infty}\int_{\mathbb{R}^{3M}} d^{3M}x^M \int_{\mathbb{R}^{3N}} d^{3N}y^N\, \phi(x^M, y^N)^*\,(H\psi)(x^M, y^N) \tag{6.57a}$$

$$= -\sum_{M,N}\int dx^M\, dy^N\, \phi(x^M, y^N)^* \sum_{j=1}^{M} \frac{\hbar^2}{2m_x}\nabla^2_{\boldsymbol{x}_j}\psi(x^M, y^N)$$

$$-\sum_{M,N}\int dx^M\, dy^N\, \phi(x^M, y^N)^* \sum_{k=1}^{N} \frac{\hbar^2}{2m_y}\nabla^2_{\boldsymbol{y}_k}\psi(x^M, y^N)$$

$$+\sum_{M,N}\int dx^M\, dy^N\, \phi(x^M, y^N)^*\, N\, E_0\, \psi(x^M, y^N)$$

$$+\sum_{M}\sum_{N=0}^{\infty}\int dx^M\, dy^N\, \phi(x^M, y^N)^*\, g\sqrt{N+1}$$

$$\times \sum_{j=1}^{M} \psi^{(M,N+1)}(x^M, (y^N, \boldsymbol{x}_j))$$

$$+ \sum_M \sum_{N=1}^{\infty} \int dx^M \, dy^N \, \phi(x^M, y^N)^* \, \frac{g}{\sqrt{N}}$$

$$\times \sum_{j=1}^{M} \sum_{k=1}^{N} \delta^3(\boldsymbol{x}_j - \boldsymbol{y}_k) \, \psi^{(M,N-1)}(x^M, y^N \setminus \boldsymbol{y}_k) \,. \qquad (6.57b)$$

The first two lines can be transformed by integrating by parts twice on each line into

$$- \sum_{M,N} \sum_{j=1}^{M} \int dx^M \, dy^N \left(\tfrac{\hbar^2}{2m_x} \nabla^2_{\boldsymbol{x}_j} \phi(x^M, y^N)^* \right) \psi(x^M, y^N)$$

$$- \sum_{M,N} \sum_{k=1}^{N} \int dx^M \, dy^N \left(\tfrac{\hbar^2}{2m_y} \nabla^2_{\boldsymbol{y}_k} \phi(x^M, y^N)^* \right) \psi(x^M, y^N) \,. \qquad (6.58)$$

In the last two lines of (6.57b), we integrate out $\boldsymbol{y}_k$ against the δ function to obtain

$$\sum_M \sum_{N=1}^{\infty} \frac{g}{\sqrt{N}} \sum_{j=1}^{M} \sum_{k=1}^{N} \int dx^M \, d^3\boldsymbol{y}_1 \cdots d^3\boldsymbol{y}_{k-1} \, d^3\boldsymbol{y}_{k+1} \cdots d^3\boldsymbol{y}_N$$

$$\times \phi(x^M, \boldsymbol{y}_1 \dots \boldsymbol{y}_{k-1}, \boldsymbol{x}_j, \boldsymbol{y}_{k+1} \dots \boldsymbol{y}_N)^* \, \psi^{(M,N-1)}(x^M, \boldsymbol{y}_1 \dots \boldsymbol{y}_{k-1}, \boldsymbol{y}_{k+1} \dots \boldsymbol{y}_N), \qquad (6.59)$$

then reorder the arguments of ϕ so $\boldsymbol{y}_k = \boldsymbol{x}_j$ becomes the last one, rename $\boldsymbol{y}_1 \dots \boldsymbol{y}_{k-1}, \boldsymbol{y}_{k+1} \dots \boldsymbol{y}_N$ into $\boldsymbol{y}_1 \dots \boldsymbol{y}_{N-1}$, and observe that $\sum_k$ has N equal terms to obtain

$$\sum_M \sum_{N=1}^{\infty} \frac{g}{\sqrt{N}} \sum_{j=1}^{M} N \int dx^M \, d^3\boldsymbol{y}_1 \cdots d^3\boldsymbol{y}_{N-1}$$

$$\times \phi(x^M, \boldsymbol{y}_1 \dots \boldsymbol{y}_{N-1}, \boldsymbol{x}_j)^* \, \psi^{(M,N-1)}(x^M, \boldsymbol{y}_1 \dots \boldsymbol{y}_{N-1}). \qquad (6.60)$$

Rename $N \to N+1$ to obtain

$$\sum_M \sum_{N=0}^{\infty} g\sqrt{N+1} \sum_{j=1}^{M} \int dx^M \, d^3\boldsymbol{y}_1 \cdots d^3\boldsymbol{y}_N$$

$$\times \phi(x^M, \boldsymbol{y}_1 \dots \boldsymbol{y}_N, \boldsymbol{x}_j)^* \, \psi^{(M,N)}(x^M, \boldsymbol{y}_1 \dots \boldsymbol{y}_N). \qquad (6.61)$$

Doing the same steps backward with the fourth and fifth line of (6.57b) and putting together the results, we obtain that

$$\langle \phi | H \psi \rangle = \langle H \phi | \psi \rangle \,, \tag{6.62}$$

which is what we wanted to show. □

We have expressed the state vector $\psi \in \mathscr{H}$ in a particle-position representation, a representation I have adopted particularly from Schweber (1961) [22] and Nelson (1964) [19]. We will consider another representation, the field representation, in Sect. 6.5.2.

6.2.4 Creation and Annihilation Operators

The Hamiltonian (6.55) is a prototypical example of a Hamiltonian with particle creation. It provides a unitary evolution on the Hilbert space $\mathscr{F}_- \otimes \mathscr{F}_+$ and thus, together with the Born rule $\rho = |\psi|^2$, a time-dependent probability distribution on the configuration space $\Gamma(\mathbb{R}^3) \times \Gamma(\mathbb{R}^3)$. In the following sections, we will deal with how to set up versions of Bohmian mechanics, GRW theory, and many-worlds for such a Hamiltonian. We will also look into the problem that the formula (6.55) for H does not make mathematical sense as it stands, a problem known as the *ultraviolet divergence problem* of H. But before, we will define some notation and terminology that is useful for dealing with Hamiltonians similar to (6.55).

On (bosonic or fermionic) Fock space $\mathscr{F} = \mathscr{F}_\pm$, the *particle number operator* (or just *number operator*) N is defined by

$$N\psi(y^n) = n\psi(y^n) \,. \tag{6.63}$$

Equivalently, the number operator is the multiplication operator that multiplies by the particle number function on configuration space, $q \mapsto \#q$. For $\mathscr{H} = \mathscr{F}_- \otimes \mathscr{F}_+ = \mathscr{F}_x \otimes \mathscr{F}_y$ as in the emission–absorption model of Sect. 6.2.3, we have an x-number operator N_x and a y-number operator N_y,

$$N_x \psi^{(M,N)} = M \psi^{(M,N)} \,, \quad N_y \psi^{(M,N)} = N \psi^{(M,N)} \,. \tag{6.64}$$

(No confusion should arise from the fact that we use the symbol N also for the number of y-particles.)

On Fock space $\mathscr{F} = \Gamma_\pm(L^2(\mathbb{R}^3))$ and for any given $f \in L^2(\mathbb{R}^3)$, the *creation operator* $a^\dagger(f)$ is defined by

$$\left(a^\dagger(f)\psi\right)^{(n)}(y^n) = \frac{1}{\sqrt{n}} \sum_{k=1}^{n} (\pm 1)^k \, f(\boldsymbol{y}_k)\, \psi^{(n-1)}(y^n \setminus \boldsymbol{y}_k) \,. \tag{6.65}$$

It is understood that $(a^\dagger(f)\psi)^{(0)} = 0$. Notice that the right-hand side of (6.65) is (anti-)symmetric in y^n if $\psi^{(n-1)}$ is and $\pm = +\ (-)$. In fact, (6.65) can be equivalently expressed as

$$a^\dagger(f)\psi^{(n)} = \pm\sqrt{n} \text{ times the (anti-)symmetrization of } f(\boldsymbol{y}_1)\,\psi^{(n-1)}(\boldsymbol{y}_2 \ldots \boldsymbol{y}_n)\,. \tag{6.66}$$

Put differently, when applied to a wave function ψ concentrated in one sector, say $n-1$, $a^\dagger(f)$ yields a wave function concentrated in the n-sector, and up to a factor $\pm\sqrt{n}$ and (anti-)symmetrization, this sector is the wave function of one particle with wave function f and $n-1$ particles with wave function $\psi^{(n-1)}$; one says that $a^\dagger(f)$ creates a further particle with wave function f. The generalization to arbitrary $\mathscr{H}_1$ instead of $L^2(\mathbb{R}^3)$ is straightforward.

The *annihilation operator* $a(f)$ is defined by

$$\big(a(f)\psi\big)^{(n)}(y^n) = (\pm 1)^{n+1}\sqrt{n+1}\int_{\mathbb{R}^3} d^3\boldsymbol{y}_{n+1}\, f(\boldsymbol{y}_{n+1})^*\,\psi^{(n+1)}(y^n, \boldsymbol{y}_{n+1})\,. \tag{6.67}$$

When applied to a wave function ψ concentrated in one sector, say $n+1$, $a(f)$ yields a wave function concentrated in the n-sector; one says that $a(f)$ annihilates a particle.

As the notation suggests, $a^\dagger(f)$ is the adjoint of $a(f)$. Indeed, the computation is analogous to that of (6.57)–(6.62):

$$\langle a^\dagger(f)\phi|\psi\rangle = \sum_{n=1}^{\infty}\int_{\mathbb{R}^{3n}} dy^n \frac{1}{\sqrt{n}}\sum_{k=1}^{n}(\pm 1)^k\, f(\boldsymbol{y}_k)^*\,\phi^{(n-1)}(y^n\setminus \boldsymbol{y}_k)^*\,\psi^{(n)}(y^n) \tag{6.68a}$$

$$= \sum_{n=1}^{\infty}\frac{1}{\sqrt{n}}\sum_{k=1}^{n}(\pm 1)^k\int_{\mathbb{R}^{3n}} dy^n\,\phi^{(n-1)}(y^n\setminus \boldsymbol{y}_k)^*\, f(\boldsymbol{y}_k)^*$$
$$\times\,(\pm 1)^{n-k}\psi^{(n)}(y^n\setminus \boldsymbol{y}_k, \boldsymbol{y}_k) \tag{6.68b}$$

since the permutation $(\boldsymbol{y}_1\ldots\boldsymbol{y}_n) \to (\boldsymbol{y}_1\ldots\boldsymbol{y}_{k-1}, \boldsymbol{y}_{k+1}\ldots\boldsymbol{y}_n, \boldsymbol{y}_k)$ requires $n-k$ neighbor exchanges. Rename $(\boldsymbol{y}_1\ldots\boldsymbol{y}_{k-1}, \boldsymbol{y}_{k+1}\ldots\boldsymbol{y}_n, \boldsymbol{y}_k) \to (\boldsymbol{y}_1\ldots\boldsymbol{y}_n)$ and $n \to n+1$ to obtain

$$\sum_{n=0}^{\infty}\frac{1}{\sqrt{n+1}}\sum_{k=1}^{n+1}(\pm 1)^{n+1}\int_{\mathbb{R}^{3n}} dy^n\,\phi^{(n)}(y^n)^*\int_{\mathbb{R}^3} d^3\boldsymbol{y}_{n+1}\, f(\boldsymbol{y}_{n+1})^*\,\psi^{(n+1)}(y^n, \boldsymbol{y}_{n+1})\,. \tag{6.68c}$$

Since $\sum_k$ contains $n+1$ equal terms, this can be written as

$$\sum_{n=0}^{\infty}(\pm 1)^{n+1}\sqrt{n+1}\int\limits_{\mathbb{R}^{3n}} dy^n\,\phi^{(n)}(y^n)^*\int\limits_{\mathbb{R}^3} d^3\boldsymbol{y}_{n+1}\,f(\boldsymbol{y}_{n+1})^*\,\psi^{(n+1)}(y^n,\boldsymbol{y}_{n+1}) \tag{6.68d}$$

$$= \langle\phi|a(f)\psi\rangle\,, \tag{6.68e}$$

which is what we wanted to show.

Exercise 6.4 Verify that for any orthonormal basis $\{f_i\}$ of $\mathscr{H}_1$,

$$\sum_i a^\dagger(f_i)\,a(f_i) = N\,, \tag{6.69}$$

the number operator.

It is often useful to consider not only L^2 functions for f, but also generalized eigenfunctions such as e^{ikx} or Dirac delta functions. Specifically, if

$$f(\boldsymbol{y}) = \delta^3(\boldsymbol{y}-\boldsymbol{x})\,, \tag{6.70}$$

then it is common to write $a^\dagger(\boldsymbol{x})$ for $a^\dagger(f)$ and $a(\boldsymbol{x})$ for $a(f)$, so

$$a^\dagger(\boldsymbol{x})\,\psi(y^n) = \frac{1}{\sqrt{n}}\sum_{k=1}^{n}(\pm 1)^k\,\delta^3(\boldsymbol{y}_k-\boldsymbol{x})\,\psi^{(n-1)}(y^n\setminus \boldsymbol{y}_k) \tag{6.71a}$$

$$a(\boldsymbol{x})\,\psi(y^n) = (\pm 1)^{n+1}\sqrt{n+1}\,\psi^{(n+1)}(y^n,\boldsymbol{x})\,. \tag{6.71b}$$

Another remark concerns the first two lines of (6.55), called the *free Hamiltonian*. They consist of applying the same 1-particle operator $-(\hbar^2/2m_x)\nabla^2$ to every x-particle and the same 1-particle operator $-(\hbar^2/2m_y)\nabla^2 + E_0$ to every y-particle. More abstractly, if H_1 is a given operator on $\mathscr{H}_1$, then the associated free Hamiltonian on $\Gamma_\pm(\mathscr{H}_1)$ is usually denoted $d\Gamma(H_1)$,

$$(d\Gamma(H_1)\psi)^{(n)} = \left(\sum_{k=1}^{n} I^{\otimes(k-1)}\otimes H_1\otimes I^{\otimes(n-k)}\right)\psi^{(n)} \tag{6.72a}$$

$$= \sum_{k=1}^{n} H_{1,k}\psi^{(n)}\,, \tag{6.72b}$$

where $H_{1,k}$ means H_1 acting on the variable $\boldsymbol{y}_k$. The operator $d\Gamma(H_1)$ is often called the "second quantization" of H_1, a name that sounds mysterious and is perhaps misleading but common.[4]

6.2.5 Ultraviolet Divergence

The expression (6.55) for the Hamiltonian actually does not make mathematical sense, for the following reason.

Every self-adjoint operator A in a Hilbert space $\mathscr{H}$ is a linear mapping $A : \mathscr{D} \to \mathscr{H}$, where $\mathscr{D}$, the domain of A, is a dense subspace of $\mathscr{H}$. Now $H\psi$ as given by the right-hand side of (6.55) contains a Dirac delta function and therefore cannot be expected to lie in $\mathscr{H} = \mathscr{F}_- \otimes \mathscr{F}_+$, which contains square-integrable functions on $\Gamma_o(\mathbb{R}^3) \times \Gamma_o(\mathbb{R}^3)$.

Another qualitative way of understanding the root of the problem is to note that according to (6.55), the wave function of a newly created y-particle is $\delta^3(\boldsymbol{y} - \boldsymbol{x}_j)$, which is a function of infinite energy: Indeed, the Fourier transform of $\delta^3(\boldsymbol{y})$ is constant, and thus

$$\Big\langle \delta^3(\boldsymbol{y}) \Big| -\Delta \Big| \delta^3(\boldsymbol{y}) \Big\rangle = \text{const.} \int_{\mathbb{R}^3} d^3\boldsymbol{k}\, \boldsymbol{k}^2 = \infty. \tag{6.73}$$

Since the integral in (6.73) would exist if we integrated, not over $\mathbb{R}^3$, but over a ball $\{\boldsymbol{k} \in \mathbb{R}^3 : |\boldsymbol{k}| < \Lambda\}$ with finite radius $\Lambda > 0$, the divergence problem is connected to large (rather than small) values of $|\boldsymbol{k}|$ and therefore called an *ultraviolet (UV) divergence*. (Because ultraviolet light is electromagnetic radiation with larger values of $|\boldsymbol{k}|$ than visible light.)

The UV divergence problem comes up not only for the specific example provided by our emission–absorption model (6.55) but, in one form or another, for most Hamiltonians involving particle creation; it was noticed early on in the history of quantum theory (Oppenheimer 1930 [20]).

While the Hamiltonian $H = -\Delta + V$ of non-relativistic quantum mechanics as in (1.7) with Coulomb and Newton potential (1.5) has been rigorously proven to exist,[5] such proofs for Hamiltonians with particle creation are a challenge for

---

[4] Here is how this expression came about historically. Around 1930, people wanted to arrive at the equations of quantum electrodynamics by "quantizing" Maxwell's equations, the equations of classical electrodynamics. When difficulties arose with that, they tried other PDEs first as an exercise, including the 1-particle Schrödinger equation, although that does not even belong to any classical theory. The result of the procedure of "quantizing" the Schrödinger equation was the bosonic Fock space as the Hilbert space and (6.72) as the Hamiltonian—which is nothing but the free (i.e., interactionless) Hamiltonian H_0 for many non-relativistic quantum particles. Since the Schrödinger equation was regarded as the result of "quantizing" Newton's equation of motion, H_0 was the "second quantization." Big name, little substance.

[5] This is known as Kato's theorem; see, e.g., Theorem X.16 in Reed and Simon (1975) [21].

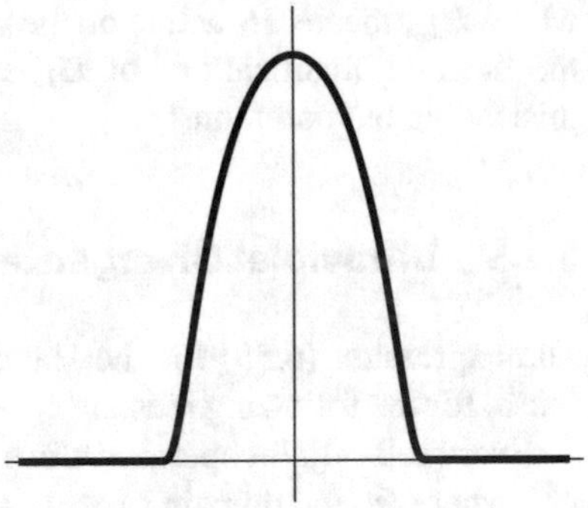

Fig. 6.1 A simple choice of a function φ approximating the Dirac delta function, here shown in 1 dimension. It is smooth (C^∞), vanishes outside the ball of radius 1, and thus is square integrable; the particular function shown is $\varphi(\boldsymbol{x}) = \exp(1/(|\boldsymbol{x}|^2 - 1))$ for $|\boldsymbol{x}| < 1$ and $\varphi(\boldsymbol{x}) = 0$ for $|\boldsymbol{x}| \geq 1$

mathematicians and still largely missing. The question is not just one of finding a proof but also one of finding a precise definition for which operator should be meant by a formula such as (6.55). We will get to know one approach, based on so-called interior-boundary conditions, in Sect. 6.4.

The UV divergence is absent if space is not continuous but discrete. In fact, one could well imagine that there exist only finitely many space points, or only finitely many in every region that we normally regard as having finite volume; space could be a lattice, perhaps not a periodic but a rather irregular one. Of course, the number of points in a cubic meter would have to be very large to create the impression of a continuum. On the other hand, there is no empirical evidence for discreteness of space.

The UV divergence would also be absent if elementary particles such as electrons were not point particles but extended, if they had a continuous charge distribution over (say) a ball of positive radius $R > 0$. Again, there is no empirical evidence for a nonzero diameter of the electron, but let us follow this thought in a little more detail. Suppose the charge distribution has density function $\varphi : \mathbb{R}^3 \to \mathbb{R}$; we may imagine that φ is a kind of approximation of the δ^3 function with a bump shape as in Fig. 6.1. (The shape of φ is, of course, a matter of fantasy since, as mentioned, there is no empirical evidence for it.)

Then φ is itself square-integrable, and replacing $\delta^3 \to \varphi$ leads to the following Hamiltonian H_φ instead of (6.55):

$$(H_\varphi \psi)^{(M,N)}(x^M, y^N)$$

$$= -\sum_{j=1}^{M} \frac{\hbar^2}{2m_x} \nabla^2_{\boldsymbol{x}_j} \psi^{(M,N)}(x^M, y^N) \tag{6.74}$$

$$- \sum_{k=1}^{N} \frac{\hbar^2}{2m_y} \nabla^2_{\boldsymbol{y}_k} \psi^{(M,N)}(x^M, y^N) + N E_0\, \psi^{(M,N)}(x^M, y^N)$$

$$+ g\sqrt{N+1}\sum_{j=1}^{M}\int_{\mathbb{R}^3} d^3y_{N+1}\, \varphi(\boldsymbol{y}_{N+1} - \boldsymbol{x}_j)\, \psi^{(M,N+1)}(x^M, (y^N, \boldsymbol{y}_{N+1}))$$

$$+ \frac{g}{\sqrt{N}}\sum_{j=1}^{M}\sum_{k=1}^{N} \varphi(\boldsymbol{x}_j - \boldsymbol{y}_k)\, \psi^{(M,N-1)}(x^M, y^N \setminus \boldsymbol{y}_k)\,. \tag{6.75}$$

Compared to the original expression (6.55), the first two lines are unchanged, the last line has δ^3 replaced by φ, and in the third line, instead of inserting $\boldsymbol{x}_j$ for $\boldsymbol{y}_{N+1}$ we have integrated $\psi^{(M,N+1)}$ against φ in the variable $\boldsymbol{y}_{N+1}$. For square-integrable φ, H_φ can be proved to be self-adjoint with standard techniques. Since $\varphi(\cdot - \boldsymbol{x}_j)$ can be thought of as the wave function of a newly created y-particle, its average energy is

$$\langle\varphi| - \Delta|\varphi\rangle = \int_{\mathbb{R}^3} d^3\boldsymbol{k}\, |\hat{\varphi}(\boldsymbol{k})|^2\, \boldsymbol{k}^2 \tag{6.76}$$

with $\hat{\varphi}$ the Fourier transform of φ; this expression is finite for φ as in Fig. 6.1. Another popular simple choice of φ is that $\hat{\varphi}(\boldsymbol{k}) = 1_{|\boldsymbol{k}|<\Lambda}$, which amounts to cutting off the integration in (6.73) at radius $|\boldsymbol{k}| = \Lambda$. For such reasons, the function φ is called a *cutoff function*, and the replacement $\delta^3 \to \varphi$ (and $H \to H_\varphi$) an *ultraviolet cutoff*. Such a cutoff is often used as a replacement for the ill-defined expression (6.55) when it is desired to study a Hamiltonian that exists mathematically.

Finally, another and more convincing way of solving the problem of UV divergence, or at least of this *particular* kind of UV divergence we encounter in the emission–absorption model (6.55), is provided by interior-boundary conditions and described in Sect. 6.4.

6.3 Particle Creation as Such

Let us consider again a Bohmian picture of particle creation. For the emission–absorption model introduced in Sect. 6.2.3, we may imagine that the y-world lines can begin and end in emission and absorption events as depicted in Fig. 6.2, while the x-world lines extend over all times.

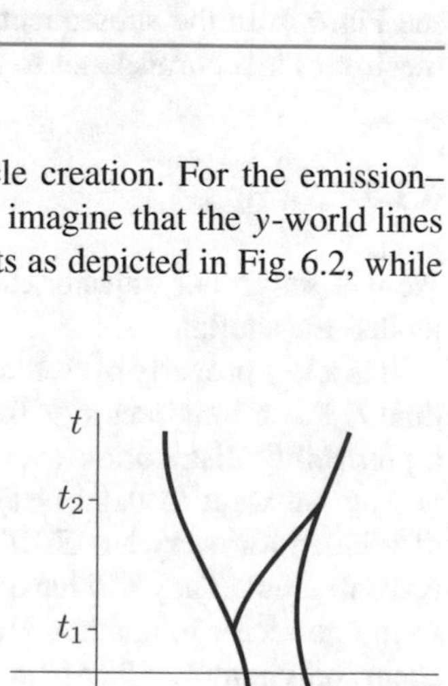

Fig. 6.2 Space-time diagram of emission of a particle by a second particle at t_1 and absorption by a third particle at t_2

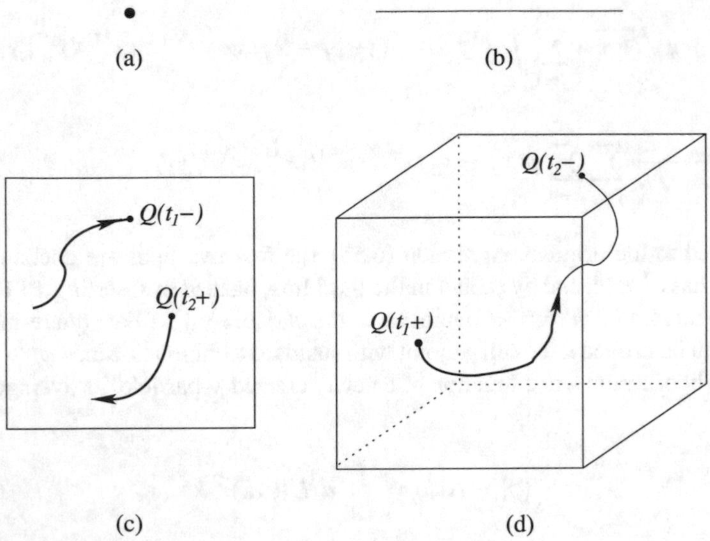

Fig. 6.3 The first four sectors of the configuration space (6.25) of a variable number of particles, for simplicity for ordered configurations in space dimension $d = 1$: (**a**) the 0-particle sector has a single element, the empty configuration; (**b**) the 1-particle sector is a copy of physical space; (**c**) the 2-particle sector; and (**d**) the 3-particle sector. In addition, the configuration curve corresponding to Fig. 6.2 is drawn (Reprinted from [7] with permission by the American Physical Society)

Let us translate this picture into a curve $t \mapsto Q(t)$ in the configuration space of a variable number of particles. The configuration Q moves in the 2-particle sector of configuration space up to the time t_1 of emission, at which Q jumps to the 3-particle sector; Q moves around there up to the time t_2 of absorption, at which it jumps back to the 2-particle sector. From then on, it moves around in the 2-particle sector; see Fig. 6.3. In the subsequent section, we will explore an extension of Bohmian mechanics that corresponds to these pictures, known as Bell's jump process.

6.3.1 Jumps

We will set up our trajectories for the emission–absorption model; those for other models are similar.

It is a key property of Bohmian mechanics that $Q(t)$ is $|\psi_t|^2$-distributed at every time t. For a unit vector ψ from the Hilbert space $\mathscr{H}$ defined in (6.50), $|\psi|^2$ is a probability distribution over the configuration space $\mathcal{Q}$ defined in (6.49). That is why we want to define trajectories $t \mapsto Q(t) \in \mathcal{Q}$ such that $Q(t)$ is $|\psi_t|^2$ distributed for every $t \in \mathbb{R}$. It is clear from Fig. 6.3 that these trajectories will not be continuous: They will have jumps because there is no continuous way of getting from one sector to another. However, only certain types of discontinuities should occur: only jumps, either from the (M, N)-sector to the $(M, N+1)$-sector or vice versa. In particular, there should be time intervals during which the configuration curve is continuous and the particle number is constant. At the end points of these intervals, the configuration can jump, but only by the creation or annihilation of a y-particle, as in

$$(x^M, y^N) \to (x^M, y^N \cup \{\boldsymbol{y}\}) \quad \text{or} \quad (x^M, y^N) \to (x^M, y^N \setminus \{\boldsymbol{y}\}) \,. \tag{6.77}$$

Now I will argue that some of these jumps should be *stochastic*, as opposed to deterministic. If we want the post-jump configuration to be continuously distributed, such as $|\psi|^2$-distributed, then a deterministic jump to a higher sector would not contain enough randomness: If the pre-jump configuration $Q_{t_1} \in \mathcal{Q}^{(M,N)}$ is continuously distributed, then it contains $3M + 3N$ degrees of randomness, and if the jump is deterministic, then the post-jump configuration $Q_{t_2} \in \mathcal{Q}^{(M,N+1)}$ still has $3M+3N$ degrees of randomness—its distribution is concentrated in a $3M+3N$-dimensional subset of the $3M + 3N + 3$ dimensions of $\mathcal{Q}^{(M,N+1)}$. This would not work well with having $|\psi|^2$ distribution at all times. The obvious way out is to make the upward jumps stochastic: The destination $\boldsymbol{y}$ and the time t of the jump together contain at least three degrees of new randomness. The configuration curve $t \mapsto Q(t)$ is then no longer determined by the initial configuration $Q(0)$; rather, it is a *stochastic process*.

If the fundamental law of motion is stochastic, this means that nature makes random decisions in the middle of the time evolution, that the evolution is intrinsically random. Some authors felt that a good physical theory should be deterministic because everything that happens in nature must have a sufficient reason. Some authors felt that determinism was essential to Bohm-type theories. "Of all people, were not the Bohmians the ones who rejected orthodox quantum mechanics for its randomness?" they might say. No, that is a misunderstanding; the Bohmians rejected orthodox quantum mechanics for its lack of coherence and clarity. It so happens that the simplest theory of trajectories for quantum mechanics of N particles, Bohmian mechanics, is deterministic, but that does not mean that determinism is indispensable. In fact, Bohm himself studied stochastic variants of Bohmian mechanics, and Bell proposed the stochastic jump process we are about to discuss.

6.3.2 Bell's Jump Process

Suppose that at any time t, nature makes a random decision about whether the configuration should jump, and if so, where it should jump. The laws of physics would then have to specify the probabilities for jumping and for the destination. The simplest possibility would be that the probabilities depend only on the present configuration $Q_t = Q_{t-}$ and not on the prior history of how the configuration curve got to that point. A stochastic process with this property is called a *Markov process*; the Wiener process (Brownian motion), the Poisson process, and the GRW process are Markov processes.

The probability of a jump within $[t, t+dt]$ should be infinitesimal, proportional to dt, or else the probability of *not jumping* would be 0 in every interval $[t, t+\varepsilon]$ with $\varepsilon > 0$. The proportionality constant, the probability per time, is called the *jump rate*. It can vary with t and Q_t. Let the rate of jumping from $q' = Q_t$ to $q \in \mathcal{Q}$ have density $\sigma_t(q', q)$ in the variable q, taken relative to the measure μ on $\mathcal{Q}$ that represents volume in $\mathcal{Q}$; that is, for $B \subseteq \mathcal{Q}$, the rate of jumping to anywhere in B, given that the present configuration is q', is

$$\int_{q\in B} \mu(dq)\, \sigma_t(q', q)\,. \tag{6.78}$$

In the following, we write $\sigma_t(q' \to q)$ instead of $\sigma_t(q', q)$ to remind us of the meaning of the variables.

We have in mind that between the jumps, the configuration should follow a continuous Bohmian trajectory within one sector. But let us first consider the math of a process that is a *pure jump process*, i.e., a Markov process $t \mapsto Q_t$ in a space $\mathcal{Q}$ that does not move between the jumps. Let $\rho_t(q)$ denote the density function (relative to μ) of the distribution of the random point Q_t. The jump rates will determine how the probability distribution of Q_t gets transported, and this can be expressed as an equation for the time evolution of ρ_t, known as a *transport equation* or *balance equation*:

$$\frac{\partial \rho}{\partial t}(t, q) = \int_{q'\in\mathcal{Q}} \mu(dq')\Big(\rho_t(q')\,\sigma_t(q' \to q) - \rho_t(q)\,\sigma_t(q \to q')\Big)\,. \tag{6.79}$$

It can be understood as follows: The first term represents the *gain* in $\rho_t(q)$ due to jumps from q' to q, the second the *loss* at q due to jumps from q to q'. I like to imagine ρ_t as a distribution of sand; the amount of probability (or in my picture, of sand) at time t in an infinitesimal volume element dq is $\mu(dq)\,\rho_t(q)$; since each grain of sand in dq' has probability $\sigma_t(q' \to q)\,\mu(dq)\,dt$ to jump to dq within time dt, the amount of sand transported from dq' to dq within dt is $\rho_t(q')\,\sigma_t(q' \to q)\,\mu(dq)\,\mu(dq')\,dt$. That yields the gain term. Exchanging $q \leftrightarrow q'$ yields the loss term. Integration over all q' yields (6.79).

In Bohmian mechanics with a fixed number of particles, probability was transported in such a way that $\rho_t(q) = |\psi_t(q)|^2$ for all q and t, and we want to choose the jump rates σ in such a way that the same relation holds in our case with a variable number of particles. If we demand that the configuration moves, between the jumps, according to an ODE

$$\frac{dQ_t}{dt} = v_t(Q_t) \tag{6.80}$$

with the *velocity field* v_t being a time-dependent vector field on $\mathcal{Q}$ yet to be determined, then the balance equation for ρ becomes

$$\frac{\partial \rho}{\partial t}(t,q) = -\mathrm{div}\big(\rho_t(q)\, v_t(q)\big) + \int\limits_{q'\in\mathcal{Q}} \mu(dq')\Big(\rho_t(q')\,\sigma_t(q'\to q) - \rho_t(q)\,\sigma_t(q\to q')\Big)\,, \tag{6.81}$$

where div = $\nabla\cdot$ means the divergence of a vector field. The analogous equation without jumps is the continuity equation (1.96) that governs the transport of probability by the solutions of an ODE.

Now $\rho_t = |\psi_t|^2$ will follow for all times if $\rho_t = |\psi_t|^2$ at one time implies $\partial_t \rho_t = \partial_t |\psi_t|^2$, i.e., if the balance equation (6.81) coincides with the equation for $\partial_t|\psi_t|^2$ that follows from the Schrödinger equation. For the emission–absorption model with UV cutoff φ as in (6.75), the latter reads

$$\frac{\partial |\psi|^2}{\partial t}(x^M, y^N) = \tfrac{2}{\hbar}\mathrm{Im}\big[\psi^*(H_\varphi \psi)\big] \tag{6.82}$$

$$\begin{aligned} &= -\sum_{j=1}^{M} \nabla_{\boldsymbol{x}_j}\cdot \Big(|\psi|^2 \tfrac{\hbar}{m_x}\mathrm{Im}\frac{\nabla_{\boldsymbol{x}_j}\psi}{\psi}\Big) - \sum_{k=1}^{N}\nabla_{\boldsymbol{y}_k}\cdot\Big(|\psi|^2\tfrac{\hbar}{m_y}\mathrm{Im}\frac{\nabla_{\boldsymbol{y}_k}\psi}{\psi}\Big) \\ &\quad + \tfrac{2g\sqrt{N+1}}{\hbar}\sum_{j=1}^{M}\int\limits_{\mathbb{R}^3} d^3\tilde{\boldsymbol{y}}\;\mathrm{Im}\Big[\psi^*(x^M,y^N)\,\varphi(\tilde{\boldsymbol{y}}-\boldsymbol{x}_j)\,\psi(x^M,(y^N,\tilde{\boldsymbol{y}}))\Big] \\ &\quad + \tfrac{2g}{\hbar\sqrt{N}}\sum_{j=1}^{M}\sum_{k=1}^{N}\mathrm{Im}\Big[\psi^*(x^M,y^N)\,\varphi(\boldsymbol{x}_j-\boldsymbol{y}_k)\,\psi(x^M,y^N\setminus \boldsymbol{y}_k)\Big]. \end{aligned} \tag{6.83}$$

Now the first line of (6.83) is exactly $-\mathrm{div}(|\psi|^2\, v_t)$ if we set

$$v_t(x^M,y^N) = \begin{pmatrix} \frac{\hbar}{m_x}\mathrm{Im}[\psi^{-1}\nabla_{\boldsymbol{x}_1}\psi] \\ \vdots \\ \frac{\hbar}{m_y}\mathrm{Im}[\psi^{-1}\nabla_{\boldsymbol{y}_1}\psi] \\ \vdots \end{pmatrix}, \tag{6.84}$$

that is, if we use the usual Bohmian equation of motion (1.77).

Now we want to match the other two lines of (6.83) with the integral in (6.81). Let us write $H_\varphi = H_{\mathrm{free}} + H_{\mathrm{inter}}$ with H_{free} given by the first two lines of (6.75) and the interaction Hamiltonian H_{inter} by the last two lines of (6.75). Then H_{free} leads to the first line of (6.83) (the divergence term) and H_{inter} to the last two lines. It seems fitting that H_{inter}, which is responsible for creation and annihilation of y-particles, should correspond to jumps in the configuration process.

To make the consideration more transparent, let us write H_{inter} as an abstract *integral operator*

$$(H_{\text{inter}}\psi)(\hat{q}) = \int\limits_{\hat{q}'\in\mathcal{Q}_o} \mu_o(d\hat{q}')\, K(\hat{q},\hat{q}')\,\psi(\hat{q}') \tag{6.85}$$

with $\mathcal{Q}_o = \Gamma_o(\mathbb{R}^3)\times\Gamma_o(\mathbb{R}^3)$ the ordered configuration space. The function $K(\hat{q},\hat{q}')$ is called the *integral kernel*. In our case, by (6.75),

$$\begin{aligned} K(&x^M, y^N, x'^{M'}, y'^{N'}) \\ &= g\sqrt{N+1}\,\delta_{MM'}\,\delta_{N+1,N'}\,\delta^{3M}\big(x^M - x'^{M'}\big) \\ &\times \sum_{j=1}^{M}\sum_{k=1}^{N'} \varphi(\boldsymbol{x}_j - \boldsymbol{y}'_k)\,\delta^{3N}\big(y^N - (y'^{N'}\setminus \boldsymbol{y}'_k)\big) \\ &+ \frac{g}{\sqrt{N}}\delta_{MM'}\,\delta_{N-1,N'}\,\delta^{3M}\big(x^M - x'^{M'}\big) \\ &\times \sum_{j=1}^{M}\sum_{k=1}^{N} \varphi(\boldsymbol{x}_j - \boldsymbol{y}_k)\,\delta^{3N'}\big((y^N\setminus \boldsymbol{y}_k) - y'^{N'}\big) \end{aligned} \tag{6.86}$$

The self-adjointness of H_{inter} has its roots in the property

$$K(\hat{q}',\hat{q}) = K(\hat{q},\hat{q}')^* \tag{6.87}$$

of the kernel (recall that $\varphi^* = \varphi$). The contribution of H_{inter} to Eq. (6.83) for $\partial_t|\psi_t|^2$ is of the form

$$\tfrac{2}{\hbar}\text{Im}\big[\psi^*(\hat{q})(H_{\text{inter}}\psi)(\hat{q})\big] = \tfrac{2}{\hbar}\text{Im}\Bigg[\psi^*(\hat{q}) \int\limits_{\hat{q}'\in\mathcal{Q}_o} \mu_o(d\hat{q}')\,K(\hat{q},\hat{q}')\,\psi(\hat{q}')\Bigg] \tag{6.88}$$

$$= \int\limits_{\hat{q}'\in\mathcal{Q}_o} \mu_o(d\hat{q}')\,\tfrac{2}{\hbar}\text{Im}\Big[\psi^*(\hat{q})K(\hat{q},\hat{q}')\,\psi(\hat{q}')\Big]. \tag{6.89}$$

Comparison to the integral in (6.81) suggests that

$$\rho_t(q')\,\sigma_t(q'\to q) - \rho_t(q)\,\sigma_t(q\to q') = \tfrac{2}{\hbar}\text{Im}\Big[\psi^*(\hat{q})K(\hat{q},\hat{q}')\,\psi(\hat{q}')\Big], \tag{6.90}$$

where q is the unordered version of $\hat{q}$ and q' that of $\hat{q}'$. The quantity on the left-hand side of (6.90) possesses the following interpretation: It is the amount of probability (per volume) that gets transported (per time) from q' to q minus that from q to q'; it is called the *net current between* q' *and* q, abbreviated $J(q', q)$. By its definition, it is anti-symmetric, $J(q, q') = -J(q', q)$, just like the right-hand side of (6.90) by virtue of (6.87).

Exercise 6.5 (Minimal Jump Rates) Consider six real numbers $J_{12}, J_{21}, \sigma_{12}, \sigma_{21}, \rho_1, \rho_2$. Suppose $J_{21} = -J_{12}$ and $\rho_1 > 0, \rho_2 > 0$ are given. We look for values $\sigma_{12} \geq 0$ and $\sigma_{21} \geq 0$ such that

$$\rho_1 \sigma_{12} - \rho_2 \sigma_{21} = J_{12} \,. \tag{6.91}$$

Show that the solutions are exactly the pairs σ_{12}, σ_{21} of the form

$$\sigma_{ij} = \sigma_{\min,ij} + \frac{a}{\rho_i} \tag{6.92}$$

with arbitrary $a \geq 0$ and

$$\sigma_{\min,ij} = \frac{J_{ij}^+}{\rho_i} \,, \tag{6.93}$$

where

$$x^+ = \max\{x, 0\} = x \, 1_{x>0} \tag{6.94}$$

denotes the positive part of $x \in \mathbb{R}$ (see Fig. 6.4). Thus, $\sigma_{\min,ij}$ are the smallest possible values. (The relation $x^+ - (-x)^+ = x$ can be useful.)

As a consequence, the minimal jump rates for given $J(q', q)$ are

$$\sigma_{\min}(q' \to q) = \frac{J(q', q)^+}{\rho(q')} \,. \tag{6.95}$$

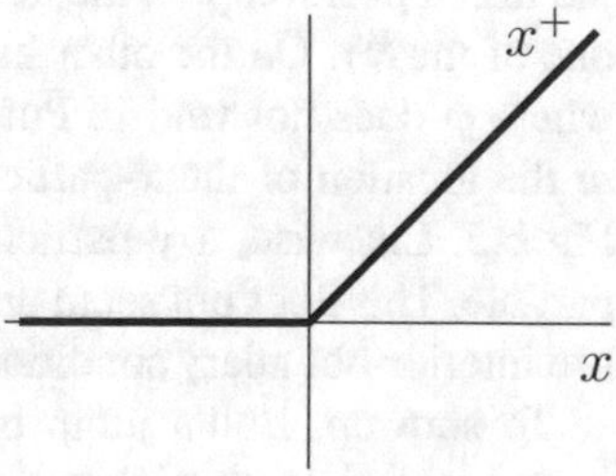

Fig. 6.4 Graph of the x^+ function defined in (6.94)

They have the property that of the two jumps $q' \to q$ and $q \to q'$, only one is allowed at any time.

Taking as $J(q', q)$ the right-hand side of (6.90), we obtain that it is possible to choose the jump rates $\sigma(q' \to q)$ so that $Q_t \sim |\psi_t|^2$ for all t, and there is a distinguished choice, the minimal jump rate:

$$\sigma(q' \to q) = \tfrac{2}{\hbar} \frac{\mathrm{Im}^+\big[\psi^*(\hat{q})\, K(\hat{q}, \hat{q}')\, \psi(\hat{q}')\big]}{|\psi(\hat{q}')|^2}, \tag{6.96}$$

where $\hat{q}$ is any ordering of q and $\hat{q}'$ any of q'. The process with these rates is *Bell's jump process*. Bell (1986) [1] proposed it for a discretized version of the theory on a lattice (so that also the motion of particles consists of jumps to neighboring lattice sites). The process in the continuum was studied by Dürr et al. (2003 [6], 2004 [7], 2005 [8,9]). For the concrete kernel of H_{inter}, the jump rates are explicitly

$$\sigma\Big((x^M, y^N) \to (x^M, y^N \cup \boldsymbol{y})\Big)$$
$$= \frac{2g\sqrt{N+1}}{\hbar} \frac{\mathrm{Im}^+\big[\psi^*(x^M, y^N, \boldsymbol{y}) \sum_j \varphi(\boldsymbol{x}_j - \boldsymbol{y})\, \psi(x^M, y^N)\big]}{|\psi(x^M, y^N)|^2} \tag{6.97a}$$
$$\sigma\Big((x^M, y^N) \to (x^M, y^N \setminus \boldsymbol{y}_k)\Big)$$
$$= \frac{2g}{\hbar\sqrt{N}} \frac{\mathrm{Im}^+\big[\psi^*(x^M, y^N \setminus \boldsymbol{y}_k) \sum_j \varphi(\boldsymbol{x}_j - \boldsymbol{y}_k)\, \psi(x^M, y^N)\big]}{|\psi(x^M, y^N)|^2}. \tag{6.97b}$$

Why should we choose the minimal jump rates? It is a natural, distinguished choice; it leads to no more stochasticity than necessary (in fact it may seem odd to allow both jumps $q' \to q$ and $q \to q'$ at the same time); the minimal rates can be expressed as a formula in ψ; if we apply the same jump rate formula to a particle moving on a lattice by jumps, then we obtain Bohm's equation of motion in the continuum limit (Vink 1993 [29]); and they converge to the jump rate associated with an interior-boundary condition (see Sect. 6.4) in a suitable limit of removing the ultraviolet cutoff in which φ approaches a delta function.

Here is another observation about Bell's jump process. Suppose φ has compact support of radius R, i.e., that $\varphi(\boldsymbol{x}) = 0$ for $|\boldsymbol{x}| \geq R$. It follows from (6.97a) that the initial position $\boldsymbol{y}$ of the newly created y-particle has to lie in the R-ball around one of the $\boldsymbol{x}_j$. On the other hand, within this ball, the position could be any place where φ does not vanish. Put differently, the new y-particle does not get created *at* the location of the x-particle, but merely *near* it, at a distance of order R; see Fig. 6.5. Likewise, a y-particle can get absorbed at distance up to R from the x-particle. This does not seem physically plausible, and this will be different when we use interior-boundary conditions in Sect. 6.4.

To sum up, Bell's jump process is an extension of Bohmian mechanics that includes particle creation and annihilation. It is a Markov process that is piecewise

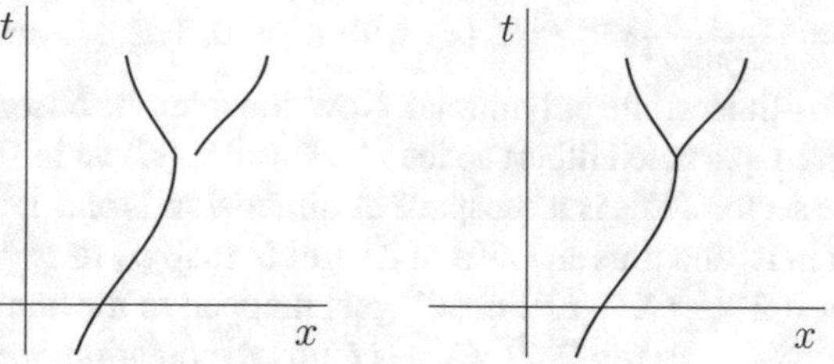

Fig. 6.5 Space-time diagram of Bell's jump process; a y-particle gets emitted by an x-particle. Left: With UV cutoff φ, the y-particle gets created at small but nonzero distance from the x-particle. Right: With interior-boundary condition, the y-particle gets created at the location of the x-particle

deterministic and piecewise continuous. The continuous pieces are solutions of Bohm's equation of motion. The stochastic jumps represent particle creation and annihilation; during these jumps, the other particles do not move. The configuration Q_t is $|\psi_t|^2$-distributed in $\mathcal{Q}$ at every time t. We have described it for the emission–absorption model, but it can easily be adapted to other models (Dürr et al. 2005 [8]).

6.3.3 Virtual Particles

The expression "virtual particles" is used in the literature in several meanings; one, which I will briefly discuss in this section, can be characterized as the use of a Hilbert space of particles without an ontology of particles; another one, which I will not discuss, concerns contributions to Feynman diagrams (i.e., operators that appear as terms in the Dyson series expansion of the time evolution) that could not appear for free particles.

Bell's jump process has provided us with a picture of particle creation and annihilation on the level of the ontology that goes along with a Fock space and creation and annihilation operators acting on it. Now suppose we consider a physical system together with a theory about the ontology (e.g., a version of Bohmian mechanics) such that the ontology does not involve particle creation or annihilation; but, suppose further, we consider a unitary isomorphism U between the system's Hilbert space $\mathscr{H}$ and a Fock space $\mathscr{F}$, for example, because the system's Hamiltonian H, when translated to $\mathscr{F}$, UHU^{-1}, assumes a simple form, say, one that can be nicely expressed in terms of creation and annihilation operators. Then for the analysis of H we might naturally talk about operators in Fock space and may even consider Bell's jump process for UHU^{-1}, while we know that the true ontology involves no particle creation; so the particles in Bell's jump process are not the actual particles, and that is why they are called *virtual particles*.

For example, consider a single spinless particle (say, in Bohmian mechanics) in a harmonic oscillator potential in 1d, $\mathscr{H} = L^2(\mathbb{R}, \mathbb{C})$ and $H = -\frac{1}{2}\partial_x^2 + \frac{1}{2}x^2$ (setting $\hbar$, m and the oscillator constant ω_0 to 1). It turns out that the eigenfunctions of H are $\phi_n = \frac{1}{\sqrt{2^n n! \pi^{1/4}}} e^{-x^2/2} H_n(x)$ with $n = 0, 1, 2, \ldots$ and $H_n = e^{x^2/2}(x - \partial_x)^n e^{-x^2/2}$ the n-th Hermite polynomial. Now consider the bosonic Fock space over a 1-dimensional 1-particle Hilbert space $\mathbb{C}$, $\mathscr{F} = \Gamma_+(\mathbb{C})$ as in (6.45); for each $n \geq 0$, its n-particle sector $\mathscr{F}^{(n)}$ is a subspace of dimension 1, and $1^{\otimes n} = 1 \otimes 1 \otimes \cdots \otimes 1$ is a unit vector in it (and thus an ONB of it). Let U map ϕ_n to $1^{\otimes n}$. Then the operator $2^{-1/2}(x + \partial_x) = 2^{-1/2}(X + iP)$ on $\mathscr{H}$ gets mapped to the annihilation operator a on $\mathscr{F}$, and $2^{-1/2}(x - \partial_x) = 2^{-1/2}(X - iP)$ to the creation operator $a^\dagger$, and H to $a^\dagger a + \frac{1}{2}$, which according to (6.69) is the number operator on $\mathscr{F}$ plus one half. The particles corresponding to the Fock space, the virtual particles, are called excitons or quanta; they do not move, but there can be different numbers of them.

Another example of virtual particles is provided by *phonons*, the representation in Fock space of vibrations of the lattice formed by the atoms of a solid body.

6.3.4 GRW Theory and Many-Worlds in Fock Space

There are natural extensions of the various GRW and many-worlds theories to Fock space, although the meaning of Fock space in terms of configurations gets lost in these theories as configurations play no role in them.

GRW Here is the extension of GRWf [27]. Consider for simplicity a single species, so the Hilbert space is a Fock space, $\mathscr{H} = \mathscr{F}_\pm$, and suppose we are given a Hamiltonian H on $\mathscr{H}$. For $\boldsymbol{x} \in \mathbb{R}^3$, let

$$N(\boldsymbol{x}) = a^\dagger(\boldsymbol{x})\, a(\boldsymbol{x}) \tag{6.98}$$

be the particle number density operator at $\boldsymbol{x}$. The total number operator is

$$N = \int_{\mathbb{R}^3} d^3\boldsymbol{x}\, N(\boldsymbol{x}) \,. \tag{6.99}$$

Define the flash rate density operator as λ times the smeared-out number density operator,

$$\Lambda(\boldsymbol{x}) = \lambda\, (N * g_\sigma)(\boldsymbol{x}) = \frac{\lambda}{(2\pi\sigma^2)^{3/2}} \int_{\mathbb{R}^3} d^3\boldsymbol{y}\, e^{-(\boldsymbol{x}-\boldsymbol{y})^2/2\sigma^2}\, N(\boldsymbol{y}) \tag{6.100}$$

with λ and σ the usual GRW constants. Connecting to Example 5.4, where the joint distribution of the flashes for GRWf in $\mathbb{R}^{3N}$ was expressed in terms of a POVM, I will define the theory on Fock space by defining the POVM G that governs the joint distribution of the flashes. The full POVM is the unique extension $n \to \infty$ of the POVM G_n governing the *first n flashes*. It is given by

$$G_n\Big(d^3\boldsymbol{x}_1 \times dt_1 \times \cdots \times d^3\boldsymbol{x}_n \times dt_n\Big) = K_n^\dagger\, K_n\, d^3\boldsymbol{x}_1\, dt_1 \cdots d^3\boldsymbol{x}_n\, dt_n \tag{6.101}$$

with

$$K_n(\boldsymbol{x}_1, t_1, \ldots, \boldsymbol{x}_n, t_n)$$
$$= 1_{0<t_1<\ldots<t_n} \times \Lambda(\boldsymbol{x}_n)^{1/2}\, W_{t_n - t_{n-1}} \Lambda(\boldsymbol{x}_{n-1})^{1/2}\, W_{t_{n-1}-t_{n-2}} \cdots \Lambda(\boldsymbol{x}_1)^{1/2}\, W_{t_1} \tag{6.102}$$

and

$$W_t = \exp\Big(- iHt/\hbar - \lambda N t/2\Big) \tag{6.103}$$

with N the total number operator. The W_t operators play a role of time evolution operators, but they are not unitary because if we write the exponent in the form $-iAt/\hbar$, then

$$A = H - i\hbar\lambda N/2 \tag{6.104}$$

is not self-adjoint (the first term is, the second is not because of the factor i). In the special case in which N commutes with H (i.e., if H involves no particle creation or annihilation), W_t can be factorized into $W_t = \exp(-iHt/\hbar)\, \exp(-\lambda N t/2)$, but this will not be true in general.

Proposition 6.2 *Suppose that*[6]

$$W_t^\dagger W_t \to 0 \quad \text{as } t \to \infty\,. \tag{6.105}$$

Then G_n is a POVM.

Proof G_n is positive because its density is of the form $K_n^\dagger K_n$. In order to show that it is normalized, it suffices to show that

$$R := \int_{\mathbb{R}^3} d^3\boldsymbol{x} \int_0^\infty dt\, W_t^\dagger\, \Lambda(\boldsymbol{x})\, W_t = I\,. \tag{6.106}$$

Indeed,

$$\int_{\mathbb{R}^3} d^3\boldsymbol{x}\, \Lambda(\boldsymbol{x}) = \frac{\lambda}{(2\pi\sigma^2)^{3/2}} \int_{\mathbb{R}^3} d^3\boldsymbol{x} \int_{\mathbb{R}^3} d^3\boldsymbol{y}\, e^{-(\boldsymbol{x}-\boldsymbol{y})^2/2\sigma^2}\, N(\boldsymbol{y}) \tag{6.107a}$$

$$= \frac{\lambda}{(2\pi\sigma^2)^{3/2}} \int_{\mathbb{R}^3} d^3\boldsymbol{y}\, N(\boldsymbol{y}) \int_{\mathbb{R}^3} d^3\boldsymbol{x}\, e^{-(\boldsymbol{x}-\boldsymbol{y})^2/2\sigma^2} \tag{6.107b}$$

---

[6] I think this is the case as soon as the vectors in the 0-particle sector $\mathscr{F}_\pm^{(0)}$ of Fock space are not eigenvectors of H, but I could not find a proof.

$$= \frac{\lambda}{(2\pi\sigma^2)^{3/2}} \int_{\mathbb{R}^3} d^3\boldsymbol{y}\, N(\boldsymbol{y}) \int_{\mathbb{R}^3} d^3\boldsymbol{x}\, e^{-\boldsymbol{x}^2/2\sigma^2} \tag{6.107c}$$

$$= \lambda \int_{\mathbb{R}^3} d^3\boldsymbol{y}\, N(\boldsymbol{y}) = \lambda N\,. \tag{6.107d}$$

Thus,

$$R = \int_0^\infty dt\, W_t^\dagger\, \lambda N\, W_t \tag{6.108a}$$

$$= -\int_0^\infty dt\, W_t^\dagger \Big(iH/\hbar - \lambda N/2 - iH/\hbar - \lambda N/2 \Big) W_t \tag{6.108b}$$

$$= -\int_0^\infty dt \left(\frac{dW_t^\dagger}{dt} W_t + W_t^\dagger \frac{dW_t}{dt} \right) \tag{6.108c}$$

$$= -\int_0^\infty dt\, \frac{d}{dt} \Big(W_t^\dagger W_t \Big) \tag{6.108d}$$

$$= -\lim_{t\to\infty} W_t^\dagger W_t + W_0^\dagger W_0 = I\,. \tag{6.108e}$$

□

This completes the simplest definition of a GRWf theory with particle creation.

Many-Worlds For many-worlds theories, the generalization to Fock space is obvious: for Sm, replace (3.74) by

$$m(\boldsymbol{x}, t) = \langle \psi_t | M(\boldsymbol{x}) | \psi_t \rangle \tag{6.109}$$

with mass density operators

$$M(\boldsymbol{x}) = m\, N(\boldsymbol{x}) \tag{6.110}$$

or, for several particle species,

$$M(\boldsymbol{x}) = \sum_i m_i\, N_i(\boldsymbol{x})\,. \tag{6.111}$$

For Bell's many-worlds theories, we only need to use the appropriate $|\psi|^2$ Born distribution on the configuration space $\mathcal{Q}$ of a variable number of particles.

Fock Space, Configurations, and Ontology The situation of Fock space shines a new light on why a primitive ontology is needed, as well as (somewhat paradoxically) on why one might think otherwise. While wave functions in quantum mechanics of a fixed number N of particles are often thought of as functions on the configuration space $\mathbb{R}^{3N}$, it is less common to think of vectors in Fock space as functions on configuration space $\mathcal{Q}$; it is more common to think of them as vectors in some abstract Hilbert space on which certain operators such as H and $a(f)$ act. And if we do not talk about configurations any more, we may get the impression that configurations (or anything like it, any primitive ontology) are not applicable or irrelevant. But if we stop considering configuration space, and if we do not have a primitive ontology, then it becomes more obscure than before how the theory is connected to physical reality.

Let me consider first an extreme example: Suppose a theory involves only an abstract Hilbert space $\mathscr{H}$ and a self-adjoint Hamiltonian H in $\mathscr{H}$, and nothing else. Then the physical content is obscure, as all relevant Hilbert spaces (such as $L^2(\mathbb{R}^d, \mathbb{C}^k)$, Fock spaces thereof, and tensor products thereof) have countably infinite dimension, any two abstract Hilbert spaces of countably infinite dimension are unitarily isomorphic, and any pairs $(\mathscr{H}, H)$ and $(\mathscr{H}', H')$ are unitarily equivalent as soon as H and H' have the same generalized eigenvalues (with multiplicities). But a Hamiltonian H with particle creation might well havc the same generalized eigenvalues (with multiplicities) as some 1d Schrödinger Hamiltonian $-\partial_x^2 + V(x)$ for suitable potential V. Or the 3d hydrogen Hamiltonian $\boldsymbol{p}^2 + 1/|\boldsymbol{x}|$ has the same generalized eigenvalues (with multiplicities) as $\boldsymbol{x}^2 + 1/|\boldsymbol{p}|$. So it seems clear that $(\mathscr{H}, H)$ does not provide the full information about a physical system.

Now in Fock space, one uses further operators, the $a^\dagger(\boldsymbol{x})$ and $a(\boldsymbol{x})$. Actually, specifying $\mathscr{H}$, H, and the $a(\boldsymbol{x})$ provides all of the mathematical information needed for computing, from any $\psi \in \mathbb{S}(\mathscr{H})$, all we could ask for: Born's distribution for any observable, Bohm's velocities, Bell's jump rates, GRW's flash distribution, and Schrödinger's m function. But that, of course, assumes that we are given laws for the primitive ontology, which then provide a clear possibility of physical reality. If we do not say what the primitive ontology is, then we do not have a complete picture of physical reality. Let me illustrate this point again through a creation myth: God has chosen an abstract Hilbert space $\mathscr{H}$ of countably infinite dimension, a self-adjoint operator H in $\mathscr{H}$, a unit vector ψ_0 in $\mathscr{H}$, and operators $a(\boldsymbol{x})$ satisfying the canonical (anti-)commutation relations (6.189) or (6.190). He makes his decisions known to the demiurgs and tells them to build the universe according to his plan. But the demiurgs ask, "what do you want us to make?"

6.4 Interior-Boundary Conditions

Interior-boundary conditions (IBCs) are a kind of boundary condition that can be used for defining Hamiltonians without ultraviolet (UV) divergence problem. This approach can be applied to the emission–absorption model of Sect. 6.2.3, among others, to obtain a well-defined Hamiltonian without UV cutoff. In this section, I explain how the trick works.

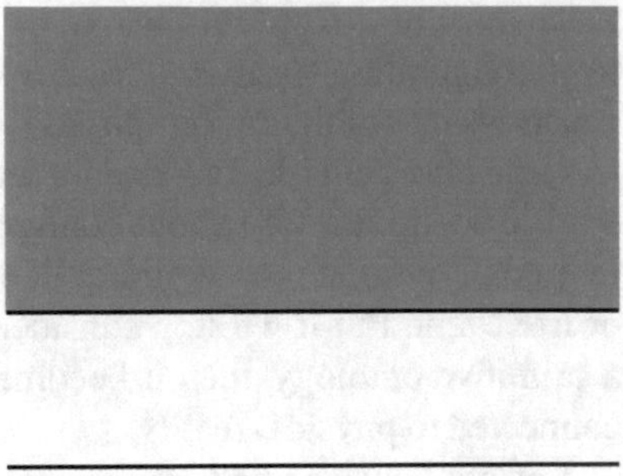

Fig. 6.6 A configuration space with boundary, consisting of a half plane and a line disjoint from it

As mentioned already, the UV problem would be absent if space were discrete (such as $\mathbb{Z}^3$) or if electrons had a positive radius, but there is no empirical evidence for either. (Protons are known to have a radius of 8.4×10^{-16} m; however, they are not elementary but consist of several particles, viz., three quarks. Quarks are believed to be point particles, and the proton radius represents half the average distance between the quarks inside a proton.)

The IBC approach allows for the definition of a Hamiltonian in continuous space $\mathbb{R}^3$ for particles of radius 0. IBCs were first proposed by Marcos Moshinsky (1921–2009) in 1951 [18] as an effective description of nuclear reactions,[7] then independently by Georgii and Tumulka (2004) [12] for fundamental creation and annihilation processes, and more recently by Teufel and Tumulka (2015 [26], 2016 [25]) as a way out of the UV divergence problem described in Sect. 6.2.5. Rigorous existence and self-adjointness of the IBC Hamiltonian for the emission–absorption model was proven by Lampart (2019) [14].

Before we apply an IBC to particle creation in Sect. 6.4.2, we consider it first for a simple example of a configuration space with a boundary.

6.4.1 What an IBC Is

For getting acquainted with IBCs, we consider a simple example, starting from an artificial choice of a configuration space $\mathcal{Q} = \mathcal{Q}^{(1)} \cup \mathcal{Q}^{(2)}$ that is the disjoint union of two connected components ("sectors") $\mathcal{Q}^{(1)}$ and $\mathcal{Q}^{(2)}$, where $\mathcal{Q}^{(1)} = \mathbb{R}$ is a line and $\mathcal{Q}^{(2)} = \{(x, y) \in \mathbb{R}^2 : y \geq 0\} =: \mathbb{R}^2_{\geq}$ is a half plane, as depicted in Fig. 6.6.

A concept of volume is given by the measure μ on $\mathcal{Q}$ defined by

$$\mu(S) = \mathrm{vol}_1(S \cap \mathcal{Q}^{(1)}) + \mathrm{vol}_2(S \cap \mathcal{Q}^{(2)}) \tag{6.112}$$

[7] In a prior publication [15], I claimed that Landau and Peierls (1930) [16] had the first IBC. I take back that claim; while their Hamiltonian can be expressed in terms of an IBC, Landau and Peierls did not express it this way.

for any $S \subseteq \mathcal{Q}$, where vol_d means volume in $\mathbb{R}^d$. We will consider a quantum theory with configuration space $\mathcal{Q}$. The wave function ψ is a function $\mathcal{Q} \to \mathbb{C}$, and the Hilbert space is

$$\mathscr{H} = L^2(\mathcal{Q}, \mathbb{C}, \mu) = L^2(\mathbb{R}, \mathbb{C}, \mathrm{vol}_1) \oplus L^2\left(\mathbb{R}^2_{\geq}, \mathbb{C}, \mathrm{vol}_2\right) = \mathscr{H}^{(1)} \oplus \mathscr{H}^{(2)} \tag{6.113}$$

with inner product

$$\langle\psi|\phi\rangle = \int_{\mathbb{R}} dx\, \psi^{(1)}(x)^*\, \phi^{(1)}(x) + \int_{\mathbb{R}} dx \int_0^\infty dy\, \psi^{(2)}(x, y)^*\, \phi^{(2)}(x, y)\,. \tag{6.114}$$

The configuration space $\mathcal{Q}$ has a *boundary*

$$\partial\mathcal{Q} = \left\{(x, y) \in \mathbb{R}^2 : y = 0\right\}. \tag{6.115}$$

That is, $\mathcal{Q}^{(2)}$ has a boundary, while the boundary of $\mathcal{Q}^{(1)}$ is empty. Any non-boundary point is called an *interior point*.

Before we define the Hamiltonian and write an IBC, let us talk a bit about the motion of the configuration Q_t, although IBCs can also be considered without trajectories. Suppose we are given time-dependent vector fields $v^{(1)}$ on $\mathcal{Q}^{(1)}$ and $v^{(2)}$ on $\mathcal{Q}^{(2)}$. Q_t can be in $\mathcal{Q}^{(1)}$ or $\mathcal{Q}^{(2)}$; when in $\mathcal{Q}^{(n)}$ $(n = 1, 2)$, it moves according to

$$\frac{dQ}{dt} = v^{(n)}(t, Q_t)\,. \tag{6.116}$$

When moving in $\mathcal{Q}^{(2)}$, the configuration could hit the boundary $\partial\mathcal{Q}$ at a point $(x, 0)$. When that happens, we postulate that it jumps to the point in $\mathcal{Q}^{(1)}$ with coordinate x. After the jump, the configuration will move in $\mathcal{Q}^{(1)}$ according to (6.116). Since we want the theory to be time reversal invariant, also the reverse history should be possible: that the configuration moves in $\mathcal{Q}^{(1)}$ for a while, then jumps from $x \in \mathcal{Q}^{(1)}$ to $(x, 0) \in \partial\mathcal{Q}$, and then moves in $\mathcal{Q}^{(2)}$ according to (6.116). However, this is only possible if $v^{(2)}(t, x, 0)$ has y-component ≥ 0; if the y-component were < 0, then there is a solution of (6.116) *ending* at $(x, 0)$ at time t but none *beginning* at $(x, 0)$ at time t. We denote the components by $v^{(2)} = (v_x^{(2)}, v_y^{(2)})$.

Suppose the jumps from x to $(x, 0)$ occur stochastically with jump rate $\sigma_t\big(x \to (x, 0)\big)$, so that $(Q_t)_{t\in\mathbb{R}}$ is a Markov process. By the previous remark, we need to assume that

$$\sigma_t\big(x \to (x, 0)\big) = 0 \text{ whenever } v_y^{(2)}(t, x, 0) < 0. \tag{6.117}$$

It then follows that the probability density of Q_t relative to μ, given by the functions $\rho^{(1)}(t,x)$ and $\rho^{(2)}(t,x,y)$, both nonnegative with normalization

$$\int_{\mathbb{R}} dx\, \rho^{(1)}(t,x) + \int_{\mathbb{R}} dx \int_0^\infty dy\, \rho^{(2)}(t,x,y) = 1 \tag{6.118}$$

for every t, evolves according to

$$\frac{\partial \rho^{(1)}(t,x)}{\partial t} = -\partial_x\Big(\rho^{(1)}(t,x)\, v^{(1)}(t,x)\Big) - 1_{v_y^{(2)}(t,x,0)<0}\, \rho^{(2)}(t,x,0)\, v_y^{(2)}(t,x,0) - \rho^{(1)}(t,x)\, \sigma_t\big(x \to (x,0)\big) \tag{6.119a}$$

$$\frac{\partial \rho^{(2)}(x,y)}{\partial t} = -\sum_{i=x,y} \partial_i\Big(\rho^{(2)}(t,x,y)\, v_i^{(2)}(t,x,y)\Big) \quad \text{for } y > 0. \tag{6.119b}$$

Here, the first term in (6.119) represents transport within $\mathcal{Q}^{(1)}$, the second the gain due to jumps coming from $\partial\mathcal{Q}$ (it is the current across the boundary, i.e., the component of the current orthogonal to the boundary), and the third the loss due to the stochastic jumps from $\mathcal{Q}^{(1)}$ to $\partial\mathcal{Q}$. Note that although (6.119b) looks like the usual probability transport equation in $\mathbb{R}^2$, the existence of the boundary changes the situation; in particular, the total amount of probability in $\mathcal{Q}^{(2)}$, $\int dx \int dy\, \rho^{(2)}(x,y)$, need not be conserved; it may shrink due to current into the boundary or grow due to current out of the boundary.

Of course, we have in mind that, once we will have set up a Schrödinger equation for ψ consisting of the sectors $\psi^{(1)}(x)$ and $\psi^{(2)}(x,y)$, the probability density should be

$$\rho^{(1)}(t,x) = \big|\psi^{(1)}(t,x)\big|^2 \tag{6.120a}$$

$$\rho^{(2)}(t,x,y) = \big|\psi^{(2)}(t,x,y)\big|^2 \tag{6.120b}$$

and the velocity

$$v^{(1)}(t,x) = \frac{j^{(1)}(t,x)}{|\psi^{(1)}(t,x)|^2} \tag{6.121a}$$

$$v^{(2)}(t,x,y) = \frac{j^{(2)}(t,x,y)}{|\psi^{(2)}(t,x,y)|^2} \tag{6.121b}$$

with the current

$$j^{(1)}(t,x) = \tfrac{\hbar}{m}\, \mathrm{Im}\big[\psi^{(1)}(t,x)^* \partial_x \psi^{(1)}(t,x)\big] \tag{6.122a}$$

$$j^{(2)}(t,x,y) = \tfrac{\hbar}{m}\, \mathrm{Im}\big[\psi^{(2)}(t,x,y)^* \nabla \psi^{(2)}(t,x,y)\big]. \tag{6.122b}$$

When there is current out of the boundary, i.e., when

$$j_y^{(2)}(t,x,0) > 0\,, \tag{6.123}$$

then the only way to maintain (6.120b) is to ensure that (6.123) is the rate at which probability arrives at $(x,0) \in \partial\mathcal{Q}$. That is, taking (6.117) into account, the jump rate has to be

$$\sigma_t\big(x \to (x,0)\big) = \frac{j_y^{(2)}(t,x,0)^+}{|\psi^{(1)}(t,x)|^2}\,. \tag{6.124}$$

This formula is very similar to Bell's jump rate formula (6.96). With this rate it then follows that the second and third terms in (6.119a) add up to

$$-1_{v_y^{(2)}(t,x,0)<0}\,\rho^{(2)}(t,x,0)\,v_y^{(2)}(t,x,0) - \rho^{(1)}(t,x)\,\sigma_t\big(x \to (x,0)\big) = -j_y^{(2)}(t,x,0)\,. \tag{6.125}$$

In other words, the gain minus loss terms for $\rho^{(1)}(x)$ are together equal to the current into the boundary at $(x,0) \in \mathcal{Q}^{(2)}$. This makes intuitive sense.

So, we want to set up the Schrödinger equation in such a way that it implies the following desired balance equation for $|\psi|^2$:

$$\frac{\partial|\psi^{(1)}|^2}{\partial t}(t,x) = -\partial_x j^{(1)}(t,x) - j_y^{(2)}(t,x,0) \tag{6.126a}$$

$$\frac{\partial|\psi^{(2)}|^2}{\partial t}(t,x,y) = -\operatorname{div} j^{(2)}(t,x,y) \quad \text{for } y > 0\,. \tag{6.126b}$$

We expect the Schrödinger equation to have the form

$$i\hbar\frac{\partial\psi^{(1)}}{\partial t}(t,x) = -\frac{\hbar^2}{2m}\partial_x^2\psi^{(1)}(t,x) + \text{further term} \tag{6.127a}$$

$$i\hbar\frac{\partial\psi^{(2)}}{\partial t}(t,x,y) = -\frac{\hbar^2}{2m}\left(\partial_x^2 + \partial_y^2\right)\psi^{(2)}(t,x,y) \quad \text{for } y > 0\,, \tag{6.127b}$$

which leads to

$$\frac{\partial|\psi^{(1)}|^2}{\partial t}(t,x) = -\partial_x j^{(1)}(t,x) + \tfrac{2}{\hbar}\operatorname{Im}\big[\psi^{(1)}(t,x)^* \times \text{the further term}\big] \tag{6.128a}$$

$$\frac{\partial|\psi^{(2)}|^2}{\partial t}(t,x,y) = -\operatorname{div} j^{(2)}(t,x,y) \quad \text{for } y > 0\,. \tag{6.128b}$$

So, we get both desired divergence terms, and agreement between (6.126) and (6.128) only requires that

$$\tfrac{2}{\hbar}\,\mathrm{Im}\big[\psi^{(1)}(t,x)^* \times \text{the further term}\big] = -j_y^{(2)}(t,x,0) \tag{6.129}$$

$$= -\tfrac{\hbar}{m}\,\mathrm{Im}\big[\psi^{(2)}(t,x,0)^*\,\partial_y\psi^{(2)}(t,x,0)\big]\,. \tag{6.130}$$

We can well imagine that the further term involves a constant prefactor times $\partial_y\psi^{(2)}(t,x,0)$, but then there is still a mismatch between $\psi^{(1)}$ on the left and $\psi^{(2)}$ on the right. Now comes a crucial step.

We impose on ψ the *interior-boundary condition* (IBC)

$$\psi^{(2)}(x,0) = \alpha\,\psi^{(1)}(x) \tag{6.131}$$

with some real constant α. The condition relates the value of ψ at any boundary point $(x,0) \in \mathcal{Q}^{(2)}$ to the value of ψ at an interior point $x \in \mathcal{Q}^{(1)}$ in another sector.

It is now easy to see from (6.130) how we can choose the further term in order to make the pieces fit together, viz., as $-\frac{\hbar^2\alpha}{2m}\partial_y\psi^{(2)}(t,x,0)$. Put differently, the Hamiltonian acts on wave functions satisfying the IBC (6.131) according to

$$(H\psi)^{(1)}(x) = -\tfrac{\hbar^2}{2m}\partial_x^2\psi^{(1)}(x) - \tfrac{\hbar^2\alpha}{2m}\,\partial_y\psi^{(2)}(x,0) \tag{6.132a}$$

$$(H\psi)^{(2)}(x,y) = -\tfrac{\hbar^2}{2m}\Big(\partial_x^2+\partial_y^2\Big)\psi^{(2)}(x,y) \quad \text{for } y>0. \tag{6.132b}$$

Now (6.130) is satisfied and equivariance follows: The jump process with motion (6.116) and jump rate (6.124) are $|\psi_t|^2$ distributed at every time t, if so at time 0. It also follows that $\|\psi_t\| = 1$ for all t, if so at time 0.

We have already encountered in Sect. 3.7.4 the situation that in a space $\mathcal{Q}$ with a boundary, the definition of the time evolution of ψ requires the specification of a boundary condition in addition to the specification of how H acts on ψ (say, as the negative Laplacian). Then the *domain* of H consists of functions satisfying the boundary condition. Indeed, it can be shown that (6.131) and (6.132) together define a time evolution, as it can be shown that they define a self-adjoint operator H (Teufel and Tumulka 2016 [25]).

6.4.2 Configuration Space with Two Sectors

We will now apply this approach to particle creation. We consider a version of the emission–absorption model of Sect. 6.2.3, simplified by assuming the x-particles do not move but just sit at fixed locations. Thus, the wave function depends only on the positions of the y-particles. To simplify further, we consider just 1 x-particle located at the origin $\mathbf{0} \in \mathbb{R}^3$, and we focus first just on two sectors, the 0-y-particle sector and the 1-y-particle sector. Our goal is to define a Hamiltonian for which the y-particle can be created or annihilated only at one point, the origin.

Hilbert Space

Let us begin with the configuration space. It is then of the form

$$\mathcal{Q} = \mathcal{Q}^{(0)} \cup \mathcal{Q}^{(1)}, \tag{6.133}$$

where $\mathcal{Q}^{(0)}$ contains a single element, the configuration with an x-particle at $\mathbf{0}$ and 0 y-particles, and $\mathcal{Q}^{(1)}$ contains the configurations with the x-particle at $\mathbf{0}$ and 1 y-particle located at any $\boldsymbol{y} \in \mathbb{R}^3$. For ease of notation, we write only the y-configuration, so we write the element of $\mathcal{Q}^{(0)}$ as $\emptyset$ and an element of $\mathcal{Q}^{(1)}$ as $\boldsymbol{y}$; thus, $\mathcal{Q} = \{\emptyset\} \cup \mathbb{R}^3$. The volume measure μ on $\mathcal{Q}$ is defined, for any $S \subseteq \mathcal{Q}$, as

$$\mu(S) = \begin{cases} 1 + \mathrm{vol}_3(S \cap \mathcal{Q}^{(1)}) & \text{if } \emptyset \in S \\ \mathrm{vol}_3(S \cap \mathcal{Q}^{(1)}) & \text{if } \emptyset \notin S. \end{cases} \tag{6.134}$$

The wave function ψ is a function $\mathcal{Q} \to \mathbb{C}$, and the Hilbert space is

$$\mathscr{H} = L^2(\mathcal{Q}, \mathbb{C}, \mu) = \mathbb{C} \oplus L^2(\mathbb{R}^3, \mathbb{C}, \mathrm{vol}_3) = \mathscr{H}^{(0)} \oplus \mathscr{H}^{(1)}. \tag{6.135}$$

Correspondingly, ψ can be written as $\psi = (\psi^{(0)}, \psi^{(1)})$, where $\psi^{(0)} = \psi(\emptyset)$ is a complex number and $\psi^{(1)}$ a function $\mathbb{R}^3 \to \mathbb{C}$. The inner product in $\mathscr{H}$ is

$$\langle\psi|\phi\rangle = \psi^{(0)*}\phi^{(0)} + \int_{\mathbb{R}^3} d^3\boldsymbol{y}\, \psi^{(1)}(\boldsymbol{y})^*\, \phi^{(1)}(\boldsymbol{y}). \tag{6.136}$$

Spherical Coordinates

It will be useful to describe $\mathbb{R}^3$ in spherical coordinates $(r, \vartheta, \varphi) \in [0, \infty) \times [0, \pi] \times [0, 2\pi)$ according to

$$\boldsymbol{x} = \Big(r \sin\vartheta \cos\varphi, r \sin\vartheta \sin\varphi, r \cos\vartheta\Big). \tag{6.137}$$

Then the origin corresponds to $r = 0$, and the volume element $d^3\boldsymbol{x}$ to $r^2 \sin\vartheta\, d\vartheta\, d\varphi$. We will also write

$$\boldsymbol{\omega} = \Big(\sin\vartheta \cos\varphi, \sin\vartheta \sin\varphi, \cos\vartheta\Big) \tag{6.138}$$

for the unit vector in the direction of $\boldsymbol{x}$, or the parametrization of $\mathbb{S}^2 = \mathbb{S}(\mathbb{R}^3)$ through ϑ and φ with area element $d^2\boldsymbol{\omega} = \sin\vartheta\, d\vartheta\, d\varphi$. (We write $d^2\boldsymbol{\omega}$ for both the infinitesimal 2d set and its area.) Correspondingly, we can write

$$\langle\psi|\phi\rangle = \psi^{(0)*}\phi^{(0)} + \int_0^\infty dr \int_{\mathbb{S}^2} d^2\boldsymbol{\omega}\, r^2\, \psi^{(1)}(r\boldsymbol{\omega})^*\, \phi^{(1)}(r\boldsymbol{\omega}). \tag{6.139}$$

In the following, it will be useful to think of the surface $r = 0$ in the coordinate space $[0, \infty) \times \mathbb{S}^2$ as "the boundary" $\partial\mathcal{Q}$, although in physical space it is not a surface but just the point $\mathbf{0} \in \mathcal{Q}^{(1)}$, and to consider the probability current through this surface.

It will be advantageous to switch back and forth between spherical and Cartesian coordinates, and it will be easiest to define the quantities ρ, j, and v in *Cartesian coordinates* according to

$$\rho^{(1)}(t, x, y, z) = \left|\psi^{(1)}(t, x, y, z)\right|^2 \tag{6.140a}$$

$$\boldsymbol{j}^{(1)}(t, x, y, z) = \tfrac{\hbar}{m}\,\mathrm{Im}\big[\psi^{(1)}(t, x, y, z)^*\nabla\psi^{(1)}(t, x, y, z)\big] \tag{6.140b}$$

$$\boldsymbol{v}^{(1)}(t, x, y, z) = \frac{\boldsymbol{j}^{(1)}}{|\psi|^2}(t, x, y, z) \tag{6.140c}$$

together with $\rho^{(0)}(t) = |\psi^{(0)}(t)|^2$. Then the amount of probability contained in the spherical volume element $dr \times d^2\boldsymbol{\omega}$ of $[0, \infty) \times \mathbb{S}^2$ at time t is $r^2\rho^{(1)}(t, r\boldsymbol{\omega})\, dr\, d^2\boldsymbol{\omega}$. We will also need to consider the current through a surface element $\{r\} \times d^2\boldsymbol{\omega}$ of constant radial coordinate. This current is defined as the amount of probability per time passing through the surface element times ± 1 depending on whether it passes outward or inward, and it can be computed in terms of $\boldsymbol{j}^{(1)}$ (the Cartesian current vector field) according to

$$\boldsymbol{\omega} \cdot \boldsymbol{j}^{(1)}(t, r\boldsymbol{\omega})\, d^2\boldsymbol{\omega} \tag{6.141}$$

using the fact that $\boldsymbol{\omega}$ is also the unit vector in the radial direction and therefore $\boldsymbol{\omega} \cdot \boldsymbol{j}$ the radial component of $\boldsymbol{j}$.

Probability Transport

After these preparations, let us consider how the configuration Q_t moves. Between jumps, it can be in $\mathcal{Q}^{(0)}$ or $\mathcal{Q}^{(1)}$. When it is in $\mathcal{Q}^{(0)}$ then, as $\mathcal{Q}^{(0)}$ has only one element, Q_t does not move. When it is in $\mathcal{Q}^{(1)}$, $Q_t = \boldsymbol{Q}_t \in \mathbb{R}^3$, then the configuration follows the velocity vector field $\boldsymbol{v}^{(1)}$ on $\mathcal{Q}^{(1)}$,

$$\frac{d\boldsymbol{Q}_t}{dt} = \boldsymbol{v}^{(1)}(t, \boldsymbol{Q}_t)\,. \tag{6.142}$$

When the trajectory hits the boundary $r = 0$, the configuration jumps to $\emptyset \in \mathcal{Q}^{(0)}$. It stays there for a while and then, at a random time, jumps to a point $(r = 0, \boldsymbol{\omega})$ on the boundary $\{0\} \times \mathbb{S}^2$. Although all points $(0, \boldsymbol{\omega})$ in the coordinate space $[0, \infty) \times \mathbb{S}^2$ correspond to the same point $\mathbf{0}$ in physical space, it is important to keep track of the $\boldsymbol{\omega}$ because different trajectories begin, in spherical coordinates, at different points $(0, \boldsymbol{\omega})$; see Fig. 6.7; equivalently, different trajectories can begin, in Cartesian coordinates, at $\mathbf{0}$ at the same time t, and in order to specify which trajectory the process Q_t jumps to, we need to specify the direction $\boldsymbol{\omega}$ in which the trajectory

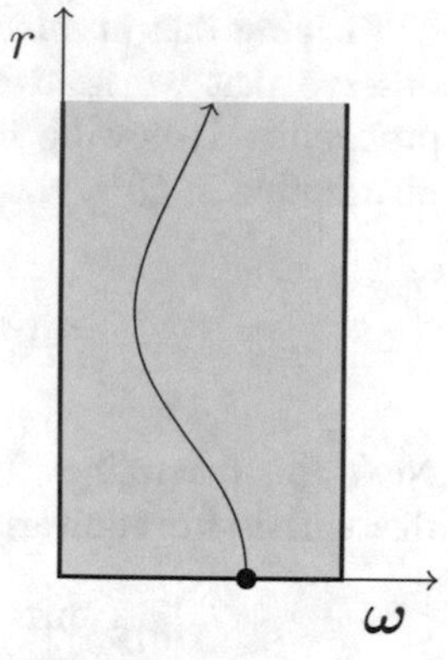

Fig. 6.7 A trajectory in $\mathcal{Q}^{(1)}$, represented in spherical coordinates, with only one of the two angles ϑ, φ of $\boldsymbol{\omega}$ drawn. The shaded region represents the admissible values of r, φ, ϑ. The trajectory begins at $r = 0$ at a particular value of $\boldsymbol{\omega}$ (Reprinted from [11] with permission by Springer)

begins. It will be convenient to speak of jumping to the boundary point $(0, \boldsymbol{\omega})$ in spherical coordinates.

The rate of jumping to a surface element $\{0\} \times d^2\boldsymbol{\omega}$ will be denoted $\sigma_t\big(\emptyset \to (0, \boldsymbol{\omega})\big)\, d^2\boldsymbol{\omega}$. The probability balance equation analogous to (6.119) reads

$$\frac{\partial \rho^{(0)}}{\partial t}(t) = -\lim_{r \searrow 0} \int_{\mathbb{S}^2} d^2\boldsymbol{\omega}\, r^2\, \rho^{(1)}(t, r\boldsymbol{\omega})\, \boldsymbol{\omega} \cdot \boldsymbol{v}^{(1)}(t, r\boldsymbol{\omega})\, 1_{\boldsymbol{\omega}\cdot \boldsymbol{v}^{(1)}(t, r\boldsymbol{\omega})<0}$$
$$- \rho^{(0)}(t) \int_{\mathbb{S}^2} d^2\boldsymbol{\omega}\, \sigma_t\big(\emptyset \to (0, \boldsymbol{\omega})\big) \tag{6.143a}$$

$$\frac{\partial \rho^{(1)}}{\partial t}(t, \boldsymbol{y}) = -\nabla_{\boldsymbol{y}} \cdot \Big(\rho^{(1)}(t, \boldsymbol{y})\, \boldsymbol{v}^{(1)}(t, \boldsymbol{y})\Big) \quad \text{for } \boldsymbol{y} \neq 0. \tag{6.143b}$$

Another observation at this point is that we should expect $\rho^{(1)}(\boldsymbol{y})$ to diverge as $\boldsymbol{y} \to \mathbf{0}$ like $1/|\boldsymbol{y}|^2$. That is because as trajectories approach $r = 0$, volumes shrink like r^2 while probabilities remain the same, so probability density goes up like $1/r^2$. Here is a simple example illustrating this phenomenon. Consider the time-independent velocity vector field $\boldsymbol{v}^{(1)}(\boldsymbol{y}) = -\boldsymbol{y}/|\boldsymbol{y}|$ of unit magnitude pointing radially inward, and the associated motion

$$\frac{d\boldsymbol{Q}}{dt} = -\frac{\boldsymbol{y}}{|\boldsymbol{y}|}. \tag{6.144}$$

The solutions $\boldsymbol{Q}(t)$ move radially inward at unit speed and reach the origin after finite time (in fact, the time needed is $|\boldsymbol{Q}(0)|$). Now consider an ensemble of trajectories with $\boldsymbol{Q}(0)$ uniformly distributed in a spherical shell with inner radius r_0 and thickness dr. After time t with $0 < t < r_0$, they will be uniformly distributed in a shell with radius $r_t = r_0 - t$ and equal thickness dr. Since the volume $4\pi r^2\, dr$ goes down with decreasing r, the density goes up like $1/r^2$. This suggests that relevant wave functions $\psi^{(1)}$ will diverge at $\mathbf{0}$ like $1/r$ and $\rho^{(1)}$ and $\boldsymbol{j}^{(1)}$ like $1/r^2$.

Keeping that in mind, we will follow similar steps as in Sect. 6.4.1. First, we observe that we need a specific jump rate to ensure that the right amount of probability is flowing in through the boundary $r = 0$ for maintaining the $|\psi|^2$ distribution in $\mathcal{Q}^{(1)}$, viz., (Dürr et al. 2020 [11])

$$\sigma_t\big(\emptyset \to (0, \boldsymbol{\omega})\big) = \lim_{r \searrow 0} \frac{r^2 \boldsymbol{\omega} \cdot \boldsymbol{j}^{(1)}(t, r\boldsymbol{\omega})^+}{|\psi^{(0)}(t)|^2}. \tag{6.145}$$

Next, this determines the desired balance equation for $|\psi|^2$ analogous to (6.126) that will ensure equivariance:

$$\frac{\partial |\psi^{(0)}|^2}{\partial t}(t) = -\lim_{r \searrow 0} r^2 \int_{\mathbb{S}^2} d^2\omega\, \boldsymbol{\omega} \cdot \boldsymbol{j}^{(1)}(t, r\boldsymbol{\omega}) \tag{6.146a}$$

$$\frac{\partial |\psi^{(1)}|^2}{\partial t}(t, \boldsymbol{y}) = -\nabla_{\boldsymbol{y}} \cdot \boldsymbol{j}^{(1)}(t, \boldsymbol{y}) \quad \text{for } \boldsymbol{y} \neq \boldsymbol{0}. \tag{6.146b}$$

Now we come to the need for an IBC. Equation (6.146b) can be obtained in the usual way from

$$(H\psi)^{(1)}(\boldsymbol{y}) = -\tfrac{\hbar^2}{2m} \nabla_{\boldsymbol{y}}^2 \psi^{(1)}(\boldsymbol{y}) + E_0\, \psi^{(1)}(\boldsymbol{y}) \quad \text{for } \boldsymbol{y} \neq \boldsymbol{0}, \tag{6.147}$$

but since the Schrödinger equation implies

$$\frac{\partial |\psi^{(0)}|^2}{\partial t} = \tfrac{2}{\hbar} \operatorname{Im}\big[\psi^{(0)*} (H\psi)^{(0)}\big], \tag{6.148}$$

(6.146a) could only come out of the Schrödinger equation if

$$\tfrac{2}{\hbar} \operatorname{Im}\big[\psi^{(0)*} (H\psi)^{(0)}\big] = -\lim_{r \searrow 0} r^2 \int_{\mathbb{S}^2} d^2\omega\, \boldsymbol{\omega} \cdot \boldsymbol{j}^{(1)}(t, r\boldsymbol{\omega}) \tag{6.149}$$

$$= -\tfrac{\hbar}{m} \lim_{r \searrow 0} r^2 \int_{\mathbb{S}^2} d^2\omega\, \boldsymbol{\omega} \cdot \operatorname{Im}\big[\psi^{(1)*} \nabla \psi^{(1)}\big](t, r\boldsymbol{\omega}), \tag{6.150}$$

which creates the difficulty, as did (6.126a), that the right-hand side involves $\psi^{(1)}$ quadratically, whereas the left-hand side involves $\psi^{(0)}$; compare (6.130). Even if $(H\psi)^{(0)}$ involves $\psi^{(1)}$, this is still not quadratic in $\psi^{(1)}$, unless we require a relation between $\psi^{(1)}$ on the boundary and $\psi^{(0)}$—an IBC.

Hamiltonian and IBC

Since $\psi^{(1)}$ should be expected to diverge like $1/r$, we cannot speak of the value of $\psi^{(1)}$ on the boundary, but we can speak of the coefficient of the $1/r$ asymptote, which is the left-hand side of the following IBC:

$$\lim_{\boldsymbol{y}\to\boldsymbol{0}} |\boldsymbol{y}|\, \psi^{(1)}(\boldsymbol{y}) = \alpha\, \psi^{(0)} \tag{6.151}$$

with constant $\alpha \in \mathbb{R}$.

If we can expand $\psi^{(1)}(r\boldsymbol{\omega})$ in powers of r,

$$\psi^{(1)}(r\boldsymbol{\omega}) = \sum_{k=-1}^{\infty} c_k(\boldsymbol{\omega})\, r^k \tag{6.152}$$

(starting at $k = -1$ because $\psi^{(1)}$ should diverge like $1/r$), then the IBC demands that

$$c_{-1}(\boldsymbol{\omega}) = \alpha\, \psi^{(0)}\,. \tag{6.153}$$

In particular, $c_{-1}(\boldsymbol{\omega})$ is independent of $\boldsymbol{\omega}$. (Actually, a calculation shows that both $c_{-1}(\boldsymbol{\omega})$ and $c_0(\boldsymbol{\omega})$ must be independent of $\boldsymbol{\omega}$ anyway or else $\nabla^2\psi^{(1)}$ could not be square-integrable; see Remark 7 in Dürr et al. (2020) [11].) The relation (6.150) needed for equivariance can be expressed in terms of the coefficients of the series (6.152) as

$$\mathrm{Im}\big[c_{-1}^*(H\psi)^{(0)}\big] = -\tfrac{4\pi\hbar^2\alpha}{2m}\mathrm{Im}[c_{-1}^* c_0] \tag{6.154}$$

because $\boldsymbol{\omega}\cdot\nabla = \partial_r$ and

$$\psi = c_{-1}r^{-1} + c_0 + O(r) \tag{6.155a}$$

$$\partial_r\psi = -c_{-1}r^{-2} + c_1(\boldsymbol{\omega}) + O(r) \tag{6.155b}$$

$$\psi^*\partial_r\psi = -|c_{-1}|^2 r^{-3} - c_0^* c_{-1} r^{-2} + O(r^{-1}) \tag{6.155c}$$

$$\mathrm{Im}[\psi^*\partial_r\psi] = -\mathrm{Im}\left[c_0^* c_{-1}\right] r^{-2} + O(r^{-1})\,. \tag{6.155d}$$

This suggests

$$(H\psi)^{(0)} = -\tfrac{4\pi\hbar^2\alpha}{2m} c_0 \tag{6.156}$$

or

$$(H\psi)^{(0)} = -\tfrac{\hbar^2\alpha}{2m}\lim_{r\searrow 0}\int_{\mathbb{S}^2} d^2\boldsymbol{\omega}\, \partial_r\Big(r\psi^{(1)}(r\boldsymbol{\omega})\Big)\,. \tag{6.157}$$

So we have arrived at a Hamiltonian.

Delta Contribution

One more point requires explanation. Let me mention a relation widely used in classical electrostatics:

$$\nabla_{\mathbf{y}}^2 \frac{1}{|\mathbf{y}|} = -4\pi\,\delta^3(\mathbf{y})\,. \tag{6.158}$$

(There it means that the Coulomb potential $V = 1/4\pi\varepsilon_0|\mathbf{y}|$ is a solution of the Poisson equation

$$\nabla^2 V = -\varepsilon_0^{-1}\,\rho_{\text{charge}} \tag{6.159}$$

with charge density corresponding to a unit point charge.) We verify it in Exercise 6.13. As a consequence,

$$\nabla_{\mathbf{y}}^2 \psi^{(1)}(\mathbf{y}) = -4\pi\alpha\,\psi^{(0)}\,\delta^3(\mathbf{y}) + \text{remainder}\,, \tag{6.160}$$

where the remainder contains no Dirac delta at $\mathbf{0}$. When we defined $(H\psi)^{(1)}$ in (6.147), we meant only the part at $\mathbf{y} \neq \mathbf{0}$—*without the delta function*. Put differently, we need to subtract the appropriate multiple of the delta function:

$$(H\psi)^{(1)}(\mathbf{y}) = -\tfrac{\hbar^2}{2m}\nabla_{\mathbf{y}}^2\psi^{(1)}(\mathbf{y}) + E_0\,\psi^{(1)}(\mathbf{y}) - \tfrac{2\pi\alpha\hbar^2}{m}\delta^3(\mathbf{y})\,\psi^{(0)}\,. \tag{6.161}$$

If we introduce the abbreviation

$$g = -\frac{2\pi\hbar^2\alpha}{m}\,, \tag{6.162}$$

then the Hamiltonian can be written as

$$(H\psi)^{(0)} = \tfrac{g}{4\pi}\int_{\mathbb{S}^2} d^2\boldsymbol{\omega}\,\lim_{r\searrow 0}\partial_r\Big(r\psi^{(1)}(r\boldsymbol{\omega})\Big) \tag{6.163a}$$

$$(H\psi)^{(1)}(\mathbf{y}) = -\tfrac{\hbar^2}{2m}\nabla_{\mathbf{y}}^2\psi^{(1)}(\mathbf{y}) + E_0\,\psi^{(1)}(\mathbf{y}) + g\,\delta^3(\mathbf{y})\,\psi^{(0)} \tag{6.163b}$$

with IBC

$$\lim_{\mathbf{y}\to\mathbf{0}}|\mathbf{y}|\,\psi^{(1)}(\mathbf{y}) = -\tfrac{mg}{2\pi\hbar^2}\,\psi^{(0)}\,. \tag{6.164}$$

Note that (6.163) is remarkably similar to the original, UV divergent Hamiltonian (6.55). However, it can be shown (Lampart et al. 2018 [15]):

Theorem 6.1 *The operator H given by* (6.163) *is well defined and self-adjoint on a domain consisting of wave functions satisfying the IBC* (6.164) *(although not all functions in the domain can be written as a power series of the form* (6.152)*).*

6.4.3 All Sectors

Let me drop the restriction to only two sectors (the 0-sector and the 1-sector) and briefly report what the model looks like in the more general case with infinitely many sectors, but still with a single x-particle fixed at the origin. Again, we write only the y-configuration, so the configuration space is

$$\mathcal{Q} = \Gamma(\mathbb{R}^3) = \bigcup_{N=0}^{\infty} {}^N\mathbb{R}^3 \tag{6.165}$$

with boundary

$$\partial\mathcal{Q} = \{q \in \mathcal{Q} : \mathbf{0} \in q\}. \tag{6.166}$$

The Hilbert space is the bosonic Fock space

$$\mathscr{H} = \Gamma_+\big(L^2(\mathbb{R}^3, \mathbb{C})\big). \tag{6.167}$$

Hamiltonian

The IBC says that for every $N \in \mathbb{N}$ and $k \in \{1, \ldots, N\}$,

$$\lim_{\mathbf{y}_k \to \mathbf{0}} |\mathbf{y}_k|\, \psi(y^N) = -\frac{mg}{2\pi\hbar^2\sqrt{N+1}}\, \psi(y^N \setminus \mathbf{y}_k), \tag{6.168}$$

and the expression for the corresponding Hamiltonian H_{IBC} then reads

$$\begin{aligned}(H_{IBC}\psi)(y^N) = &-\sum_{k=1}^{N} \tfrac{\hbar^2}{2m}\nabla^2_{\mathbf{y}_k}\psi^{(N)}(y^N) + N\,E_0\,\psi^{(N)}(y^N)\\ &+ \frac{g\sqrt{N+1}}{4\pi}\int_{\mathbb{S}^2} d^2\boldsymbol{\omega}\, \lim_{r\searrow 0}\frac{\partial}{\partial r}\Big(r\psi^{(N+1)}(y^N, r\boldsymbol{\omega})\Big)\\ &+ \frac{g}{\sqrt{N}}\sum_{k=1}^{N}\delta^3(\mathbf{y}_k)\,\psi^{(N-1)}(y^N \setminus \mathbf{y}_k).\end{aligned} \tag{6.169}$$

Here is a rigorous result about H_{IBC} (Lampart et al. 2018 [15]):

Theorem 6.2 *For every $E_0 \geq 0$ and $g \in \mathbb{R}$, there is a dense domain of ψs in $\mathscr{H}$ satisfying the IBC (6.168) on which H_{IBC} defined by (6.169) is well defined and self-adjoint.*

Jump Process

The process $(Q_t)_{t\in\mathbb{R}}$ for the configuration looks as follows (Dürr et al. 2020 [11]). Between jumps, the configuration moves deterministically according to Bohm's equation of motion,

$$\frac{d\boldsymbol{Q}_k}{dt} = \frac{\hbar}{m}\,\mathrm{Im}\frac{\nabla_{\boldsymbol{y}_k}\psi}{\psi}(t, Q_t)\,. \tag{6.170}$$

When one of the y-particles reaches the origin, it disappears,

$$Q_{t+} = Q_{t-} \setminus \{\boldsymbol{0}\}\,. \tag{6.171}$$

And a new y-particle gets emitted in direction $\boldsymbol{\omega} \in \mathbb{S}^2$ (or at the asymptotic position $0\boldsymbol{\omega}$) with rate

$$\sigma\Big(y^N \to y^N \times 0d^2\boldsymbol{\omega}\Big) = \lim_{r\to 0}\frac{\hbar}{m}\frac{\mathrm{Im}^+\big[r^2\psi(y^N, r\boldsymbol{\omega})^*\,\partial_r\psi(y^N, r\boldsymbol{\omega})\big]}{|\psi(y^N)|^2}d^2\boldsymbol{\omega}\,. \tag{6.172}$$

Again, Q_t is $|\psi_t|^2$ distributed at all t.

Ground State

For $E_0 > 0$, the Hamiltonian has a unique ground state, which is

$$\psi_{\min}(\boldsymbol{y}_1, \ldots, \boldsymbol{y}_N) = \mathcal{N}\frac{1}{\sqrt{N!}}\left(-\frac{gm}{2\pi\hbar^2}\right)^N\prod_{k=1}^N\frac{e^{-\sqrt{2mE_0}|\boldsymbol{y}_k|/\hbar}}{|\boldsymbol{y}_k|} \tag{6.173}$$

with normalization constant $\mathcal{N}$ independent of N and eigenvalue

$$E_{\min} = \frac{g^2m\sqrt{2mE_0}}{2\pi\hbar^3}\,. \tag{6.174}$$

Let me point out the aspects of (6.173) relevant to us: First, in the ground state of the model, every y-particle has the same 1-particle wave function (the fraction after the product sign). Second, this 1-particle wave function is of the form e^{-cr}/r

with some constant c and $r = |\boldsymbol{y}|$; this function diverges at the origin but is still square-integrable because

$$\int_{\mathbb{R}^3} d^3\boldsymbol{y} \left| \frac{e^{-c|\boldsymbol{y}|}}{|\boldsymbol{y}|} \right|^2 = 4\pi \int_0^\infty dr\, r^2 \frac{e^{-2cr}}{r^2} < \infty. \tag{6.175}$$

Third, the 1-particle wave function tends to 0 at large distances from the origin, so y-particles are more likely to be near the origin (i.e., near the x-particle) than far from it. Fourth and more basically, even in the ground state, the $|\psi|^2$ distribution is not concentrated in a single sector; that is, there is a nontrivial probability distribution over different numbers of y-particles. In particular, the number of y-particles has positive probability to be nonzero. That is, in the ground state, the x-particle is surrounded by a cloud of y-particles. Likewise, it is expected that also in the true theory of quantum electrodynamics, the electron is surrounded by a cloud of photons. The electron together with its photon cloud is called a *dressed electron*. The same phenomenon is expected to take place for Higgs bosons in the place of photons. As a consequence of this and the relativistic relation

$$E = mc^2 \tag{6.176}$$

between energy and mass, the observed mass of the electron equals the sum of the electron's fundamental mass ("bare mass") and c^{-2} times the ground state energy of its boson cloud; it is even a serious possibility that the bare mass of the electron (or any other particle) is zero, and the observed mass arises completely from the surrounding cloud, to which the Higgs bosons are believed to contribute particularly.

Here is another remark about this model. When an analogous model is set up for two x-particles at fixed locations in space at a distance of $R > 0$, and the ground state energy analogous to (6.174) is computed, the result is

$$E_{\min} = a - b\frac{e^{-cR}}{R} \tag{6.177}$$

with positive constants a, b, c. Now imagine, as a heuristic reasoning, that the two x-particles move slowly (e.g., because they are very heavy compared to the y-particles), and that in every new configuration of the xs the wave function in the y-Hilbert space will quickly go to the ground state. Then it would feel to the x-particle as if they moved in a potential, a configuration-dependent potential energy, called an *effective potential*. And this potential, as a function of the distance $R = |\boldsymbol{x}_1 - \boldsymbol{x}_2|$, is given by (6.177); a potential of this form (where the constant a can of course be dropped) is called a *Yukawa potential*. It turns out that in the limit $E_0 \to 0$, the constant c tends to 0; in this limit the Yukawa potential approaches the Coulomb potential $-b/R$. It is believed that something like this mechanism is the explanation of the Coulomb potential: that the ground state energy of the photons emitted and absorbed by two electrons at distance R varies with R like $1/R$.

6.5 A Brief Look at Quantum Field Theory

When speaking of "a quantum field theory," one refers to quantum electrodynamics (i.e., the theory of electrons and photons), or to quantum chromodynamics (i.e., the theory of quarks and gluons), or to an extension of those, or to any model considered as a (simplified) variant of those. The relevant extensions concern the so-called weak interaction (i.e., the theory of W and Z bosons) and the Higgs boson. Suitably combined, quantum electrodynamics and quantum chromodynamics with weak interaction and the Higgs boson form the *standard model of particle physics*. In addition, "quantum field theory" is also the name of the field of physics in which quantum field theories are relevant.

A simple example of a quantum field theory (QFT) is provided by the emission–absorption model (6.55). This example suggests the following:

Hypothesis *Quantum field theory is essentially quantum mechanics with creation and annihilation of particles, usually in relativistic space-time.*

Although it is far from settled whether this hypothesis gives the correct picture, I think this is quite possible, even likely, although many other authors think otherwise (e.g., Wallace 2022 [30]).

Be that as it may, I should say explicitly that the problems of the Copenhagen interpretation, such as the quantum measurement problem, certainly do not get better but rather persist in QFT, and that nonlocality is equally valid in QFT. (When one says of a QFT that it be local, then the statement refers to a different meaning of the word "local," usually that it is interaction local, i.e., that the Hamiltonian contains no interaction terms between spacelike separated regions.) And then there are problems in QFT that were not present in quantum mechanics.

6.5.1 Problems of Quantum Field Theory

Let me try to formulate some of the further problems (maybe related to each other) that arise in QFT:

(i) The problem of UV divergence that we have already encountered in Sect. 6.2.5
(ii) The problem of possible creation of infinitely many particles in finite time
(iii) The problem of finding the correct position operators (the correct POVM) for photons (see Sect. 7.3.9)
(iv) The problem of the status and nature of the Dirac sea; more or less equivalently, the problem of finding the correct position operators (the correct POVM) for

positrons,[8] which of course is interlocked with the correct position operators for electrons[9]

(v) In curved space-time, the problem of which subspace should be filled by the Dirac sea

(vi) The problem of whether superluminal signaling is strictly impossible or just unlikely or difficult

All of these problems are largely open questions. Some of the problems get better if we consider, instead of arbitrary times, only $t = -\infty$ and $t = +\infty$. For example, wave functions tend to become locally plane waves as $t \to \pm\infty$, and for local plane waves it is known how to compute the correct Born distribution for photon positions, while it is not known for arbitrary wave functions. For another example, the time evolution from $t = -\infty$ to $t = +\infty$, which is essentially the same as the *scattering matrix* S, can sometimes be defined even if the Hamiltonian H is ill defined. Moreover, even if S is ill defined, it can often be expanded, by means of formal (i.e., possibly ill-defined) calculation, into a series known as the Dyson series, and even if the series does not converge, at least the terms in the series are sometimes well defined, and the sum of the first few terms of the series sometimes yields excellent agreement with experimental data. Moreover, even if the terms in the Dyson series are ill defined, they can sometimes be split into several contributions, some of which are finite and some of which are believed to be unphysical (i.e., should not actually be there), in such a way that leaving out the unphysical contributions yields terms that agree well with experimental data. All this could perhaps be cleared up if we knew the correct Hamiltonian.

6.5.2 Field Ontology vs. Particle Ontology

Now let us leave these problems aside, limit ourselves to the non-relativistic case, and ignore the UV divergence. Even then there is disagreement about whether QFT should be thought of as a theory of particles or as a theory of fields, and that concerns particularly the ontology.

The fields mentioned in the name of quantum field theory are the field operators $\hat{\phi}(\boldsymbol{x})$. They are operators on the Hilbert space $\mathscr{H}$ containing the state vector, and usually related to the creation and annihilation operators, for example, for a bosonic species of particles by

$$\hat{\phi}(\boldsymbol{x}) = a(\boldsymbol{x}) + a^\dagger(\boldsymbol{x})\,. \tag{6.178}$$

[8] For discussion, see my paper [28].

[9] Usually, in QFT one works with field operators and leaves position operators aside. But even if particles did not exist, and even if position operators for microscopic configurations of individual particles did not exist, we would still need position operators (or something replacing them) at least for macroscopic objects such as pointers of measurement instruments. How else would we justify from fundamental laws that certain quantum states are states of pointers pointing in certain directions?

The field operators play a key role, and many relevant calculations can be carried out in terms of the field operators. The Wightman axioms, a certain formalization of quantum field theory, are about the field operators.

The existence of field operators may suggest a field representation of the quantum state vector $|\Psi\rangle \in \mathscr{H}$ and a field ontology for a Bohm-type theory. The latter means that the primitive ontology, the ontology beyond $|\Psi\rangle$, is not given by particles with actual positions but by an *actual field* $\Phi(t, \boldsymbol{x})$ (with values in the real or complex scalars or vectors or tensors for every time $t \in \mathbb{R}$ and every location $\boldsymbol{x} \in \mathbb{R}^3$). The actual field at time t can be represented by a point in the *space of field configurations* $\mathcal{Q}_{\text{field}}$, which is a class of functions on $\mathbb{R}^3$, for example, all continuous functions $\phi : \mathbb{R}^3 \to \mathbb{R}$. One may then expect that the quantum state $|\Psi\rangle$ can be represented by a function $\Psi : \mathcal{Q}_{\text{field}} \to \mathbb{C}$ on the configuration space, called a *wave functional* (because "functional" is a word for a mapping, especially a nonlinear one, whose argument is a function); some authors write $\Psi[\phi]$ for a functional of the function ϕ. One may expect that the Hamiltonian for Ψ involves functional derivatives $\delta/\delta\phi(\boldsymbol{x})$, for example,

$$H\Psi[\phi] = -\int_{\mathbb{R}^3} d^3\boldsymbol{x}\, \frac{\delta^2\Psi}{\delta\phi(\boldsymbol{x})^2} \tag{6.179}$$

in analogy to the negative Laplacian operator on $\mathbb{R}^d$,

$$H\psi(x_1, \ldots, x_d) = -\sum_{i=1}^{d} \frac{\partial^2\psi}{\partial x_i^2}\,. \tag{6.180}$$

On may expect further that the actual field configuration Φ is guided by the wave functional $\Psi[\phi]$ in a way analogous to how, in N-particle Bohmian mechanics, the actual particle configuration Q is guided by $\psi(q)$:

$$\frac{\partial\Phi}{\partial t}(\boldsymbol{x}) = \text{Im}\left[\Psi[\phi]^{-1}\frac{\delta\Psi}{\delta\phi(\boldsymbol{x})}\right]_{\phi=\Phi} \tag{6.181}$$

in analogy to

$$\frac{dQ_i}{dt} = \frac{\hbar}{m}\,\text{Im}\left[\psi(q)^{-1}\frac{\partial\psi}{\partial q_i}\right]_{q=Q}. \tag{6.182}$$

Finally, one may expect that the representation of $|\Psi\rangle$ as a functional Ψ on $\mathcal{Q}_{\text{field}}$ is the *field representation* that diagonalizes the field operators $\hat{\phi}(\boldsymbol{x})$, i.e., turns them into multiplication operators

$$\Big(\hat{\phi}(\boldsymbol{x})\,\Psi\Big)[\phi] = \phi(\boldsymbol{x})\,\Psi[\phi] \tag{6.183}$$

in analogy to

$$\Big(X_i\, \psi\Big)(x_1, \ldots, x_d) = x_i\, \psi(x_1, \ldots, x_d) \tag{6.184}$$

for the position operators X_i in N-particle quantum mechanics with ψ the position representation of the quantum state $|\psi\rangle$. The field representation can be thought of as the representation relative to a certain generalized ONB, the joint eigenbasis of all field operators.

The field ontology has been studied and advocated in particular by Struyve (2010) [24] and Sebens (2022) [23]. Bohm (1952) [3] himself and Bohm and Hiley (1993) [4] have actually advocated the field ontology for the electromagnetic field (replacing photons), while they suggested a particle ontology for electrons. I see the following reasons for preferring a particle ontology to a field ontology:

(i) *Fermions.* Any multiplication operators such as (6.183) commute with each other, but the field operators associated with a fermionic Fock space do not—they anti-commute. This suggests that the field ontology does not work for fermionic Fock spaces.

(ii) *No volume in infinite dimension.* $\dim \mathcal{Q}_{\text{field}} = \infty$, and while $\mathbb{R}^d$ has a natural notion of volume (which mathematicians call the Lebesgue measure), ∞-dimensional spaces do not have a reasonable notion of volume. For example, if a cube of side length 1 therein were assumed to have volume 1, then a cube of side length α would have volume ∞ for every $\alpha > 1$ and volume 0 for every $\alpha < 1$. Almost all relevant sets would have either volume ∞ or volume 0. As a consequence, there is no reasonable notion of probability density, as any density refers to volume. In particular, for a wave functional $\Psi : \mathcal{Q}_{\text{field}} \to \mathbb{C}$, $|\Psi|^2$ is a function on $\mathcal{Q}_{\text{field}}$ but not a measure, so it does not define a probability distribution. In other words, there is no Born rule, and without a Born rule, it is unclear how the theory would make any empirical predictions at all.[10]

(iii) *Standard applications.* The way QFT is usually applied and compared to experiments is by means of a particle representation at $t = \pm\infty$, not by fields.

(iv) *Appearances.* We observe particle tracks in cloud chambers and spots on photographic plates, and we have photon counters. So it appears like there should be particles, not fields.[11]

(v) *Macroscopic objects.* They seem to consist of particles, not fields. For example, we can say how many particles an ordinary cup of coffee consists of: about 10^{26} electrons and 6 times as many quarks.

---

[10] I am not saying that my reasons (i) and (ii) are watertight no-go proofs. Maybe these obstacles can somehow be overcome, just like the UV divergence problem could. But I am not aware of solutions to these problems. Problem (ii) also arises for systems with infinitely many particles.

[11] Of course, appearances can be misleading. But the point remains that the particle ontology is kind of obvious, and the field ontology kind of eccentric.

(vi) *Continuity with quantum mechanics.* The particle ontology is theoretically attractive as N-particle quantum mechanics then is not completely wrong but contained as the special case without creation and annihilation.

Exercises

Exercises 6.1 and 6.2 can be found in Sect. 6.1.1, Exercise 6.3 in Sect. 6.2.2, Exercise 6.4 in Sect. 6.2.4, and Exercise 6.5 in Sect. 6.3.2.

Exercise 6.6 (Essay Question) Describe the concepts of field ontology and particle ontology.

Exercise 6.7 (Wedge Product) The wedge product of the N 1-particle wave functions $\psi_1, \ldots, \psi_N \in L^2(\mathbb{R}^3, \mathbb{C}^d)$ in this order is defined as the anti-symmetrized tensor product,

$$\psi_1 \wedge \psi_2 \wedge \cdots \wedge \psi_N := P_{\text{anti}}(\psi_1 \otimes \psi_2 \otimes \cdots \otimes \psi_N) . \tag{6.185}$$

Show that its value at a configuration $(\boldsymbol{x}_1, \ldots, \boldsymbol{x}_N)$ and spin indices $s_1, \ldots, s_N$ is given by the determinant

$$(\psi_1 \wedge \cdots \wedge \psi_N)_{s_1 \ldots s_N}(\boldsymbol{x}_1 \ldots \boldsymbol{x}_N) = \det\big(\psi_{i,s_j}(\boldsymbol{x}_j)\big)_{ij} \tag{6.186}$$

$$= \begin{vmatrix} \psi_{1s_1}(\boldsymbol{x}_1) & \psi_{1s_2}(\boldsymbol{x}_2) & \ldots & \psi_{1s_N}(\boldsymbol{x}_N) \\ \psi_{2s_1}(\boldsymbol{x}_1) & \psi_{2s_2}(\boldsymbol{x}_2) & \ldots & \psi_{2s_N}(\boldsymbol{x}_N) \\ \vdots & \vdots & & \vdots \\ \psi_{Ns_1}(\boldsymbol{x}_1) & \psi_{Ns_2}(\boldsymbol{x}_2) & \ldots & \psi_{Ns_N}(\boldsymbol{x}_N) \end{vmatrix} \tag{6.187}$$

called the *Slater determinant.*

Exercise 6.8 (More About Wedge Products) Show that if $\psi_1, \ldots, \psi_N$ form an ONB of the subspace $S = \text{span}\{\psi_1, \ldots, \psi_N\} \subset L^2(\mathbb{R}^3, \mathbb{C}^d)$, and if $\phi_1, \ldots, \phi_N$ is another ONB of S, then

$$\phi_1 \wedge \cdots \wedge \phi_N = e^{i\theta}\psi_1 \wedge \cdots \wedge \psi_N \tag{6.188}$$

for some $\theta \in [0, 2\pi)$.

Exercise 6.9 (Canonical Commutation Relations) Verify the *canonical commutation relations* for creation and annihilation operators in bosonic Fock space $\Gamma_+(\mathscr{H}_1)$,

$$[a(f), a(g)] = 0\,, \quad [a^\dagger(f), a^\dagger(g)] = 0\,, \tag{6.189a}$$

$$[a^\dagger(f), a(g)] = \langle f|g\rangle_{\mathscr{H}_1} \quad \forall f, g \in \mathscr{H}_1 \tag{6.189b}$$

and the *canonical anti-commutation relations* for creation and annihilation operators in fermionic Fock space $\Gamma_-(\mathscr{H}_1)$,

$$\{a(f), a(g)\} = 0\,, \quad \{a^\dagger(f), a^\dagger(g)\} = 0\,, \tag{6.190a}$$

$$\{a^\dagger(f), a(g)\} = \langle f|g\rangle_{\mathscr{H}_1} \quad \forall f, g \in \mathscr{H}_1\,. \tag{6.190b}$$

Here, $[A, B] = AB - BA$ is the commutator and $\{A, B\} = AB + BA$ the anti-commutator.

Exercise 6.10 (Occupation Number Representation) Let $\{|j\rangle : j \in J\}$ be an ONB of the (1-particle) Hilbert space $\mathscr{H}_1$, where J is an ordered, finite, or countably infinite index set. For every "occupation number list," we define the "occupation number states" as follows. In the bosonic case: For every mapping $n : J \to \mathbb{N} \cup \{0\}$ with $\sum_{j\in J} n_j = N < \infty$, define

$$|n\rangle := P_{\mathrm{sym}} \bigotimes_{\substack{j\in J\\ n_j\neq 0}} |j\rangle^{\otimes n_j} \in (\mathscr{H}_1^{\otimes N})_{\mathrm{sym}} \subset \mathscr{F}_+ = \Gamma_+(\mathscr{H}_1)\,. \tag{6.191}$$

In the fermionic case: For every mapping $n : J \to \{0, 1\}$ with $\sum_{j\in J} n_j = N < \infty$, define

$$|n\rangle := P_{\mathrm{anti}} \bigotimes_{\substack{j\in J\\ n_j\neq 0}} |j\rangle \in (\mathscr{H}_1^{\otimes N})_{\mathrm{anti}} \subset \mathscr{F}_- = \Gamma_-(\mathscr{H}_1)\,. \tag{6.192}$$

(a) In each case, show that the $|n\rangle$ form an ONB of $\mathscr{F}_\pm$.

(b) Let $n'_j = n_j + 1$ and $n'_k = n_k$ for $k \neq j$; let $n''_j = n_j - 1$ and $n''_k = n_k$ for $k \neq j$. Show that

$$a^\dagger(|j\rangle)|n\rangle = \begin{cases} \sqrt{n_j+1}\,|n'\rangle & \text{in the bosonic case} \\ (-1)^{\sum_{k<j} n_k}\,|n'\rangle & \text{in the fermionic case with } n_j = 0 \\ 0 & \text{in the fermionic case with } n_j = 1 \end{cases} \tag{6.193}$$

$$a(|j\rangle)|n\rangle = \begin{cases} \sqrt{n_j}\,|n''\rangle & \text{if } n_j > 0 \\ 0 & \text{if } n_j = 0. \end{cases} \tag{6.194}$$

Exercise 6.11 (Fock Space of a Sum)

(a) Find a bijection between $\Gamma(A_1 \cup A_2)$ and $\Gamma(A_1) \times \Gamma(A_2)$ for disjoint sets A_1, A_2 and Γ as in (6.23).

(b) Find a unitary isomorphism between $\Gamma_\pm(\mathscr{H}_1 \oplus \mathscr{H}_2)$ and $\Gamma_\pm(\mathscr{H}_1) \otimes \Gamma_\pm(\mathscr{H}_2)$ for $\oplus$ the orthogonal sum, arbitrary $\mathscr{H}_1$, $\mathscr{H}_2$, and $\Gamma_\pm$ the Fock space as in (6.45).

Exercise 6.12 (Time Reversal) It may seem that every non-deterministic stochastic process breaks time reversal invariance because whenever nature makes a random decision, the outcome of the decision is fixed afterward but not before. However, this is not correct. Let W be the space of all possible histories according to some theory, and μ^ψ a probability distribution over W. Let $T(h)$ be the time reverse of the history h. Suppose that $T(h) \in W$ for every $h \in W$, and that whenever h has distribution μ^ψ, $T(h)$ has distribution $\mu^{\tilde\psi}$ for some $\tilde\psi$. Then we say that the theory is time reversal invariant. Explain why Bell's jump process is time reversal invariant with $\tilde\psi(q) = \psi(q)^*$. To this end, compute how much probability gets transported from dq' to dq within dt according to both ψ and $\tilde\psi$, and use (6.87).

Exercise 6.13 (Laplacian of the Coulomb Potential) In order to verify the relation $\nabla_y^2 \frac{1}{|y|} = -4\pi\delta^3(y)$ mentioned in (6.158), proceed as follows.

(a) Let $e_r : \mathbb{R}^3 \setminus \{\mathbf{0}\} \to \mathbb{R}^3$ be the radial unit vector field $e_r(x) = x/|x|$. Show that for a radial scalar function $f(x) = g(|x|)$, $\nabla f = g'(|x|)e_r$ for $x \neq \mathbf{0}$.

(b) Show that a radial vector field $v(x) = h(|x|)e_r$ has divergence $\nabla \cdot v = h'(|x|) + \frac{2}{|x|}h(x)$ for $x \neq \mathbf{0}$.

(c) Conclude that $\nabla^2 g(|x|) = g''(|x|) + \frac{2}{|x|}g'(|x|)$ for $x \neq 0$.

(d) Conclude that $\nabla^2 \frac{1}{|x|} = 0$ for $x \neq 0$.

(e) In order to verify that the contribution at the origin is $-4\pi\delta^3$, use the Ostrogradsky–Gauss integral theorem to show that for every $\varepsilon > 0$ and every smooth function $\varphi : \mathbb{R}^3 \to \mathbb{R}$,

$$\int_{B_\varepsilon(\mathbf{0})} d^3\boldsymbol{x}\, \varphi(\boldsymbol{x})\, \nabla^2 \frac{1}{|\boldsymbol{x}|} = -\int_{B_\varepsilon(\mathbf{0})} d^3\boldsymbol{x}\, \nabla\varphi(\boldsymbol{x}) \cdot \nabla \frac{1}{|\boldsymbol{x}|} + \int_{\partial B_\varepsilon(\mathbf{0})} d^2\boldsymbol{x}\, \varphi(\boldsymbol{x})\, \boldsymbol{e}_r \cdot \nabla \frac{1}{|\boldsymbol{x}|} \tag{6.195}$$

$$\xrightarrow{\varepsilon \to 0} -4\pi\varphi(\mathbf{0}), \tag{6.196}$$

where $B_\varepsilon(\mathbf{0}) = \{\boldsymbol{x} \in \mathbb{R}^3 : |\boldsymbol{x}| < \varepsilon\}$ is the ball of radius ε around the origin.

References

1. J.S. Bell, Beables for quantum field theory. Physics Reports **137**, 49–54 (1986). Reprinted as chapter 19 of [2]. Also reprinted in: *Quantum Implications: Essays in Honour of David Bohm*. ed. by F.D. Peat and B.J. Hiley (Routledge, London, 1987), p. 227
2. J.S. Bell, *Speakable and Unspeakable in Quantum Mechanics*. (Cambridge University Press, Cambridge, 1987)
3. D. Bohm, A suggested interpretation of the quantum theory in terms of "hidden" variables II. Phys. Rev. **85**, 180–193 (1952)
4. D. Bohm, B.J. Hiley, *The Undivided Universe: An Ontological Interpretation of Quantum Theory* (Routledge, London, 1993)
5. C. Dove, E. Squires, Symmetric versions of explicit wavefunction collapse models. Found. Phys. **25**, 1267–1282 (1995)
6. D. Dürr, S. Goldstein, R. Tumulka, N. Zanghì, Trajectories and particle creation and annihilation in quantum field theory. J. Phys. A Math. Gen. **36**, 4143–4149 (2003). https://arxiv.org/abs/quant-ph/0208072
7. D. Dürr, S. Goldstein, R. Tumulka, N. Zanghì, Bohmian mechanics and quantum field theory. Phys. Rev. Lett. **93**, 090402 (2004). https://arxiv.org/abs/quant-ph/0303156. Reprinted in [10]
8. D. Dürr, S. Goldstein, R. Tumulka, N. Zanghì, Bell-type quantum field theories. J. Phys. A Math. Gen. **38**, R1–R43 (2005). https://arxiv.org/abs/quant-ph/0407116
9. D. Dürr, S. Goldstein, R. Tumulka, N. Zanghì, Quantum Hamiltonians and stochastic jumps. Commun. Math. Phys. **254**, 129–166 (2005). https://arxiv.org/abs/quant-ph/0303056
10. D. Dürr, S. Goldstein, N. Zanghì, *Quantum Physics Without Quantum Philosophy* (Springer, Heidelberg, 2013)
11. D. Dürr, S. Goldstein, S. Teufel, R. Tumulka, N. Zanghì, Bohmian trajectories for Hamiltonians with interior-boundary conditions. J. Stat. Phys. **180**, pp. 34–73 (2020). http://arxiv.org/abs/1809.10235
12. H.-O. Georgii, R. Tumulka, Some jump processes in quantum field theory, in *Interacting Stochastic Systems*, ed. by J.-D. Deuschel, A. Greven (Springer, Berlin, 2004), pp. 55–73. http://arxiv.org/abs/math.PR/0312326
13. S. Goldstein, Stochastic mechanics and quantum theory. J. Stat. Phys. **47**, 645–667 (1987)
14. J. Lampart, A nonrelativistic quantum field theory with point interactions in three dimensions. Ann. Henri Poincaré **20**, 3509–3541 (2019). http://arxiv.org/abs/1804.08295

15. J. Lampart, J. Schmidt, S. Teufel, R. Tumulka, Particle creation at a point source by means of interior-boundary conditions. Math. Phys. Anal. Geom. **21**, 12 (2018). http://arxiv.org/abs/1703.04476
16. L. Landau, R. Peierls, Quantenelektrodynamik im Konfigurationsraum. Zeitschrift für Physik **62**, 188–200 (1930). English translation: Quantum electrodynamics in configuration space. Pages 71–82 in R.H. Dalitz and R. Peierls (eds): *Selected Scientific Papers of Sir Rudolf Peierls With Commentary* (World Scientific, Singapore, 1997)
17. J.M. Leinaas, J. Myrheim, On the theory of identical particles. Il Nuovo Cimento **37B**, 1–23 (1977)
18. M. Moshinsky, Boundary conditions for the description of nuclear reactions. Phys. Rev. **81**, 347–352 (1951)
19. E. Nelson, Interaction of nonrelativistic particles with a quantized scalar field. J. Math. Phys. **5**, 1190–1197 (1964)
20. J.R. Oppenheimer, Note on the theory of the interaction of field and matter. Phys. Rev. **35**, 461–477 (1930)
21. M. Reed, B. Simon, *Methods of Modern Mathematical Physics*, vol. 2 (Academic Press, New York, 1975)
22. S. Schweber, *An Introduction to Relativistic Quantum Field Theory* (Row, Peterson and Company, Ogden, 1961)
23. C. Sebens, *The Fundamentality of Fields* (2022). http://arxiv.org/abs/2202.09425
24. W. Struyve, Pilot-wave theory and quantum fields. Rep. Prog. Phys. **73**, 106001 (2010). http://arxiv.org/abs/0707.3685
25. S. Teufel, R. Tumulka, Avoiding Ultraviolet Divergence by Means of Interior–Boundary Conditions, in *Quantum Mathematical Physics: A Bridge between Mathematics and Physics*, ed. by F. Finster, J. Kleiner, C. Röken, J. Tolksdorf (Birkhäuser, Basel, 2016), pp. 293–311. https://arxiv.org/abs/1506.00497
26. S. Teufel, R. Tumulka, Hamiltonians without ultraviolet divergence for quantum field theories. Quantum Studies: Mathematics and Foundations **8**, 17–35 (2021). https://arxiv.org/abs/1505.04847
27. R. Tumulka, On Spontaneous wave function collapse and quantum field theory. Proc. R. Soc. A **462**, 1897–1908 (2006). http://arxiv.org/abs/quant-ph/0508230.
28. R. Tumulka, *Positron Position Operators I. A Natural Option*. Ann. Phys. **443**, 168988 (2022). http://arxiv.org/abs/2111.12304
29. J.C. Vink, Quantum mechanics in terms of discrete beables, Phys. Rev. A **48**, 1808–1818 (1993)
30. D. Wallace, The quantum theory of fields, in *The Routledge Companion to the Philosophy of Physics*, ed. by E. Knox, A. Wilson (Routledge, London, 2022), pp. 275–295. http://philsci-archive.pitt.edu/15296/

Relativity

7

We will focus here on special relativity and leave general relativity aside, except for a few remarks. After a brief introduction to relativity and to relativistic versions of the Schrödinger equation, we turn to the main question in this chapter, how the various theories of quantum mechanics can be generalized to relativistic space-time. In classical physics, the principles of relativity required certain modifications of the fundamental laws, and these modifications made the theories relativistic. But classical theories are local, and it turns out that for nonlocal theories, it is far from obvious how one should define which of them count as relativistic. Of course, they should use the relativistic metric, but that alone is not a strong requirement. Of course, they should be Lorentz invariant; however, as I will explain, while for local theories the requirement of Lorentz invariance narrows down the possible theories very much, Lorentz invariance is actually a very weak requirement for nonlocal theories that does not even exclude superluminal signaling. As we will see, there are several different properties that a nonlocal theory may have or not have and that capture some aspects of "being relativistic." It seems fair to say that some theories are more relativistic and others less so, which then leads to another question we will discuss: How much of relativity is in the true laws of nature?

7.1 Brief Introduction to Relativity

We begin with a precursor of relativity.

7.1.1 Galilean Relativity

A *space-time point* is a pair $(t, \boldsymbol{x})$ of a time t and a location $\boldsymbol{x}$ in physical 3-space. The set of all space-time points is called the *space-time*. A *Galilean boost* is a

R. Tumulka, *Foundations of Quantum Mechanics*, Lecture Notes in Physics 1003,
https://doi.org/10.1007/978-3-031-09548-1_7

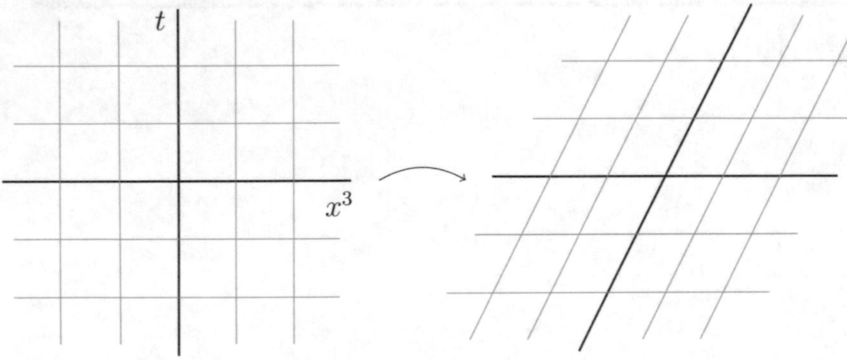

Fig. 7.1 Galiean boost as in (7.1) (only t and x^3 axes shown)

mapping of space-time to itself of the form

$$t' = t \tag{7.1a}$$

$$x_1' = x_1 + v_1 t \tag{7.1b}$$

$$x_2' = x_2 + v_2 t \tag{7.1c}$$

$$x_3' = x_3 + v_3 t \tag{7.1d}$$

with some fixed vector $\boldsymbol{v} = (v_1, v_2, v_3) \in \mathbb{R}^3$; see Fig. 7.1. It can also be regarded as associating new coordinates $(t', \boldsymbol{x}')$ with the space-time point $(t, \boldsymbol{x})$. The intuitive meaning is that the primed coordinate system is moving uniformly relative to the unprimed one with velocity $\boldsymbol{v}$. A (proper) *Galilean transformation* is any composition of a Galilean boost, a rotation $(t, \boldsymbol{x}) \mapsto (t, R\boldsymbol{x})$ with $R \in SO(3)$, and a space-time translation $(t, \boldsymbol{x}) \mapsto (t + t_0, \boldsymbol{x} + \boldsymbol{x}_0)$; the Galilean transformations correspond to all uniformly moving Cartesian coordinate systems and form a group called the *Galilean group*.

As we have checked in Exercise 1.18, both Newtonian and Bohmian mechanics are invariant under Galilean transformations. This means that for every solution of Newtonian (respectively, Bohmian) mechanics, thought of as N paths in space-time, the result of a Galilean transformation yields another solution (with a possibly different wave function in Bohmian mechanics). As a consequence, it is in principle impossible to determine empirically how fast any object (such as the Earth) is moving. On how to understand this issue, there have been two classic views: One view, often associated with Isaac Newton, claims that there is a fact about which space point at time t_1 is the same point as a given space point at time t_2, and therefore, there is a fact about the velocity at which an object (such as Earth) is actually moving, even though this velocity cannot be measured empirically. Another view, often associated with Gottfried Wilhelm Leibniz, claims that there are no facts in nature about the absolute velocity of an object, only about relative velocities (hence the name "relativity"); correspondingly, there is no fact about the identity of space points at different times. We may imagine space-time as a four-dimensional real affine space[1] $\mathscr{A}$ equipped with an equivalence class of affine-linear functions $t : \mathscr{A} \to \mathbb{R}$ ("time"), where equivalence means that two functions differ by addition of a real constant (so there is no fact about which time is time 0), and the structure of a Euclidean space on each level set of t ("time slice") such that every translation of $\mathscr{A}$ is an isometry on each time slice. Ontologically, this means that "the 3d physical space" at different times is really different spaces consisting of different points; I like to imagine that nature creates new points at each instant of time. In this picture, the 3d physical space does not exist by itself; rather, space-time exists as a 4d set $\mathscr{A}$, a *Galilean space-time*. The Galilean invariance of Newtonian and Bohmian mechanics then guarantees that these theories make sense in a Galilean space-time.

7.1.2 Minkowski Space

In contrast to Newtonian and Bohmian mechanics, Maxwell's equations, the fundamental equations governing the electromagnetic field according to classical electrodynamics, are not invariant under Galilean transformations. However, they are invariant under a similar family of linear transformations $\mathbb{R}^4 \to \mathbb{R}^4$, the *Lorentz transformations*, which we will define presently.

Einstein's Principle of Special Relativity (1905) *The true symmetry of space-time in nature is given by the Lorentz transformations, not the Galilean transformations. All laws of nature are invariant under Lorentz transformations.*

Definition 7.1 *Minkowski space* is a 4d real vector space M equipped with a symmetric bilinear form $(\cdot, \cdot) : M \times M \to \mathbb{R}$ (called a pseudo inner product or the *metric*) such that in a suitable basis b_0, b_1, b_2, b_3, the product of $x = x^0 b_0 + x^1 b_1 + x^2 b_2 + x^3 b_3$ and $y = y^0 b_0 + y^1 b_1 + y^2 b_2 + y^3 b_3$ is given by

$$(x, y) = x^0 y^0 - x^1 y^1 - x^2 y^2 - x^3 y^3 \tag{7.2}$$

$$= \sum_{\mu,\nu=0}^{3} x^\mu y^\nu \eta_{\mu\nu} \tag{7.3}$$

[1] An *affine space* is sometimes defined as an *affine subspace* of a vector space $\mathscr{V}$, which is a subset of the form $\boldsymbol{v} + \mathscr{U}$ with $\boldsymbol{v} \in \mathscr{V}$ and $\mathscr{U}$ a subspace of $\mathscr{V}$. Here, we define an affine space as a set $\mathscr{A}$ together with a family $\mathscr{V}$ of bijections $\mathscr{A} \to \mathscr{A}$ ("translations") such that $\mathscr{V}$ is a vector space; the addition in $\mathscr{V}$ is the composition of mappings; and for any $a_1, a_2 \in \mathscr{A}$ there is a unique $\boldsymbol{v} \in \mathscr{V}$ with $\boldsymbol{v}(a_1) = a_2$. The dimension of $\mathscr{A}$ is that of $\mathscr{V}$.

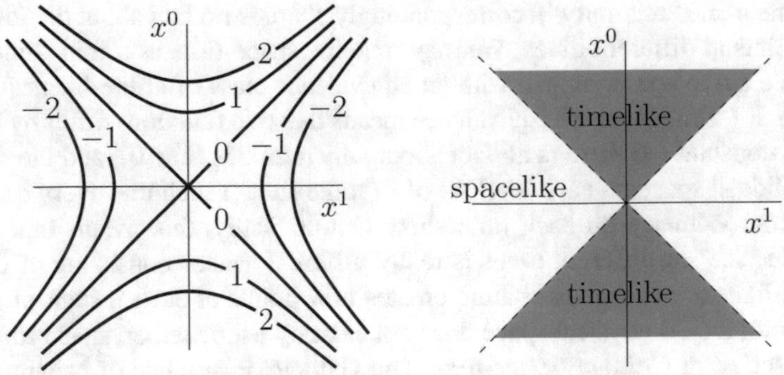

Fig. 7.2 1+1 dimensions of Minkowski space. Left: Level sets of the function $(x, x) = x^\mu x^\nu \eta_{\mu\nu}$. Right: x with $x^\mu x^\nu \eta_{\mu\nu} > 0$ are called timelike, with $x^\mu x^\nu \eta_{\mu\nu} = 0$ lightlike (dashed), and with $x^\mu x^\nu \eta_{\mu\nu} < 0$ spacelike. Two points x, y are said to be *spacelike separated* if and only if $x - y$ is spacelike

with

$$\eta = \begin{pmatrix} 1 & & & \\ & -1 & & \\ & & -1 & \\ & & & -1 \end{pmatrix}. \tag{7.4}$$

(See also Fig. 7.2.) Such a basis is called a *Lorentz frame*.

While an inner product is required to be positive definite, the Minkowski metric is *indefinite*. The connection with space-time is that $x^0 = ct$, where t is the time coordinate and c is the speed of light. Henceforth, we will use units in which $c = 1$. Following a common notation, we will often write x^μ for the vector x; this is somewhat illogical because x^μ would actually mean the μ-th component of the vector x (much like, if f is a function, then $f(x)$ is a number), but since I am telling you now that I will use this *abstract index notation*, no problem arises as long as you can understand from the context whether x^μ means the vector x or its μ-th component.

Einstein Sum Convention As a shorthand notation, we omit sum signs over every index that appears once as an upper and once as a lower index; every such index is understood to be summed over from 0 to 3. For example, we write

$$x^\mu y^\nu \eta_{\mu\nu} \text{ instead of } \sum_{\mu,\nu=0}^{3} x^\mu y^\nu \eta_{\mu\nu}\,. \tag{7.5}$$

Definition 7.2 A *Lorentz transformation* is a linear mapping $\Lambda : M \to M$ that preserves the metric,

$$(\Lambda x, \Lambda y) = (x, y) \quad \forall x, y \in M\,. \tag{7.6}$$

Alternatively, it is the coordinate expression of such a mapping relative to a Lorentz frame, i.e., a matrix $\Lambda^{\mu}{}_{\nu}$ such that

$$\Lambda^{\mu}{}_{\lambda}\eta_{\mu\nu}\Lambda^{\nu}{}_{\rho} = \eta_{\lambda\rho} \tag{7.7}$$

or, in usual matrix multiplication notation,

$$\Lambda^t \eta \Lambda = \eta \tag{7.8}$$

(where t denotes the transpose matrix). These matrices form a group called the *Lorentz group* or $O(1,3)$. The *proper Lorentz group* $O_+^{\uparrow}(1,3)$ is a subgroup containing the *proper Lorentz transformations* (also called the restricted Lorentz transformations), i.e., those that involve neither a reflection of the time direction nor a reflection of space (equivalently, those with $\Lambda^0{}_0 > 0$ and $\det \Lambda > 0$).

Example 7.1 The *Lorentz boost* in the x^3 direction with *rapidity* ξ is the matrix

$$(\Lambda^{\mu}{}_{\nu}) = \begin{pmatrix} \cosh\xi & & & \sinh\xi \\ & 1 & & \\ & & 1 & \\ \sinh\xi & & & \cosh\xi \end{pmatrix}, \tag{7.9}$$

shown in Fig. 7.3. It is an easy exercise to verify (7.8), so it is a Lorentz transformation, in fact a proper one. The rapidity is roughly the analog of the angle in a rotation matrix.

Exercise 7.1 (Rapidity) Show that when we compose Lorentz boosts in the x^3 direction, their rapidities add.

The Lorentz transformations are the symmetries of Minkowski space. It is a simple consequence of the definition that every Lorentz transformation maps a Lorentz frame to another Lorentz frame. Conversely, if a linear mapping $\Lambda : M \to M$ maps any Lorentz frame to another Lorentz frame, then Λ is a Lorentz transformation. It turns out that there is a 6-parameter family of Lorentz transformations (and thus of Lorentz frames), $\dim O(1,3) = 6$. A Lorentz frame is sometimes called an "observer" in the literature, but I do not recommend that terminology.

Let us draw some consequences of Einstein's principle of relativity. The analog of the Newtonian view (though I am not claiming that Newton would have thought

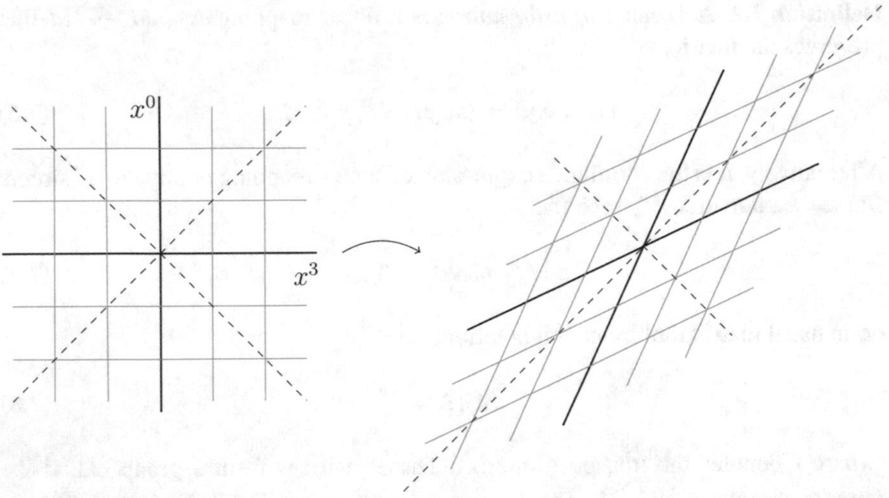

Fig. 7.3 Lorentz boost in x^3 direction (only x^0 and x^3 axes shown, lightlike vectors dashed)

that) would be that there is one Lorentz frame in which space is truly at rest (the "absolute rest frame"), but we cannot find out which frame that is. After all, any solution could be mapped to a different Lorentz frame. The analog of the Leibnizian view, which seems particularly convincing given Einstein's principle of relativity, would be that there is no absolute rest frame. That was, in fact, Einstein's view. Now a crucial difference to the Galilean case is that different Lorentz frames disagree about which space-time points are simultaneous. So, as a consequence of Einstein's view, there is no fact about whether two space-time points are simultaneous, or in fact about the temporal order of two spacelike separated points. That is a rather radical and counterintuitive view, but not an impossible one. According to it, space and time as such do not exist; only space-time exists, and it is an affine Minkowski space.[2] In particular, space-time is regarded as (what is called) a *block universe*: time does not pass in a literal sense; instead, all space-time points are equally real although not equally accessible to us here-now; the words "past" and "future" refer to *relations* between space-time points, not to events that exist no more or not yet.

7.1.3 Dual Space

The dual space of a real vector space V is the vector space V' of all linear mappings $V \to \mathbb{R}$. Elements of V' are often called *covectors*. In matrix notation, elements of V can be written as column vectors and elements of V' as row vectors. Now

[2] That is, an affine space whose associated vector space is equipped with a Minkowski metric. By a slight abuse of notation, we will call both the affine and the vector space M.

we often regard row vectors simply as a different way of arranging the entries of a column vector, so why do we make a big deal now out of the difference between row and column vectors? Here is why. An inner product $\langle\cdot|\cdot\rangle$ on V defines an isomorphism $\varphi : V \to V'$ by $v \mapsto \langle v|\cdot\rangle$ that allows us to identify V with V'. Given an orthonormal basis $\{b_i\}$ of V, the dual basis $\{\hat{b}_j\}$ defined by $\hat{b}_j(b_i) = \delta_{ij}$ is just $\hat{b}_j = \varphi(b_j)$, and the components of $\varphi(v)$ relative to $\{\hat{b}_j\}$ are the same as the components of v relative to $\{b_i\}$ (just arranged as a row vector); that is why there is usually no need to distinguish between V and V'. Now that is different if we use a metric that is indefinite: Although it still defines an isomorphism $\varphi : M \to M'$ by $x \mapsto (x, \cdot)$, if $\{b_\mu\}$ is a Lorentz frame, $(b_\mu, b_\nu) = \eta_{\mu\nu}$, then the dual basis $\hat{b}_\mu$ is *not* the same as $\varphi(b_\mu)$, and the components of $\varphi(x)$ relative to $\{\hat{b}_\mu\}$ are *not* the same as the components of x relative to $\{b_\mu\}$. The components of x relative to $\{b_\mu\}$ are traditionally denoted x^μ (and called the "contravariant components of x"), and those of $\varphi(x)$ relative to $\{\hat{b}_\mu\}$ are x_μ (and called the "covariant components of x"); they are related according to

$$x_\mu = \eta_{\mu\nu} x^\nu , \tag{7.10}$$

so some components have a different sign. It follows that

$$(x, y) = x^\mu y_\mu , \tag{7.11}$$

and this notation for the Minkowski product is more common than (x, y); we will also use it henceforth. By symmetry of the metric,

$$x^\mu y_\mu = x_\mu y^\mu . \tag{7.12}$$

Moreover, the inverse relation to (7.10) is obtained by multiplying by the inverse matrix of η, which has the same entries as η and is therefore called $\eta^{\mu\nu}$,

$$x^\nu = \eta^{\nu\mu} x_\mu . \tag{7.13}$$

$\eta^{\mu\nu}$ also shows up as the coefficients of the metric on M': If, for two covectors $u, v \in M'$, we define their Minkowski product (u, v) as that of their associated vectors $\varphi^{-1}(u)$ and $\varphi^{-1}(v)$,

$$(u, v) := \left(\varphi^{-1}(u), \varphi^{-1}(v)\right), \tag{7.14}$$

then

$$(u, v) = u_\mu v_\nu \eta^{\mu\nu} . \tag{7.15}$$

Furthermore, $\{\hat{b}_\mu\}$ is then a Lorentz frame in M'. (Sometimes, $\hat{b}_\mu$ is denoted b^μ.)

The gradient of a function $f : M \to \mathbb{R}$ is by its nature a covector field. That is because one can think of the derivative of f at x as a linear mapping $u : M \to \mathbb{R}$ such that $f(x) + u(y - x)$ provides the best affine-linear (i.e., first order) approximation to $f(y)$. The covector nature of the gradient can be expressed by saying that the components $\partial f/\partial x^\mu$ form the covariant components of a vector; that is why they are abbreviated as $\partial_\mu f$.

7.1.4 Arc Length

A piecewise smooth curve $[s_1, s_2] \to M : s \mapsto X(s)$ is *future-timelike* if dX/ds is everywhere future-timelike (and at points where two smooth pieces meet, both one-sided derivatives are future-timelike). Likewise for past-timelike; a timelike curve is one that is either future-timelike or past-timelike. Its *invariant length* or *proper length* or *arc length* is

$$\int_{s_1}^{s_2} ds \sqrt{\frac{dX^\mu}{ds}\frac{dX_\mu}{ds}}\,. \tag{7.16}$$

Example 7.2 If $x - y$ is future-timelike, then the straight line segment from y to x has invariant length $\sqrt{(x^\mu - y^\mu)(x_\mu - y_\mu)}$.

Arc Length Rule *Invariant length is time along the curve as measured by ideal clocks.*

Derivation from the Principle of Relativity Fix a Lorentz frame $F = \{b_\mu\}$. An ideal clock at rest in F should measure the coordinate time of F. An ideal clock moving along the straight line from x to y, where $y - x$ is timelike, is at rest in another Lorentz frame $F' = \{b'_\mu\}$. By the principle of relativity, it should measure the coordinate time of F', which agrees with the invariant length of the straight line from x to y. Thus, for a timelike polygonal chain, its duration as measured by ideal clocks should be its invariant length. Since every timelike curve can be approximated to arbitrary accuracy by a timelike polygonal chain, the same should be true for any timelike curve. □

Example 7.3 Of two twins, one stays at $x^1 = 0 = x^2 = x^3$ of a Lorentz frame from time $x^0 = 0$ to time $x^0 = 20$, while the other travels from $(0, 0, 0, 0)$ to $(10, 0, 0, 9)$ to $(20, 0, 0, 0)$; see Fig. 7.4. Say, the unit of time is years and the unit of distance light years. The stay-at-home twin has aged by 20 years, as the proper length of his world line between $(0, 0, 0, 0)$ and $(20, 0, 0, 0)$ is 20. The traveling twin has only aged 8.72 years, as that is the invariant length of his world line: the timelike distance from $(0, 0, 0, 0)$ to $(10, 0, 0, 9)$ is $\sqrt{10^2 - 9^2} = \sqrt{19} \approx 4.36$, and the one from $(10, 0, 0, 9)$ to $(2, 0, 0, 0)$ is the same. The phenomenon that the

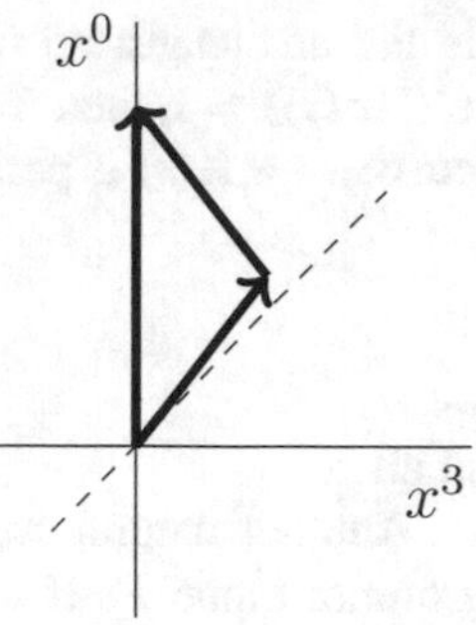

Fig. 7.4 World lines of the two twins

traveling twin is younger is called *time dilation*. It allows for time travel, but only toward the future.

This example is sometimes called the "twin paradox" although there is not much paradoxical about it; it is just surprising. Here is a variant that appears more paradoxical, although it is also a contradiction-free mathematical fact: Consider two Lorentz frames, $F = \{b_\mu\}$ and $F' = \{b'_\mu\}$. Suppose that near every space-time point there is a clock showing the coordinate time of F and another clock showing that of F'. Then at any fixed 3-location of F (i.e., along any parallel to the time axis of F, or along any straight line in direction b_0), the F'-clocks are slower than the F-clocks (i.e., x'^0 increases more slowly than x^0). But by symmetry, the F-clocks are slower than the F'-clocks along any straight line in direction b'_0.

Proposition 7.1 *For any timelike curve from x to y,*

$$\textit{the invariant length is } \leq \sqrt{(x^\mu - y^\mu)(x_\mu - y_\mu)}, \tag{7.17}$$

and equality holds if and only if the curve is the straight line segment. (The straight line is the longest curve connecting two timelike separated points, in contrast to Euclidean space, where the straight line is the shortest curve connecting two given points.)

Arc Length Parameterization It can be convenient to parameterize a future-timelike curve by arc length. If $s \mapsto X(s)$ is any parameterization, then

$$\tau(s_0) = \int_0^{s_0} ds \sqrt{\frac{dX^\mu}{ds}\frac{dX_\mu}{ds}} \tag{7.18}$$

is the arc length up to s_0. If the function τ^{-1} is the inverse of the function τ, $\tau^{-1}(\tau(s)) = s$, then $X(\tau^{-1}(\cdot))$ is the arc length parameterization. A future-timelike curve $s \mapsto X(s)$ is parameterized by arc length if and only if

$$\frac{dX^\mu}{ds}\frac{dX_\mu}{ds} = 1 \tag{7.19}$$

at all s.

Another natural choice of parameterization is by coordinate time in a given Lorentz frame F: If a curve $s \mapsto X(S)$ is future-timelike, then, since dX/ds is future-timelike, we have that $dX^0/ds > 0$, so $t(s) := X^0(s)$ is a strictly increasing function of s, so it possesses an inverse t^{-1}, $t^{-1}(t(s)) = s$, and $X(t^{-1}(\cdot))$ is the parameterization by coordinate time in F.

7.1.5 Index Contraction

If u, v, w are vectors in M, then expressions such as $u^\mu v^\nu$, $u_\mu v_\nu$, $u^\mu v_\nu$, $u^\lambda v^\mu w^\nu$, and $u_\lambda v^\mu w_\nu$ are elements of the tensor product spaces $M \otimes M$, $M' \otimes M'$, $M \otimes M'$, $M \otimes M \otimes M$, and $M' \otimes M \otimes M'$, respectively. General elements of these spaces are called *tensors* and denoted by symbols such as $T^{\mu\nu}$, $T_{\mu\nu}$, $T^\mu{}_\nu$, $T^{\lambda\mu\nu}$, $T_\lambda{}^\mu{}_\nu$, so the positions of the indices indicate which tensor space they are from. Since the pseudo inner product $(\cdot,\cdot) : M \times M \to \mathbb{R}$ extends uniquely to a linear mapping $M \otimes M \to \mathbb{R}$ given by $T^{\mu\nu} \mapsto T^{\mu\nu}\eta_{\mu\nu}$, this linear mapping is a Lorentz-invariant operation, commonly written as $T^{\mu\nu} \mapsto T^\mu_{\ \mu}$ in analogy to $x^\mu y_\mu = x^\mu y^\nu \eta_{\mu\nu}$. The Lorentz-invariant isomorphism $M \to M'$ defined by the pseudo inner product $x^\mu \mapsto x_\mu = x^\nu \eta_{\mu\nu}$ ("lowering an index") can be used to define Lorentz-invariant isomorphisms $M \otimes M \to M \otimes M'$ ($T^{\mu\nu} \mapsto T^\mu{}_\nu = T^{\mu\lambda}\eta_{\lambda\nu}$) and $M \otimes M \to M' \otimes M$ ($T^{\mu\nu} \mapsto T_\mu{}^\nu = T^{\lambda\nu}\eta_{\lambda\mu}$); "any index can be lowered." Using $\eta^{\mu\nu}$, any index can be raised. The operation $M \otimes M' \to \mathbb{R}$, $T^\mu{}_\nu \mapsto T^\mu{}_\mu$ is Lorentz invariant as well; it is called the *contraction* of the index μ with the index ν. Lowering an index and subsequently contracting it with an upper index is the same as applying $T^{\mu\nu} \mapsto T^{\mu\nu}\eta_{\mu\nu}$ mentioned above.

Example 7.4 The second derivatives of a scalar field $f : M \to \mathbb{R}$ form a tensor $T_{\mu\nu} = \partial_\mu \partial_\nu f$. The Lorentz-invariant operation of raising one index and contracting it with the other yields the scalar field

$$\partial^\mu \partial_\mu f = \partial_\mu \partial^\mu f = \eta^{\mu\nu}\partial_\mu \partial_\nu f = c^{-2}\frac{\partial^2 f}{\partial t^2} - \Delta f\,. \tag{7.20}$$

The mapping $f \mapsto \partial^\mu \partial_\mu f$ (also for complex-valued f) is known as the *D'Alembertian operator* denoted by

$$\Box := \partial^\mu \partial_\mu\,, \tag{7.21}$$

and the equation

$$\Box f = 0 \tag{7.22}$$

as the *D'Alembert equation* or *wave equation*.

7.1.6 Classical Electrodynamics as a Paradigm of a Relativistic Theory

According to classical electrodynamics, the world consists of space-time, an electromagnetic field, and N particles. Space-time is an affine Minkowski space M, and each particle $i \in \{1, \ldots, N\}$ has a path in M called its *world line* that is a timelike curve $\tau \mapsto X_i(\tau)$, which we assume to be parameterized by arc length τ. (It does not matter in the following where we choose the origin of the parameterization.) Each world line obeys the equation of motion ("relativistic Newtonian equation of motion with Lorentz force")

$$m_i \frac{d^2 X_{i\mu}}{d\tau^2} = e_i \frac{dX_i^\nu}{d\tau} F_{\mu\nu}(X_i(\tau)), \tag{7.23}$$

where $m_i > 0$ is a constant called the *rest mass* of particle i, $e_i \in \mathbb{R}$ is a constant called the *electric charge* of particle i, and the function $F : M \to M \otimes M$ is called the *electromagnetic field*. (The tensor product is taken over the reals, not the complex numbers.) It is anti-symmetric,

$$F_{\nu\mu} = -F_{\mu\nu} \tag{7.24}$$

(in the language of differential forms, it is a 2-form on M) and governed by *Maxwell's equations* (1865)

$$\partial_\lambda F_{\mu\nu} + \partial_\mu F_{\nu\lambda} + \partial_\nu F_{\lambda\mu} = 0 \tag{7.25a}$$

$$\partial^\mu F_{\mu\nu} = 4\pi J_\nu , \tag{7.25b}$$

where the vector field J is called the *charge current density* and given by

$$J^\nu(x) = \sum_{i=1}^{N} e_i \int d\tau' \, \delta^4\big(x - X_i(\tau')\big) \, \frac{dX_i^\nu}{d\tau}(\tau'). \tag{7.26}$$

This completes the definition of classical electrodynamics.

A basic problem with this theory is that its definition is inconsistent—it does not possess any solutions. That is because for a charge current that is concentrated on curves such as (7.26), every solution $F_{\mu\nu}$ of the Maxwell equations (7.25) will diverge on these curves, in fact at the rate

$$F_{\mu\nu}(x) \sim \frac{1}{r^2} \tag{7.27}$$

with

$$r^2 = (x^1 - X_i^1(\tau))^2 + (x^2 - X_i^2(\tau))^2 + (x^3 - X_i^3(\tau))^2 \tag{7.28}$$

with τ so chosen that $x^0 = X_i^0(\tau)$. At the same time, the equation of motion (7.23) demands that we evaluate $F_{\mu\nu}$ exactly at $x = X_i(\tau)$, where it is not defined.

I give a brief discussion of this problem in Appendix A.2. For our present purposes, let us ignore it and pretend that classical electrodynamics was a well defined theory. It helps that each of the defining equations (7.23), (7.25), and (7.26) by itself is consistent: we could solve (7.23) for $X_i(\cdot)$ if a smooth field $F_{\mu\nu}$ were given and (7.25) and (7.26) if the $X_i(\cdot)$ were given. As initial conditions for (7.23), we can specify on the $\{x^0 = 0\}$ hyperplane in M the positions X_i and the tangent direction to the world line (which amounts to specifying the velocity at this time, or $dX_i/d\tau$); as initial conditions for (7.25), we can specify $F_{\mu\nu}$ on the same hyperplane (although the Maxwell equations (7.25) imply that the initial data have to satisfy certain constraint conditions).

Each of the defining equations (7.23), (7.25), and (7.26) is Lorentz invariant: if it is true in one Lorentz frame, it is true in every Lorentz frame. In particular, a Lorentz transform of a solution is again a solution. The theory is *deterministic*: the intial data determine the field $F_{\mu\nu}$ and the world lines $X_i(\cdot)$ on all of M.

The theory is *local*: if we solve (7.23) for two different choices of a (given) field $F_{\mu\nu}$ that differ only in a region $A \subset M$ in the future of $\{x^0 = 0\}$, but with the same initial data for X_i and $dX_i/d\tau$, then the world lines of the solutions will differ only in the future of A. Here,

$$\text{future}(A) = \bigcup_{x \in A} \text{future}(x) \tag{7.29}$$

means the *relativistic future* or *future light cone* of A, where future(x) contains all vectors y with $y^0 \geq x^0$ and $y - x$ timelike or lightlike (including x itself); see Fig. 7.5. Likewise, if we solve Maxwell's equations (7.25) for two different choices of the charge current J that differ only in A in the future of $\{x^0 = 0\}$, but with the

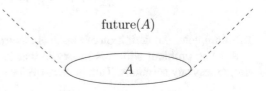

Fig. 7.5 The future of a space-time region A

same initial data, then the solutions will differ only in the future of A. We conclude that events cannot influence each other at spacelike separation, which means the theory is local.

In particular, a no-signaling theorem holds in this theory. It also turns out that world lines always stay timelike (so particles cannot move faster than light), that energy and momentum are conserved and cannot be transported faster than light. Moreover, nothing in the theory defines any concept of simultaneity-at-a-distance or temporal order of spacelike separated points.

This theory has strongly influenced the ideas of physicists of what a relativistic theory should be like. However, it is not obvious how many of its properties will carry over to quantum theories, as we know they must be *nonlocal*. It is remarkably difficult and subtle to give a general definition of what it means for a nonlocal theory to be relativistic; this will be a recurrent theme in this chapter. In addition, a concept of temporal order of spacelike separated points arises as soon as we consider *general relativity*, as I will discuss in Sect. 7.5.4.

7.1.7 Cauchy Surfaces

Definition 7.3 A *Cauchy surface* is a subset Σ of M which is intersected by every inextendible timelike-or-lightlike curve exactly once.

Intuitively, and leaving aside some technical subtleties, a Cauchy surface is a spacelike surface that extends to infinity; see Fig. 7.6. We will mostly use only *smooth* surfaces.

Example 7.5 A *hyperplane* is a 3d affine subspace P of M. It is called *spacelike* if and only if every $x, y \in P$ with $x \neq y$ are spacelike separated. Then there exists a Lorentz frame in which P is horizontal, i.e., $P = \{x^0 = \text{const.}\}$. Every spacelike hyperplane is a Cauchy surface.

For calculations, we often need to parameterize a surface Σ by a mapping $\Phi : B \to \Sigma$ on a subset B of $\mathbb{R}^3$. For example, if Σ is defined as the graph of

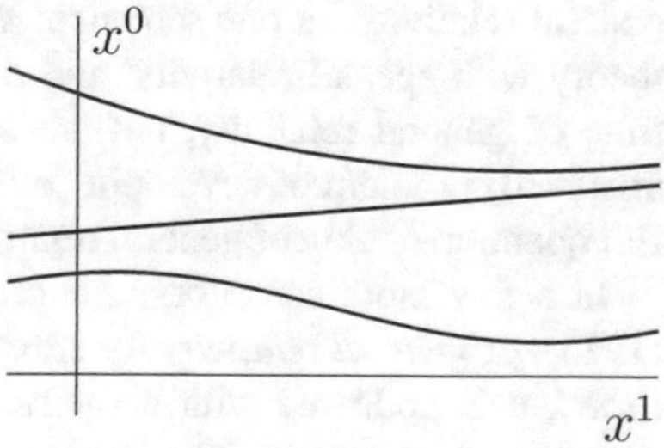

Fig. 7.6 Three examples of Cauchy surfaces

a function $T(x^1, x^2, x^3)$, $\Sigma = \{(T(\boldsymbol{x}), \boldsymbol{x}) : \boldsymbol{x} \in \mathbb{R}^3\}$, then $\Phi(\boldsymbol{x}) = (T(\boldsymbol{x}), \boldsymbol{x})$ is a parameterization.

Like surfaces in Euclidean space, Cauchy surfaces in Minkowski space have a natural measure of surface area, denoted by $V(d^3x)$ in the following, where d^3x is a 3d surface element. That is, we can integrate a function $f : \Sigma \to \mathbb{R}$ in the form

$$\int_\Sigma V(d^3x)\, f(x)\,. \tag{7.30}$$

Technically, V is a measure on the σ-algebra of Borel subsets of Σ. It can be expressed explicitly as follows.

Proposition 7.2 *In any bijective parameterization $\Phi : \mathbb{R}^3 \to \Sigma$ (so $x = \Phi(\boldsymbol{y})$),*

$$\int_\Sigma V(d^3x)\, f(x) = \int_{\mathbb{R}^3} d^3\boldsymbol{y}\, \sqrt{-\det({}^3g(\boldsymbol{y}))}\, f(\Phi(\boldsymbol{y}))\,, \tag{7.31}$$

where ${}^3g$ is the Riemann metric on Σ,

$${}^3g_{ij} = \frac{\partial \Phi^\mu}{\partial y^i}\frac{\partial \Phi^\nu}{\partial y^j}\eta_{\mu\nu} \tag{7.32}$$

for $i, j \in \{1, 2, 3\}$.

Remark 7.1 Also for *timelike* 3-surfaces (i.e., with one timelike and two spacelike dimensions), there is a natural and frame-independent surface area measure, which we will also denote by $V(d^3x)$. In coordinates, it is given by the same expression as in (7.31) but without the minus sign under the square root because the determinant is positive.

7.1.8 Outlook on General Relativity

Let me briefly put our considerations into perspective concerning special versus general relativity. In one sentence: We focus here on the combination of quantum theory with special relativity, and much can be generalized to the curved space-time of general relativity, but we stay away from the much harder (and as yet unsolved) problem of developing a theory of quantum gravity, which would be the full quantum analog of general relativity.

In a few more sentences: According to Einstein's general theory of relativity (1915), or *general relativity* for short, space-time is not actually Minkowski space-time but is endowed with a metric $g_{\mu\nu}(x)$ that varies from point to point, i.e., depends on x; one speaks of a *curved* space-time. At every x, $g_{\mu\nu}(x)$ is a bilinear form on the tangent space T_x at x; this is analogous to a curved surface (such as a sphere) in Euclidean 3-space, for which we can consider at every point x the tangent space T_x and the inner product on it, which is just the inner product of Euclidean 3-space restricted to T_x. A key difference is that for a surface in Euclidean space, the inner product on each T_x is positive definite (with the consequence that there exists a basis of T_x for which the matrix of the metric is the unit matrix), whereas for curved space-time, it is equivalent to the metric of Minkowski space (with the consequence that there exists a basis of T_x for which the matrix of the metric is η as in (7.4)). A manifold with a positive definite metric on each tangent space is called a *Riemannian manifold*; one with a metric equivalent to Minkowski space on each tangent space is called a *pseudo-Riemannian* or *Lorentzian manifold*.

General relativity also claims that gravity is nothing but the effect of space-time curvature, and it provides an equation, the *Einstein field equation*, that prescribes the curvature in terms of the mass current density. As a consequence, the metric of space-time itself depends on the distribution and motion of the matter in space-time. The Einstein field equation thus defines a classical field theory when combined with a suitable (classical) equation of motion for the matter.

Here, we will sometimes add remarks on how our considerations generalize to curved space-time, but we will always regard the metric as given.

7.2 Relativistic Schrödinger Equations

There are several relativistic versions of the Schrödinger equation that need separate discussion. That is because for different spins (and for zero versus nonzero mass), the equations look quite different, and because they do not all work equally well when it comes to formulating Born's rule. We focus first on the 1-particle equations before we turn to the many-particle case in Sect. 7.4.

7.2.1 The Klein-Gordon Equation

Fourier Transform

All Lorentz transformations have determinant ± 1. As a consequence, a subset $B \subset M$ of Minkowski space M will be represented in all Lorentz frames by subsets of $\mathbb{R}^4$ of equal volume. This volume we define as the *space-time volume* $\mathrm{vol}_4(B)$ of B; it is Lorentz invariant or, put differently, defined without the need to choose any coordinates.

As a consequence of *that*, Fourier transformation in all 4 space-time variables is a Lorentz-invariant operation, defined for $\psi : M \to \mathbb{C}$ by

$$\widehat{\psi}(k^\mu) = \frac{1}{4\pi^2} \int_M \mathrm{vol}_4(d^4x)\, e^{ik_\mu x^\mu}\, \psi(x^\mu)\,. \tag{7.33}$$

Here, we have used abstract index notation in two ways: first, we have written k^μ for the 4-vector (k^0, k^1, k^2, k^3) and, second, the notation $k^\mu x_\mu$, although it looks as if we had chosen a basis and were considering components relative to that basis, just means the Minkowski product and does not require the choice of any basis. In the following, we will just write d^4x instead of $\mathrm{vol}_4(d^4x)$. The inverse transformation reads

$$\psi(x^\mu) = \frac{1}{4\pi^2}\int_M d^4x\, e^{-ik_\mu x^\mu}\, \widehat{\psi}(k^\mu)\,. \tag{7.34}$$

Dispersion Relation

Consider the free 1-particle Schrödinger equation

$$i\hbar\partial_t\psi(t,\boldsymbol{x}) = -\frac{\hbar^2}{2m}\nabla_{\boldsymbol{x}}^2\psi(t,\boldsymbol{x})\,. \tag{7.35}$$

Recall that Fourier transformation translates the derivative operator ∂_{x^j} into multiplication by $-ik_j$; see (2.17). After expressing t as x^0/c and taking the Fourier transform in $x^0,\ldots,x^3$, (7.35) reads

$$\hbar ck^0\widehat{\psi}(k^0,k^1,k^2,k^3) = \frac{\hbar^2}{2m}\sum_{i=1}^{3}k_i^2\widehat{\psi}(k^0,k^1,k^2,k^3) \tag{7.36}$$

or, with $\boldsymbol{k} = (k^1,k^2,k^3)$,

$$\left(\hbar ck^0 - \frac{\hbar^2}{2m}\boldsymbol{k}^2\right)\widehat{\psi}(k^\mu) = 0\,. \tag{7.37}$$

Since a product vanishes whenever one of the factors does, this equation says that $\widehat{\psi}(k^\mu)$ can be nonzero only on the surface in $\mathbb{R}^4$ where the bracket vanishes, just as a delta function is nonzero only at a point; that is, $\widehat{\psi}$ must involve a factor $\delta(k^0 - k^0(\boldsymbol{k}))$ with

$$k^0(\boldsymbol{k}) = \frac{\hbar}{2mc}\boldsymbol{k}^2\,. \tag{7.38}$$

A relation between k^0 and $\boldsymbol{k}$ is called a *dispersion relation*.[3]

Now, (7.36) and (7.38) are not Lorentz invariant: if postulated in one Lorentz frame, they will fail in another. But there is an obvious Lorentz-invariant expression similar to $\boldsymbol{k}^2$, and that is $k^\mu k_\mu$. This leads us to the dispersion relation

$$\frac{\hbar^2}{2m}k^\mu k_\mu = 0\,. \tag{7.39}$$

This equation is clearly Lorentz invariant.

---

[3] A dispersion relation is often expressed as a relation between energy and momentum. That is equivalent, with $\hbar ck^0$ called the energy and $\hbar\boldsymbol{k}$ the momentum.

Exercise 7.2 (Lorentz Invariant Polynomials) Show that all Lorentz invariant polynomials of second order are of the form $\alpha k^\mu k_\mu + \beta$ for some (possibly complex) constants α, β.

So, writing $\gamma := -\beta/\alpha$, let us consider the more general dispersion relation

$$k^\mu k_\mu = \gamma \tag{7.40}$$

or, solving for k^0,

$$k^0 = \pm\sqrt{\gamma + \boldsymbol{k}^2}\,. \tag{7.41}$$

Since we need real (rather than complex) k^0, γ better be real and ≥ 0.

Now we can make a comparison with the dispersion relation (7.38) of the Schrödinger equation in the following way. We would expect the Schrödinger equation to come out in a non-relativistic limit, something like $c \to \infty$. So, let us assume there is a Lorentz frame in which all velocities are small. Small velocities are usually associated with small magnitudes of $\boldsymbol{k}$, and since

$$\sqrt{1+x} \approx 1 + \frac{x}{2} \quad \text{for } |x| \ll 1\,, \tag{7.42}$$

we can, at least for $\gamma \neq 0$, approximate the dispersion relation (7.41) by

$$k^0 \approx \pm\sqrt{\gamma}\Big(1 + \frac{\boldsymbol{k}^2}{2\gamma}\Big) = \pm\sqrt{\gamma} \pm \frac{1}{2\sqrt{\gamma}}\,\boldsymbol{k}^2 \tag{7.43}$$

for small $|\boldsymbol{k}|$ (in fact, for $|\boldsymbol{k}| \ll \sqrt{\gamma}$). The addition of a constant in a dispersion relation does not matter for most purposes, as replacing $k^0 \to k^0 + C$ corresponds to replacing $\psi(t, \boldsymbol{x}) \to e^{-iCct}\psi(t, \boldsymbol{x})$. But after dropping the additive constant, (7.43) agrees with (7.38) for

$$\pm = + \quad \text{and} \quad \gamma = \frac{m^2c^2}{\hbar^2}\,. \tag{7.44}$$

Inserting this in (7.40) leads us to the dispersion relation

$$k^\mu k_\mu = \frac{m^2c^2}{\hbar^2}\,. \tag{7.45}$$

The Klein-Gordon Equation

Multiplying by $\widehat{\psi}(k^\mu)$ and taking the inverse Fourier transform yields

$$-\,\partial^\mu \partial_\mu \psi(x^\mu) = \tfrac{m^2c^2}{\hbar^2}\psi(x^\mu)\,, \tag{7.46}$$

or

$$-\Box\psi = \frac{m^2c^2}{\hbar^2}\psi \tag{7.47}$$

in the notation of (7.21). This is the Klein-Gordon equation. For mass 0, it reduces to the D'Alembert equation (7.22).

The Klein-Gordon equation can also be considered for, apart from scalar-valued ($\mathbb{C}$-valued) functions, vector fields, spinor fields, or tensor fields, e.g.,

$$\Box\psi^\mu = -\frac{m^2c^2}{\hbar^2}\psi^\mu\,. \tag{7.48}$$

Example 7.6 Every solution $F_{\mu\nu}$ of the Maxwell equations (7.25) without sources, $J_\nu = 0$, satisfies the massless Klein-Gordon equation

$$\Box F_{\mu\nu} = 0\,. \tag{7.49}$$

Indeed,

$$\partial^\lambda\partial_\lambda F_{\mu\nu} \overset{(7.25a)}{=} \partial^\lambda\Big(-\partial_\mu F_{\nu\lambda} - \partial_\nu F_{\lambda\mu}\Big) \tag{7.50a}$$

$$= -\partial_\mu\partial^\lambda F_{\nu\lambda} - \partial_\nu\partial^\lambda F_{\lambda\mu} \tag{7.50b}$$

$$= \partial_\mu\partial^\lambda F_{\lambda\nu} - \partial_\nu\partial^\lambda F_{\lambda\mu} \tag{7.50c}$$

$$\overset{(7.25b)}{=} 4\pi\,\partial_\mu J_\nu - 4\pi\,\partial_\nu J_\mu = 0. \tag{7.50d}$$

Positive Energy Solutions

One respect in which the Klein-Gordon equation and the Schrödinger equation do not match is that the former is of second order in time, while the latter is of first order in time. So, the Klein-Gordon equation is not, as we might have expected, a Schrödinger equation

$$i\hbar\frac{\partial\psi}{\partial t} = H\psi \tag{7.51}$$

with a particular Hamiltonian H. However, there is a further twist to this story, as I will explain now: there is a way in which the Klein-Gordon equation does fit the format (7.51).

Consider a PDE of the form (7.51) with a Hamiltonian H that is a polynomial in the derivative operators, $H = P(\partial^1, \partial^2, \partial^3)$ with P a polynomial in three variables. Then the Fourier transform in space and time of the PDE reads

$$\hbar ck^0\widehat{\psi}(k^\mu) = P(i\boldsymbol{k})\,\widehat{\psi}(k^\mu) \tag{7.52}$$

corresponding to the dispersion relation

$$\hbar c k^0 = P(i\boldsymbol{k}). \tag{7.53}$$

Since the Hamiltonian is the energy operator, $P(i\boldsymbol{k})$ can be regarded as the energy of the plane wave $e^{i\boldsymbol{k}\cdot\boldsymbol{x}}$ with wave vector $\boldsymbol{k}$. Negative energies are often regarded as unphysical. For (7.41), this suggests using the positive square root,

$$k^0 = +\sqrt{m^2c^2/\hbar^2 + \boldsymbol{k}^2} \tag{7.54}$$

or

$$i\hbar\partial_t\psi = \sqrt{m^2c^4 - c^2\hbar^2\nabla^2}\,\psi \tag{7.55}$$

or (7.51) with

$$H = \sqrt{m^2c^4 - c^2\hbar^2\nabla^2}\,. \tag{7.56}$$

Here, the square root of a positive operator T is defined as the unique positive operator R such that $R^2 = T$.[4] (Note that $-\nabla^2$ is a positive operator.) Equation (7.55) is also called the *Klein-Gordon equation of first order*. It is widely used as the relativistic Schrödinger equation for a spin-0 particle. (While it is not clear whether spin-0 particles exist in nature, they are often considered in theoretical studies as the simplest possible case.)

Operators in $L^2(\mathbb{R}^3)$ that act on the 3d Fourier transform of ψ by multiplication by a polynomial in $\boldsymbol{k}$ are differential operators. Operators in $L^2(\mathbb{R}^3)$ that act on the 3d Fourier transform of ψ by multiplication by *some function* in $\boldsymbol{k}$ are called *pseudo-differential operators*; (7.56) is an example. A relevant difference between differential and other pseudo-differential operators is that differential operators are *local* and pseudo-differential operators in general are not. This means the following: if I know $\psi(\boldsymbol{x})$ not for all $\boldsymbol{x} \in \mathbb{R}^3$ but only in some open set $A \subset \mathbb{R}^3$, and if D is a differential operator, then I still know $D\psi(\boldsymbol{x})$ everywhere in A. Put differently, if $\phi \neq \psi$ but they agree on A, then also $D\phi$ and $D\psi$ agree on A. This is not so with pseudo-differential operators R: for computing $(R\psi)(\boldsymbol{x})$, it is not sufficient to know ψ in a neighborhood of $\boldsymbol{x}$; rather, $(R\psi)(\boldsymbol{x})$ will depend on the whole function ψ,

[4] Equivalently, one can say that it is the square root function $\sqrt{\cdot} : [0,\infty) \to [0,\infty)$ applied to the operator T (viz., to the generalized eigenvalues of T). The square root of a diagonal matrix with non-negative diagonal entries $d_1, \ldots, d_n$ is the diagonal matrix with diagonal entries $\sqrt{d_1}, \ldots, \sqrt{d_n}$. The square root of a multiplication operator M_f multiplying by a non-negative function f is $M_{\sqrt{f}}$. The square root of $UM_fU^\dagger$ with unitary U is $UM_{\sqrt{f}}U^\dagger$. Thus, (7.56) is the operator acting on the 3d Fourier transform of ψ like multiplication by $\sqrt{m^2c^4 + c^2\hbar^2\boldsymbol{k}^2}$.

also in places remote from $\boldsymbol{x}$. For a Hamiltonian H, it seems more plausible that H should be a local operator.

Another problem with the Klein-Gordon Hamiltonian (7.56) is that it is not clear, for a wave function governed by it, what the Born rule for position should look like, and how the probability current should be computed. We will discuss that problem in Sect. 7.3.8.

7.2.2 Two-Spinors and Four-Vectors

In Sect. 2.3 we have introduced spinors. Now we discuss how they connect with Minkowski space. We will often say "2-spinor" for a spinor because it has two complex components, in contrast to the Dirac spinors used in the Dirac equation, which have four complex components. From now on, we will work in units with

$$c = 1 \quad \text{and} \quad \hbar = 1. \tag{7.57}$$

Two-Spinors and Three-Vectors

Let us begin by reviewing a few relevant facts from Sect. 2.3 from a different angle: To every 2-spinor $\phi \in S$, there corresponds a 3-vector (i.e., a vector in physical space, as opposed to 4-vectors from space-time), given by $\phi^* \boldsymbol{\sigma} \phi$. Rotations $R \in SO(3)$ act on spinors as follows. Suppose R is the rotation through the angle ϑ around the axis through the unit vector $\boldsymbol{n} \in \mathbb{R}^3$ (with the sense of rotation given by the right-hand rule). If $\boldsymbol{n}\times$ denotes the matrix of the linear mapping $\boldsymbol{v} \mapsto \boldsymbol{n} \times \boldsymbol{v}$,

$$\boldsymbol{n}\times = \begin{pmatrix} 0 & -n_3 & n_2 \\ n_3 & 0 & -n_1 \\ -n_2 & n_1 & 0 \end{pmatrix}, \tag{7.58}$$

then R can be expressed as

$$R = \exp(\vartheta \boldsymbol{n}\times)\,. \tag{7.59}$$

Under the rotation R, the spinor ϕ gets rotated in spin space S according to the unitary matrix

$$M(R) = \pm \exp(i\sigma_{\boldsymbol{n}}\vartheta/2) \tag{7.60}$$

with $\sigma_{\boldsymbol{n}} = \boldsymbol{n} \cdot \boldsymbol{\sigma}$ the Pauli matrix in the $\boldsymbol{n}$ direction. Since $\sigma_{\boldsymbol{n}}$ is self-adjoint, $M(R)$ is unitary. The appearance of $\vartheta/2$ has to do with the fact that $M(R)$ rotates by half the angle. The $\pm$ sign means that two matrices $M \in U(2)$, differing by a sign, can be associated with $R \in SO(3)$. Roughly speaking, the 2-spinors (or the matrices $M(R)$) form a representation of $SO(3)$, but because of the sign ambiguity, a correct statement would be that the $M(R)$ form a *projective* representation (meaning a representation up to scalar prefactors). Here is another approach to dealing with the sign ambiguity: the *universal covering group*. As a manifold, $SO(3)$ is *multiply connected* ("topologically folded"), a property we discuss in more detail in a different context in Appendix A.1. We can "unfold" $SO(3)$ and obtain the universal covering group $\widehat{SO(3)}$ and a group homomorphism $\varphi : \widehat{SO(3)} \to SO(3)$. In fact, $\widehat{SO(3)}$ is canonically identified with $SU(2)$, and every R has two pre-images under φ, $M(R)$, and $-M(R)$. Roughly speaking, rotating by 360° does not bring back the identity but $-I \in SU(2)$. We can then say that the action of rotations on spinors is given by the appropriate representation of $\widehat{SO(3)}$.

Action of Lorentz Transformations

Now let us turn to relativistic space-time. Lorentz transformations also act on 2-spinors. For any proper Lorentz transformation Λ, you apply to the spinor a certain matrix $M(\Lambda)$, which is unique up to sign. Consider again the Lorentz boost Λ of (7.9) in Example 7.1; it can be expressed as

$$\Lambda = \exp(\xi \begin{pmatrix} 0&0&0&1 \\ 0&0&0&0 \\ 0&0&0&0 \\ 1&0&0&0 \end{pmatrix}). \tag{7.61}$$

It acts on S according to[5]

$$M(\Lambda) = \pm \begin{pmatrix} e^{\xi/2} & \\ & e^{-\xi/2} \end{pmatrix} = \pm \exp(\tfrac{1}{2}\xi \begin{pmatrix} 1 & \\ & -1 \end{pmatrix}). \tag{7.62}$$

Noticing that this matrix is not unitary, you might worry that probability conservation will break down. I can give away already that it will not, despite the fact that $M(\Lambda)$ is not unitary.

Let us express this in terms of group representations. As a manifold, the Lorentz group $O(1, 3)$ has four connected components: One consists of those Lorentz transformations that (among other things) reverse time, one of those that reflect space, one of those that do both, and one of those that do neither. The latter one, the connected component of the identity, forms a subgroup of $O(1, 3)$ called the "proper Lorentz group" $\mathscr{L}_0 = O^{\uparrow}_{+}(1, 3)$. Lorentz transformations Λ that reverse time have negative entry $\Lambda^0{}_0$, the others positive; Λ that either reverse time or reflect space but not both have negative determinant; those that do both or neither have positive determinant. The 2-spinors (or the matrices $M(\Lambda)$) form a projective representation of $\mathscr{L}_0$. Alternatively, we can say that the $M(\Lambda)$ form a representation of the universal covering group $\widehat{\mathscr{L}_0}$ of $\mathscr{L}_0$.

[5] For derivation see Eq. (1.2.37) in Penrose and Rindler (1984) [38].

Proposition 7.3 *$\widehat{\mathscr{L}_0}$ is canonically identified with $SL(2, \mathbb{C})$ (the "special linear group" over $\mathbb{C}$, i.e., the group of complex 2×2 matrices with determinant 1).*

For proof and detailed discussion, see pages 16–18 and 45–46 in Penrose and Rindler (1984) [38] or Ch. 8 in Sexl and Urbantke (2001) [41]. As the representation is not unitary, the usual inner product in spin space S is not Lorentz invariant, it depends on the Lorentz frame. For this reason, we will mostly not use this inner product, and correspondingly not regard S as a Hilbert space. However, as we will see, there is another kind of product in S that is Lorentz invariant but not positive definite and not even symmetric.

Conjugate Vector Space

But before we get there, I need to introduce the conjugate spin space. For every complex vector space V, one can define the *conjugate vector space* $\overline{V}$. Before I give a definition, let me describe the intuition behind it. Suppose that for every element $v \in V$, we can form the *conjugate of* v, Cv. While the conjugate of a function $\psi \in L^2(\mathbb{R}^d)$ is another element of $L^2(\mathbb{R}^d)$, we now need to consider the case that Cv does not belong again to V but to a different space $\overline{V}$. The mapping C will be conjugate-linear ("anti-linear"), i.e.,

$$C(\lambda v) = \lambda^* C v \ \forall \lambda \in \mathbb{C} \quad \text{and} \quad C(v + v') = C(v) + C(v') \ \forall v, v' \in V\,. \tag{7.63}$$

Furthermore, C will be bijective; in particular, $\overline{V}$ contains *only* the conjugates of elements of V. Such a mapping C is called an "anti-isomorphism."

Here are two ways of defining $\overline{V}$ for given V. According to the first, more abstract, definition, we say that $\overline{V}$ is any complex vector space equipped with an anti-isomorphism $C : V \to \overline{V}$. It follows that the pair $(\overline{V}, C)$ is uniquely determined up to isomorphism: For any complex vector space W and anti-isomorphism $K : V \to W$, there exists an isomorphism $J : \overline{V} \to W$ such that $K = JC$. The second, more constructive, definition is based on the idea of labeling every element $\overline{v} \in \overline{V}$ by the v with $Cv = \overline{v}$ and simply writing v for $\overline{v}$. Thus, according to this definition, $\overline{V}$ is the same set as V but a different vector space: while addition is the same as in V, the scalar multiplication is different; for $\lambda \in \mathbb{C}$ and $v \in \overline{V} = V$, scalar multiplication in $\overline{V}$ is $\lambda^* v$ when expressed in terms of the scalar multiplication of the vector space V.

In the following, it will not matter which definition we use. As a consequence of both, $\overline{\overline{V}}$ is canonically identified with V. The choice of a basis B in V also provides a basis $\overline{B}$ of $\overline{V}$. And instead of Cv, we will mostly write $\overline{v}$.

For spin space S (a 2d complex vector space), we will have to consider its conjugate $\overline{S}$. Lorentz transformations act differently on $\overline{S}$ than on S, viz., according to the conjugate representation.

We note further that if ϕ has complex coefficients c_1 and c_2 in the basis B, then $\overline{\phi}$ has coefficients c_1^* and c_2^* in the basis $\overline{B}$. The components of $\phi \in S$ relative to B are henceforth denoted by ϕ^A, those of $\chi \in \overline{S}$ relative to $\overline{B}$ by $\chi^{A'}$.

Relation to 4-Vectors

Consider the tensor product $S \otimes \overline{S}$ over the complex numbers. An element has four complex components $u^{AA'}$. Define the "adjoint" mapping as conjugation of each factor, following by the permutation of the two factors; it is a mapping $S \otimes \overline{S} \to S \otimes \overline{S}$. The components of the adjoint element are exactly the entries of the adjoint (conjugate-transpose) matrix. An element of $S \otimes \overline{S}$ is called *Hermitian* if it is equal to its adjoint. The Hermitian elements form a real subspace of $S \otimes \overline{S}$ of dimension 4, henceforth denoted $\mathrm{Herm}(S \otimes \overline{S})$, and they transform like 4-vectors:

Proposition 7.4 *The representation of* $\widehat{\mathscr{L}_0}$ *on Minkowski space* M *is equivalent to that on* $\mathrm{Herm}(S \otimes \overline{S})$.

For proof and more detail, see again Penrose and Rindler (1984) [38] or Sexl and Urbantke (2001) [41]. As a consequence, the Hermitian elements can be canonically identified with M. That is, we can always translate Hermitian elements $u^{AA'}$ of $S \otimes \overline{S}$ into 4-vectors u^μ and back; roughly speaking, the index μ is equivalent to the index pair AA'. The translation is an isomorphism of real vector spaces $M \to S \otimes \overline{S}$, and the coefficients $\frac{1}{\sqrt{2}}\sigma_\mu^{AA'}$ of this linear mapping relative to the given bases (a Lorentz frame in M, $|\uparrow\rangle$, and $|\downarrow\rangle$ in S for this Lorentz frame and their conjugates in $\overline{S}$) are (up to the $\sqrt{2}$) given by $\sigma_0 = I$ the unit matrix and $\sigma_1, \sigma_2, \sigma_3$ the Pauli matrices.

For every 2-spinor ϕ, the expression $\sqrt{2}\,\phi^A\,\overline{\phi}^{A'}$ is Hermitian and corresponds to the 4-vector

$$j^\mu = (\phi^*\phi, \phi^*\sigma\phi) \tag{7.64}$$

with the notation

$$\phi^*\psi = \phi_1^*\psi_1 + \phi_2^*\psi_2 \,. \tag{7.65}$$

The association $\phi \mapsto j$ is Lorentz invariant because $u^{AA'} \mapsto u^\mu$ is.

Proposition 7.5 *j is always future-lightlike.*

Proof This follows from $j^0 = \phi^*\phi \geq 0$ and (writing all spinor indices downstairs as in Chap. 2.3)

$$j^\mu j_\mu = (j^0)^2 - \sum_{i=1}^{3}(j^i)^2 = (\phi^*\phi)^2 - \sum_{i=1}^{3}(\phi^*\sigma^i\phi)^2 \tag{7.66a}$$

$$\begin{aligned} &= (|\phi_1|^2 + |\phi_2|^2)^2 - (\phi_1^*\phi_2 + \phi_2^*\phi_1)^2 \\ &\quad - (-i\phi_1^*\phi_2 + \phi_2^* i\phi_1)^2 - (|\phi_1|^2 - |\phi_2|^2)^2 \end{aligned} \tag{7.66b}$$

$$= |\phi_1|^4 + 2|\phi_1|^2|\phi_2|^2 + |\phi_2|^4 - (\phi_1^*\phi_2)^2 - 2|\phi_1|^2|\phi_2|^2 - (\phi_2^*\phi_1)^2$$

$$+ (\phi_1^*\phi_2)^2 - 2|\phi_1|^2|\phi_2|^2 + (\phi_2^*\phi_1)^2 - |\phi_1|^4 + 2|\phi_1|^2|\phi_2|^2 - |\phi_2|^4 \tag{7.66c}$$

$$= 0 \,. \tag{7.66d}$$

□

That fits together with Sect. 2.3: The 3-vector associated with ϕ is exactly the space part of j, i.e., $\boldsymbol{j} = (j^1, j^2, j^3)$. $SO(3)$ is a subgroup of $\mathscr{L}_0$, and the representation of $SO(3)$ is just the restriction of the representation of $\mathscr{L}_0$ to the subgroup.

We also understand more now about why the inner product (7.65) we used in Chap. 2.3 is not Lorentz invariant: $\phi^*\phi$ is the time component j^0 of a 4-vector, rather than a scalar.

Lorentz-Invariant Product

There is a Lorentz-invariant operation between two 2-spinors ϕ and χ, an anti-symmetric scalar product

$$\phi^1\chi^2 - \phi^2\chi^1 \tag{7.67}$$

(no conjugation!), usually written as

$$\phi^A\chi^B\varepsilon_{AB} \tag{7.68}$$

with the matrix

$$(\varepsilon_{AB}) = \begin{pmatrix} 0 & 1 \\ -1 & 0 \end{pmatrix}, \tag{7.69}$$

known as the *Levi-Civita spinor* or *epsilon spinor*, and the *sum convention*

$$S^A T_A = \sum_{A=1}^{2} S^A T_A \,, \tag{7.70}$$

according to which every index that appears in both the upper and the lower position is to be summed over. (The need for upper and lower indices arises again from the fact that the scalar product is not the Euclidean/Hermitian inner product $\sum_\alpha x_\alpha^* y_\alpha$, as explained in Sect. 7.1.3.)

Here is a way of seeing that the operation (7.67) is Lorentz invariant: This expression equals the determinant of the 2×2 matrix A with columns ϕ and χ. If the entries were real, this quantity would mean the signed area of the parallelogram spanned by ϕ and χ. Now matrices in $SL(2)$ have determinant 1 and thus leave areas unchanged; for matrices from $SL(2, \mathbb{C})$, it remains true that they preserve $\det(A)$.

There is also a corresponding epsilon spinor in $\overline{S}$, obtained by taking the conjugate of the epsilon spinor in S and denoted by $\varepsilon_{A'B'}$ (although a logical notation would be $\overline{\varepsilon}_{A'B'}$). Now consider the tensor product of the two epsilon spinors,

$$\varepsilon_{AB}\,\varepsilon_{A'B'}\,, \tag{7.71}$$

an object with four indices. It turns out to be Hermitian in the index pair AA' and again in the index pair BB'. Thus, it can be identified with a tensor $T_{\mu\nu}$, in fact

$$T_{\mu\nu} = \eta_{\mu\nu}\,, \tag{7.72}$$

the Minkowski metric. So the epsilon spinor plays a role in S roughly similar to the metric in Minkowski space.

Spinor indices can be raised and lowered using the epsilon spinors,

$$\phi_A = \phi^B\,\varepsilon_{BA} \quad \text{and} \quad \phi^A = \varepsilon^{AB}\,\phi_B\,. \tag{7.73}$$

However, the order is relevant, as the "see-saw rule" makes clear:

$$\phi^A\chi_A = -\phi_A\chi^A\,, \text{ as opposed to } x^\mu y_\mu = x_\mu y^\mu\,. \tag{7.74}$$

The Dirac 4-spinor that occurs in the Dirac equation can be regarded as a pair consisting of a 2-spinor and a conjugate 2-spinor, $(\phi^A, \chi^{A'})$.

7.2.3 The Weyl Equation

Since every 4-vector u^μ can be translated into $u^{AA'}$ and u_μ into $u_{AA'}$, we can write ∂_μ as $\partial_{AA'}$. Explicitly, it is the following matrix of differential operators:

$$(\partial_{AA'}) = \frac{1}{\sqrt{2}}\begin{pmatrix} \partial_0 + \partial_3 & \partial_1 - i\partial_2 \\ \partial_1 + i\partial_2 & \partial_0 - \partial_3 \end{pmatrix}. \tag{7.75}$$

The simplest possible Lorentz-invariant equation for a spinor field ψ^A, that is, for $\psi : M \to S$, then reads

$$\partial_{AA'}\psi^A = 0 \tag{7.76}$$

or (dropping the unnecessary factor $\sqrt{2}$)

$$\sigma^\mu_{AA'}\partial_\mu\psi^A = 0\,. \tag{7.77}$$

This is known as the *Weyl equation*. In Hamiltonian form (i.e., solved for the time derivative), it reads

$$i\frac{\partial \psi(t,\boldsymbol{x})}{\partial t} = H\psi(t,\boldsymbol{x}) \tag{7.78a}$$

$$= i\boldsymbol{\sigma}\cdot\nabla\psi(t,\boldsymbol{x}) \tag{7.78b}$$

with $\nabla = (\partial_x, \partial_y, \partial_z)$. We will turn to the Dirac equation in the next section; we can mention already that the Dirac equation with mass zero corresponds to two decoupled equations, the Weyl equation and the conjugate Weyl equation.

Proposition 7.6 *The Weyl Hamiltonian* $i\boldsymbol{\sigma}\cdot\nabla$ *(defined, say, on the space of smooth functions* $\mathbb{R}^3 \to \mathbb{C}^2$ *with compact support) possesses a unique self-adjoint extension.*

This follows, e.g., from the results of Chernoff (1973) [13].

Relation to the Klein-Gordon Equation

Proposition 7.7 *Every solution of the Weyl equation also satisfies the massless second-order Klein-Gordon equation* $\Box\psi = 0$.

For this reason, the wave function of the Weyl equation belongs to a particle of mass 0 and spin $\frac{1}{2}$.

Proof One way of seeing this is to take $-i$ times the time derivative of (7.78). Note that

$$(\boldsymbol{\sigma}\cdot\nabla)^2 = \sum_{i,j=1}^{3} \sigma^i\sigma^j\partial_i\partial_j\,. \tag{7.79}$$

Using $\partial_i\partial_j = \partial_j\partial_i$ and renaming $i \leftrightarrow j$, we can rewrite the last expression as

$$\sum_{i,j=1}^{3} \sigma^j\sigma^i\partial_i\partial_j\,. \tag{7.80}$$

Taking the average of (7.79) and (7.80) yields

$$(\boldsymbol{\sigma}\cdot\nabla)^2 = \sum_{i,j=1}^{3} \tfrac{1}{2}(\sigma^i\sigma^j + \sigma^j\sigma^i)\partial_i\partial_j\,. \tag{7.81}$$

Since

$$\sigma^i\sigma^j + \sigma^j\sigma^i = 2\delta^{ij}\,I\,, \tag{7.82}$$

it follows that

$$(\sigma \cdot \nabla)^2 = I\Delta\,, \tag{7.83}$$

as claimed.

An *alternative proof* makes use of the fact that for every symmetric tensor $T_{\mu\nu} = T_{\nu\mu}$, the rank-2 spinor

$$s_{AB} := T_{AA'BB'}\varepsilon^{A'B'} \tag{7.84}$$

can be expressed as

$$s_{AB} = \tfrac{1}{2}\varepsilon_{AB}T^{\mu}{}_{\mu}\,. \tag{7.85}$$

Indeed, every spinor $T_{A'B'}$ (possibly with further indices) can be decomposed into a symmetric part $T_{(A'B')} = \frac{1}{2}T_{A'B'} + \frac{1}{2}T_{B'A'}$ and an anti-symmetric part $T_{[A'B']} = \frac{1}{2}T_{A'B'} - \frac{1}{2}T_{B'A'}$. Since $\varepsilon^{A'B'}$ is anti-symmetric, only the anti-symmetric part $T_{[A'B']}$ contributes to s_{AB}. But since $T_{AA'BB'}$ is symmetric against the simultaneous exchange $A \leftrightarrow B$ and $A' \leftrightarrow B'$, $T_{[A'B']}$ is anti-symmetric in the index pair AB. But every anti-symmetric R_{AB} is a multiple of ε_{AB}, with the prefactor given by $\frac{1}{2}\varepsilon^{CD}R_{CD}$. Finally,

$$\varepsilon^{CD}\varepsilon^{C'D'}T_{CC'DD'} = \eta^{CC'DD'}T_{CC'DD'} = \eta^{\mu\nu}T_{\mu\nu} = T^{\mu}{}_{\mu}\,, \tag{7.86}$$

which yields (7.85).

The relation (7.85) remains valid when T has a further spin index as in $T_{\mu\nu}{}^{C}$. Now apply this to

$$T_{\mu\nu}{}^{C} := \partial_\mu\partial_\nu\psi^C\,. \tag{7.87}$$

Then

$$\partial_{AA'}\partial_{BB'}\psi^C\,\varepsilon^{A'B'} = \tfrac{1}{2}\varepsilon_{AB}\partial^\mu\partial_\mu\psi^C\,. \tag{7.88}$$

Now set $C = A$ and contract, so

$$\partial_{AA'}\partial_{BB'}\psi^A\,\varepsilon^{A'B'} = \tfrac{1}{2}\varepsilon_{AB}\partial^\mu\partial_\mu\psi^A = \tfrac{1}{2}\Box\psi_B\,. \tag{7.89}$$

On the other hand, starting from the Weyl equation (7.76), taking the $\partial_{BB'}$ derivative and contracting with $\varepsilon^{A'B'}$,

$$0 = \partial_{BB'}\partial_{AA'}\psi^A\,\varepsilon^{A'B'} = \partial_{AA'}\partial_{BB'}\psi^A\,\varepsilon^{A'B'} = \tfrac{1}{2}\Box\psi_B\,, \tag{7.90}$$

so also $0 = \Box\psi^A$. □

7.2.4 The Dirac Equation

The fact that the Weyl equation implies the mass-zero Klein-Gordon equation signals that it can be correct only for mass-zero particles. But electrons have nonzero mass. To implement that, we would like to put a term of the form "mass times wave function" on the right-hand side of the Weyl equation (7.76). However, the term on the left-hand side, $\partial_{AA'}\psi^A$, has an index A', so it takes values in the conjugate spin space $\overline{S}$, not in S, so the equation "$\partial_{AA'}\psi^A = m\psi^A$" is not meaningful. But a nonzero mass *can* be implemented if we assume that the wave function consists of a pair, $\psi = (\phi^A, \chi_{A'})$, of a spinor field ϕ^A and a conjugate spinor field $\chi_{A'}$. (It does not matter much whether the indices are upstairs or downstairs since they can be lowered or raised if needed using the epsilon spinors as in (7.73).) Then we can write, with constant $m \geq 0$ called the rest mass,

$$i\sqrt{2}\,\partial_{AA'}\phi^A = m\chi_{A'} \tag{7.91a}$$

$$i\sqrt{2}\,\partial^{AA'}\chi_{A'} = m\phi^A \tag{7.91b}$$

or, in slightly different notation,

$$i\sigma^{\mu}_{AA'}\partial_{\mu}\phi^A = m\chi_{A'} \tag{7.92a}$$

$$i\sigma_{\mu}^{AA'}\partial^{\mu}\chi_{A'} = m\phi^A\,. \tag{7.92b}$$

This pair of equations is the *Dirac equation* (Dirac 1928 [20]). It is usually written in a more compact form for $\psi : M \to D$ with the *Dirac spin space*

$$D := S \oplus \overline{S}' \tag{7.93}$$

(where the prime means dual space), or in coordinates $\psi : \mathbb{R}^4 \to \mathbb{C}^4$, as

$$i\gamma^{\mu}\partial_{\mu}\psi(x) = m\psi(x) \tag{7.94}$$

with 4×4 matrices

$$\gamma^0 = \begin{pmatrix} 0 & 1 \\ 1 & 0 \end{pmatrix} \tag{7.95a}$$

$$\gamma^i = \begin{pmatrix} 0 & -\sigma_i \\ \sigma_i & 0 \end{pmatrix}\,. \tag{7.95b}$$

Here, 0 means $\left(\begin{smallmatrix} 0 & 0 \\ 0 & 0 \end{smallmatrix}\right)$, 1 means $\left(\begin{smallmatrix} 1 & 0 \\ 0 & 1 \end{smallmatrix}\right)$, and σ_i the i-th Pauli matrix.

The specific form (7.95) of the gamma matrices is linked to a certain basis of D called the *Weyl representation*, consisting of a certain basis of 2d spin space S and the conjugate dual basis. Other choices of basis in D will lead to different components of the gamma matrices, but the abstract form (7.94) remains valid, as does the following relation known as the *Clifford relation*:

$$\gamma^\mu \gamma^\nu + \gamma^\nu \gamma^\mu = 2\eta^{\mu\nu} I . \tag{7.96}$$

In particular, $(\gamma^0)^{-1} = \gamma^0$. In Hamiltonian form, the Dirac equation reads

$$i\frac{\partial \psi}{\partial t} = -i\boldsymbol{\alpha} \cdot \nabla \psi + \beta m \psi \tag{7.97}$$

with matrices

$$\alpha^i = \gamma^0 \gamma^i \quad \text{and} \quad \beta = \gamma^0 , \tag{7.98}$$

given in the Weyl representation by (see, e.g., Thaller 1992 [44, p. 36]):

$$\boldsymbol{\alpha} = \begin{pmatrix} \sigma & 0 \\ 0 & -\sigma \end{pmatrix} \tag{7.99a}$$

$$\beta = \begin{pmatrix} 0 & 1 \\ 1 & 0 \end{pmatrix} . \tag{7.99b}$$

So, the Dirac equation is a PDE of the form of a general Schrödinger equation, and indeed it defines a unitary time evolution for ψ:

Proposition 7.8 *The Dirac Hamiltonian $-i\boldsymbol{\alpha} \cdot \nabla + \beta m$ (on the space of smooth functions $\mathbb{R}^3 \to \mathbb{C}^4$ with compact support) possesses a unique self-adjoint extension.*

For the proof, see Chernoff (1973) [13] or Thaller (1992) [44, Sec. 1.4.4].

Relation to the Klein-Gordon Equation

Proposition 7.9 *Every solution of the Dirac equation also satisfies the second-order Klein-Gordon equation $\Box\psi = -m^2\psi$.*

Proof Apply $-i\gamma^\nu \partial_\nu$ to both sides of the Dirac equation to obtain

$$\gamma^\nu \partial_\nu \gamma^\mu \partial_\mu \psi(x) = -im\gamma^\nu \partial_\nu \psi(x) . \tag{7.100}$$

Since γ^μ does not depend on x, we can exchange the order of ∂_ν and γ^μ on the left-hand side (while we may not exchange γ^ν and γ^μ, as these two matrices may not commute) yielding

$$\gamma^\nu \gamma^\mu \partial_\nu \partial_\mu \psi(x) . \tag{7.101}$$

Since $\partial_\mu \partial_\nu = \partial_\nu \partial_\mu$, the left-hand side of (7.100) can be rewritten as

$$\gamma^\nu \gamma^\mu \partial_\mu \partial_\nu \psi(x) . \tag{7.102}$$

Since μ and ν are summed over (contracted), they can be renamed $\mu \leftrightarrow \nu$, yielding

$$\gamma^\mu \gamma^\nu \partial_\nu \partial_\mu \psi(x) . \tag{7.103}$$

Since each of (7.101) and (7.103) is equal to the left-hand side of (7.100), so is their average,

$$\tfrac{1}{2}(\gamma^\nu \gamma^\mu + \gamma^\mu \gamma^\nu)\partial_\nu \partial_\mu \psi(x) . \tag{7.104}$$

By the Clifford relation (7.96), this expression equals

$$\eta^{\mu\nu} \partial_\mu \partial_\nu \psi(x) = \Box \psi(x) . \tag{7.105}$$

So,

$$\Box \psi(x) = -im\gamma^\nu \partial_\nu \psi(x) . \tag{7.106}$$

Renaming $\nu \to \mu$ and using the Dirac equation (7.94), we obtain that

$$\Box \psi(x) = -m^2 \psi(x) , \tag{7.107}$$

as claimed. □

Lorentz Invariance

The Dirac equation is Lorentz invariant. This can be seen from its expression in 2-spinor calculus (7.91), which involves only Lorentz-invariant operations. Explicitly, Lorentz invariance means the following: for every proper Lorentz transformation Λ, there is a matrix $S(\Lambda)$, unique up to a sign, representing the action of Λ on Dirac spinors, and a Dirac spinor field transforms according to

$$\widetilde{\psi}(x) = S(\Lambda)\, \psi(\Lambda x) . \tag{7.108}$$

If ψ is a solution of the Dirac equation, then so is $\widetilde{\psi}$. This fact can also be verified directly; the key step is the relation

$$\gamma^\nu = \Lambda^\nu{}_\mu\, S(\Lambda)\gamma^\mu S(\Lambda)^{-1} . \tag{7.109}$$

In the Weyl representation, the transformation matrix $S(\Lambda)$ has the form

$$S(\Lambda) = \begin{pmatrix} M(\Lambda) & 0 \\ 0 & M(\Lambda)^* \end{pmatrix} \tag{7.110}$$

with $M(\Lambda)$ as in (7.62) and the paragraphs around it.

The inner product in $\mathbb{C}^4$,

$$\phi^* \psi = \sum_{s=1}^{4} \phi_s^* \psi_s \,, \tag{7.111}$$

is not Lorentz invariant, but the indefinite product

$$\overline{\phi}\psi := \phi^* \gamma^0 \psi \tag{7.112}$$

is Lorentz invariant. This can be seen in the Weyl representation $\psi = (\psi^A, \psi_{A'})$, where, by virtue of (7.95a),

$$\phi^* \gamma^0 \psi = (\phi^1)^* \psi_1 + (\phi^2)^* \psi_2 + (\phi_1)^* \psi^1 + (\phi_2)^* \psi_2 = \overline{\phi}^{A'} \psi_{A'} + \overline{\phi}_A \psi^A \,, \tag{7.113}$$

which is a Lorentz invariant expression. Another way of expressing the Lorentz invariance is that

$$(S(\Lambda)\phi)^* \gamma^0 = \phi^* \gamma^0 S(\Lambda)^{-1} \,. \tag{7.114}$$

7.3 Probability

We now discuss the flow of probability according to the Weyl, Dirac, and other equations, in particular with regard to Cauchy surfaces.

7.3.1 Current for the Weyl Equation

Given a solution $\psi : M \to S$ of the Weyl equation $\partial_{AA'} \psi^A = 0$, we define the 4-vector field $j^\mu(x^\nu)$ by

$$j^{AA'}(x^\nu) = \sqrt{2}\, \overline{\psi}^{A'}(x^\nu)\, \psi^A(x^\nu) \tag{7.115}$$

or, equivalently,

$$j^\mu(x^\nu) = \overline{\psi}^{A'}(x^\nu)\, \sigma^\mu_{AA'}\, \psi^A(x^\nu) \,. \tag{7.116}$$

It is called the *probability current 4-vector field*. It is analogous to the 4-vector field $(\rho, \boldsymbol{j})$ in the non-relativistic case. Since $\sigma^0_{AA'}$ is the unit matrix, we have that

$$j^0 = \sum_{A=1}^{2} |\psi^A|^2 . \tag{7.117}$$

Born's Rule for the Weyl Equation *If we measure the position of a particle governed by the Weyl equation at time t in some Lorentz frame, then the outcome is random in $\mathbb{R}^3$ with probability density*

$$\rho(\boldsymbol{x}) = j^0(t, \boldsymbol{x}) = \sum_{A=1}^{2} \big|\psi^A(t, \boldsymbol{x})\big|^2 . \tag{7.118}$$

Since every spacelike hyperplane is of the form $\{x^0 = \text{const.}\}$ in some Lorentz frame, we can also say that if we place detectors along any spacelike hyperplane P, then the probability density at $x \in P$, relative to 3-volume in P, is $n_\mu j^\mu(x)$, where n^μ is the unique future-pointing unit 4-vector orthogonal to P in the Minkowski metric (i.e., $n_\mu(x^\mu - y^\mu) = 0$ for all $x, y \in P$). In a Lorentz frame in which P is horizontal, $n^\mu = (1, 0, 0, 0)$.

Proposition 7.10

$$\partial_\mu j^\mu = 0 . \tag{7.119}$$

Proof

$$\partial_\mu j^\mu = \partial_{AA'} j^{AA'} \tag{7.120a}$$

$$= \partial_{AA'}\big(\psi^A \overline{\psi}^{A'}\big) \tag{7.120b}$$

$$= \underbrace{\big(\partial_{AA'}\psi^A\big)}_{=0 \text{ by } (7.76)} \overline{\psi}^{A'} + \psi^A \underbrace{\partial_{AA'}\overline{\psi}^{A'}}_{=0} , \tag{7.120c}$$

where the last term vanishes because the conjugate of the Weyl equation (7.76) reads

$$\partial_{AA'}\overline{\psi}^{A'} = 0 . \tag{7.121}$$

□

Equation (7.119) is a continuity equation: with the notation $\rho = j^0$, it can be rewritten as

$$\frac{\partial \rho}{\partial t} = -\nabla \cdot \boldsymbol{j} \,. \tag{7.122}$$

We will study the relation between probabilities and the vector field j^μ more closely in Sect. 7.3.3.

7.3.2 Current for the Dirac Equation

Definition 7.4 A 4-vector $x \in M$ is called *causal* if and only if it is timelike or lightlike. A causal vector is *future-causal* (past-causal) if and only if $x^0 \geq 0$ ($x^0 \leq 0$) in any (and thus every) Lorentz frame.

Exercise 7.3 (Causal Vectors) Show that the following statements about a 4-vector u^μ are equivalent:

(i) u^μ is future-causal.
(ii) For every future-causal v^μ, the Minkowski product $u^\mu v_\mu$ is ≥ 0.
(iii) For every future-lightlike v^μ, the Minkowski product $u^\mu v_\mu$ is ≥ 0.

Exercise 7.4 (Causal and Timelike Vectors) Show that if $u^\mu \neq 0$ is future-timelike and $v^\mu \neq 0$ is future-causal, then $u^\mu v_\mu > 0$.

The probability current 4-vector field for the Dirac equation is defined by

$$j^\mu = \overline{\psi} \gamma^\mu \psi \,. \tag{7.123}$$

In the Weyl representation $\psi = (\phi^A, \chi_{A'})$, it can be expressed as

$$j^{AA'} = \sqrt{2}\Big(\overline{\phi}^{A'} \phi^A + \overline{\chi}^A \chi^{A'}\Big) \tag{7.124}$$

(where the index has been raised using the ε spinor as in (7.73)). It is clearly defined in a covariant way. It has the following properties:

Proposition 7.11 *At every* $x = x^\mu \in M$,

$$j^0(x) = \sum_{s=1}^{4} \big|\psi_s(x)\big|^2 \,, \tag{7.125}$$

and $j^\mu(x)$ is future-causal. Moreover, the vector field is 4-divergence free,

$$\partial_\mu j^\mu = 0\,. \tag{7.126}$$

Proof Equation (7.125) follows from $j^0 = \overline{\psi}\gamma^0\psi = \psi^*\gamma^0\gamma^0\psi = \psi^*\psi$.

Since $\overline{\phi}^{A'}\phi^A$ is a future-lightlike vector for every $\phi \in S$, and likewise $\overline{\chi}^A\chi^{A'}$, and since a positive multiple and the sum of future-lightlike vectors is future-timelike-or-lightlike (future-causal), j^μ is future-causal.

Here is a way to verify (7.126):

$$\partial_\mu j^\mu = \partial_\mu(\psi^*\gamma^0\gamma^\mu\psi) \tag{7.127a}$$

$$= \underbrace{(\partial_\mu\psi)^*\gamma^0\gamma^\mu\psi}_{(\gamma^0\gamma^\mu\partial_\mu\psi)^*} + \psi^*\gamma^0\underbrace{\gamma^\mu\partial_\mu\psi}_{-im\psi} \tag{7.127b}$$

$$= (-im\gamma^0\psi)^*\psi - im\psi^*\gamma^0\psi \tag{7.127c}$$

$$= im\psi^*\gamma^0\psi - im\psi^*\gamma^0\psi = 0 \tag{7.127d}$$

since $\gamma^0\gamma^0 = I$, $\gamma^0\gamma^i = \alpha^i$, and γ^0 are all self-adjoint matrices. □

Born's Rule for the Dirac Equation *If we measure the position of a particle governed by the Dirac equation at time t in some Lorentz frame, then the outcome is random in $\mathbb{R}^3$ with probability density*

$$\rho(\boldsymbol{x}) = j^0(t, \boldsymbol{x}) = \sum_{s=1}^{4}\big|\psi_s(t, \boldsymbol{x})\big|^2\,. \tag{7.128}$$

7.3.3 Probability Flow

The basic properties of the Weyl and Dirac equations concerning probability can be expressed and proved particularly easily from the Bohmian viewpoint. We will formulate the statements and arguments for the Dirac equation, but everything in this section equally applies to the Weyl equation.

Equation of Motion

Consider a particle world line $X : \mathbb{R} \to M$ with $X^\mu(s)$ the space-time point corresponding to parameter value s. We regard the parametrization as arbitrary; a re-parametrization $X \circ \varphi$ with bijective $\varphi : \mathbb{R} \to \mathbb{R}$ would still represent the same physical situation. If desired, we can parametrize by arc length ("proper time") or by coordinate time of some Lorentz frame.

At the space-time point $X^\mu(s_0)$ corresponding to a particular parameter value s_0, the *velocity* corresponds to the direction (up to a sign) of the *tangent line*. Let me explain. The apparent velocity 3-vector of the particle in a particular Lorentz frame would be

$$\boldsymbol{v} = (v^1, v^2, v^3) \text{ with } v^\mu = \frac{dX^\mu}{dX^0}(s_0) = \frac{dX^\mu/ds}{dX^0/ds} \text{ for } \mu = 1, 2, 3, \tag{7.129}$$

which is a quantity that depends only on the 4-vector dX^μ/ds but remains invariant if dX^μ/ds gets changed by a scalar factor and thus remains invariant if the parametrization of X gets changed. The direction of dX^μ/ds is exactly the direction of the tangent line (except that the latter is only defined up to a sign).

We define the Bohmian trajectories to be the solutions $s \mapsto X^\mu(s)$ of the equation (Bohm 1953 [10]),

$$\frac{dX^\mu}{ds} \propto j^\mu(X^\nu(s)), \tag{7.130}$$

where s is an arbitrary curve parameter and $\propto$ means proportional to (i.e., the two vectors point in the same or opposite directions or are equal up to a scalar factor); if the curve is parametrized with s increasing toward the future, then they point in the same direction. A re-parametrization of the curve will change the length but not the direction of dX^μ/ds except possibly for the sign. As a consequence, if (7.130) is satisfied in one parametrization, then it is also in every other. Equivalently, the Bohmian trajectories are the integral curves of j^μ, i.e., the curves everywhere tangent to j.[6]

In any Lorentz frame, (7.130) can equivalently be expressed as

$$\frac{d\boldsymbol{X}}{dt} = \frac{\psi^* \boldsymbol{\alpha} \psi}{\psi^* \psi}(t, \boldsymbol{X}(t)). \tag{7.131}$$

Indeed, if we parametrize the curve by coordinate time t, i.e., so that $X^0(t) = t$, then $dX^0/dt = 1$, which fixes the proportionality constant to be $1/j^0$. Thus,

$$\frac{dX^\mu}{dt} = \frac{j^\mu}{j^0}(X^\nu(t)), \tag{7.132}$$

which yields (7.131) by setting $\mu = 1, 2, 3$ and recalling that $j^\mu = \overline{\psi}\gamma^\mu\psi$ and $\overline{\psi} = \psi^*\gamma^0$ and $\alpha^i = \gamma^0\gamma^i$. Conversely, (7.131) and $dX^0/dt = 1$ together yield (7.132), which implies (7.130).

---

[6] Usually, an *integral curve* of a vector field j in any space is defined to be a solution $s \mapsto X(s)$ of the ODE $dX/ds = j(X(s))$, and thus a *parametrized* curve. Since we are interested in *unparametrized* curves, or regard reparametrizations as equivalent, we mean the curves for which a parametrization exists such that $dX/ds = j(X(s))$; this is equivalent to (7.130).

Since j^μ is a future-causal vector, the curve $X^\mu(s)$ must be everywhere timelike-or-lightlike (i.e., its tangents are timelike or lightlike, or, equivalently, any two points on the curve are timelike or lightlike separated).

From the form (7.131), it is visible that this equation is an equation of motion, a first-order ODE: it determines the velocity from the position. Equivalently, (7.130) determines the tangent line to the curve for whichever given space-time point $X^\nu(s)$ is, and in the space-time view, to determine the tangent line of a world line amounts to determining the velocity in a Lorentz-invariant way.

Another way of expressing that (7.130) is an equation of motion is to say that there is exactly one integral curve through every space time point, more precisely, through every space time point x^ν where $j^\mu(x^\nu) \neq 0$. If at a point x^ν, $j^\mu(x^\nu) = 0$, then $\psi(x^\nu) = 0$, and such a point is called a *node* of ψ. At a node, the right-hand side of (7.130) does not define a line; it does not define a direction up to a sign; the equation of motion is ill-defined there, as was the non-relativistic Bohmian equation of motion at nodes of the wave function. Standard theorems from the general theory of ODEs confirm that there is exactly one integral curve through every x^ν that is not a node, provided the vector field j^μ is sufficiently regular (continuously differentiable with locally bounded derivatives), which will be satisfied if the initial wave function is sufficiently regular (C^2).[7]

An integral curve can end if it runs into a node; also, if we trace the curve backward from $t = 0$, it may run into a node; else it is defined from $t = -\infty$ to $t = +\infty$ in every Lorentz frame. According to a rigorous result of Teufel et al. (2004) [43], if in some frame the initial position $\boldsymbol{X}(t = 0)$ is random with probability density $|\psi_0|^2$, and if ψ_0 is sufficiently regular, then the probability is 0 that the solution $t \mapsto \boldsymbol{X}(t)$ will *ever* run into a node either toward the future or toward the past. Put differently, with probability 1 the curve $t \mapsto \boldsymbol{X}(t)$ is defined for all $t \in \mathbb{R}$ (the curve "exists globally").[8] From now on, we only consider curves that exist globally.

Note also that (7.131) is an instance of the general Bohm-type scheme

$$\frac{d\boldsymbol{X}}{dt} = \frac{\boldsymbol{j}}{\rho}(t, \boldsymbol{X}(t)), \tag{7.133}$$

as $\boldsymbol{j} = \psi^* \boldsymbol{\alpha} \psi$ are the spacelike components of $j^\mu = \overline{\psi} \gamma^\mu \psi$. Since we know that ρ and $\boldsymbol{j}$ satisfy a continuity equation, we obtain as in the non-relativistic case the following *equivariance statement*: *If* $\boldsymbol{X}(0)$ *is* $|\psi_0|^2$*-distributed, then* $\boldsymbol{X}(t)$ *is* $|\psi_t|^2$*-distributed.* But much more is true, as I will explain now.

[7] Presumably, this is true regardless of the regularity properties of ψ, but that would not follow from the standard ODE theorems.

[8] Again, this is presumably true regardless of regularity, but that has not been proven.

Surface Equivariance

Recall that X is a timelike-or-lightlike curve that, assuming global existence, is inextendible; and recall that, according to Definition 7.3 in Sect. 7.1.7, a Cauchy surface intersects every such curve exactly once.

Non-Rigorous Theorem 7.1 (Surface Equivariance) *If the initial position $X(0)$ is $|\psi_0|^2$ distributed, then on any Cauchy surface Σ, the unique point $X(\Sigma)$ at which X intersects Σ has probability distribution on Σ given by ("curved Born distribution," as Σ can be a curved surface)*

$$\mathbb{P}\Big(X(\Sigma) \in d^3x\Big) = j^\mu(x)\, n_\mu(x)\, V(d^3x) \tag{7.134}$$

with $n_\mu(x)$ the future unit normal vector to Σ and $V(d^3x)$ the 3-volume of an infinitesimal element d^3x of Σ as in Proposition 7.2.

As usual, we assume henceforth that the initial position is $|\psi_0|^2$ distributed.

Derivation We use the following version of the 4d Ostrogradski-Gauss integral theorem that is taylored to Minkowski space M: *If A is a 4d subset of M whose boundary ∂A is sufficiently regular and piecewise spacelike or timelike, if $n_\mu(x)$ is the outward unit normal vector on ∂A at $x \in \partial A$, and if F^μ is a sufficiently regular vector field on $A \cup \partial A$, then*

$$\int_A d^4x^\nu\, \partial_\mu F^\mu(x^\nu) = \int_{\partial A} V(d^3x^\nu)\, F^\mu(x^\nu)\, n_\mu(x^\nu)\,. \tag{7.135}$$

The *proof* of this theorem is best done by means of differential forms. Differential forms are briefly explained in Appendix A.6; readers not familiar with differential forms may skip this proof and assume (7.135). By means of the volume 4-form on M given by the Levi-Civita tensor or epsilon tensor $\varepsilon_{\mu_1\mu_2\mu_3\mu_4}$, a vector field F^μ can be translated into a 3-form

$$\varphi_{\mu_1\mu_2\mu_3} = \varepsilon_{\mu_1\mu_2\mu_3\mu_4}\, F^{\mu_4} \tag{7.136}$$

whose exterior derivative is related to the 4-divergence of F^μ according to

$$(d\varphi)_{\mu_1\mu_2\mu_3\mu_4} = \varepsilon_{\mu_1\mu_2\mu_3\mu_4}\partial_\mu F^\mu\,. \tag{7.137}$$

We will use the Stokes theorem for differential forms, which says that

$$\int_A d\varphi = \int_{\partial A} \varphi\,. \tag{7.138}$$

We have seen that the left-hand side agrees with that of (7.135), and we will now verify that also their right-hand sides agree. For an infinitesimal 3d parallelepiped d^3x^ν spanned by the three infinitesimal vectors $dx_1^{\mu_1}, dx_2^{\mu_2}, dx_3^{\mu_3}$, suppose first that the three vectors are mutually orthogonal in the 3-metric on the surface ∂A and none of them is lightlike. Consider the 4d ONB (Lorentz frame) B associated with these three directions together with $n^\mu(x^\nu)$; if ∂A is spacelike at x, then n^μ is the timelike basis vector, and if ∂A is timelike at x, then the 3-metric on ∂A is indefinite, and n^μ is spacelike, while one of the dx_i^μ is timelike. In this ONB, as in every ONB, the components of $\varepsilon_{\mu_1\mu_2\mu_3\mu_4}$ are $+1$ if $\mu_1\mu_2\mu_3\mu_4$ is an even permutation of 0123, -1 if odd, and 0 if not a permutation. Thus,

$$\varphi_{\mu_1\mu_2\mu_3}\, dx_1^{\mu_1}\, dx_2^{\mu_2}\, dx_3^{\mu_3} = \varepsilon_{\mu_1\mu_2\mu_3\mu_4}\, dx_1^{\mu_1}\, dx_2^{\mu_2}\, dx_3^{\mu_3}\, F^{\mu_4} \tag{7.139}$$

will only have contributions from the component of F^μ in the direction of n^μ, so we can replace in the last expression F^{μ_4} by $n^{\mu_4}\, n_\mu\, F^\mu$. So, the contribution to $\int_{\partial A} \varphi$ from d^3x^ν is the product of the lengths of the dx_i^ν times $n_\mu\, F^\mu$, or $V(d^3x^\nu)\, F^\mu(x^\nu)\, n_\mu(x^\nu)$. But if this relation is true for mutually orthogonal dx_i^ν, it is also true for all directions. This completes the proof of (7.135).

Now we use (7.135) for $F = j$ for deriving (7.134). Suppose first that Σ lies in the future of the 3-surface $\{t = 0\}$. Consider a region $B_0 \subseteq \{t = 0\}$, let B_Σ be the set it gets transported to on Σ,

$$B_\Sigma = \{X(\Sigma) : X(0) \in B_0\}\,, \tag{7.140}$$

and let A be the union of all trajectories in between starting in B_0 (so A is a deformed cylinder). Then ∂A consists of three parts, the bottom B_0, the lid B_Σ, and the mantle. Since the mantle consists of trajectories, it is tangent to the trajectories and thus to j, so $n_\mu j^\mu = 0$ on the mantle—the flux of j across the mantle vanishes. Thus,

$$0 = \int_A d^4x^\nu\, \partial_\mu j^\mu(x^\nu) \tag{7.141a}$$

$$= \int_{\partial A} V(d^3x^\nu)\, j^\mu(x^\nu)\, n_\mu(x^\nu) \tag{7.141b}$$

$$= -\int_{B_0} V(d^3x^\nu)\, j^0(x^\nu) + \int_{B_\Sigma} V(d^3x^\nu)\, j^\mu(x^\nu)\, n_\mu(x^\nu) \tag{7.141c}$$

or, identifying $\{t = 0\}$ with $\mathbb{R}^3$ by means of the coordinates,

$$\int_{B_\Sigma} V(d^3x^\nu)\, j^\mu(x^\nu)\, n_\mu(x^\nu) = \int_{B_0} d^3\boldsymbol{x}\, |\psi_0(\boldsymbol{x})|^2\,. \tag{7.142}$$

Since B_0 (or B_Σ) was largely arbitrary, (7.134) follows.

If Σ lies in the past of $\{t = 0\}$, then the same calculation applies with the signs of the B_0 and B_Σ integrals reversed, which leads to (7.134) as well. If Σ lies partly in the future and partly in the past of $\{t = 0\}$, then separate consideration of these parts shows that (7.134) still holds. □

7.3.4 Evolution Between Cauchy Surfaces

The curved Born distribution (7.134) is the analog of the $|\psi|^2$ distribution on curved surfaces. Let us collect some consequences of surface equivariance.

A first consequence of surface equivariance is the converse statement: If $X(\Sigma)$ has curved Born distribution, then $\boldsymbol{X}(0)$ has Born ($|\psi_0|^2$) distribution. And then it follows further that $X(\Sigma')$ has curved Born distribution also on every other Cauchy surface Σ'. That is, *if* $X(\Sigma)$ *has curved Born distribution for one* Σ *then for every* Σ.

As another consequence, the curved Born distribution must be normalized to 1 on every Cauchy surface. This suggests the following definition of a 1-particle Hilbert space $\mathscr{H}_{1\Sigma}$ for any Cauchy surface Σ: The elements are functions $\psi : \Sigma \to D$ ("surface wave function") for which the norm

$$\|\psi\|_{1\Sigma} = \left(\int_\Sigma V(d^3x^\nu)\, \overline{\psi}(x^\nu)\, \gamma^\mu n_\mu(x^\nu)\, \psi(x^\nu) \right)^{1/2} \tag{7.143}$$

is finite, where we identify two functions differing only on a set of measure 0.

Remark 7.2 The integral on the right-hand side of (7.143) is well defined (though possibly ∞) and ≥ 0. This is because, first, it turns out that n_μ, which may not be defined at non-differentiable points (such as kinks) of Σ, is defined everywhere except on a set of measure 0 (which is irrelevant to the value of the integral in (7.143)). Second, the integrand is everywhere ≥ 0 because it is $j^\mu n_\mu$, and both vectors are future-causal; see Exercise 7.3 in Sect. 7.3.2. (Equivalently, writing $\overline{\psi}\, \gamma^\mu n_\mu\, \psi$ in coordinates as $\phi^* h \psi$ with 4×4 matrix $h = \gamma^0 \gamma^\mu n_\mu$, the statement means that h is self-adjoint and positive definite.)

Proceeding further with the definition of $\mathscr{H}_{1\Sigma}$, the obvious definition of the inner product reads

$$\langle \phi | \psi \rangle_{1\Sigma} = \int_\Sigma V(d^3x^\nu)\, \overline{\phi}(x^\nu)\, \gamma^\mu n_\mu(x^\nu)\, \psi(x^\nu)\,. \tag{7.144}$$

Remark 7.3 For ϕ, ψ with finite norm (7.143), the inner product (7.144) is well defined and finite. Indeed, the Cauchy-Schwarz inequality for $L^2(\Sigma, \mathbb{C}^4, V(d^3x))$ says that

$$\left|\int_\Sigma V(d^3x)\, u^*(x)\, v(x)\right| \leq \left(\int_\Sigma V(d^3x)\, |u(x)|^2\right)^{1/2} \left(\int_\Sigma V(d^3x)\, |v(x)|^2\right)^{1/2}. \tag{7.145}$$

Setting $u = h^{1/2}\phi$ and $v = h^{1/2}\psi$, the left-hand side is the absolute value of the right-hand side of (7.144), and the right-hand side is $\|\phi\|_{1\Sigma}\, \|\psi\|_{1\Sigma}$, so the former must be finite if both norms are.

The surface equivariance theorem then also suggests that initial data for the wave function can be specified on any Cauchy surface Σ. That is, for any $\psi_\Sigma \in \mathscr{H}_{1\Sigma}$, there is a unique solution $\psi : M \to D$ of the Dirac equation whose restriction $\psi|_\Sigma$ to Σ agrees with ψ_Σ. Restricting ψ to *another* Cauchy surface Σ' defines a time evolution $\psi_\Sigma \mapsto \psi|_{\Sigma'}$ from Σ to Σ'. Let $U_{1\Sigma}^{\Sigma'} : \mathscr{H}_{1\Sigma} \to \mathscr{H}_{1\Sigma'}$ denote the operator that maps ψ_Σ to $\psi|_{\Sigma'}$. Since $\|\psi|_{\Sigma'}\|_{1\Sigma'} = 1$ if $\|\psi_\Sigma\|_{1\Sigma} = 1$, $U_{1\Sigma}^{\Sigma'}$ preserves norms, and since $U_{1\Sigma}^{\Sigma'} U_{1\Sigma'}^{\Sigma} = I_{1\Sigma'}$, $U_{1\Sigma}^{\Sigma'}$ must be surjective, so $U_{1\Sigma}^{\Sigma'}$ is unitary. This is known to be rigorously true (Dimock 1982 [19], Deckert and Merkl 2014 [18]):

Theorem 7.1 *Between any two smooth Cauchy surfaces Σ and Σ', the 1-particle Dirac time evolution $U_{1\Sigma}^{\Sigma'} : \mathscr{H}_{1\Sigma} \to \mathscr{H}_{1\Sigma'}$ is well defined and unitary. It satisfies*

$$U_{1\Sigma}^{\Sigma} = I_{1\Sigma} \quad \text{and} \quad U_{1\Sigma'}^{\Sigma''} U_{1\Sigma}^{\Sigma'} = U_{1\Sigma}^{\Sigma''}. \tag{7.146}$$

Finally, the surface equivariance theorem suggests the

Curved Born Rule for 1 Dirac Particle *If we place detectors along a Cauchy surface Σ, then the probability distribution of the detection point is given by* (7.134).

I will say more about its justification in Sect. 7.4.2, when we consider the N-particle case.

Proposition 7.12 *For any region $B \subseteq \Sigma$, multiplication by 1_B is a projection operator P_B on $\mathscr{H}_{1\Sigma}$. The P_B form a PVM on Σ acting on $\mathscr{H}_{1\Sigma}$, and the Born distribution for this PVM is the curved Born distribution on the right-hand side of* (7.134).

Proof Obviously, $P_B^2 = P_B$, and P_B maps $\mathscr{H}_{1\Sigma}$ to $\mathscr{H}_{1\Sigma}$ (functions stay square-integrable). In order to verify that P_B is self-adjoint, note that

$$\langle\phi|P_B\psi\rangle_{1\Sigma} = \int_B V(d^3x^\nu)\, \overline{\phi}(x^\nu)\, \gamma^\mu n_\mu(x^\nu)\, \psi(x^\nu) = \langle P_B\phi|\psi\rangle_{1\Sigma}\,. \tag{7.147}$$

Obviously, $P_\Sigma = I$, and σ-additivity follows in the same way as in the non-relativistic case. □

7.3.5 Propagation Locality

Another consequence of the surface equivariance for the Dirac equation is propagation locality: the wave function propagates at most at the speed of light. If the initial wave function ψ_0 is concentrated in a region A in $\mathbb{R}^3$ (i.e., vanishes outside of A), then ψ_t is concentrated in the $|t|$-neighborhood of A,

$$B_{|t|}(A) = \bigcup_{\boldsymbol{x}\in A} B_{|t|}(\boldsymbol{x})\,. \tag{7.148}$$

More generally, if we specify initial data ψ_Σ for the wave function on some Cauchy surface Σ, and if ψ_Σ is concentrated in a region $A \subset \Sigma$, then the solution $\psi : M \to D$ of the Dirac equation with initial datum ψ_Σ is concentrated in future$(A) \cup$ past(A) (with the relativistic future of a set as in Fig. 7.5, and likewise for the past). Equivalently, $U^{\Sigma'}_{1\Sigma}\psi_\Sigma$ is concentrated in

$$B_{\Sigma'}(A) = \Sigma' \cap (\text{future}(A) \cup \text{past}(A)) \tag{7.149}$$

for every Cauchy surface Σ'. This can be seen from surface equivariance together with the fact that the trajectories are timelike or lightlike: If $\|\psi_\Sigma\|_{1\Sigma} = 1$ and all trajectories begin in A, then they can intersect Σ' only in $B_{\Sigma'}(A)$. But the flux of j^μ through $B_{\Sigma'}(A)$ is 1 by equivariance, i.e.,

$$\int_{B_{\Sigma'}(A)} V(d^3x)\, j^\mu\, n_\mu = 1. \tag{7.150}$$

By unitarity of $U^{\Sigma'}_{1\Sigma}$, $\|\psi|_{\Sigma'}\|_{1\Sigma'} = 1$, so $\psi|_{\Sigma'}$ must vanish almost everywhere (i.e., everywhere except on a set of measure 0) outside of $B_{\Sigma'}(A)$.

A further consequence concerns the dependence on initial data. Suppose initial data are specified on a Cauchy surface Σ, and $x^\nu \in M$ is a point not on Σ, then $\psi(x^\nu)$ depends only on initial data in $B_\Sigma(\{x^\nu\})$. In particular, if $\Sigma = \{t = 0\}$, then $\psi(t, \boldsymbol{x})$ depends only on initial data in the ball in $\mathbb{R}^3$ of radius $|t|$ around $\boldsymbol{x}$. Put differently, if we change the initial data outside that region, then $\psi(t, \boldsymbol{x})$ respectively $\psi(x^\nu)$ will not change. That is because the difference between the two wave functions vanishes outside $A = \Sigma \setminus B_\Sigma(\{x^\nu\})$ and is another solution of the Dirac equation, so it vanishes outside future$(A) \cup$ past(A); in particular, it vanishes at x^ν.

7.3.6 External Fields

The effect of an external electromagnetic field can be incorporated in the (first or second order) Klein-Gordon equation, Weyl equation, and Dirac equation by replacing the derivative operator ∂_μ by

$$\partial_\mu + ieA_\mu(x)\,, \tag{7.151}$$

where e is the particle's electric charge and $A_\mu = (A_0, -\boldsymbol{A})$ is the electromagnetic 4-vector potential with $\boldsymbol{A}$ the magnetic vector potential as in (2.51);[9] A_μ is related to $F_{\mu\nu}$ according to

$$F_{\mu\nu} = \partial_\mu A_\nu - \partial_\nu A_\mu\,. \tag{7.152}$$

For example, the first-order Klein-Gordon equation with A_μ reads

$$i\partial_t\psi(x) = \sqrt{m^2 - (\nabla - ie\boldsymbol{A}(x))^2}\,\psi + eA_0(x)\,\psi(x)\,, \tag{7.153}$$

where the radicand is a positive operator because $m^2 \geq 0$ and $-(\nabla - ie\boldsymbol{A}(x))^2 = \sum_{j=1}^3 R_j^\dagger R_j$ is positive with $R_j = R_j^\dagger = i\partial_j + eA_j(t,\cdot)$.

In the presence of A_μ, no change in the formulas (7.116) and (7.123) for j^μ is needed, but the trajectories will be different in the presence of $A_\mu \neq 0$ because the wave function will be different.

Exercise 7.5 (Gauge Transformation) The replacement, for an arbitrary (sufficiently smooth) function $\theta : M \to \mathbb{R}$,

$$\psi(x) \to \tilde{\psi}(x) = e^{ie\theta(x)}\psi(x) \quad\text{and}\quad A_\mu(x) \to \tilde{A}_\mu(x) = A_\mu(x) - \partial_\mu\theta(x)\,, \tag{7.154}$$

is called a *gauge transformation*.

(a) Verify that

$$\big(\partial_\mu + ie\tilde{A}_\mu(x)\big)\tilde{\psi} = e^{ie\theta(x)}\big(\partial_\mu + ieA_\mu(x)\big)\psi(x)\,. \tag{7.155}$$

(b) Verify that the Weyl and Dirac equations with A_μ are invariant under gauge transformations, i.e., that $\tilde{\psi}$ solves the Weyl (Dirac) equation with $\tilde{A}_\mu$ if ψ solves it with A_μ.

---

[9] The sign in (7.151) may appear inconsistent with that in (2.50), but it is not. Keep in mind that, in any Lorentz frame, the first component of $\nabla\psi$ is the same as $\partial_1\psi = -\partial^1\psi$, while the first component of $\boldsymbol{A}$ is the same as $A^1 = -A_1$.

(c) Verify that for both the Weyl and Dirac equations with A_μ, the current 4-vector field j^μ is invariant under gauge transformations.

Remark 7.4 These properties of *gauge invariance* allow for a deeper interpretation in terms of vector bundles. A *vector bundle* is a family of vector spaces S_x indexed by (in our case) a space-time point $x \in M$. For example, when space-time is curved, then the one spin space S must be replaced with an x-dependent spin space S_x. A function $\psi : M \to \cup_x S_x$ with the property $\psi(x) \in S_x$ (as the Weyl wave function will satisfy) is called a *cross-section* of the vector bundle $\cup_x S_x$. Since $\psi(x + dx)$ and $\psi(x)$ lie in different spaces S_{x+dx} and S_x, one cannot form their difference $\psi(x + dx) - \psi(x)$, so for taking derivatives, one needs a further prescription for how to identify S_{x+dx} and S_x for infinitesimal dx; such a prescription is called a *gauge connection* or *connection* and provides a *covariant derivative operator*. It turns out that different electromagnetic fields then correspond to different gauge connections and gauge transformations to an x-dependent change of basis in S_x.

Remark 7.5 In the presence of A_μ, the time evolution $\mathscr{H}_{1\Sigma} \to \mathscr{H}_{1\Sigma'}$ is still unitary (the definition of the Hilbert spaces is unchanged), and (7.146) and propagation locality still hold.

7.3.7 Non-Relativistic Limit

For every $\boldsymbol{k} \in \mathbb{R}^3$, the free Dirac equation has four linearly independent plane wave solutions

$$\psi(x^\nu) = u \exp(ik_\mu x^\mu) \tag{7.156}$$

with x-independent $u \in D$, two linearly independent ones with $k^0 = \sqrt{m^2 + \boldsymbol{k}^2}$ and two with $k^0 = -\sqrt{m^2 + \boldsymbol{k}^2}$. Contributions with negative energy (i.e., positive k^0) are often regarded as a problem which the concept of the Dirac sea should deal with. I will not get into this issue here (but see my paper [55]).

If $m > 0$ and, in a certain Lorentz frame, only $\boldsymbol{k}$ values with $|\boldsymbol{k}| \ll m$ contribute significantly to ψ (as would correspond to speeds $\ll c$), one can obtain a non-relativistic limit (informally, $c \to \infty$) of the Dirac equation. Here, I only report the result; see, e.g., Bohm and Hiley (1993) [11, Sec. 10.4] for the calculation. If only contributions with positive energy (i.e., negative k^0) are considered, then only two degrees of freedom per $\boldsymbol{k}$ (and thus per $\boldsymbol{x}$) remain, which can be regarded as the ϕ^A components of $\psi = (\phi^A, \chi_{A'})$, and in the non-relativistic limit, one obtains the Pauli equation (2.112) plus an extra constant term in the Hamiltonian given by m, the rest mass of the electron. It makes sense to assume that in nature the rest mass contributes to the Hamiltonian but had been dropped in the non-relativistic theory.

In this limit, the Bohmian equation of motion (7.131) for a Dirac particle becomes (Bohm and Hiley 1993 [11, Sec. 10.4])

$$\frac{d\boldsymbol{X}}{dt} = \frac{\hbar}{m}\text{Im}\frac{\psi^*\nabla\psi}{\psi^*\psi} + \frac{\hbar}{m}\frac{\nabla\times(\psi^*\boldsymbol{\sigma}\psi)}{\psi^*\psi} \tag{7.157}$$

with $\psi : \mathbb{R}^4 \to S$ the (2-component) wave function of the Pauli equation and $\times$ the cross product in $\mathbb{R}^3$, so $\nabla\times$ is the curl as in (2.51) (and with $\hbar$ put back in the right place). That is, we almost get the equation of motion (2.120) we had considered with the Pauli equation, but we get an extra term, the second term in (7.157).

The extra term in the equation of motion corresponds to an extra term $\frac{\hbar}{m}\nabla\times(\psi^*\boldsymbol{\sigma}\psi)$ in the current $\boldsymbol{j}$. However, it does not show up in the continuity equation because the latter involves only div $\boldsymbol{j}$, and the divergence of a curl always vanishes. In particular, both alternative equations of motion, (2.120) and (7.157), make the $|\psi|^2$ distribution equivariant. As a consequence, we have two different versions of Bohmian mechanics, with different trajectories, but they make exactly the same empirical predictions, they are *empirically equivalent*. That is because if we wanted to use any experiment for testing the two theories against each other, then the outcome of the experiment would be indicated by pointer positions, or ink in lab books, or ink in journal publications, in sum by the particle configuration at some final time t after the completion of the experiment. But in both theories, the configuration at time t would be $|\psi_t|^2$-distributed, so it cannot be concentrated in different regions of configuration space in the two theories. Thus, the question of whether the additional term in (7.157) exists in nature is empirically undecidable—another limitation to knowledge.

However, it appears convincing that the term should be there because the relativistic equation (7.130) is the more fundamental one and appears very natural. And if (7.130) is correct, then (7.157) will hold in the non-relativistic regime. We see now that we should have used (7.157) all along; one lesson here is that we may have to revise our theories in light of more general frameworks. But at least, the conclusions we have drawn from (2.120) are equally valid for (7.157), except for the exact shape of the critical surface (separatrix) in Figs. 2.4 and 2.6. So, we have not made a serious mistake.

7.3.8 Probability and the Klein-Gordon Equation

For spin-0 wave functions, the only available relativistic Schrödinger equation is the Klein-Gordon equation, either of first or second order. But with these equations, it is not obvious what the formula for the Born rule should be. On the other hand, it is not clear whether spin-0 wave functions even exist in nature.[10] In this section, I discuss the issue of the spin-0 Born rule.

Psi Squared

Spin-0 wave functions are scalar-valued, $\psi : M \to \mathbb{C}$. As a consequence, $|\psi|^2$ is a scalar, too. That is, it is Lorentz invariant: for any point $x \in M$ (which will have different coordinates in different Lorentz frames), $|\psi(x)|^2$ has the same value in every Lorentz frame. However, probability density should be the 0 component of a 4-vector[11] and therefore transform in a nontrivial way when we change the frame. So, the relation $\rho = |\psi|^2$ cannot be true for spin-0 wave functions. This was different for Dirac wave functions: there, ψ had several components which transformed in a way that would ensure $|\psi|^2$ transforms like the 0 component of a 4-vector.

Here is another problem with $\rho = |\psi|^2$ that arises specifically for the *first-order* Klein-Gordon equation (7.55): propagation locality is violated. That is, if ψ evolves according to (7.55) and $\psi_{t=0}$ has compact support in $\mathbb{R}^3$, then at arbitrarily small times $t > 0$, ψ_t will have nonzero tails to infinity in $\mathbb{R}^3$ (Thaller 1992 [44, Sec. 1.8.1]). (In contrast, propagation locality is satisfied for the *second-order* Klein-Gordon equation: if both $\psi(t = 0)$ and $\partial_t \psi(t = 0)$ are concentrated in a region $A \subset \mathbb{R}^3$, then ψ_t is concentrated in $B_{|t|}(A)$. However, the second-order Klein-Gordon equation includes contributions of negative energy and is therefore widely avoided.)

The Klein-Gordon Current

Another proposal is to abandon $\rho = |\psi|^2$ and assume instead the formula

$$j^\mu = -\mathrm{Im}\big[\psi^* \partial^\mu \psi\big] \tag{7.158}$$

(perhaps times some prefactor) for the probability current, called the *Klein-Gordon current*. In the presence of an external electromagnetic field A_μ, it reads

$$j^\mu = -\mathrm{Im}\big[\psi^* (\partial^\mu + ieA^\mu)\psi\big]\,. \tag{7.159}$$

[10] The only spin-0 particle species in the standard model of particle physics is the Higgs boson, which might turn out to consist of several particles. Also, while all other particle species fit into families reminiscent of the periodic table, the Higgs boson stands out as the only member of its family; this suggests we may not have discovered the full structure of the Higgs boson yet.

[11] Here is a justification of this statement based on differential forms (see Appendix A.6 for a brief introduction). Along any Cauchy surface Σ, we could place detectors and would detect the particle in a random point $X \in \Sigma$. Its probability distribution on Σ should attribute a number in $[0, 1]$ to every (measurable) subset B of Σ. Since 3-forms are the kinds of mathematical objects that can, by their nature, be integrated over a 3-set such as B, probability density should correspond to a 3-form $J_{\lambda\mu\nu}$. But this is mathematically equivalent to a vector field j^μ according to $J_{\lambda\mu\nu} = \varepsilon_{\lambda\mu\nu\rho} j^\rho$. On a horizontal surface $\{t = \text{const.}\}$, the density relative to ordinary 3-volume is then given by $j^0(x)$.

It will become clear in this section and Sect. 7.6.3 (where we discuss the N-particle case) that there is reason for reservations about this formula; but let us analyze step by step what is and what is not problematical about it and why.

Exercise 7.6 (The Klein-Gordon Current is Divergence-Free)

(a) Verify that the Klein-Gordon equation of second order, $-\partial^\mu\partial_\mu\psi = m^2\psi$ (and thus a fortiori also that of first order), implies that

$$\partial_\mu j^\mu = 0 \tag{7.160}$$

for j^μ given by (7.158).

(b) Verify that in the presence of (real) A_μ, the Klein-Gordon equation

$$-\eta^{\mu\nu}(\partial_\mu + ieA_\mu)(\partial_\nu + ieA_\nu)\psi = m^2\psi \tag{7.161}$$

implies $\partial_\mu j^\mu = 0$ for j^μ given by (7.159).

A property of the Klein-Gordon current that differs from the Dirac current (7.123) and the Weyl current (7.115) is that the Klein-Gordon current is not everywhere future-causal. It can be spacelike or even past-timelike. Thus, so can be the integral curves, the solutions of the associated equation of motion

$$\frac{dX^\mu}{ds} \propto j^\mu(X(s)) \tag{7.162}$$

proposed explicitly for (7.158) and (7.159) by de Broglie (1960) [17, Sec. III.7]; see Fig. 7.7. In particular, they can be $\bigcup$- or $\bigcap$-shaped.[12] Correspondingly, an integral curve may intersect a Cauchy surface repeatedly (or not at all).

Some authors (e.g., Bohm and Hiley 1993 [11]) have criticized that j^0 is sometimes negative and thus cannot be interpreted as probability density. However, this is not really a problem, for the following reason (Tumulka 2001 [47]). If we consider a future-causal current j^μ (as for the Dirac equation), so we have a probability distribution over the integral curves $s \mapsto X(s)$, and if we consider a piece B of 3-surface that is *timelike* (i.e., contains a timelike dimension and two spacelike ones), then a curve X may intersect B repeatedly. If n_μ is a continuous unit normal vector field on B (whose choice amounts to the choice of an *orientation* of B), then X can cross B *in the sense of* n^μ (i.e., $n_\mu\, dX^\mu/ds > 0$) or *against the sense of* n^μ (i.e., $n_\mu\, dX^\mu/ds < 0$). Let us call the former a *positive crossing* and the latter a *negative crossing*. If X crosses B multiple times, the number of positive

[12] It is tempting to think that a $\bigcup$-shaped trajectory represents the creation of a pair of particles. However, a pair of particles should be governed by a 2-particle wave function $\psi(x_1, x_2)$, whereas the trajectory we are considering is governed by a 1-particle wave function $\psi(x)$. In particular, there is no candidate in sight for Born's rule, which would have to relate the probability density $\rho(x_1, x_2)$ to the wave function.

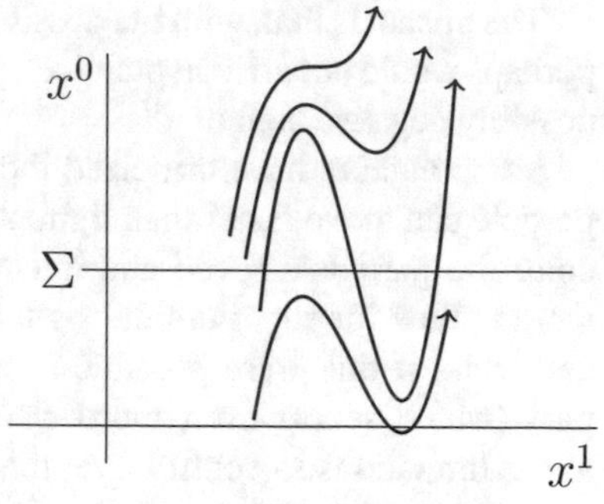

Fig. 7.7 Integral curves of the Klein-Gordon current (7.158) are sometimes spacelike and sometimes past-timelike; they can intersect a Cauchy surface Σ repeatedly

crossings minus the number of negative crossings is called the number of signed crossings $N(X, B) \in \mathbb{Z}$. And then

$$\int_B V(d^3x)\, j^\mu(x)\, n_\mu(x) = \mathbb{E}\, N(X, B) \tag{7.163}$$

yields the *expected number of signed crossings*. The same consideration can be applied to current vector field j^μ that is not future-causal and a 3-set B that is a piece of a Cauchy surface with future unit normal n_μ: the flux of j through B is not the probability of crossing B but the expected number of signed crossings through B. Locally and up to a sign, this agrees with intersection probability, so the probability that the random integral curve X intersects the 3-surface element d^3x is

$$V(d^3x)\, |j^\mu(x)\, n_\mu(x)|\,. \tag{7.164}$$

That this can be carried out consistently is the content of the following.

Theorem 7.2 (Tumulka 2001 [47]) *Any divergence-free vector field j^μ on Minkowski space gives rise to a measure $\mathbb{P}$ on the set of integral curves of j^μ within the wandering set of j^μ, which is the set of space-time points x such that the integral curve X_y through every y in a neighborhood of x is non-recurrent in both time directions; a* recurrent curve *is one that returns infinitely often to every neighborhood of some point it has traversed. If the total weight of the measure $\mathbb{P}$ is 1, it defines a random trajectory X, and for any oriented piece B of 3-surface lying within the wandering set, the expected number of signed crossings of X through B equals*

$$\mathbb{E}\, N(X, B) = \int_B V(d^3x)\, j^\mu(x)\, n_\mu(x) \tag{7.165}$$

with n_μ denoting the (properly oriented) unit normal.[13]

[13] The formula (7.165) is not applicable to points x where B is lightlike. There, a more careful formulation in terms of differential forms is needed.

The upshot is that, with the possible exception of recurrent integral curves (which perhaps would not arise in practice), probabilities are indeed well defined despite the possibly negative sign of j^0.

Some authors have criticized the use of the Klein-Gordon current because the particle can move faster than light. Also this is not really a problem. If Alice could control a particle's speed and intentionally shoot it to Bob faster than light, that would allow for superluminal signaling (which seems impossible in our universe) and even, if this were possible in any Lorentz frame, for sending signals to the past (which seems even more clearly impossible). However, we know already about limitations to control over the position of a Bohmian particle within a wave packet (see Sect. 5.1.5), so the problem need not arise, and we know already that propagation locality holds for the second-order Klein-Gordon equation, which suggests that the problem cannot arise.

However, a theory with particle world lines that can turn around in time as in Fig. 7.7 is certainly eccentric, implausible, and hard to believe. The implausibility will stand out even more clearly in the many-particle case, which we will discuss in Sect. 7.6.3.

7.3.9 The Maxwell Equation as the Schrödinger Equation for Photons

The photon has spin 1 and mass 0, and there is essentially just one relativistic free Schrödinger equation for such a particle; it is mathematically equivalent to the free Maxwell equations (7.25) for a complexified Maxwell field $F_{\mu\nu}$. So we know what a *photon wave function* is, but there is something that is not known, and that is what the *Born rule* for a single photon would say. We would expect that there is some formula for the probability current j^μ that takes the same place for the Maxwell equations as (7.123) for the Dirac equation or (7.115) for the Weyl equation, but no convincing candidate for such a formula is known—the Born rule for photons is an open question. I discuss this issue in this section.

The problem is not created by having mass 0, but by having spin 1. We do have a formula (7.171) that is valid in a special regime and covers most (perhaps all) applications in present-day experiments, but we would expect a general law that is applicable under all circumstances, and this law is not known. We also have another formula (7.180) that appears to be a good approximation, but for theoretical reasons, that I will explain does not seem convincing as a fundamental law.

A law for j^μ is particularly important for the Bohmian approach because once the current j^μ is known, it is straightforwardly possible to define the trajectories as the integral curves of j^μ, or

$$\frac{dX^\mu}{ds} \propto j^\mu(X^\nu(s)), \tag{7.166}$$

and if $\partial_\mu j^\mu = 0$, then equivariance follows. Without a current j^μ, we also lack a law of motion. Some authors have contemplated Bohm-type theories without photon particles: Bohm (1952b) [9] proposed a field ontology replacing photons, and Bell (1986) [4] proposed that, while fermions are particles, bosons are not associated with any primitive ontology at all; they exist only in the wave function.

But even so, the question remains how to compute the probability of a photon detection event. We have photon counters, so it makes sense to ask for the probability that a certain counter will be triggered in a certain run of an experiment. We can see interference patterns made up by spots on the screen corresponding to individual photons, so there is a nontrivial distribution of these spots in space. For such detection probabilities, there should be a formula, regardless of whether photons have trajectories; in fact, regardless of whether we want a Bohm-type theory or prefer some other theory of quantum physics. The lack of a Born rule for photons is a gap in any theory of quantum physics. Some authors have suggested that no Born rule exists for photon positions, but I do not see how that could go together with the fact that in our experiments, we routinely detect photons in different positions. So I tend to think that there is a formula that we just have not discovered yet.

So let me discuss some known proposals.

Locally Plane Waves

There is a convincing way of defining j^μ if the photon wave function $F_{\mu\nu}$ is a plane wave or at least a locally plane wave (i.e., a function such that every point has a small neighborhood on which the function is to a good degree of approximation a plane wave): Suppose

$$A_\mu(x) = a_\mu\, e^{-ik_\lambda x^\lambda}, \quad F_{\mu\nu}(x) = -i(a_\mu k_\nu - a_\nu k_\mu)e^{-ik_\lambda x^\lambda} \tag{7.167}$$

with future-lightlike k^μ and $k^\mu a_\mu = 0$ to fulfill the free Maxwell equation $\partial^\mu F_{\mu\nu} = 0$. A situation with many photons, all having the same wave function $F_{\mu\nu}$, should represent a classical regime with electromagnetic field $F_{\mu\nu}$ (or something closely related). Classically, the energy-momentum density is given, up to a positive constant factor, by the tensor field

$$T_{\mu\nu} = -\mathrm{Re}\big[F^*_{\mu\lambda} F_\nu{}^\lambda\big] + \tfrac{1}{4}\eta_{\mu\nu} F^*_{\lambda\rho} F^{\lambda\rho} \tag{7.168}$$

(with $\eta_{\mu\nu}$ the Minkowski metric), which for the plane wave (7.167) amounts to

$$T_{\mu\nu} = -a^*_\lambda\, a^\lambda\, k_\mu k_\nu\,. \tag{7.169}$$

If each photon has momentum $\hbar k^\mu$ according to the de Broglie relation, and if photon number density is proportional to the probability density j^ν for each photon, then the energy-momentum density should be

$$T_{\mu\nu} = \hbar k_\mu j_\nu \tag{7.170}$$

up to a positive constant factor. So, we can obtain j^μ by comparison with (7.169),

$$j^\mu = -a^*_\lambda a^\lambda \, k^\mu \tag{7.171}$$

up to a positive constant factor. (Note that $a^*_\lambda a^\lambda \leq 0$, so j^μ is future-lightlike.)

For the Maxwell equation as for the non-relativistic Schrödinger equation, wave functions evolving freely become locally plane waves as $t \to \pm\infty$. Thus, if we limit ourselves to considering the time evolution from $t = -\infty$ to $t = +\infty$ (as in scattering theory), it can presumably be assumed of all wave functions that they are locally plane waves. It may also be the case in most optical or quantum optical experiments that, by the time of detection, the wave function has become a locally plane wave, so (7.171) may suffice for most practical applications. But what is the general law?

The Poynting Vector

One of the simplest expressions quadratic in $F_{\mu\nu}$ is $T_{\mu\nu}$, the tensor that in classical electrodynamics means the energy-momentum tensor.[14] It is symmetric, $T_{\nu\mu} = T_{\mu\nu}$, and divergence-free in the sense that

$$\partial_\mu T^{\mu\nu} = 0\,. \tag{7.172}$$

In any Lorentz frame, T^{00} would represent the energy density and the Poynting vector $P^i = T^{i0}$ the energy current. Let me combine them into the 4-vector $P^\mu = T^{\mu 0}$. The idea of taking

$$j^\mu = P^\mu = T^{\mu 0} \tag{7.173}$$

was reportedly [42, 57] first proposed by Einstein (1924) and then considered by Slater (1924) [42], Bialynicki-Birula (1994) [8], and (in a slightly different way) Kiessling and Tahvildar-Zadeh (2018) [31]. From (7.172) it follows by setting $\nu = 0$ that $\partial_\mu j^\mu = 0$ (which would represent the local conservation of energy in classical electrodynamics).

Proposition 7.13 *P^μ is future-causal.*

[14] The degree of simplicity is not as clearly visible from the expression (7.168) for $T_{\mu\nu}$ as it is in spinor components: a real $F_{\mu\nu}$ can be expressed through a symmetric spinor φ^{AB} (Penrose and Rindler 1984 [38, Eq. (5.1.39)]), and $\overline{\varphi}^{A'B'}\varphi^{AB}$ yields $T^{\mu\nu}$ up to a prefactor of 2π (Penrose and Rindler 1984 [38, Eq. (5.2.4)]).

Proof A 4-vector P^μ is future-causal if and only if its Minkowski product $P^\mu v_\mu$ with every future-lightlike vector v^μ is ≥ 0; see Exercise 7.3 in Sect. 7.3.2. Since every future-lightlike vector v^μ is of the form $v_{AA'} = \overline{\phi}_{A'}\phi_A$ for some spinor $\phi \in S$, it suffices to show that $P^{AA'}\overline{\phi}_{A'}\phi_A \geq 0$ for every $\phi \in S$. Let $n_\mu = (1, 0, 0, 0)$ be the timelike basis vector of the Lorentz frame; then $n_{BB'}$ is $2^{-1/2}$ times the identity matrix, and $P^\mu = T^{\mu\nu} n_\nu$ or (by Footnote 14)

$$P^{AA'} = \frac{1}{2\pi}\overline{\varphi}^{A'B'}\varphi^{AB} n_{BB'} \tag{7.174a}$$

$$= \frac{1}{2\pi\sqrt{2}}\left(\overline{\varphi}^{A'1}\varphi^{A1} + \overline{\varphi}^{A'2}\varphi^{A2}\right), \tag{7.174b}$$

so

$$P^{AA'}\overline{\phi}_{A'}\phi_A = \frac{1}{2\pi\sqrt{2}}\left(\overline{\varphi}^{A'1}\overline{\phi}_{A'}\varphi^{A1}\phi_A + \overline{\varphi}^{A'2}\overline{\phi}_{A'}\varphi^{A2}\phi_A\right) \tag{7.175a}$$

$$= \frac{1}{2\pi\sqrt{2}}\left(\left|\varphi^{A1}\phi_A\right|^2 + \left|\varphi^{A2}\phi_A\right|^2\right) \geq 0. \tag{7.175b}$$

□

Here are the problems with the proposal $j^\mu = T^{\mu 0}$:

- *Lack of Lorentz invariance.* The equation $j^\mu = T^{\mu 0}$, if true in one Lorentz frame, will not be true in another. This suggests to assume that a certain timelike direction (and thus a certain future-timelike unit vector u_μ) is preferred in nature (maybe related to the time foliation; see Sect. 7.5.4) and that the equation really means $j^\mu = T^{\mu\nu} u_\nu$. But then another problem arises:
- *Visibly preferred frame.* It would seem that we can empirically determine, for every wave function $F_{\mu\nu}$, the full current vector field j^μ. After all, we could prepare many photons with the same wave function, group them into subensembles, and for each measure the probability distribution over some Cauchy surface Σ. If for every spacelike 3-surface element d^3x the value $V(d^3x)\, j^\mu(x)\, n_\mu(x)$ is known, j^μ can be reconstructed. But then we know j^μ and $T^{\mu\nu}$, so we can reconstruct u_ν; that is, there would be an empirically distinguished frame (which is different from the situation with the time foliation), and this does not seem to be correct in our universe.
- *Classical regime.* The Poynting vector disagrees with the current (7.171) that seems to be correct for plane waves. Let me spell out what is at the center of this disagreement. One would expect that classical electrodynamics is still valid in a certain regime and that this regime prevails when the number of photons is large, there is no entanglement, and all photons have the same wave function $F_{\mu\nu}$. One would expect further that in this situation, the classical field $F_{\mu\nu}$ is just the photon wave function $F_{\mu\nu}$. (The fact that the photon wave function is complex

could be dissolved by assuming it has only positive energies and supplementing the negative energy part to be the conjugate of the positive energy part, which would make $F_{\mu\nu}$ real.) But then the proposal $j^\mu = P^\mu$ does not fit with known rules of classical electrodynamics, as the vector j^μ would then mean the *photon number current* and j^0 the photon number density, whereas P^μ is the *energy current* and $P^0 = T^{00}$ the energy density. In fact, for a plane wave (7.167), each photon would carry the energy $\hbar k^0$, so energy density should be $\hbar k^0$ times the photon number density.

- *Conformal Killing vector.* In curved space-time, u_μ would have to be a vector *field*, and local conservation of probability would correspond to the equation $\nabla_\mu j^\mu = 0$ with ∇_μ the covariant derivative; since it is known that $\nabla_\mu T^{\mu\nu} = 0$, what is needed for obtaining $\nabla_\mu j^\mu = 0$ is that

$$\nabla_\mu u_\nu + \nabla_\nu u_\mu - \tfrac{1}{2} g_{\mu\nu} \nabla_\lambda u^\lambda = 0 \tag{7.176}$$

(since $T_{\mu\nu}$ is symmetric and traceless but otherwise arbitrary). This condition happens to be known as the condition for u_μ to be a "conformal Killing vector field," which is a vector field whose flow map consists of conformal isometries of space-time, a certain kind of symmetries. Generic curved space-times have no symmetries and thus no conformal Killing vector fields. I conclude that the $j^\mu = P^\mu$ approach breaks down in almost every curved space-time.

One Over Root omega

The Poynting vector is quadratic in $F_{\mu\nu}$ and contains one factor of k^0 too many. Since $F_{\mu\nu}$ consists of first derivatives of A_μ, an expression quadratic in A_μ should be expected to contain one factor of k^0 too few. This observation suggests to use a quantity in between A_μ and $F_{\mu\nu}$. Roland Good, Jr. (1957) [27] proposed to Fourier transform $F_{\mu\nu}$ (in all 4 space-time variables), multiply by

$$(k^0)^{-1/2} = 1/\sqrt{\omega} \tag{7.177}$$

(ω is a common notation for k^0, and we are assuming here only contributions with $k^0 > 0$), and Fourier transform back; call the resulting field $\tilde{F}_{\mu\nu}$; in another notation,

$$\tilde{F}_{\mu\nu} = (-\nabla^2)^{-1/4} F_{\mu\nu} \,. \tag{7.178}$$

Then an expression quadratic in $\tilde{F}_{\mu\nu}$ has the right number of factors k^0. Specifically, this is so for the Poynting-type vector $\tilde{P}^\mu = \tilde{T}^{\mu 0}$ with

$$\tilde{T}_{\mu\nu} = -\mathrm{Re}\big[\tilde{F}^*_{\mu\lambda} \tilde{F}_\nu{}^\lambda\big] + \tfrac{1}{4} \eta_{\mu\nu} \tilde{F}^*_{\lambda\rho} \tilde{F}^{\lambda\rho} \,, \tag{7.179}$$

formed from $\tilde{F}_{\mu\nu}$ just like $T_{\mu\nu}$ from $F_{\mu\nu}$. The proposal is to take

$$j^\mu = \tilde{P}^\mu \,. \tag{7.180}$$

Exercise 7.7 (One Over Root Omega) Verify that for a plane wave (7.167) with future-lightlike k^μ, (7.180) agrees with (7.171).

Here are the problems with this current:

- *Lack of Lorentz invariance.* Multiplication by $(k^0)^{-1/2}$ is an operation that depends on the Lorentz frame.
- *Visibly preferred frame.* The problem of the Poynting vector P^μ with an empirically distinguished frame gets inherited.
- *Nonlocal.* $(-\nabla^2)^{-1/4}$ is a nonlocal operator; that is, $\tilde{F}_{\mu\nu}$ in an open set A cannot be computed from $F_{\mu\nu}$ in A. In fact, the probability density $\rho(x)$ at a point x would depend on the wave function $F_{\mu\nu}(y)$ at arbitrarily large distances from x. That does not seem plausible. Note also that this is very different from the kind of nonlocality in Bell's theorem.
- *Curved space-time.* In a generic curved space-time $\mathscr{M}$, there are no plane waves, and since Fourier transformation is the expansion in plane waves, there is no Fourier transform of functions on $\mathscr{M}$. Thus, (7.180) is not defined in $\mathscr{M}$.

The Kappa Operator

Recall that every 4-vector k^μ can be translated into a spinor $k^{AA'} = \frac{1}{\sqrt{2}}\sigma_\mu^{AA'} k^\mu$. If k^μ is future-lightlike, then $k^{AA'}$ is the square of a spinor $\kappa \in S$,

$$k^{AA'} = \overline{\kappa}^{A'} \kappa^A \tag{7.181}$$

(e.g., Penrose and Rindler 1984 [38, Eq. (3.1.32)]). The free Maxwell field $F_{\mu\nu}$ has dispersion relation $k^0 = \pm|\boldsymbol{k}|$ (see Example 7.6); that is, the 4d Fourier transform $\widehat{F}_{\mu\nu}$ is concentrated on the light cone; equivalently, only lightlike k^μ occur. If we limit ourselves to positive energies, $k^0 > 0$, then k^μ can be written in the form (7.181). Now κ^A is not uniquely determined by (7.181); it is determined only up to a phase; $\tilde{\kappa}^A = e^{i\theta}\kappa^A$ will have the same square as κ^A, and this is in fact the only freedom. So suppose we have chosen a κ^A for every future-lightlike k^μ. Then this defines an operator on $F_{\mu\nu}$ as follows: Fourier transform, multiply by κ^A, and Fourier transform back. This operator (which we will also, in a slight abuse of notation, denote by κ^A) is defined in a covariant way (we never chose a Lorentz frame); it can be applied to any kind of field (e.g., a scalar field or a vector field such as A_μ) that obeys the D'Alembert equation (or massless Klein-Gordon equation) and comprises only contributions of positive energy; and it has the property

$$\overline{\kappa}^{A'} \kappa^A = i\partial^{AA'}, \tag{7.182}$$

which means it is something like a half derivative. This fits with the idea, discussed above, of using a quantity of which $F_{\mu\nu}$ is half a derivative; we could try to take half a derivative of A_μ.

Indeed, a (possibly complex) A_μ can, through a suitable gauge transformation, be made sure to satisfy the *Lorenz gauge condition*

$$\partial_\mu A^\mu = 0\,. \tag{7.183}$$

It then follows from the Maxwell equations (7.25) that

$$\Box A_\mu = 0\,. \tag{7.184}$$

So, assuming only positive energies, the κ operator is defined on A_μ, and we can define the *electromagnetic spinor potentials*

$$U^A = \overline{\kappa}_{A'}\, A^{AA'}\,, \tag{7.185a}$$

$$V^{A'} = \kappa_A\, A^{AA'}. \tag{7.185b}$$

They have the property that

$$F_{\mu\nu} = F_{AA'BB'} = \tfrac{i}{2}(\kappa_A U_B + \kappa_B U_A)\varepsilon_{A'B'} + \tfrac{i}{2}(\overline{\kappa}_{A'} V_{B'} + \overline{\kappa}_{B'} V_{A'})\varepsilon_{AB} \tag{7.186}$$

(and every complex positive-energy solution $F_{\mu\nu}$ of the Maxwell equations can be written in this way). They also have the property of being gauge invariant: if $\tilde{A}_\mu = A_\mu - \partial_\mu\theta$, then

$$\tilde{U}^A = \overline{\kappa}_{A'}\, \tilde{A}^{AA'} = \overline{\kappa}_{A'}\, A^{AA'} + i\overline{\kappa}_{A'}\,\overline{\kappa}^{A'}\kappa^A\theta = U^A \tag{7.187}$$

because $\overline{\kappa}_{A'}\,\overline{\kappa}^{A'} = 0$, and likewise $\tilde{V}^{A'} = V^{A'}$.

So, U^A and $V^{A'}$ lie, with regard to differentiation, between A_μ and $F_{\mu\nu}$, and this suggests defining the current

$$j^{AA'} = \sqrt{2}\Big(\overline{U}^{A'} U^A + \overline{V}^A V^{A'}\Big)\,. \tag{7.188}$$

For plane waves, this current agrees with (7.171), and it has the advantage over Good's current (7.180) of being defined in a Lorentz invariant way. Here are the problems with this current:

- *Not unique.* It depends on the choice of the phase of κ^A for every future-lightlike k^μ, different choices of the phase will lead to different vector fields j^μ, and there does not seem to be a distinguished choice.
- *Nonlocal.* As for Good's current (7.180), this j^μ depends on $F_{\mu\nu}$ (or A_μ) at arbitrary distances.
- *Curved space-time.* As for Good's current (7.180), this construction depends on Fourier transformation and thus is not defined in an obvious way in curved space-time.

Desiderata

I can formulate some desiderata[15] of the missing formula for j^μ for photons [54]:

1. The expression is quadratic in A_μ and its derivatives.
2. The expression is local, i.e., $j^\nu(x)$ depends only on A_μ and its derivatives *at* x.
3. j^μ is future-causal.
4. $\partial_\mu j^\mu = 0$ if A_μ obeys the free Maxwell equations.
5. For a plane wave, j^μ agrees with (7.171) up to a constant factor.
6. No choices need to be made, i.e., if the law requires a special gauge or Lorentz frame, then it also specifies this gauge or Lorentz frame.
7. The law can be generalized to curved space-time.

None of the proposals for j^μ that I am aware of satisfy all of the properties above, so I am hesitant to accept any of them. In the subsequent sections, we will focus on spin-$\frac{1}{2}$ particles and leave photons aside.

7.4 Many Particles

So far we have only considered 1-particle wave functions. Let us now turn to the case of N particles.

7.4.1 Multi-Time Wave Functions

In non-relativistic quantum mechanics, a wave function is a function ψ of time t and of N locations in space, one for each particle:

$$\psi = \psi(t, \boldsymbol{x}_1, \ldots, \boldsymbol{x}_N)\,. \tag{7.189}$$

That is, particle number j is considered at space-time point $(t, \boldsymbol{x}_j)$, so the N particles together are considered at a configuration of N space-time points lying on the 3d surface $\{t = \text{const.}\}$, which is horizontal in the given Lorentz frame. The obvious relativistic analog that does not refer to any Lorentz frame would be a function ϕ of N space-time points,

$$\phi = \phi(x_1, \ldots, x_N) = \phi(t_1, \boldsymbol{x}_1, \ldots, t_N, \boldsymbol{x}_N)\,, \tag{7.190}$$

where $x_j \in M$ has coordinates $(t_j, \boldsymbol{x}_j) \in \mathbb{R}^4$ in a given Lorentz frame. The key difference to ψ is that ϕ depends on N time variables, one for each particle; it is therefore called a *multi-time wave function*. In particular, ψ is a function of $3N + 1$

[15] = desired things (Latin).

variables and ϕ of $4N$ variables. A comprehensive discussion of multi-time wave functions is given in Lienert et al. (2020) [32].

The single-time wave function ψ can be recovered from ϕ when we set all time coordinates equal,

$$\psi(t, \boldsymbol{x}_1, \ldots, \boldsymbol{x}_N) = \phi(t, \boldsymbol{x}_1, \ldots, t, \boldsymbol{x}_N). \tag{7.191}$$

We may expect that ϕ will be defined, not on the set M^N of *all* space-time configurations, but on the set of *spacelike* configurations,

$$\mathscr{S}_N = \big\{(x_1, \ldots, x_N) \in M^N \,:\, \forall i \neq j \,:\, x_i \times x_j \text{ or } x_i = x_j\big\}, \tag{7.192}$$

where $x \times y$ means that x and y are spacelike separated.

An obvious way of defining a time evolution for ϕ is to provide a PDE for each time coordinate,

$$\begin{aligned} i\frac{\partial\phi}{\partial t_1} &= H_1\phi\,, \\ &\vdots \\ i\frac{\partial\phi}{\partial t_N} &= H_N\phi\,. \end{aligned} \tag{7.193}$$

We call a system of PDEs of this form *multi-time equations* and the operators H_j appearing on the right-hand sides the *partial Hamiltonians*. It follows from (7.191) and the chain rule that

$$i\frac{\partial\psi}{\partial t}(t, \boldsymbol{x}_1, \ldots, \boldsymbol{x}_N) = i\frac{\partial}{\partial t}\phi(t, \boldsymbol{x}_1, \ldots, t, \boldsymbol{x}_N) \tag{7.194}$$

$$= i\sum_{j=1}^{N}\frac{\partial\phi}{\partial t_j}(t, \boldsymbol{x}_1, \ldots, t, \boldsymbol{x}_N) \tag{7.195}$$

$$= \sum_{j=1}^{N} H_j\phi(t, \boldsymbol{x}_1, \ldots, t, \boldsymbol{x}_N)\,, \tag{7.196}$$

so the Hamiltonian H governing the evolution of ψ is the sum of the partial Hamiltonians,

$$H = \sum_{j=1}^{N} H_j\Big|_{t_1=\ldots=t_N=t}\,. \tag{7.197}$$

As a simple example of multi-time equations, consider N non-interacting Dirac particles with wave function $\phi : M^N \to (\mathbb{C}^4)^{\otimes N}$, $\phi = \phi_{s_1 \ldots s_N}(x_1, \ldots, x_N)$, governed by

$$\begin{aligned} i\frac{\partial\phi}{\partial t_1} &= -i\boldsymbol{\alpha}_1 \cdot \nabla_1\phi + m_1\beta_1\phi\,, \\ &\;\vdots \\ i\frac{\partial\phi}{\partial t_N} &= -i\boldsymbol{\alpha}_N \cdot \nabla_N\phi + m_N\beta_N\phi\,, \end{aligned} \tag{7.198}$$

where $\boldsymbol{\alpha}_j$ and β_j mean the $\boldsymbol{\alpha}$ and β matrices acting on the index s_j, $m_j \geq 0$ is the mass of particle j, and $\nabla_j = (\partial_{j1}, \partial_{j2}, \partial_{j3})$ with notation

$$\partial_{j\mu}\phi := \frac{\partial\phi}{\partial x_j^\mu} \tag{7.199}$$

for $j \in \{1, \ldots, N\}$ and $\mu \in \{0, 1, 2, 3\}$. Equivalently, we can write (7.198) in the covariant form

$$\begin{aligned} i\gamma_1^\mu\partial_{1\mu}\phi &= m_1\phi\,, \\ &\;\vdots \\ i\gamma_N^\mu\partial_{N\mu}\phi &= m_N\phi\,, \end{aligned} \tag{7.200}$$

which makes clear that this system of PDEs can be formulated without even choosing a Lorentz frame. As initial data, one can specify ϕ on the set where $t_1 = 0, \ldots, t_N = 0$ arbitrarily as a (square-integrable) function ϕ_0 of the $3N$ variables $\boldsymbol{x}_1, \ldots, \boldsymbol{x}_N$; ϕ is then uniquely determined on all of the $4N$-dimensional set M^N by

$$\phi(t_1, \cdot, \ldots, t_N, \cdot) = e^{-iH_N t_N} \cdots e^{-iH_1 t_1}\phi_0 \tag{7.201}$$

with H_j the 1-particle Dirac operator acting on particle j as on the right-hand side of the j-th equation in (7.198). This is the same freedom of choice of initial data as in the single-time picture for ψ_t. Conversely, (7.201) is actually a solution of (7.198). In particular, we see that in the non-interacting case, ϕ can be defined on M^N, i.e., also on non-spacelike configurations.

The case with interaction is a bit more complicated, in two ways. First, a system of multi-time equations (7.193) can be inconsistent, that is, it may fail to possess a solution for (some or every) nonzero initial condition ϕ_0. Here is a heuristic outline of the reason. Since, for $j \neq k \in \{1, \ldots, N\}$,

$$-\partial_{j0}\partial_{k0}\phi = -\partial_{k0}\partial_{j0}\phi\,, \tag{7.202}$$

the multi-time equations (7.193) entail that

$$i\,\partial_{j0}(H_k\phi) = i\,\partial_{k0}(H_j\phi)\,. \tag{7.203}$$

Since

$$\partial_{j0}(H_k\phi) = (\partial_{j0}H_k)\phi + H_k\partial_{j0}\phi \tag{7.204}$$

(where the $\partial_{j0}H_k$ term reflects the situation that the operator H_k may depend on the variable x_j^0), we obtain that

$$(i\partial_{j0}H_k)\phi + H_kH_j\phi = (i\partial_{k0}H_j)\phi + H_jH_k\phi\,. \tag{7.205}$$

Since this condition must be true at every $(t_1, \boldsymbol{x}_1, \ldots, t_N, \boldsymbol{x}_N)$ and does not involve any time derivatives of ϕ any more, and since we could take any fixed $t_1, \ldots, t_N$ as the "initial" times and choose any function of $\boldsymbol{x}_1, \ldots, \boldsymbol{x}_N$ as initial data there, the operator $(i\partial_{j0}H_k) + H_kH_j$ must agree with the operator $(i\partial_{k0}H_j)\phi + H_jH_k$ on arbitrary functions of $\boldsymbol{x}_1, \ldots, \boldsymbol{x}_N$, that is, they must agree as operators,

$$i\partial_{j0}H_k + H_kH_j = i\partial_{k0}H_j + H_jH_k\,. \tag{7.206}$$

But this *consistency condition* will be met by some choices of the partial Hamiltonians and not by others. For example, if the partial Hamiltonians are time-independent, then (7.206) demands that H_j and H_k commute—a condition met by some operators and not by others. In short, we cannot choose the H_j arbitrarily; to make the multi-time equations (7.193) consistent, we need H_j that satisfy the consistency condition (7.206). For non-interacting particles as in (7.198), the H_j are time-independent and commute with each other, so the consistency condition (7.206) is satisfied; in fact, we have already seen in (7.201) that all initial data possess solutions, i.e., that the system (7.198) is consistent. But the consistency condition limits the possibilities for choosing interaction terms; for example, more or less, all interaction potentials are excluded.

I said there are two complications with interaction. The first was the consistency condition. The second is that, in a relativistic context, interaction potentials seem impossible because they would entail instantaneous interaction and thus violate interaction locality and no-signaling, except for point interactions (zero-range interactions) as described in Sect. 3.7.5. Two further ways of implementing relativistic interaction are provided by interaction between space-time points x, y that are lightlike separated and by interaction through the creation and annihilation of particles. And in fact, all three kinds of relativistic interaction can be implemented with consistent multi-time equations on the set of spacelike configurations; see Lienert et al. (2020) [32] and references therein.

7.4.2 Surface Wave Functions

For a single particle, we considered wave functions sometimes as functions on space-time M and sometimes as functions on a Cauchy surface Σ. Likewise for N-particle wave functions: we can consider them as functions ϕ on the set $\mathscr{S}_N$ of spacelike configurations or as functions ψ_Σ associated with a Cauchy surface Σ. For definiteness, let us consider Dirac particles. The relation between ϕ and ψ_Σ is straightforward: a given $\phi : \mathscr{S}_N \to D^{\otimes N}$ can be restricted to configurations on Σ, that is,

$$\psi_\Sigma(x_1, \ldots, x_N) = \phi(x_1, \ldots, x_N) \quad \forall x_1, \ldots, x_N \in \Sigma\,, \tag{7.207}$$

yielding a function $\psi_\Sigma : \Sigma^N \to D^{\otimes N}$. Here, we used that every configuration on Σ is spacelike. We call ψ_Σ a *surface wave function*. Conversely, a family of ψ_Σ for every Cauchy surface Σ fits together as a multi-time wave function if for every configuration x^N that lies on both Σ and Σ', $\psi_\Sigma(x^N) = \psi_{\Sigma'}(x^N)$. Then, multi-time wave functions can be translated into surface wave functions and vice versa.

For fixed Σ and N, the possible ψ_Σ lie in a Hilbert space $\mathscr{H}_{N\Sigma}$ that can be defined in either of two ways:

- as the space of functions $\psi : \Sigma^N \to D^{\otimes N}$ (modulo changes on sets of measure 0) that are (anti-symmetric for fermions and) square-integrable in the sense $\langle\psi|\psi\rangle < \infty$, where the inner product is defined as

$$\langle\psi|\chi\rangle_{N\Sigma} = \int_\Sigma V(d^3x_1)\cdots\int_\Sigma V(d^3x_N)\,\overline{\psi}(x_1\ldots x_N) \\ \times [\gamma^{\mu_1} n_{\mu_1}(x_1)\otimes\cdots\otimes\gamma^{\mu_N} n_{\mu_N}(x_N)]\,\chi(x_1\ldots x_N) \tag{7.208}$$

  with $\overline{\psi} = \psi^*\gamma^0\otimes\cdots\otimes\gamma^0$.
- as the tensor product of N copies of the 1-particle Hilbert space $\mathscr{H}_{1\Sigma}$ as defined in (7.143) and (7.144) in Sect. 7.3.4 (or the anti-symmetric subspace thereof).

On any Cauchy surface Σ, a $\psi_\Sigma \in \mathscr{H}_{N\Sigma}$ can serve as initial datum for multi-time equations for ϕ. By evaluating ϕ on Σ', we can define a time evolution $U^{\Sigma'}_{N\Sigma} : \mathscr{H}_{N\Sigma} \to \mathscr{H}_{N\Sigma'}$. For the free multi-time Dirac evolution (7.200) or the one including an external electromagnetic field, $U^{\Sigma'}_{N\Sigma}$ is unitary and in fact is the tensor product of N copies of the 1-particle evolution operator $U^{\Sigma'}_{1\Sigma}$ as in Theorem 7.1 in Sect. 7.3.4 (restricted to the anti-symmetric subspace). The factorization of $U^{\Sigma'}_{N\Sigma}$ for N particles into a tensor product of N operators indicates the absence of interaction between the particles. However, for multi-time equations with interaction, $U_{N\Sigma}^{\Sigma'}$ should still be unitary and satisfy the composition law (7.146).

Let us turn to how multi-time or surface wave functions are related to detection probabilities.

Curved Born Rule for N Dirac Particles *If we place detectors along a Cauchy surface Σ and let them act on a system of N Dirac particles, then the observed configuration $Q \in \Sigma^N$ has distribution*

$$\begin{aligned}\mathbb{P}\Big(Q \in d^3x_1 \times \cdots \times d^3x_N\Big) \\ &= V(d^3x_1) \cdots V(d^3x_N)\, \overline{\psi_\Sigma}(x_1 \ldots x_N) \\ &\quad \times [\gamma^{\mu_1} n_{\mu_1}(x_1) \otimes \cdots \otimes \gamma^{\mu_N} n_{\mu_N}(x_N)]\, \psi_\Sigma(x_1 \ldots x_N)\,. \end{aligned} \tag{7.209}$$

Note that the probability distribution on the right-hand side (the N-particle curved Born distribution) coincides with the Born distribution associated with the PVM $P_{N\Sigma} := P \otimes \cdots \otimes P$ (N factors of the 1-particle PVM P as in Proposition 7.12) on Σ^N acting on $\mathscr{H}_{1\Sigma}^{\otimes N}$ (and, thus, for permutation-invariant sets, on $\mathscr{H}_{N\Sigma}$).

There is an aspect about the curved Born rule that needs explanation. One cannot simply *postulate* the curved Born rule for every Σ. If a law determines in one Lorentz frame the probability distribution of the configuration at every time t, then it also determines the distribution of the outcomes of any experiment, regardless of whether the experiment refers to detectors along some non-horizontal surface Σ. Thus, once we assume the Born rule on the horizontal surfaces in one Lorentz frame, the curved Born rule is *either false or a theorem.* In fact, it is a theorem. To formulate it, it is convenient to use the Hilbert space $\mathscr{H}_\Sigma = \oplus_{N=0}^{\infty}\mathscr{H}_{N\Sigma}$ of a variable number of particles and the configuration space $\Gamma(\Sigma)$ of a variable number of particles as in (6.23).

Theorem 7.3 (Lienert et al. 2017 [33]) *Suppose we are given a unitary time evolution $U_\Sigma^{\Sigma'} : \mathscr{H}_\Sigma \to \mathscr{H}_{\Sigma'}$ satisfying $U_\Sigma^\Sigma = I_\Sigma$ and $U_{\Sigma'}^{\Sigma''} U_\Sigma^{\Sigma'} = U_\Sigma^{\Sigma''}$, interaction locality, and propagation locality. Assume the horizontal Born rule for configuration measurements at t and the horizontal collapse rule (that after the configuration was found to lie in a set B in configuration space, the wave function gets projected to B). For a setup of horizontal detectors converging to Σ, the detection distribution converges to the curved Born distribution.*

A similar result using a different model of the detectors has been established in Lill et al. (2021) [34].

Remark 7.6 (Tomonaga-Schwinger Equation) When we add interaction to the free time evolution, then there is a possibility of using just one Hilbert space $\mathscr{H}_0$ (say, referring to a fixed surface Cauchy Σ_0) by transporting everything to $\mathscr{H}_0$ by means

Fig. 7.8 Two Cauchy surfaces that are infinitesimally close

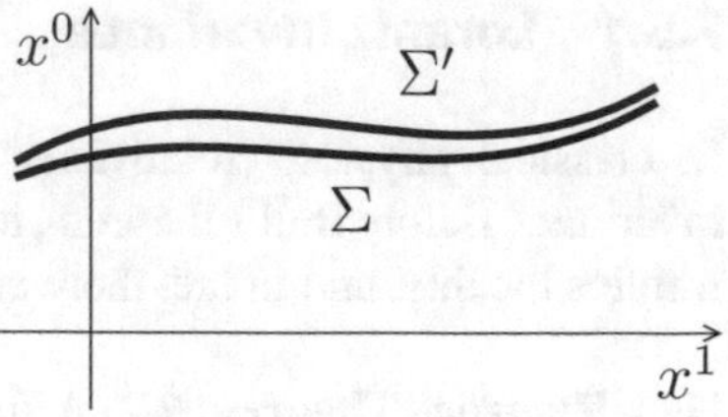

of the free time evolution $F_\Sigma^{\Sigma'}$ ("interaction picture"): then

$$\widetilde{\psi}_\Sigma = F_\Sigma^{\Sigma_0} \psi_\Sigma \tag{7.210}$$

will be an element of $\mathscr{H}_0$ that is an image of, or represents, ψ_Σ; $\widetilde{\psi}_\Sigma$ is called the Tomonaga-Schwinger wave function. It evolves according to

$$\widetilde{\psi}_{\Sigma'} = \widetilde{U}_\Sigma^{\Sigma'}\, \widetilde{\psi}_\Sigma \tag{7.211}$$

with unitary operators $\widetilde{U}_\Sigma^{\Sigma'} : \mathscr{H}_0 \to \mathscr{H}_0$ given by

$$\widetilde{U}_\Sigma^{\Sigma'} = F_{\Sigma'}^{\Sigma_0}\, U_\Sigma^{\Sigma'}\, F_{\Sigma_0}^{\Sigma}\,. \tag{7.212}$$

For explicit calculations of the time evolution, Tomonaga (1946) [46] and Schwinger (1948) [40] have introduced the following form of writing the evolution from a surface Σ to a surface Σ' infinitesimally close to Σ as in Fig. 7.8:

$$i\,(\widetilde{\psi}_{\Sigma'} - \widetilde{\psi}_\Sigma) = \int_\Sigma^{\Sigma'} d^4x\, \mathcal{H}_I(x)\, \widetilde{\psi}_\Sigma\,. \tag{7.213}$$

This equation is known as the *Tomonaga-Schwinger equation*. The operator-valued function $\mathcal{H}_I(x)$ is called the interaction Hamiltonian density.

7.5 Which Theories Count as Relativistic?

As I will explain in this section, a local theory is relativistic whenever it is Lorentz invariant, while for a nonlocal theory, Lorentz invariance is a very weak condition that does not seem to capture the idea of being relativistic. For nonlocal theories, there are different gradations of being relativistic. Even before I discuss the relevant modifications for Bohmian mechanics and GRW theory in relativistic space-time in detail in Sects. 7.6 and 7.8, I can give away that Bohmian mechanics is less relativistic than GRW theory (though still a serious possibility). I will describe in what way it is less relativistic and discuss what to conclude about the nature of reality.

7.5.1 Lorentz Invariance

In classical physics, "relativistic" is often taken as synonymous with "Lorentz invariant." Before Bell's theorem, it was also widely believed that Lorentz invariance implies locality, and in fact there are reasons for this belief.

Non-Rigorous Theorem 7.2 *A field theory that is governed by a PDE, which allows for external sources (or external fields or localized external parameters) and is Lorentz invariant, must be local.*

Derivation If the PDE (think of the wave equation or the second-order Klein-Gordon equation) contains terms that experimentalists can control and choose (within certain options) at whim (i.e., with free will) (which is what I called external sources or fields or parameters), then we can consider solutions ϕ and ϕ' of the PDE with the same initial conditions (say, on the surface $\{x^0 = 0\}$ in some frame) and different external parameters (in the region $\{x^0 > 0\}$ in that frame). If the two choices of parameters differ only in a bounded space-time region A, then ϕ and ϕ' will agree up to the first time of A, $t(A) = \inf\{x^0 : x \in A\}$. By Lorentz invariance, this is true in every Lorentz frame, so ϕ and ϕ' agree everywhere outside the future of A as in Fig. 7.5. But that means events in A do not influence the values of ϕ outside the future of A; in other words, the theory is local. □

Now let us consider what Lorentz invariance means for a *particle theory*. For simplicity, let us consider a fixed number N of particles. In space-time M, a particle is represented by its world line (a connected 1d subset or curve) $L \subset M$. So, a possible history h for an ontology of N particles is an N-tuple of world lines, $h = (L_1, \ldots, L_N)$. An abstract characterization of such a theory may be taken to consist of a set W of possible histories.[16] A Lorentz transformation $\Lambda : M \to M$ maps every space-time point to another space-time point, so it also maps every world line L to another world line $\Lambda_* L$ and thus every N-tuple of world lines to another N-tuple of world lines, that is, every possible history h to another possible history $\Lambda_* h$. The obvious meaning of the statement that *the theory is Lorentz invariant* is that the set W is closed under Λ_*, i.e., $\Lambda_* h \in W$ for every $h \in W$.

There is an obvious way, then, of making any such theory Lorentz invariant: simply form the closure W_* of any given W under all Lorentz transformations,

$$W_* := \{\Lambda_* h : h \in W, \Lambda \in O(1,3)\}\,. \tag{7.214}$$

[16] Alternatively, we might consider other types of theories, described by a set W of possible histories together with a probability distribution μ over W or described by a family Ω of possible values of a further variable ω (e.g., a wave function) together with, for every $\omega \in \Omega$, a set W_ω of possible histories and a probability distribution μ_ω over W_ω. But all these further complications make no difference to the point I want to make, so we can focus on the simplest case given by a set W of possible histories.

For example, take W to be the solutions of Newtonian mechanics as in (1.59), a $6N$-parameter family of histories. Of course, Newtonian mechanics is a paradigm of a non-relativistic theory; with its instantaneous interaction via the Coulomb potential and the Newtonian gravity, it should be possible to send messages arbitrarily fast, in particular superluminally. W is not closed under Lorentz transformations, so let us consider its closure W_*. The theory corresponding to W_*, what is it like? It says that those histories can physically occur that obey Newton's equation of motion (1.59) in *some* Lorentz frame F. It allows for arbitrarily fast signaling in F; it does not entail that fast-moving clocks slow down, that fast-moving particles get heavier, that fast-moving rods contract, or that particle cannot move superluminally, or any relativistic effect, or any relativistic correction to the equation of motion, provided we consider the motion in F. Inhabitants of a universe governed by this theory would not even come up with the idea of relativity. This theory is a clear-cut example of a theory that is *not* relativistic. And yet, it is Lorentz invariant. It is because it leaves open *which* Lorentz frame is the one in which the equation of motion (1.59) holds.

This example shows why for a *general* theory, Lorentz invariance is a very weak property that does not capture what we mean by "relativistic." For a *local* theory, in contrast, I cannot think of a case in which the theory is Lorentz invariant but would not be regarded as relativistic. So it seems that for local theories, Lorentz invariance is a good criterion for being relativistic.

7.5.2 Other Relativistic Properties

Apart from Lorentz invariance (which we have seen is not sufficient for being relativistic) and locality (which we have seen is violated in nature), there are a number of other properties of theories related to "being relativistic." Two space-time regions $A, B \subset M$ are *spacelike separated* if and only if every $a \in A$ is spacelike separated from every $b \in B$.

1. Interaction locality: The Hamiltonian contains no interaction terms between spacelike separated regions.
2. Propagation locality: Wave functions do not propagate superluminally.
3. No superluminal signaling: It is impossible for agents to send signals between spacelike separated regions.
4. Independent Lorentz invariance: If regions $A, B \subset M$ are spacelike separated, if Λ, Λ' are Lorentz transformations, and if $\Lambda(A), \Lambda'(B)$ are also spacelike separated, then for every solution X of the primitive ontology (local beables) of the theory, there is a solution $\tilde{X}$ such that

$$\tilde{X} \cap \Lambda(A) = \Lambda(X \cap A) \text{ and } \tilde{X} \cap \Lambda'(B) = \Lambda'(X \cap B)\,. \tag{7.215}$$

(The notation applies directly to X that is a *subset* of M such as world lines; for a field, $X \cap A$ should be understood as the restriction of the field X to A.) That is, the theory is invariant under two Lorentz transformations applied to spacelike separated regions.

5. Microscopic parameter independence (MPI): If regions $A, B \subset M$ are spacelike separated, then the probability distribution of the primitive ontology X obeys

$$\mathbb{P}(X \cap A|\Phi_A, \Phi_B, \lambda) = \mathbb{P}(X \cap A|\Phi_A, \lambda) \tag{7.216}$$

for external fields Φ and any further variables λ of the theory. Here, Φ_A means the restriction of Φ to A, i.e., the external fields in A. So, (7.216) expresses that the distribution of the local beables X in A does not depend on external fields in B.

These properties are not necessarily independent of each other; for example, the no-signaling property may require some version of interaction locality and propagation locality. Further properties that sometimes get mentioned: a unitary representation of the proper Lorentz group $O_+^\uparrow(1, 3)$ on the relevant Hilbert space (which amounts to a version of Lorentz invariance for wave functions) or that, in a quantum field theory, the field operators associated with spacelike separated space-time points x, y either commute or anti-commute, $\phi(x)\phi(y) = \pm\phi(y)\phi(x)$ (which is usually closely related to interaction locality and thus the no-signaling property).

7.5.3 Relativistic Quantum Theories Without Observers

Now, can quantum theories without observers be made relativistic? Let us focus first on Bohmian mechanics and GRW theory (I will make some remarks on orthodox quantum mechanics in Sect. 7.7.1 and on many-worlds in Sect. 7.9.1).

GRW-type theories violate propagation locality because at a collapse of a 1-particle wave function, one wave packet may disappear, while another grows in size; but we have already seen that even in non-relativistic GRW theories, this violation cannot be used for superluminal signaling, and I will argue that this violation does not constitute a violation of relativity. We will see in Sect. 7.8 that GRW theory can be so adjusted as to satisfy properties 1 and 3–5, whereas suitable versions of Bohmian mechanics satisfy 1–4 but not 5. I will conclude that GRW theory can be made more relativistic than Bohmian mechanics. At the same time, I would not take for granted that property 5 is satisfied in nature.

The situation of Bohmian mechanics in relativistic space-time is as follows. If a preferred foliation (slicing) of space-time into spacelike surfaces as in Fig. 7.9 is permitted, then there is a simple, convincing analog of Bohmian mechanics that I will describe in detail in Sect. 7.6. I call this foliation the *time foliation* $\mathcal{F}$ and the version of Bohmian mechanics based on it $\mathrm{BM}_{\mathcal{F}}$. Without a time foliation, no version of Bohmian mechanics is known that would make predictions anywhere near the quantum formalism. Should such a version be found in the future, it might

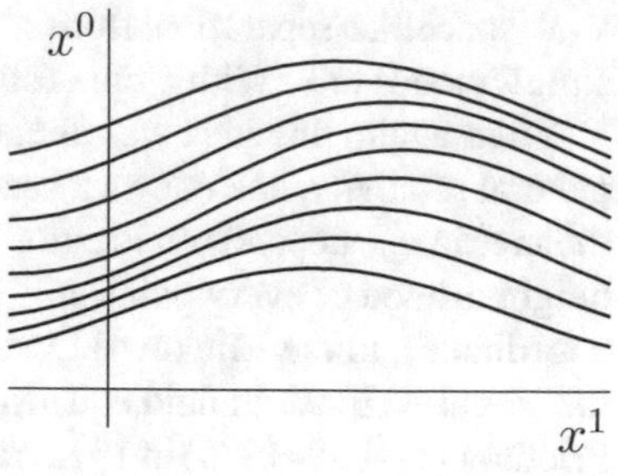

Fig. 7.9 A spacelike foliation

be more convincing than $\mathrm{BM}_{\mathcal{F}}$, but I have studied the question closely, and I have no hope that such a version can be found.

7.5.4 The Time Foliation

A foliation means in general a partition

$$M = \bigcup_{\Sigma \in \mathcal{F}} \Sigma \tag{7.217}$$

into disjoint submanifolds (here, into Cauchy surfaces). Every space-time point lies on exactly one surface ("time leaf") belonging to the foliation. The idea is that nature knows which foliation, among all mathematically possible ones, is the time foliation $\mathcal{F}$, that there is something in physical reality that is mathematically represented by $\mathcal{F}$. It turns out that while the particle world lines, according to $\mathrm{BM}_{\mathcal{F}}$, depend on $\mathcal{F}$, the predicted probability distribution of the outcome of any experiment does not. As a consequence, inhabitants of a universe governed by $\mathrm{BM}_{\mathcal{F}}$ cannot empirically determine $\mathcal{F}$. In fact, they cannot obtain any information from their observed outcomes of experiments about which foliation is $\mathcal{F}$; one can say that $\mathcal{F}$ is invisible to them.

The time foliation $\mathcal{F}$ defines a temporal order for spacelike separated points x, y because we can ask whether the time leaf through x is before, after, or identical to the time leaf through y. I will describe in Sect. 7.6 how this temporal order, and in particular the notion of simultaneity-at-a-distance corresponding to x and y lying on the same time leaf, is relevant to setting up a Bohm-type law of motion.

This notion of simultaneity-at-a-distance does not imply a return to a pre-relativistic idea of passing time; on the contrary, it is fully compatible with the "block universe" view of space-time that we use in relativistic physics, the view that past, present, and future space-time points are equally real. And yet, the use of a preferred foliation seems against the spirit of relativity. Most of us, when we learned relativity, made a big effort to accept and get used to the idea that there is no simultaneity-at-a-distance, that there is no absolute time, and that every point

y at spacelike separation from x should have the same right to be regarded as simultaneous to x. With a time foliation, we would revert this idea.

Then again, this idea was already reverted in general relativity. Here is how. In general relativity, one allows a space-time with a curved metric (i.e., one for which the metric $g_{\mu\nu}$ depends on location and time in such a way that in a sufficiently small neighborhood of every point, $g_{\mu\nu}$ becomes arbitrarily close, in suitable curvilinear coordinates, to the Minkowski metric $\eta_{\mu\nu}$ as in (7.4)), and the metric is subject to the so-called Einstein field equation. As discovered by Russian physicist Alexander Friedmann (1888–1925) in 1922, reasonable models of the approximate geometry of the universe as a whole in agreement with the Einstein field equation ("cosmological models") feature an *initial singularity*, also known as the *big bang*. A singularity is a point or set of points where the curvature density is infinite; as an example from among curved 2d surfaces in 3d Euclidean space, the surface of a cube has a singularity at each corner (the curvature density is infinite because a finite amount of curvature is concentrated in a single point), whereas the sphere has no singularity (the curvature density is finite everywhere). In each of the space-times studied by Friedmann, the set of initial singular points can be regarded as a 3d spacelike Cauchy surface Σ_0. Every space-time point lies in the future of Σ_0, so time has a beginning in the sense that there exists no space-time point in the past of Σ_0. When fitted to observational data, the singularity lies approximately 13.7 billion years in the past from here-now, so that is the absolute time here-now. Some authors (e.g., Carroll and Chen 2005 [12], Ijjas and Steinhardt 2019 [29]) have proposed that, on the contrary, our space-time had no initial singularity and that curvature density was much higher but not infinite 13.7 billion years ago. But at least it is a serious and reasonable possibility that our space-time might have had an initial singularity and thus a beginning in time.

Now in every space-time $\mathscr{M}$ with an initial singularity, one can define the function $T : \mathscr{M} \to [0, \infty)$ that specifies for every space-time point $x \in \mathscr{M}$ the timelike distance from the initial singularity; for example (assuming we live in a Friedmann space-time), T(here-now) = 13.7 billion years. This function T provides a natural notion of absolute time, often called *comoving time* in cosmology, and its level sets provide a spacelike foliation $\mathcal{F}_T$ of $\mathscr{M}$—a foliation that is preferred, among all foliations of $\mathscr{M}$, by the space-time metric $g_{\mu\nu}$ itself. And this foliation might be the time foliation that $\text{BM}_\mathcal{F}$ requires, $\mathcal{F} = \mathcal{F}_T$. (If we knew that $\mathcal{F} = \mathcal{F}_T$, then we could empirically determine $\mathcal{F}$, not from observing positions of entangled Bohmian particles but from observing the large-scale geometry of space-time.) Alternatively, $\mathcal{F}$ might be defined in terms of the wave function Ψ of the universe (Dürr et al. 2014 [24]). Or, $\mathcal{F}$ might be determined by a covariant evolution law (possibly involving Ψ) from initial data (such as an initial time leaf); see my paper [50, Sec. 3.3.1] for an example of such a law.

So, a preferred foliation does not mean a violation of relativity, as relativity itself often provides a preferred foliation. On the other hand, it turns out that GRWf and GRWm can be defined in relativistic (even curved) space-time without assuming a preferred foliation. Readers who have reservations about theories with a preferred foliation should thus prefer GRW theories. But perhaps the question of what we should mean by "relativistic" is rather irrelevant, and what is relevant is that the possibility that our universe has a time foliation is worth considering.

In the next sections, I will describe $BM_{\mathcal{F}}$ and relativistic GRW theory in some detail.

7.6 Bohmian Mechanics in Relativistic Space-Time

Here is the definition of $BM_{\mathcal{F}}$, the natural analog of Bohmian mechanics in relativistic space-time when a time foliation $\mathcal{F}$ (as described in Sect. 7.5.4) is given. The equations were developed by Dürr et al. (1999) [22], and for the special case in which $\mathcal{F}$ consists of parallel hyperplanes (a Lorentz frame) by Bohm and Hiley (1993) [11]. The theory, an N-particle generalization of the 1-particle equation of motion described in Sect. 7.3.3, can also be defined in curved space-time; $\mathcal{F}$ can be an arbitrary spacelike foliation. The theory uses multi-time wave functions. We define it here for N non-interacting Dirac particles; as already remarked, the multi-time evolution can also be defined with particle creation terms, and the equation of motion we consider below can then be combined with the stochastic particle creation events of Bell's jump process as introduced in Sect. 6.3.1.

The theory defines N world lines in space-time M (for definiteness, Minkowski space-time) from initial data on some time leaf $\Sigma_0 \in \mathcal{F}$ regarded as the initial time leaf. It will satisfy an equivariance theorem saying that if the initial configuration Q_{Σ_0} is $|\phi_{\Sigma_0}|^2$ distributed, then on any $\Sigma \in \mathcal{F}$, the configuration Q_Σ is $|\phi_\Sigma|^2$ distributed. The basic idea of the theory is to provide a law of motion analogous to Bohm's in the form of an ODE that defines the velocity of each particle in terms of the wave function and the simultaneous positions of the other particles, where "simultaneous" is defined as "lying on the same leaf of $\mathcal{F}$."

7.6.1 Law of Motion

The equation of motion will specify (up to a sign) the direction of dX^μ/ds for each particle. Let $\phi : M^N \to (\mathbb{C}^4)^{\otimes N}$ be a solution of the multi-time equations (7.198) without interaction. At every $(x_1, \ldots, x_N) \in M^N$, let

$$j^{\mu_1 \ldots \mu_N}(x_1, \ldots, x_N) = \overline{\phi}(x_1, \ldots, x_N)\,[\gamma^{\mu_1} \otimes \cdots \otimes \gamma^{\mu_N}]\,\phi(x_1, \ldots, x_N)\,. \tag{7.218}$$

j, called the *probability current tensor field*, is the N-particle generalization of the vector field j^μ introduced in (7.123); it is a function of $4N$ variables and, with

N indices $\mu_k \in \{0, 1, 2, 3\}$, has 4^N components.[17] The equation of motion (the relativistic analog of Bohm's (1.77)) reads, for every $k \in \{1, \dots, N\}$,

$$\frac{dX_k^{\mu_k}}{ds}(s_0) \propto j^{\mu_1\dots\mu_N}\Big(X_1(\Sigma), \dots, X_N(\Sigma)\Big) \prod_{i \neq k} n_{\mu_i}\big(X_i(\Sigma)\big), \tag{7.219}$$

where $\Sigma \in \mathcal{F}$ is the unique time leaf containing $X_k^\mu(s_0)$, $X_i(\Sigma)$ is the unique point where the world line X_i intersects the Cauchy surface Σ (so $X_k(\Sigma) = X_k(s_0)$), and $n_\mu(x)$ is the future unit normal vector to Σ at $x \in \Sigma$.

Proposition 7.14 *The right-hand side of* (7.219) *is a future-causal 4-vector, and it is nonzero at each* $(x_1, \dots, x_N) \in M^N$ *where* ψ *is nonzero.*

Proof A 4-vector u^μ is future-causal if and only if its Minkowski product $u^\mu v_\mu$ with every future-causal vector is ≥ 0; see Exercise 7.3 in Sect. 7.3.2. So, for the first statement of the proposition, it suffices to show that for all future-causal vectors $v_1, \dots, v_N$,

$$j^{\mu_1\dots\mu_N} v_{1\mu_1} \cdots v_{N\mu_N} \geq 0\,. \tag{7.220}$$

But the quantity on the left-hand side is of the form

$$\langle\phi|R_1 \otimes \cdots \otimes R_N|\phi\rangle\,, \tag{7.221}$$

where the inner product is taken in $(\mathbb{C}^4)^{\otimes N}$, ϕ is short for $\phi(x_1, \dots, x_N)$, and $R_k = \gamma^0 v_{k\mu}\gamma^\mu$. Since the tensor product of positive operators is positive (see Exercise 5.11), it suffices to show that each R_k is positive. Now this is equivalent to $\langle\psi|R_k|\psi\rangle \geq 0$ for every $\psi \in \mathbb{C}^4$, and this quantity is just $v_{k\mu} j^\mu$ with $j^\mu = \overline{\psi}\gamma^\mu\psi$. But we know that for every $\psi \in \mathbb{C}^4$, j^μ is future-causal, and the Minkowski product $v_{k\mu} j^\mu$ between any two future-causal vectors is ≥ 0.

The proof of the last statement of the Proposition proceeds along similar lines. The 1-particle Dirac current $j^\mu = \overline{\psi}\gamma^\mu\psi$ is future-causal, and for $\psi \neq 0$ it is nonzero because $j^0 = \sum_{s=1}^4 |\psi_s|^2$; $n_\mu(x)$ is future-timelike and nonzero. Thus, by Exercise 7.4 in Sect. 7.3.2, $j^\mu n_\mu > 0$ for $\psi \neq 0$. Since $j^\mu = \psi^*\gamma^0\gamma^\mu\psi$, it follows that $R(x) = \gamma^0 n_\mu(x)\,\gamma^\mu$ is positive *definite* (rather than semi-definite), i.e., that 0 is not an eigenvalue. The same argument as in Exercise 5.11 shows that a tensor

[17] In curved space-time, $\phi(x_1, \dots, x_N)$ takes values in $D_{x_1} \otimes \cdots \otimes D_{x_1}$, where D_x is the (complex 4d) spin space associated with space-time point x, and $j(x_1, \dots, x_N)$ in $T_{x_1} \otimes \cdots \otimes T_{x_N}$, where T_x is the tangent space to the space-time manifold in x.

product of positive definite matrices is positive definite, so $R(x_1) \otimes \cdots \otimes R(x_N)$ is positive definite. Thus, at any configuration where $\psi \neq 0$,

$$j^{\mu_1 \dots \mu_N} n_{\mu_1}(x_1) \cdots n_{\mu_N}(x_N) > 0\,, \tag{7.222}$$

which implies the last statement of the Proposition. □

Remark 7.7 In the special case in which $\mathcal{F}$ consists of parallel hyperplanes, i.e., in which the time leaves are the horizontal surfaces of some Lorentz frame F, the vectors n_μ are of the form $n_\mu = (1, 0, 0, 0)$ in F. If we parametrize the world lines by coordinate time of F, (7.219) assumes the form

$$\frac{dX_k^{\mu_k}}{dt}(t_0) \propto j^{0\dots0\mu_k0\dots0}\big(X_1(t_0), \dots, X_N(t_0)\big) \tag{7.223}$$

or (since $X_k^0 = t$ then),

$$\frac{d\boldsymbol{X}_k}{dt}(t_0) = \frac{\psi^* \boldsymbol{\alpha}_k \psi}{\psi^* \psi}\big(X_1(t_0), \dots, X_N(t_0)\big) \tag{7.224}$$

with $\boldsymbol{\alpha}_k$ the $\boldsymbol{\alpha}$ matrices acting on the spin index of particle k.

Remark 7.8 Nonlocality is manifest in the law of motion (7.219), as the velocity of particle 1 depends instantaneously (relative to $\mathcal{F}$) on the position of particle 2, no matter how distant. Since that position, $X_2(\Sigma)$ with Σ the time leaf through $X_1(s_0)$ as in Fig. 7.10, will typically depend on external fields near $X_2(\Sigma)$ in the past of Σ, also the motion of X_1 at $X_1(\Sigma)$ will depend on these external fields—but only up to Σ. In particular, microscopic parameter independence (7.216) is violated. As another consequence, an agent Bob who can arrange external fields near $X_2(\Sigma)$ can influence the motion of X_1; in a sense, the influence propagates along Σ. On the other hand, we are already familiar with limitations to knowledge that can make this influence undetectable for Alice near $X_1(\Sigma)$; we will see in Sect. 7.6.2 that a no-signaling theorem is again in place.

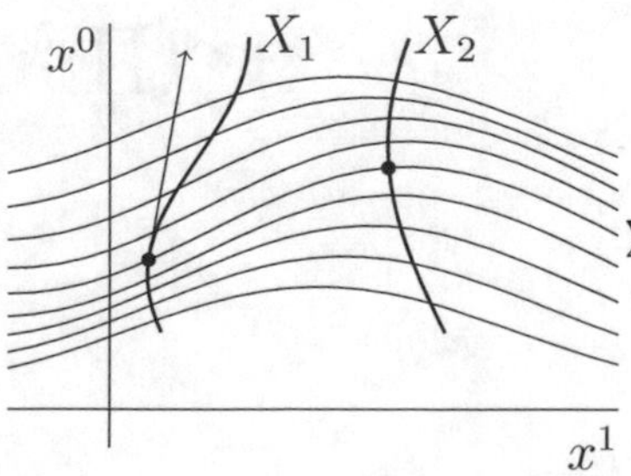

Fig. 7.10 Law of motion of $\text{BM}_{\mathcal{F}}$ for two particles: the tangent vector dX_1^μ/ds (indicated by the arrow) at $X_1(s_0) = X_1(\Sigma)$ (left dot) depends on the point $X_2(\Sigma)$ (right dot) where X_2 crosses the same time leaf

Remark 7.9 If the wave function $\phi : M^N \to (\mathbb{C}^4)^{\otimes N}$ factorizes,

$$\phi_{s_1\ldots s_N}(x_1,\ldots,x_N) = \phi_{1,s_1}(x_1)\cdots\phi_{N,s_N}(x_N) \tag{7.225}$$

(which can happen only in the absence of interaction), then the world line of any particle does not depend on the initial positions of the other particles; in fact, the world line of particle k is a solution of the 1-particle equation of motion (7.130) associated with the 1-particle wave function $\psi = \phi_k$. This behavior is parallel to the non-relativistic case; see Exercise 1.20(c): unentangled particles move independently of each other.

Proposition 7.15 *For N non-interacting Dirac particles,*

$$\partial_{k\mu_k} j^{\mu_1\ldots\mu_N}(x_1\ldots x_N) = 0 \tag{7.226}$$

for all $k \in \{1,\ldots,N\}$ *and all* $(x_1\ldots x_N) \in M^N$.

Proof We use the non-interacting multi-time Dirac equations

$$i\gamma_k^\mu\big(\partial_{x_k^\mu} + ieA_\mu(x_k)\big)\phi = m\phi \tag{7.227}$$

for real (i.e., not complex) A_μ, where γ_k^μ means γ^μ acting on the spin index of particle k, and the fact that γ^0 and $\gamma^0\gamma^\mu$ are self-adjoint in $\mathbb{C}^4$, to obtain that

$$\partial_{k\mu_k} j^{\mu_1\ldots\mu_N} = \partial_{k\mu_k}\Big(\overline{\phi}\gamma_1^{\mu_1}\cdots\gamma_N^{\mu_N}\phi\Big) \tag{7.228a}$$

$$= \partial_{k\mu_k}\Big(\phi^*\gamma_1^0\cdots\gamma_N^0\gamma_1^{\mu_1}\cdots\gamma_N^{\mu_N}\phi\Big) \tag{7.228b}$$

$$\begin{aligned}&= (\partial_{k\mu_k}\phi)^*(\gamma_1^0\gamma_1^{\mu_1})\cdots(\gamma_N^0\gamma_N^{\mu_N})\phi\\ &\quad + \phi^*(\gamma_1^0\gamma_1^{\mu_1})\cdots(\gamma_N^0\gamma_N^{\mu_N})\partial_{k\mu_k}\phi\end{aligned} \tag{7.228c}$$

$$\begin{aligned}&= \Big(\gamma_k^0 \underbrace{\gamma_k^\mu\partial_{k\mu_k}\phi}_{-im\phi - ieA_{\mu_k}(x_k)\gamma_k^{\mu_k}\phi}\Big)^*\Big(\prod_{j\neq k}\gamma_j^0\gamma_j^{\mu_j}\Big)\phi\\ &\quad + \phi^*\Big(\prod_{j\neq k}\gamma_j^0\gamma_j^{\mu_j}\Big)\Big(\gamma_k^0 \underbrace{\gamma_k^\mu\partial_{k\mu_k}\phi}_{-im\phi - ieA_{\mu_k}(x_k)\gamma_k^{\mu_k}\phi}\Big)\end{aligned} \tag{7.228d}$$

$$= i\Big(m\gamma_k^0\phi + eA_{\mu_k}(x_k)\gamma_k^0\gamma_k^{\mu_k}\phi\Big)^*\Big(\prod_{j\neq k}\gamma_j^0\gamma_j^{\mu_j}\Big)\phi$$
$$- i\phi^*\Big(\prod_{j\neq k}\gamma_j^0\gamma_j^{\mu_j}\Big)\Big(m\gamma_k^0\phi + eA_{\mu_k}(x_k)\gamma_k^0\gamma_k^{\mu_k}\phi\Big) = 0\,. \tag{7.228e}$$

□

7.6.2 Equivariance

Non-Rigorous Theorem 7.3 (Foliation Equivariance) *Suppose we are given a foliation $\mathcal{F}$ of M into Cauchy surfaces with future unit normal field n_μ and a tensor field $j^{\mu_1\dots\mu_N}(x_1,\dots,x_N)$ defined on M^N with the properties*

(i) $j^{\mu_1\dots\mu_N}v_{1\mu_1}\cdots v_{N\mu_N}\geq 0$ *for all future-causal* $v_1,\dots,v_N$
(ii) $\partial_{k\mu_k}j^{\mu_1\dots\mu_N}=0$ *for all* $k\in\{1,\dots,N\}$
(iii) $\int_{\Sigma^N}V(d^3x_1)\cdots V(d^3x_N)\,j^{\mu_1\dots\mu_N}(x_1\dots x_N)\,n_{\mu_1}(x_1)\cdots n_{\mu_N}(x_N) = 1$ *for some* $\Sigma\in\mathcal{F}$.

Then the solutions of the Bohm-type equation of motion (7.219) *are such that if on one time leaf* $\Sigma\in\mathcal{F}$, $X(\Sigma)$ *is curved Born distributed, i.e., with density*

$$j^{\mu_1\dots\mu_N}(x_1\dots x_N)\,n_{\mu_1}(x_1)\cdots n_{\mu_N}(x_N)\,, \tag{7.229}$$

then so is $X(\Sigma')$ *for every* $\Sigma'\in\mathcal{F}$.

Corollary 7.1 *For N non-interacting Dirac particles with wave function ϕ : $M^N\to D^{\otimes N}$ governed by* (7.198) *(possibly with external field A_μ), normalized so that $\|\psi_\Sigma\|_\Sigma = \|\phi|_{\Sigma^N}\|_\Sigma = 1$ for one (and thus every) Cauchy surface Σ, and $j^{\mu_1\dots\mu_N}$ given by* (7.218)*, the solutions of the Bohm-type equation of motion* (7.219) *are such that if on one time leaf $\Sigma\in\mathcal{F}$, $X(\Sigma)$ is curved Born distributed as in* (7.209)*, then so is $X(\Sigma')$ for every $\Sigma'\in\mathcal{F}$.*

As usual, we assume henceforth that the initial configuration is so distributed.

Derivation of Non-Rigorous Theorem 7.3 Let

$$\mathcal{C} = \bigcup_{\Sigma\in\mathcal{F}}\Sigma^N \tag{7.230}$$

be the set in M^N of simultaneous configurations relative to $\mathcal{F}$; it is a $3N+1$-dimensional surface (manifold) in M^N. The derivation requires the study of currents on curved manifolds, and this is best done using differential forms (see Appendix A.6). This proof is adapted from [47]. Define the $3N$-form β on M^N by

$$\beta_{K_1\ldots K_{3N}} = (-1)^{N(N+1)/2}\, \varepsilon_{K_1\ldots K_{3N},1\mu_1\ldots N\mu_N}\, j^{\mu_1\ldots\mu_N} \tag{7.231}$$

where indices $K = (k,\mu)$ run through the $4N$ dimensions of M^N and $\varepsilon(x_1\ldots x_N) = \varepsilon(x_1)\wedge\cdots\wedge\varepsilon(x_N)$ is the volume form in $4N$ dimensions.

Next we show that

$$d\beta|_{\mathcal{C}} = 0\,, \tag{7.232}$$

where d denotes the exterior derivative. (As will become clear below, this equation is a kind of continuity equation.) To verify it, choose a curvilinear coordinate system on M such that the x^0 coordinate is constant on every time leaf $\Sigma \in \mathcal{F}$. The $4N$ functions $x^{10},\ldots,x^{N3}$ then form a coordinate system on M^N; on $\mathcal{C}$, the x^{i0} functions coincide, so I will call them x^0. The coordinate vector fields on $\mathcal{C}$ (expressed as derivative operators) are then $\partial_0 = \sum_{i=1}^N \partial_{i0}|_{\mathcal{C}}$ and $\partial_{11}|_{\mathcal{C}},\ldots,\partial_{N3}|_{\mathcal{C}}$. Equation (7.232) is equivalent to

$$d\beta(\partial_0,\partial_{11},\ldots,\partial_{N3}) = 0 \tag{7.233}$$

on $\mathcal{C}$. Since

$$d\beta_{K'K_1\ldots K_{3N}} = \partial_{[K'}\beta_{K_1\ldots K_{3N}]} \tag{7.234a}$$

$$= (-1)^{N+N(N+1)/2}\, \varepsilon_{[K_1\ldots K_{3N}|1\mu_1\ldots N\mu_N}\, \partial_{|K']}j^{\mu_1\ldots\mu_N} \tag{7.234b}$$

with $[ABC|DE|FG]$ meaning anti-symmetrization in the indices $ABCFG$, we have that

$$d\beta(\partial_0,\partial_{11},\ldots,\partial_{N3}) = \sum_{i=1}^N d\beta(\partial_{i0},\partial_{11},\ldots,\partial_{N3}) \tag{7.235a}$$

$$= (-1)^{N+N(N+1)/2}\sum_{i=1}^N \varepsilon_{[11\ldots N3|1\mu_1\ldots N\mu_N}\, \partial_{|i0]}j^{\mu_1\ldots\mu_N} \tag{7.235b}$$

$$= \frac{(-1)^N}{3N+1} \sum_{i=1}^{N} \Big(\partial_{i0} j^{10\ldots N0} + \sum_{a=1}^{3} \partial_{ia} j^{10\ldots(i-1)0,ia,(i+1)0\ldots N0} \Big) \tag{7.235c}$$

$$= \frac{(-1)^N}{3N+1} \sum_{i=1}^{N} \partial_{i\mu} j^{10\ldots(i-1)0,i\mu,(i+1)0\ldots N0} = 0\,. \tag{7.235d}$$

Next, we use the analog of Theorem 7.2 on manifolds, in which the current vector field j^μ gets replaced by a differential form β and the continuity equation $\partial_\mu j^\mu = 0$ by the equation $d\beta = 0$:

Theorem 7.4 ([47]) *On a $d+1$-dimensional manifold $\mathcal{C}$, a d-form β with $d\beta = 0$ gives rise to a measure $\mathbb{P}$ on the set of integral curves of the kernels of β (which are one-dimensional wherever $\beta \neq 0$) within the wandering set. If the total weight of the measure $\mathbb{P}$ is 1, it defines a trajectory-valued random variable X, and for any oriented piece B of d-surface lying within the wandering set, the expected number of signed crossings of X through B equals*

$$\mathbb{E}\, N(X, B) = \int_B \beta\,. \tag{7.236}$$

It can be verified that the integral curves of the kernels of β are exactly the solutions of (7.219) and that

$$\int_B \beta = \int_B V(d^3x_1)\cdots V(d^3x_N)\, j^{\mu_1\ldots\mu_N}(x_1\ldots x_N)\, n_{\mu_1}(x_1)\cdots n_{\mu_N}(x_N) \tag{7.237}$$

for $B \subseteq \Sigma^N$ for some $\Sigma \in \mathcal{F}$. Assumption (i) implies that no integral curve can be recurrent, so the wandering set contains all $q \in \mathcal{C}$ with $\beta(q) \neq 0$ or, equivalently, $j(q) \neq 0$ or $\psi(q) \neq 0$. As long as none of the integral curves run into a node, or only a set of measure 0 does,[18] Assumption (iii) implies that $\mathbb{P}$ has total weight 1. Since the world lines are timelike-or-lightlike and taken to be oriented toward the future, every Cauchy surface Σ can only be crossed once and positively, so the expected number of signed crossing equals the crossing probability. It follows that the random trajectory with measure $\mathbb{P}$ is exactly the trajectory with Born distributed initial point $X(\Sigma_0)$ and intersects every Cauchy surface with Born distribution. □

[18] In fact, this is known to be the case for a dense subspace of initial wave functions and flat $\mathcal{F}$ (Teufel and Tumulka 2004 [43]).

Remark 7.10

- The foliation equivariance theorem (Non-Rigorous Theorem 7.3) is still true in curved space-time (Tumulka 2001 [47]).
- Assumption (ii) can be weakened. For example, (7.235d) shows that it suffices to have $\partial_{k\mu_k} j^{\mu_1 \ldots \mu_N} \prod_{i \neq k} n_{\mu_i}(x_i) = 0$. In fact, it suffices that (7.235d) vanishes.
- The foliation equivariance theorem (Non-Rigorous Theorem 7.3) is still true if j is defined only on the set $\mathscr{S}_N$ of spacelike configurations, provided condition (ii) gets re-interpreted in the following way. The set $\mathscr{S}_N$ as defined in (7.192) also contains configurations with two particles at the same location, $x_i = x_k$ with $i \neq k$. At such configurations, one cannot form partial derivatives such as $\partial_{k0} j^{\mu_1 \ldots \mu_N}$ because, if we vary x_k in a timelike direction and keep all other x's (including x_i) fixed, then we leave $\mathscr{S}_N$. However, it is still possible to move both x_i and x_k in the same timelike direction without leaving $\mathscr{S}_N$, so the expression $(\partial_{i0}+\partial_{k0}) j^{\mu_1 \ldots \mu_N}$ is still defined as a directional derivative. Or, taking the directional derivative in the direction of $n_\mu(x_i) = n_\mu(x_k)$ and using that we *can* move x_i and x_k independently in spacelike directions, the expression
$$\partial_{i\mu_i} j^{\mu_1 \ldots \mu_N} \prod_{j \neq i} n_{\mu_j}(x_j) + \partial_{k\mu_k} j^{\mu_1 \ldots \mu_N} \prod_{j \neq k} n_{\mu_j}(x_j) \tag{7.238}$$
is still defined at $x_i = x_k$. Instead of (ii) at $x_i = x_k$, we demand that (7.238) vanishes, while for every particle not at the same location as another, we demand $\partial_{k\mu_k} j^{\mu_1 \ldots \mu_N} = 0$ as before.
Alternatively, we may exclude collision configurations, i.e., those with $x_i = x_k$ for $i \neq k$, from $\mathcal{C}$ and keep track by means of a separate argument of the probability that a trajectory will run into a collision configuration. (Usually, it would be 0, but IBCs may be an exception.)
- If ϕ factorizes as in (7.225), then also the Born distribution factorizes. Thus in this case, not only do the particles move independently as pointed out in Remark 7.9, they are also probabilistically independent; that is, the N world lines are independently chosen Born-distributed solutions of 1-particle equations of motion of the type (7.130).

Intersection Probability and Detection Probability

The equivariance theorem 7.3 says that the world lines will intersect every *time leaf* in a Born ($|\psi|^2$)-distributed configuration, but not that the intersection with *every Cauchy surface* will be Born distributed. In fact, the intersection with a Cauchy surface $\Sigma \notin \mathcal{F}$ will in some cases deviate from the Born distribution. In fact, it has been shown using Bell's experiment and Bell's no-hidden-variable theorem that there is no probability distribution over N-tuples of timelike world lines so that the configuration would be Born distributed on *every* Cauchy surface (Berndl et al. 1996 [7]).

However, this situation is less worrisome than it may seem. One might worry that if we place detectors along a Cauchy surface $\Sigma \notin \mathcal{F}$, the distribution of the

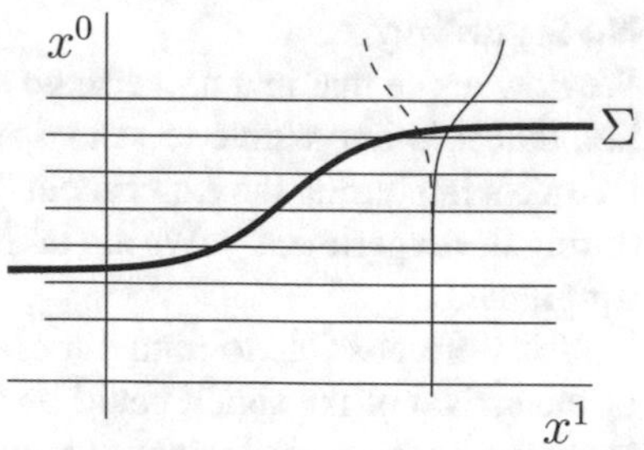

Fig. 7.11 Example of a time foliation $\mathcal{F}$ consisting of parallel hyperplanes and a Cauchy surface $\Sigma \notin \mathcal{F}$ together with a particle world line and (dashed) what the world line would have been in the absence of detectors on Σ

observed configuration might deviate from the Born distribution and thus from the prediction of the quantum formalism. However, in such a situation, we need to take into account the effects of the presence of detectors. For example, consider $\mathcal{F}$ and $\Sigma \notin \mathcal{F}$ as in Fig. 7.11.

Detectors in the leftmost part of Σ will collapse the conditional wave function of particles located at spacelike separation near the right end of the figure. Since the nonlocal influences travel along the time leaves, particles on the right can follow different trajectories before they hit Σ. Thus, although *detectors just find particles on Σ where they actually hit Σ*, they find them in a configuration that is *different from where they would have intersected Σ if no detectors had been present*—because detectors in the leftmost part of Σ influence trajectories on the right well before they hit Σ. Even if the detectors on the left did not find any particle!

And the detected configuration on Σ will be Born distributed. Here are two ways of seeing this.

1. Treating the detectors as another quantum system, we may consider the wave function Ψ of the N particles and the detectors together. Suppose $\Sigma' \in \mathcal{F}$ is so late that $\Sigma' \subset \text{future}(\Sigma)$, then $\Psi_{\Sigma'}$ is a superposition of different measurement outcomes, and the Bohmian configuration on Σ' has the corresponding $|\Psi|^2$ probability to display a particular outcome. Since recorded outcomes should be stable, the $|\Psi|^2$ weight of each outcome should be the same on all Cauchy surfaces $\Sigma'' \subset \text{future}(\Sigma)$. (In particular, since the evolution of wave functions does not depend on $\mathcal{F}$, the $|\Psi|^2$ weights will not depend on $\mathcal{F}$.) Thus, the $|\Psi|^2$ distribution of the outcomes on Σ' should agree with that on Σ, which should be the curved Born distribution. Thus, the probability of the Bohmian configuration displaying a particular outcome should agree with the curved Born distribution.
2. The effect of ideal detectors on the N-particle system can be summarized by the Born rule and collapse rule along $\mathcal{F}$. By Theorem 7.3, for $\mathcal{F}$ consisting of parallel hyperplanes and under reasonable assumptions on the unitary time evolution, the detected configuration is Born distributed. Presumably, Theorem 7.3 can be generalized to arbitrary spacelike foliations.

So, the detection probabilities on a surface Σ do not depend on $\mathcal{F}$; that is, $\mathcal{F}$ is invisible, as claimed in Sect. 7.5.4.

No Signaling

We now argue that in a universe governed by $\mathrm{BM}_{\mathcal{F}}$, a no-signaling theorem holds, i.e., that it is impossible to send a message or signal from a space-time region A ("Alice's lab during the experiment") to a spacelike separated region B ("Bob's lab during the experiment"). We are taking interaction locality and propagation locality for granted.

If it were possible to send a message or signal, then the content of the message or the arrival of the signal could be (e.g.) written down by Alice and thus encoded in the Bohmian particle configuration in A in a stable way, so it could be read off from the configuration in A from some surface Σ_0 onward, i.e., on any Cauchy surface whose past contains $\Sigma_0 \cap A$. In particular, it could be read off from the configuration in A on any *time leaf* $\Sigma \in \mathcal{F}$ whose past contains $\Sigma_0 \cap A$. Therefore, it suffices to show that for any $\Sigma \in \mathcal{F}$, the marginal distribution of the configuration in $\Sigma \cap A$ is independent of anything Bob does in B. Since by equivariance, the distribution of the Bohmian configuration on $\Sigma \in \mathcal{F}$ is the curved Born ($|\psi_\Sigma|^2$) distribution, it suffices that on any Cauchy surface, the marginal in A of the curved Born distribution is independent of anything Bob does in B. (Note that the last condition is independent of the Bohmian approach.)

Let system a consist of the particles in A and system b of those in B. Let Σ_1 and Σ_2 be two Cauchy surfaces in the relevant time range, so $\mathscr{H}_{\Sigma_i} = \mathscr{H}_{a\Sigma_i} \otimes \mathscr{H}_{b\Sigma_i}$ for both $i \in \{1, 2\}$. The configuration PVM is of the form $P_a \otimes P_b$, so the marginal distribution of the a-configuration on Σ_i is given by

$$\langle \psi_{\Sigma_i} | P_{a\Sigma_i}(\cdot) \otimes I_b | \psi_{\Sigma_i} \rangle_{\Sigma_i} = \mathrm{tr}(P_{a\Sigma_i}(\cdot)\rho_{a\Sigma_i}) \tag{7.239}$$

with $\rho_{a\Sigma_i} = \mathrm{tr}_b |\psi_{\Sigma_i}\rangle\langle\psi_{\Sigma_i}|$ the reduced density matrix of system a on Σ_i. By interaction locality, there is no interaction between a and b, so the time evolution $U_{\Sigma_1}^{\Sigma_2} : \mathscr{H}_{\Sigma_1} \to \mathscr{H}_{\Sigma_2}$ factorizes into

$$U_{\Sigma_1}^{\Sigma_2} = U_{a\Sigma_1}^{\Sigma_2} \otimes U_{b\Sigma_1}^{\Sigma_2} \tag{7.240}$$

with $U_{a\Sigma_1}^{\Sigma_2} : \mathscr{H}_{a\Sigma_1} \to \mathscr{H}_{a\Sigma_2}$ and $U_{b\Sigma_1}^{\Sigma_2} : \mathscr{H}_{b\Sigma_1} \to \mathscr{H}_{b\Sigma_2}$. It follows that the reduced density matrix of system a evolves according to U_a, that is,

$$\rho_{a\Sigma_2} = U_{a\Sigma_1}^{\Sigma_2}\, \rho_{a\Sigma_1}\, U_{a\Sigma_1}^{\Sigma_2\dagger} \tag{7.241}$$

This means two things in particular: (i) The marginal a-distribution at any time does not depend on U_b, and thus not on any external fields in place in B. (ii) ρ_a at the late time Σ_2 does not contain any information that could not be computed from ρ_a at the early time Σ_1; in particular, it does not depend on any information in the wave function ψ_{Σ_1} that is not contained in $\rho_{a\Sigma_1}$ (such as Bob's wave function, disentangled from a). Together, (i) and (ii) exhaust the possibilities of "what Bob can do" and thus represent the absence of possible messages or signals from Bob in A, which is what we claimed.

7.6.3 The Spin-0 Case

I now describe the combination of $\mathrm{BM}_{\mathcal{F}}$ with the Klein-Gordon current (7.158) and discuss reservations about this theory.

Definition of the Current Tensor

Let the multi-time wave function $\phi : M^N \to \mathbb{C}$ obey a Klein-Gordon equation with external field A_μ in every time variable,

$$\eta^{\mu\nu}\big(i\partial_{k\mu} - eA_\mu(x_k)\big)\big(i\partial_{k\nu} - eA_\nu(x_k)\big)\phi = m^2\phi \tag{7.242}$$

for $k \in \{1, \ldots, N\}$. We now define a current tensor field $j^{\mu_1\ldots\mu_N}(x_1, \ldots, x_N)$ replacing (7.218); the equation of motion is then given again by (7.219). As a preparation for the definition, we introduce for a given differential operator D (where the coefficients in front of the derivatives may be complex-valued functions of the coordinates) acting on wave functions $\phi(\xi)$ a corresponding differential operator $\overleftrightarrow{D}$ acting on density matrices $\rho(\xi, \xi')$ by

$$(\overleftrightarrow{D}\rho)(\xi, \xi') = \tfrac{1}{2}D_\xi\rho + \tfrac{1}{2}D^*_{\xi'}\rho\,, \tag{7.243}$$

where the asterisk denotes complex conjugation of the coefficients. For our purposes, $\xi = (x_1, \ldots, x_N) \in M^N$ and

$$D = D_{k\mu} := i\partial_{k\mu} - eA_\mu(x_k)\,, \tag{7.244}$$

so

$$\overleftrightarrow{D}_{k\mu}\rho = \tfrac{i}{2}\partial_{x_k^\mu}\rho - \tfrac{e}{2}A_\mu(x_k)\rho - \tfrac{i}{2}\partial_{x_k'^\mu}\rho - \tfrac{e}{2}A_\mu(x_k')\rho\,. \tag{7.245}$$

With these abbreviations, we define the current tensor by

$$j_{\mu_1\ldots\mu_N}(x_1\ldots x_N) = \overleftrightarrow{D}_{1\mu_1}\cdots\overleftrightarrow{D}_{N\mu_N}\Big(\phi(x_1\ldots x_N)\phi^*(x_1'\ldots x_N')\Big)\Big|_{\xi'=\xi}\,. \tag{7.246}$$

The basic properties of this current tensor are:

Proposition 7.16 *(i)* $j^{\mu_1\ldots\mu_N}$ *is real. (ii)* $\partial_{k\mu_k} j_k^{\mu_1\ldots\mu_N} = 0$. *(iii) For* $N = 1$, (7.246) *coincides with* (7.159).

Proof

(i) If ρ is conjugate symmetric, $\rho(\xi', \xi) = \rho(\xi, \xi')^*$, then so is $\overleftrightarrow{D}\rho$ for arbitrary D. Thus, entries at $\xi' = \xi$ must be real.

(ii) Consider applying $\leftrightarrow$ to $i D_{k\mu}$ instead of $D_{k\mu}$. We find that

$$(i D_{k\mu})^{\leftrightarrow} \rho = -\tfrac{1}{2}\partial_{x_k^\mu}\rho - \tfrac{ie}{2}A_\mu(x_k)\rho - \tfrac{1}{2}\partial_{x_k'^\mu}\rho + \tfrac{ie}{2}A_\mu(x_k')\rho\,. \tag{7.247}$$

Thus,

$$\Big((i D_{k\mu_k})^{\leftrightarrow}\rho\Big)(\xi, \xi) = -\tfrac{1}{2}\partial_{k\mu_k}\big(\rho(\xi, \xi)\big)\,. \tag{7.248}$$

Observe that always

$$(i D)^{\leftrightarrow}\overleftrightarrow{D}\rho = \tfrac{i}{4}(D_\xi D_\xi - D^*_{\xi'} D^*_{\xi'})\rho\,, \tag{7.249}$$

using that operators acting on different variables commute. Thus,

$$\eta^{\mu\nu}(i D_{k\mu})^{\leftrightarrow}\overleftrightarrow{D}_{k\nu}\rho = \tfrac{i}{4}\eta^{\mu\nu}(D_{x_k^\mu} D_{x_k^\nu}\rho - D^*_{x_k'^\mu} D^*_{x_k'^\nu}\rho)\,, \tag{7.250}$$

which vanishes for $\rho(\xi, \xi') = \phi(\xi)\phi^*(\xi')$ by virtue of (7.242). Using again that operators acting on different variables commute, it follows that

$$\eta^{\nu_k\mu_k}(i D_{k\nu_k})^{\leftrightarrow}\overleftrightarrow{D}_{1\mu_1}\cdots\overleftrightarrow{D}_{N\mu_N}\Big(\phi(\xi)\phi^*(\xi')\Big) = 0\,. \tag{7.251}$$

Combining this with (7.248) for $\rho(\xi, \xi') = \overleftrightarrow{D}_{1\mu_1}\cdots\overleftrightarrow{D}_{N\mu_N}\big(\phi(\xi)\phi^*(\xi')\big)$ yields the claim.

(iii) is immediate from the definition.

□

Trajectories

The relevant difference to the Dirac case is that the right-hand side of the equation of motion (7.219), $j^{\mu_1\ldots\mu_N}\prod_{i\neq k} n_{\mu_i}(x_i)$, is not necessarily future-causal. But we can still obtain trajectories and the equivariance of the probabilities as in the Dirac case:

Non-Rigorous Theorem 7.4 *If, in the statement of Non-Rigorous Theorem 7.3, we drop the hypothesis (i) of the future-causal property, then the $BM_{\mathcal{F}}$-equation of motion* (7.219) *still yields a family of trajectories in the* $3N+1$*-dimensional set*

$\mathcal{C} = \cup_{\Sigma \in \mathcal{F}} \Sigma^N \subset M^N$ of simultaneous configurations (and thus a family of N-tuples of trajectories in space-time M) together with a probability distribution over those trajectories in the wandering set. If the wave function is suitably normalized and $\Sigma \in \mathcal{F}$ is a time leaf, then the probability of the random trajectory intersecting the volume $d^3x_1 \times \cdots \times d^3x_N \subset \Sigma^N$ is

$$V(d^3x_1)\cdots V(d^3x_N)\left|j^{\mu_1\ldots\mu_N}(x_1\ldots x_N)\,n_{\mu_1}(x_1)\cdots n_{\mu_N}(x_N)\right|. \tag{7.252}$$

However, trajectories are not necessarily timelike-or-lightlike.

Derivation In the derivation of Non-Rigorous Theorem 7.3, we did not make use of the future-causal property when defining the d-form β or checking that $d\beta$ vanishes on $\mathcal{C}$. Theorem 7.4 thus provides the random trajectory in $\mathcal{C}$, with expected number of signed crossings given by the integral of β, and thus locally the crossing probability given by the absolute integral of β, or (7.252). (Note the parallel to (7.164).) □

Without the future-causal property, it is not automatic how big the wandering set will be; on the other hand, I see no particular reason to expect recurrent trajectories in a generic wave function.

One might have expected that the following obstruction precludes a consistent definition of trajectories and probabilities: when one particle, say k, turns around in time, what happens to another particle, say i, whose trajectory still points toward the future? Is there not a conflict between several particles in which time direction to proceed? The answer is *no*, because the sign of $n_\mu(X_k)\,dX_k^\mu/ds$ (i.e., whether the trajectory X_k points toward the future or the past at s) equals that of $j^{\mu_1\ldots\mu_N}n_{\mu_1}(x_1)\cdots n_{\mu_N}(x_N)$ and is thus independent of k, so either all trajectories point toward the future or all toward the past.

In a little more detail, the turn of the curve $s \mapsto X(s)$ in $\mathcal{C}$ occurs at a point $s = s_0$ when dX/ds is tangent to Σ^N for some time leaf Σ. This tangent direction projects, in the generic case, to N directions tangent to Σ at $X_1(s_0), \ldots, X_N(s_0)$, being the directions of the N world lines. That is, it comes out by itself that the N particles decide to turn simultaneously, see Fig. (7.12).

The (exceptional) case where this does not happen is when the tangent direction to X at s_0 has zero projection on the coordinates of particle i. In this case, the i-th world line possibly has a kink (see Fig. 7.13) and is no longer a smooth curve (although it is smoothly parametrized), because $dX_i/ds\,(s_0) = 0$. But note that even in this case, X is well-defined, and no extra law is needed for avoiding an ill-defined situation. This case occurs, e.g., for disentangled wave functions.

Time Travel

The question arises whether trajectories going back in time could be used for time travel. Well, time travel of the sort described in H.G. Wells' (1895) [56] novel *The Time Machine* is not possible because even if we could make the trajectory X of all

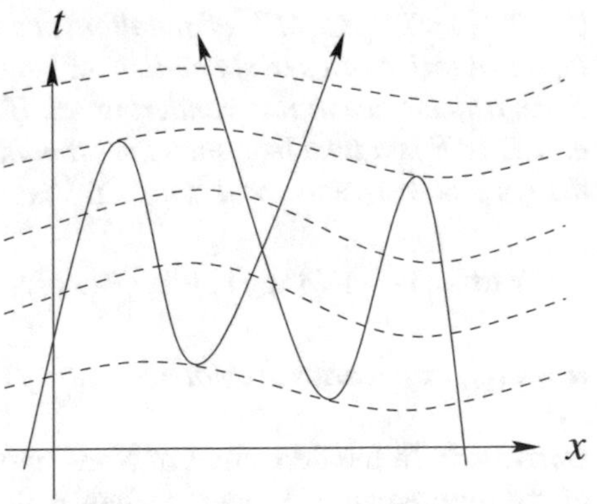

Fig. 7.12 A typical solution for a two-particle system assuming the spin-0 current tensor (7.246), with the time foliation indicated by dashed lines

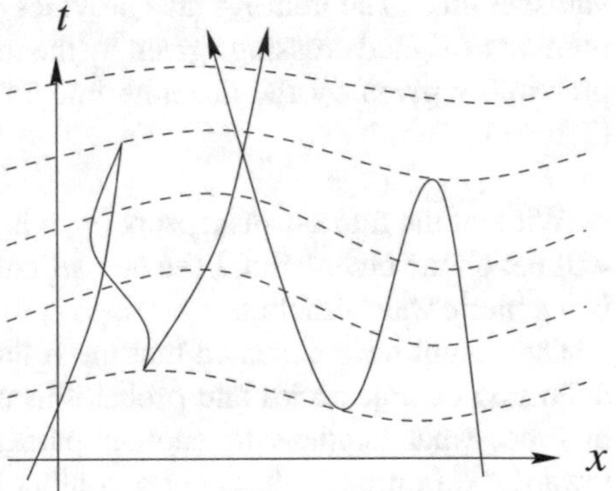

Fig. 7.13 In certain cases for the spin-0 current (7.246), a trajectory may have a kink at its turning point while other trajectories turn around in a smooth way

particles in the universe go back to the year 1895, we would not be able to remember anything from the twenty-first century because it is $\psi(1895)$ that guides its motion.

But something else, a milder sort of time machine, appears compatible with the law of motion based on the spin-0 Klein-Gordon current tensor (7.246): a machine that brings the universe back to a certain point in time and effectively erases the history that has elapsed since. More precisely, suppose the machine has two buttons named "start" and "back." Say at time t_1, you press the "start" button; from then on, the display reads "first round." If at t_2 you press the "back" button, the universe is brought back to t_1, and from the macrostate it was in at t_1, history starts anew, the only difference being that the display now reads "second round." You cannot, however, remember anything about the branch of history you have aborted, nor can anybody else. The only information available about the aborted branch of history is that at some point, the "back" button must have been pressed.

Here is how such a machine can be constructed: Find a one-particle Klein-Gordon wave function whose trajectories look qualitatively like those depicted in Fig. 7.14. This wave function is prepared at t_1. After a short evolution, at time t', a measurement is performed on whether $x < x'$ (then display "first round") or $x > x'$ (then display "second round"). The parts left and right of x' thus move into very

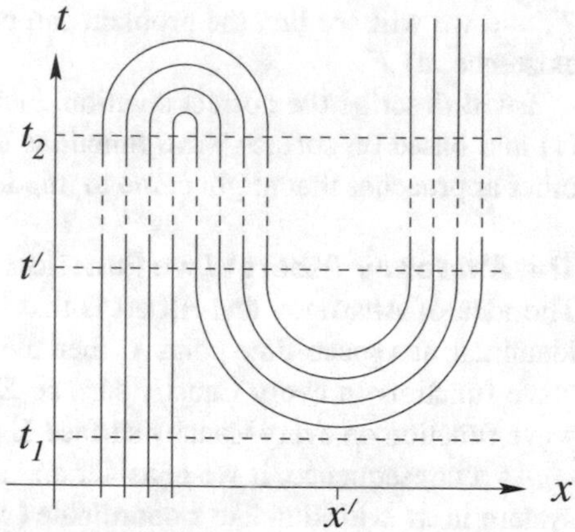

Fig. 7.14 A single-particle wave function whose trajectories look qualitatively like this can be used for the "time machine" described in the text

different regions of configuration space, but the part left of x' is kept in a coherent superposition, disentangled from all other variables. At t_2, if the "back" button is pressed, something is done to (A_μ and thus to) the wave function that will cause the future-pointing and the past-pointing trajectories to merge, as shown in the figure. Then X will go back very close along the path it came until it reaches t' and then turn toward the future again. Although the "first round" of history really existed, it does not have any influence on the "second round." If the "back" button is never pressed, then Fig. 7.14 must be modified in that the merger of trajectories never takes place. The ⋃-shaped trajectories are then disconnected from the ones starting before t_1, and they are never realized because at t_1 there was a configuration.

It is fun to think through the consequences of this law of motion, but its unusual behavior of turning around in time makes it hard to take seriously as a fundamental law of nature.

7.7 Predictions in Relativistic Space-Time

7.7.1 Is Collapse Compatible with Relativity?

The following problem would seem to arise with collapse in relativistic space-time. In non-relativistic quantum mechanics, when we collapse a wave function, we collapse it instantaneously in all of configuration space. (This is a consequence of the equation $\psi_{t+} = P_\alpha \psi_{t-}/\|P_\alpha \psi_{t-}\|$ of the projection postulate that we had as (3.16), where P_α is the appropriate projection operator.) But what counts as instantaneous depends on the Lorentz frame; that is the problem. It comes up for collapse theories such as GRW, but just as well for the quantum formalism when it prescribes collapses triggered by measurements.

A first thought might be that if we can use a time foliation $\mathcal{F}$ for the Bohmian trajectories, why not collapse the wave function instantaneously everywhere on a time leaf. However, the wave function in Bohmian mechanics never depended on

$\mathcal{F}$, and we will see that the problem can be solved without using (or assuming the existence of) $\mathcal{F}$.

I will describe the correct solution, introduced by Aharonov and Albert (1981) [1] and based on surface wave functions ψ_Σ. Afterward, I will comment on some other approaches that might come to mind.

The Aharonov-Albert Wave Function

The idea of Aharonov and Albert is that if a measurement takes place, somewhat idealized, at a space-time point X, then the wave function ψ_Σ should be a collapsed wave function on every Cauchy surface Σ with $X \in \text{past}(\Sigma)$ and an uncollapsed wave function on every Cauchy surface Σ with $X \in \text{future}(\Sigma)$.

As a consequence, if we consider an arbitrary (possibly curvilinear) coordinate system in M with timelike t coordinate ($g_{tt} > 0$) and take Σ_t to be the surface of constant t coordinate, then $\psi(t) := \psi_{\Sigma_t}$ will collapse at coordinate time $t_1 = t(X)$. (We may also imagine a surface in M that we "push" toward the future as a function of some parameter t; $\psi(t)$ collapses when Σ_t crosses X.)

More generally and in more detail, suppose that:

1. at each of the space-time points $X_1, \ldots, X_n \in M$ (of which some may be spacelike separated and some timelike or lightlike), a quantum measurement of a local observable takes place;
2. the unitary time evolution is given by operators $U_\Sigma^{\Sigma'} : \mathscr{H}_\Sigma \to \mathscr{H}_{\Sigma'}$ for any Cauchy surfaces Σ, Σ';
3. the local observable for each X_i can be represented for any Cauchy surface Σ through X_i as an operator $A_i(\Sigma) : \mathscr{H}_\Sigma \to \mathscr{H}_\Sigma$ such that
$$A_i(\Sigma') = U_\Sigma^{\Sigma'} A_i(\Sigma)\, U_{\Sigma'}^{\Sigma} \tag{7.253}$$
for every Σ' through X_i (so we can use the *Heisenberg picture* and consistently define, for some surface Σ_0 regarded as an initial surface, $A_i := U_\Sigma^{\Sigma_0} A_i(\Sigma)\, U_{\Sigma_0}^{\Sigma}$);
4. each A_i has spectral decomposition $A_i = \sum_{\alpha_i} \alpha_i P_{i\alpha_i}$;
5. for spacelike separated $X_i \times X_j$, A_i commutes with A_j.

Then, the Aharonov-Albert wave function is defined as
$$\psi_\Sigma = \mathcal{N}\, U_{\Sigma_0}^{\Sigma} \left(\prod_{i: X_i \in \text{past}(\Sigma)} P_{i\alpha_i} \right) \psi_0 \,, \tag{7.254}$$
where $\mathcal{N}$ is the normalizing factor, α_i is the outcome at X_i, the factors in the product are ordered so that i is left of j whenever $X_i \in \text{future}(X_j)$, an empty product is understood as the identity operator, and $\psi_0 \in \mathscr{H}_{\Sigma_0}$ the initial wave function. (We

may imagine that all $x_i \in \text{future}(\Sigma_0)$, although the location of Σ_0 does not actually matter since we can always use $U_\Sigma^{\Sigma'}$ to evolve the wave function to another surface.)

Exercise 7.8 (Aharonov-Albert Wave Function) Verify that all possible operators orderings in (7.254) with i left of j whenever $X_i \in \text{future}(X_j)$ yield the same ψ_Σ. (Use that if A_i and A_j commute, then so do their spectral projections.)

Put differently, the Aharonov-Albert wave function ψ_Σ is what we would obtain on $\Sigma = \{t = \text{const.}\}$ using a curvilinear coordinate system with timelike t coordinate if the X_i are ordered so that $t(X_1) \leq \ldots \leq t(X_n)$, we always evolve ψ from $\Sigma_{t(X_{i-1})}$ to $\Sigma_{t(X_i)}$ using $U_{\cdots}^{\cdots}$, and then collapse ψ on $\Sigma_{t(X_i)}$ according to the spectral projections of $A_i(\Sigma_{t(X_i)})$.

In this idealized example, it is clear that the joint probability distribution of all outcomes should be

$$\mathbb{P}\Big(Z_1 = \alpha_1, \ldots, Z_n = \alpha_n\Big) = \left\| \Big(\prod_{i=1}^{n} P_{i\alpha_i}\Big)\psi_0 \right\|^2 \tag{7.255}$$

with factors ordered so that i is left of j whenever $X_i \in \text{future}(X_j)$. As a consequence, the conditional distribution, given the outcomes up to some Cauchy surface Σ, is given by the same formula with $U_\Sigma^{\Sigma_0}\psi_\Sigma$ instead of ψ_0; that is, if $X_1, \ldots, X_j \in \text{past}(\Sigma)$ and $X_{j+1}, \ldots, X_n \in \text{future}(\Sigma)$, then

$$\mathbb{P}\Big(Z_{j+1} = \alpha_{j+1}, \ldots, Z_n = \alpha_n \Big| Z_1 = \alpha_1, \ldots, Z_j = \alpha_j\Big) = \left\| \Big(\prod_{i=j+1}^{n} P_{i\alpha_i}\Big) U_\Sigma^{\Sigma_0}\psi_\Sigma \right\|^2 . \tag{7.256}$$

This relation provides the link between the Aharonov-Albert wave function and observable probabilities.

Let me compare the Aharonov-Albert wave function to a multi-time wave function $\phi(x_1, \ldots, x_N)$. As we saw in Sect. 7.4.2, ϕ always defines a surface wave function, which we also called ψ_Σ; but that consideration involved only the unitary time evolution, not collapses. From the unitarily evolved surface wave function, we could recover ϕ by setting

$$\phi(x_1, \ldots, x_N) = \psi_\Sigma(x_1, \ldots, x_N) \tag{7.257}$$

provided that the right-hand side has the same value for every Σ containing all of $x_1, \ldots, x_N$. This condition is actually satisfied for reasonable unitary time evolutions [39, Sec. 4.3], but not for the Aharonov-Albert wave function, because of the collapses. That is why a multi-time wave function cannot be used for a time evolution with collapses, but a surface wave function can.

Another remark I should make is that the Aharonov-Albert wave function also comes out of Bohmian mechanics as the conditional wave function. As described in

Sect. 3.2.2 in the non-relativistic case, the conditional wave function $\psi(x) \propto \Psi(x, Y)$ collapses during a quantum measurement in just the way described by the projection postulate, provided x is the object configuration and Y the apparatus configuration. The same is true in $BM_{\mathcal{F}}$ on any time leaf $\Sigma \in \mathcal{F}$, but even on Cauchy surfaces Σ that do *not* belong to $\mathcal{F}$, we can consider inserting the intersection points of the Bohmian world lines of the apparatus particles with Σ into the multi-time wave function ϕ and thereby obtain a conditional wave function ψ_Σ such that conditional detection probabilities of X given Y are given by $|\psi_\Sigma(x)|^2$. Some thought shows that if outcomes get recorded stably in Y and Y does not provide more information about x than the outcomes, then the conditional wave function agrees with the Aharonov-Albert wave function.

Other Approaches to Relativistic Collapse

- (*Use the time foliation*) As I mentioned, one might first think of something like "collapsing the wave function along the time leaves." When we think more carefully about what that should mean, we arrive at ψ_Σ that agrees with the Aharonov-Albert wave function but is defined only for time leaves $\Sigma \in \mathcal{F}$. This is less general and less satisfactory than the Aharonov-Albert wave function because we also want to ask for detection probabilities along surfaces Σ that are *not* time leaves, which should be expressed through a curved Born rule such as (7.209) in terms of a wave function on Σ.
- (*The proposal of Hellwig and Kraus*) In reaction to the same problem, Hellwig and Kraus (1970) [28] proposed to collapse the wave function, for every local quantum measurement at $X \in M$, along the past light cone of X. In general, for quantum measurements of $A_1, \ldots, A_n$ at $X_1, \ldots, X_n$ as above, they associate a (Heisenberg-picture) state vector, not with every surface Σ, but with every point $x \in M$,

$$\psi_x = \mathcal{N}\Biggl(\prod_{i:X_i \in \text{past}(x)} P_{i\alpha_i}\Biggr)\psi_0 \in \mathscr{H}_{\Sigma_0}\,. \tag{7.258}$$

  One drawback of this proposal is that it involves a kind of retrocausation, as the decision, made by an observer at X_i, about which observable A_i to measure influences the reality in the coordinate past, more precisely at those points x that are spacelike separated from X_i (and therefore at an earlier time coordinate than X_i in some Lorentz frame).

  Another drawback is that there is no simple relation between ψ_x and the observable probability distribution (7.255) of the outcomes: one can define the vector ψ_x, but there is no use for it in a Born rule.

7.7.2 Tunneling Speed

We have seen for the Dirac equation that wave functions propagate no faster than light and (Bohmian) particles move no faster than light. Yet some authors (e.g., Nimtz et al. 1994 [36]) have claimed that particles (electrons, photons, or others) sometimes move faster than light when tunneling through a barrier. How can this go together with the fact that Bohm's equation of motion (7.130) for Dirac particles does not involve the potential and thus predicts subluminal speeds also inside the barrier? In this section, we look at the reasoning behind the superluminal-tunneling claim.

Consider a particle in 1d, a barrier potential $V(x) = V_0 \, 1_{0 \le x \le L}$ as in (2.47), and a wave packet ψ_{in} coming in from the left. In some of the many runs of the experiment, we measure the time T_ξ at which the particle first passes the location $x = \xi < 0$ to the left of and close (but not too close) to the barrier, in others the time T_L at which the particle passes $x = L$, given that it gets transmitted at all. (The presence of detectors may of course itself change the times at which a particle passes $x = \xi$ or $x = L$, but let us ignore this change.[19]) Then Nimtz would call

$$v = \frac{L - \xi}{\mathbb{E}(T_L|\text{transmitted}) - \mathbb{E}(T_\xi)} \tag{7.259}$$

the average tunneling speed and found it in experiments to be greater than c.

Here is a slightly different version that is a bit easier to analyze: After the passage through the barrier, the wave function will have split into two packets, $\psi_{\text{reflected}}$ and $\psi_{\text{transmitted}}$. Suppose that at time t_1, ψ_{in} starts entering the barrier region $[0, L]$ and, at t_2, $\psi_{\text{transmitted}}$ has just left $[0, L]$. Let μ_1 be the mean position of ψ_{in} at t_1 and μ_2 the mean position of $\psi_{\text{transmitted}}$ at t_2. Then a quantity analogous to v is

$$\tilde{v} = \frac{\mu_2 - \mu_1}{t_2 - t_1}, \tag{7.260}$$

which will again be greater than c for suitable parameters of the barrier and ψ_{in}.

Now note that neither v nor $\tilde{v}$ is the average velocity of those Bohmian particles that get transmitted. Since Bohmian trajectories cannot cross, any that gets transmitted must from the beginning have been to the right of any that gets reflected. Thus, it is the particles in the *front* of the packet that get transmitted; this is visible in Fig. 2.2. As a consequence, the average position at t_1 of *trajectories that will get transmitted* is greater than μ_1 (the middle of the packet), so the transmitted trajectories have less distance to travel. Likewise, for (7.259) with ξ a point that all trajectories pass, the time of passing ξ is earlier for *trajectories that will get transmitted* than for others, and thus earlier than $\mathbb{E}(T_\xi)$. Taking $\tilde{v}$ for the average velocity is like

[19] Two reasons for thinking that the change should be small are that the detectors do not have to be hard and that the current always points to the right at these times anyway.

computing the average increase of income between 2010 and 2020 by taking the average income of yacht owners in 2020 and subtracting the average income in the overall population in 2010.

That is why there is no conflict between subluminal Bohmian velocity and $v > c$ or $\tilde{v} > c$, for either experimental or theoretical values of v and $\tilde{v}$.[20]

7.8 GRW Theory in Relativistic Space-Time

One aspect that makes GRW theory particularly interesting is that it possesses a relativistic version that (unlike Bohmian mechanics) does not involve a preferred foliation into spacelike surfaces, although it is nonlocal. Such a version can also be set up in curved space-time, but we will only consider it in Minkowski space-time M.

I will describe a version of the theory that I developed in 2004 [48] for N non-interacting particles and in 2020 [53] for N interacting particles. The 2004 version, the first relativistic version of a collapse theory, made use of ideas from Bell (1987) [5]; a model for a discrete space-time lattice due to Dowker and Henson (2002) [21] can be regarded as a precursor. Philip Pearle developed a relativistic collapse process for ψ (based on CSL) already in 1990 [37], but without a primitive ontology; moreover, this process is divergent and leads to infinite energy increase in arbitrarily short time. An improved, regularized relativistic version of Pearle's approach was developed by Bedingham (2011) [2]. Another approach for setting up relativistic collapse theories was proposed by Tilloy (2017) [45].

The theory I will describe makes predictions very close to those of the quantum formalism. It is nonlocal, excludes superluminal signaling, and satisfies microscopic parameter independence over distances greater than σ. I will define it using the flash ontology and call it rGRWf for relativistic GRW theory with flash ontology, but I will also explain how it can be adapted for the matter density ontology. In the non-relativistic limit, it becomes the GRWf theory we already know from Sect. 3.3. Like the latter, rGRWf involves unitary evolution interrupted by discrete jumps that are associated with a certain jump rate λ (such as 10^{-16} sec^{-1}) and a certain collapse width σ (such as 10^{-7} m).

From the fundamental point of view, what we need to do in order to define the theory is to define the probability distribution of the primitive ontology, that is, of the flashes. The distribution will be defined using a wave function. We will also see that with every Cauchy surface Σ, we can associate a collapsed wave function

[20] If we did not know about Bohmian mechanics, it would be tempting to take the observable quantity (7.259) or (7.260) as the *definition* of the intuitive concept of average tunneling speed. This example illustrates the danger of operational definitions and thus once more the illusion of modesty of positivism. Nimtz had the bad luck of being trained in orthodox quantum mechanics, not in Bohmian mechanics.

ψ_Σ analogous to the Aharonov-Albert wave function of Sect. 7.7.1, so surface wave functions are relevant again.

7.8.1 1-Particle Case

Let us begin with the simplest case of a 1-particle wave function.

Ingredients

As mentioned, the unitary evolution will be interrupted by discrete jumps. For the construction, we regard the unitary evolution as given in the form of unitary isomorphisms $U^{\Sigma'}_{1\Sigma} : \mathscr{H}_{1\Sigma} \to \mathscr{H}_{1\Sigma'}$, as usual with the composition laws $U^{\Sigma}_{1\Sigma} = I$ and $U^{\Sigma''}_{1\Sigma'} U^{\Sigma'}_{1\Sigma} = U^{\Sigma''}_{1\Sigma}$. For example, we know from Sect. 7.3.4 that the 1-particle Dirac equation defines such an evolution. However, we make a slightly stronger assumption that concerns the surfaces that are the analogs of spheres in Minkowski space, the sets

$$\mathbb{H}_s(y) = \Big\{ x \in \text{future}(y) : (x^\mu - y^\mu)(x_\mu - y_\mu) = s^2 \Big\} \tag{7.261}$$

for any $s > 0$ and $y \in M$, the surface of constant timelike distance s from y; see Fig. 7.15.

Exercise 7.9 (Hyperboloid) Show that for any $y \in M$ and $s > 0$, $\mathbb{H}_s(y)$ is a spacelike surface but not a Cauchy surface by verifying that $t \mapsto y + (t, \sqrt{1+t^2}, 0, 0)$ is a timelike curve $\mathbb{R} \to M$ that does not intersect $\mathbb{H}_s(y)$. Why is this curve inextendible?

Henceforth, we call $\mathbb{H}_s(y)$ the *hyperboloid* of radius s based at y. From now on, we make the following.

Assumption *The 1-particle evolution operators $U^{\Sigma'}_{1\Sigma}$ are defined whenever both Σ and Σ' are either a Cauchy surface or a hyperboloid.*

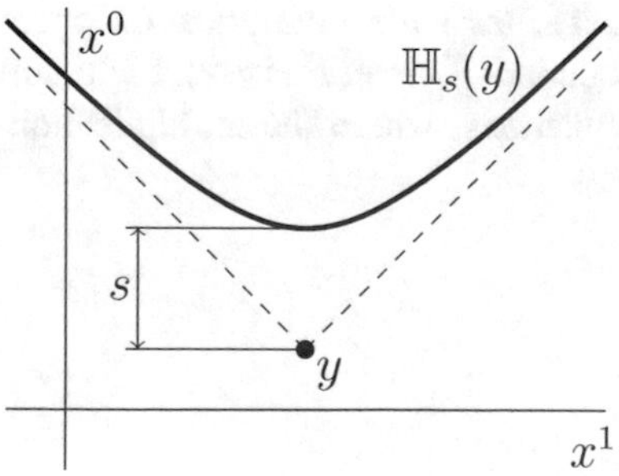

Fig. 7.15 In Minkowski space-time, the surface of constant timelike distance s from y in the future of y, $\mathbb{H}_s(y)$, has the shape of a hyperboloid that is asymptotic to the future light cone of y (dashed)

That is, we assume that we can also evolve ψ to and from hyperboloids. This assumption has been proven to be correct in [52, Lemma 17] for the free Dirac equation with $m > 0$ and some other cases (viz., the Dirac equation with $m > 0$ and an external field A_μ that is time-independent in some Lorentz frame and tends to 0 at infinity sufficiently fast).

For any Σ that is a hyperboloid or a Cauchy surface, define the *Gaussian function* $g_{\Sigma,z} : \Sigma \to \mathbb{R}$ with width σ and center $z \in \Sigma$ to be

$$g_{\Sigma,z}(x) = \mathcal{N} \exp\Big(- \frac{\text{s-dist}_\Sigma(x,z)^2}{4\sigma^2} \Big), \tag{7.262}$$

where s-dist_Σ means the spacelike distance along Σ and the normalizing factor $\mathcal{N}$ is chosen so that

$$\int_\Sigma V(d^3x)\, g_{\Sigma,z}(x)^2 = 1\,. \tag{7.263}$$

Since $\text{s-dist}_\Sigma(x,z) = \text{s-dist}_\Sigma(z,x)$, we also have that

$$\int_\Sigma V(d^3z)\, g_{\Sigma,z}(x)^2 = 1\,. \tag{7.264}$$

Definition

We are now ready to define the probability distribution of the flashes. We do this by constructing a random sequence of flashes, each inside the future light cone of the previous ones. We assume that, as part of the initial data, we are given the zeroth flash $X_0 \in M$, called the *seed flash*. Choose a random value $T_1 \sim \text{Exp}(\lambda)$ called the first waiting time. We decide that the next flash X_1 will have timelike distance T_1 from X_0; put differently, $X_1 \in \mathbb{H}_{T_1}(X_0)$. Evolve the wave function unitarily to $\Sigma := \mathbb{H}_{T_1}(X_0)$. Choose $X_1 \in \Sigma$ randomly with distribution

$$\mathbb{P}(X_1 \in d^3x_1) = \rho(x_1)\, V(d^3x_1) := \big\| g_{\Sigma,x_1} \psi_\Sigma \big\|^2\, V(d^3x_1)\,. \tag{7.265}$$

That is, we multiply ψ_Σ by the function g_{Σ,x_1}, the Gaussian centered at x_1. Abstractly, if the position observable is given as a POVM $P_{1\Sigma}$ on Σ acting on $\mathscr{H}_{1\Sigma}$, then the multiplication operator by the function g can be defined as the operator $\int_\Sigma P_{1\Sigma}(d^3x)\, g(x)$. I will write the formulas more concretely for Dirac wave functions, where the multiplication by g is literally just that—the multiplication by g.

To verify that (7.265) is a probability distribution, note that, with $n_\mu(y)$ the future unit normal vector to Σ at $y \in \Sigma$,

$$\int_\Sigma V(d^3x_1)\,\rho(x_1) = \int_\Sigma V(d^3x_1)\,\|g_{\Sigma,x_1}\psi_\Sigma\|^2 \tag{7.266a}$$

$$= \int_\Sigma V(d^3x_1)\int_\Sigma V(d^3y)\,g_{\Sigma,x_1}(y)^2\,\overline{\psi}_\Sigma(y)\gamma^\mu n_\mu(y)\psi_\Sigma(y) \tag{7.266b}$$

$$= \int_\Sigma V(d^3y)\,\overline{\psi}_\Sigma(y)\gamma^\mu n_\mu(y)\psi_\Sigma(y)\int_\Sigma V(d^3x_1)\,g_{\Sigma,x_1}(y)^2 \tag{7.266c}$$

$$\overset{(7.264)}{=} \int_\Sigma V(d^3y)\,\overline{\psi}_\Sigma(y)\gamma^\mu n_\mu(y)\psi_\Sigma(y) = \|\psi_\Sigma\|^2 = 1\,. \tag{7.266d}$$

Once we have chosen X_1, collapse the wave function to

$$\psi'_\Sigma = \frac{g_{\Sigma,X_1}\psi_\Sigma}{\|g_{\Sigma,X_1}\psi_\Sigma\|^2}. \tag{7.267}$$

That is, we collapse by multiplying with a Gaussian along the hyperboloid Σ. Now repeat the procedure with X_1 in the place of X_0 and ψ' in the place of ψ. That defines the joint distribution of a sequence $X_0, X_1, X_2, \ldots$ of flashes; see Fig. 7.16.

This construction is a rather straightforward analog of the non-relativistic flash process using the relativistic notion of timelike distance instead of the non-relativistic time difference.

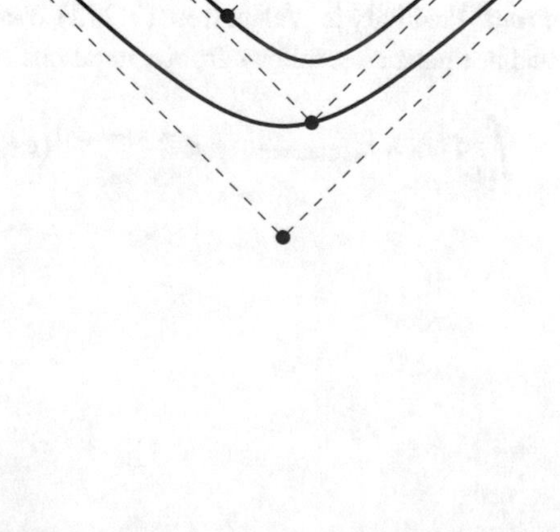

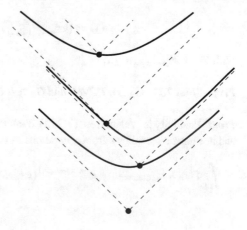

Fig. 7.16 A realization of the flashes $X_0, \ldots, X_3$ (dots), along with their future light cones (dashed) and the hyperboloids $\mathbb{H}_{T_k}(X_k)$ containing the next flash

POVM

Writing $|x - y|$ for the timelike distance between x and y, i.e.,

$$|u| = \sqrt{u^\mu u_\mu}\,, \tag{7.268}$$

x_0 for X_0, and Σ_1 for $\mathbb{H}_{|x_1-x_0|}(x_0)$, we can write the distribution of X_1 in space-time as

$$\mathbb{P}(X_1 \in d^4x_1) = \lambda\, 1_{x_1 \in \text{future}(x_0)} e^{-\lambda|x_1-x_0|} \left\| g_{\Sigma_1,x_1} U^{\Sigma_1}_{1\Sigma_0} \psi_0 \right\|^2 d^4x_1 \tag{7.269a}$$

$$= \langle \psi_0 | G_1(d^4x_1) | \psi_0 \rangle \tag{7.269b}$$

with POVM

$$G_1(d^4x_1) = 1_{x_1 \in \text{future}(x_0)} e^{-\lambda|x_1-x_0|} U^{\Sigma_0}_{1\Sigma_1} g^2_{\Sigma_1,x_1} U^{\Sigma_1}_{1\Sigma_0}\, \lambda\, d^4x_1\,. \tag{7.270}$$

More generally, the joint distribution of the first n flashes is

$$\mathbb{P}(X_1 \in d^4x_1, \ldots, X_n \in d^4x_n) = \langle \psi_0 | G_n(d^4x_1 \times \cdots \times d^4x_n) | \psi_0 \rangle \tag{7.271}$$

with POVM G_n on M^n acting on $\mathscr{H}_{1\Sigma_0}$ given by

$$G_n(d^4x_1 \times \cdots \times d^4x_n) =$$
$$\left(\lambda^n \prod_{k=1}^n 1_{x_k \in \text{future}(x_{k-1})} e^{-\lambda|x_k-x_{k-1}|} \right) L(\underline{x})^\dagger\, L(\underline{x})\, d^4x_1 \cdots d^4x_n\,, \tag{7.272}$$

where $\underline{x} = (x_1, \ldots, x_n)$ and $L(\underline{x}) : \mathscr{H}_{1\Sigma_0} \to \mathscr{H}_{1\Sigma_n}$ is defined by

$$L(\underline{x}) = \left(g_{\Sigma_n x_n} U^{\Sigma_n}_{1\Sigma_{n-1}}\right) \cdots \left(g_{\Sigma_1 x_1} U^{\Sigma_1}_{1\Sigma_0}\right) \tag{7.273}$$

with $\Sigma_k = \mathbb{H}_{|x_k - x_{k-1}|}(x_{k-1})$.

Proposition 7.17 *G_n is a POVM, and G_{n-1} is its marginal, $G_{n-1}(B) = G_n(B \times M)$.*

Proof Positivity is evident from (7.272), σ-additivity because it is the integral of a density, and normalization follows from n iterations of the relation

$$\int_M d^4x_n\, \lambda\, 1_{x_n \in \text{future}(x_{n-1})}\, e^{-\lambda|x_n-x_{n-1}|} \left(g_{\Sigma_n x_n} U^{\Sigma_n}_{1\Sigma_{n-1}}\right)^\dagger \left(g_{\Sigma_n x_n} U^{\Sigma_n}_{1\Sigma_{n-1}}\right) = I \tag{7.274}$$

in $\mathscr{H}_{1\Sigma_{n-1}}$. To verify this relation for given $(x_0, \ldots, x_{n-1})$, we use the so-called co-area formula

$$\int_{\text{future}(y)} d^4x\, f(x, y) = \int_0^\infty ds \int_{\mathbb{H}_s(y)} V(d^3x)\, f(x, y) \tag{7.275}$$

and the elementary relation

$$\int_0^\infty ds\, \lambda\, e^{-\lambda s} = 1\,. \tag{7.276}$$

Thus,

$$\int_M d^4x_n\, \lambda\, 1_{x_n \in \text{future}(x_{n-1})}\, e^{-\lambda |x_n - x_{n-1}|} (g_{\Sigma_n x_n} U^{\Sigma_n}_{1\Sigma_{n-1}})^\dagger (g_{\Sigma_n x_n} U^{\Sigma_n}_{1\Sigma_{n-1}})$$

$$= \int_0^\infty ds\, \lambda\, e^{-\lambda s} \int_{\mathbb{H}_s(x_{n-1})} V(d^3x_n)\, (g_{\Sigma_n x_n} U^{\Sigma_n}_{1\Sigma_{n-1}})^\dagger (g_{\Sigma_n x_n} U^{\Sigma_n}_{1\Sigma_{n-1}}) \tag{7.277a}$$

$$= \int_0^\infty ds\, \lambda\, e^{-\lambda s}\, U^{\Sigma_{n-1}}_{1\mathbb{H}_s(x_{n-1})} \Bigg(\int_{\mathbb{H}_s(x_{n-1})} V(d^3x_n)\, g^2_{\Sigma_n x_n} \Bigg) U^{\mathbb{H}_s(x_{n-1})}_{1\Sigma_{n-1}} \tag{7.277b}$$

$$\overset{(7.264)}{=} \int_0^\infty ds\, \lambda\, e^{-\lambda s}\, U^{\Sigma_{n-1}}_{1\mathbb{H}_s(x_{n-1})}\, I_{\mathbb{H}_s(x_{n-1})}\, U^{\mathbb{H}_s(x_{n-1})}_{1\Sigma_{n-1}} \tag{7.277c}$$

$$= \int_0^\infty ds\, \lambda\, e^{-\lambda s}\, I_{\Sigma_{n-1}} = I_{\Sigma_{n-1}}\,. \tag{7.277d}$$

The relation $G_{n-1}(B) = G_n(B \times M)$ also follows from (7.274). □

The joint distribution of *all* (infinitely many) flashes $X_1, X_2, \ldots$ is also given by a POVM $G^{(1)}$ (on the space $M^{\mathbb{N}}$ of sequences) of which each G_n is a marginal; this follows from the fact that G_{n-1} is a marginal of G_n together with some technical steps carried out in [51].

7.8.2 The Case of N Non-Interacting Particles

Before I define the theory for interacting particles in Sect. 7.8.3, let us focus on the simpler case without interaction.

Definition

For N non-interacting particles, we set $\mathscr{H}_{N\Sigma} = \mathscr{H}_{1\Sigma}^{\otimes N}$ and use the POVM

$$G^{(N)} = G^{(1)}_{x_{10}} \otimes \cdots \otimes G^{(1)}_{x_{N0}} \tag{7.278}$$

on $(M^{\mathbb{N}})^N$ acting on $\mathscr{H}_{N\Sigma}$, where $G^{(1)}_{x_0}$ is the POVM for 1 particle with seed flash x_0. We write elements of $(M^{\mathbb{N}})^N$ as collections $F = (x_{ik})$ with $i \in \{1, \ldots, N\}$ and $k \in \mathbb{N} = \{1, 2, \ldots\}$. The existence of $G^{(N)}$ is provided by Proposition 6.1. Given the seed flashes x_{i0} for all i, the flash process is then defined as having probability distribution

$$\mathbb{P}(F \in B) = \langle \psi_0 | G^{(N)}(B) | \psi_0 \rangle \tag{7.279}$$

for all B and initial wave function $\psi_0 \in \mathbb{S}(\mathscr{H}_{\Sigma_0})$. (Since $U^{\Sigma'}_{N\Sigma}$ can be used for transporting a wave function to any other surface, it is irrelevant where Σ_0 is chosen.) This completes the definition.

Properties

(i) *Non-interacting.* The non-interacting nature is already indicated by the fact that for each i, separate operators $U^{\Sigma'}_{1\Sigma}$ are used. In the following, it will be convenient to call the different values of i the "particles" although there are no particles in the ontology.

(ii) *Independence.* If ψ_0 factorizes, $\psi_0 = \psi_{10} \otimes \cdots \otimes \psi_{N0}$, then the measure (7.279) factorizes into a product of N measures, so the flash process consists of N independent flash processes, one for each i. However, if ψ_0 does not factorize, then the flashes for one particle will not be independent from those of the others.

(iii) *Spacelike separation.* While flashes belonging to one particle are timelike separated from each other, flashes belonging to different particles can be spacelike separated. The fact that they are not independent already indicates the nonlocal nature of the theory.

(iv) *Non-relativistic limit.* In the non-relativistic limit $c \to \infty$, the hyperboloids become horizontal planes, the timelike distance become differences of time coordinate, spacelike distance along Σ become distance of space coordinates, and $g_{\Sigma,z}$ becomes the usual Gaussian with center z. Putting these pieces together, the distribution of the flashes becomes that of non-relativistic GRWf.

(v) *Experiments.* Although the model as described so far is non-interacting, experiments such as Bell's can be considered in it as follows. Arbitrary external electromagnetic fields can be considered also in this non-interacting model (even, if needed, different ones for different particles). A Stern-Gerlach magnet will split the wave packet in two, and usually the interaction with a detector forces the two wave packets to collapse to a single one within a fraction of a second. But if the wave function happened to collapse already before the interaction with the detector, then that collapse would pre-determine the outcome of the detection. So, if we consider theoretically leaving the split wave function for hundreds of millions of years, such a collapse would occur with probability close to 1, and if the two particles were separated by a billion light years, they would remain spacelike separated throughout. That is why this model gives an answer to what would happen in a Bell experiment. Of course, the prediction is very, very close to that of the quantum formalism as discussed in Chap. 4, so that Bell's theorem yields another way to see that rGRWf is nonlocal.

(vi) *Who influences whom?* Nonlocality means that for space-time regions $A \times B$, events in A influence events in B *or vice versa*, but Bell's theorem does not tell us which way the influence went. In Bohmian mechanics with a time foliation $\mathcal{F}$, if A is earlier than B relative to $\mathcal{F}$ (i.e., if $A \subset \text{past}(\Sigma)$ and $B \subset \text{future}(\Sigma)$ for some $\Sigma \in \mathcal{F}$), then external fields in A influence trajectories in B, while external fields in B do not influence trajectories in A, so there is a fact about which way the influence went: from A to B. That is different in rGRWf: there is no fact about which way the influence went.

For example, consider Bell's experiment. The wave function gets split into four packets corresponding to the outcomes $\uparrow\uparrow$, $\uparrow\downarrow$, $\downarrow\uparrow$, and $\downarrow\downarrow$, and the first flash on each side will decide about the outcome. The fundamental laws of the theory prescribe the joint distribution density $\rho(x_{11}, x_{21})$ of the two flashes X_{11} and X_{21} (with particle $i = 1$ in A and particle $i = 2$ in B); it is a correlated distribution that depends on the external fields in both A and B. Random values with this joint distribution could be created by first choosing X_{11} with the marginal distribution

$$\rho(X_{11} = x_{11}) = \int_M d^4x_{21}\, \rho(x_{11}, x_{21}) \tag{7.280}$$

and then X_{21} with the conditional distribution

$$\rho(X_{21} = x_{21}|X_{11} = x_{11}) = \rho(x_{11}, x_{21}) \Big/ \int_M d^4y_{21}\, \rho(x_{11}, y_{21})\,. \tag{7.281}$$

Alternatively, they could be created by first choosing X_{21} with the appropriate marginal distribution

$$\rho(X_{21} = x_{21}) = \int_M d^4x_{11}\, \rho(x_{11}, x_{21}) \tag{7.282}$$

and then X_{11} with the conditional distribution

$$\rho(X_{11} = x_{11}|X_{21} = x_{21}) = \rho(x_{11}, x_{21}) \Big/ \int_M d^4y_{11}\, \rho(y_{11}, x_{21})\,. \tag{7.283}$$

Now the claim is that the theory does not have to specify which of the two ways nature uses and that nature does not have to use either of the two ways. Nature can just choose X_{11} and X_{21} jointly with joint distribution $\rho(x_{11}, x_{21})$, period. (This possibility becomes even more evident in the interacting version of rGRWf that can be applied to a whole universe instead of just the two particles.) Thus, there is no need for a direction of influence; it is enough if the fundamental physical theory prescribes the joint distribution.

(vii) *Microscopic parameter independence (MPI).* This is the name of a property defined in (7.216) expressing that the distribution of the flashes in the space-time region A does not depend on external fields in region $B \bigtimes A$ and thus that the distribution of flashes in past(Σ) does not depend on the external fields in future(Σ) for any Cauchy surface Σ. For simplicity, consider two particles, $i = 1$ in A and $i = 2$ in B. Then the marginal POVM of the flashes for $i = 1$,

$$G^{(2)}(B \times M^{\mathbb{N}}) = G^{(1)}_{x_{10}}(B) \otimes G^{(1)}_{x_{20}}(M^{\mathbb{N}}) = G^{(1)}_{x_{10}}(B) \otimes I, \tag{7.284}$$

does not depend on the external fields at work on particle $i = 2$. As a consequence, also the marginal distribution of the flashes for $i = 1$,

$$\mathbb{P}(X_{11} \in d^4x_{11}, X_{12} \in d^4x_{12}, \ldots) = \mathrm{tr}\Big[\rho_1\, G^{(1)}_{x_{10}}(d^4x_{11} \times d^4x_{12} \times \cdots)\Big] \tag{7.285}$$

with $\rho_1 = \mathrm{tr}_2|\psi_0\rangle\langle\psi_0|$ the reduced density matrix of particle $i = 1$ on the initial surface Σ_0, does not depend on the external fields at work on particle $i = 2$, as required by MPI.

It follows further that no signals can be sent from B to A using external fields acting on particle $i = 2$ in B and their effects on flashes of particle $i = 1$ in A. The no-signaling theorem can be strengthened further by noting that (7.285) does not depend either on any feature of the wave function ψ_0 that is not encoded in ρ_1 (such as a factor disentangled from particle $i = 1$, or such as messages that observers consisting of other particles in B may want to send). However, there is a subtlety. Particle $i = 1$ could have nonzero wave function in both A and B. If A and B touch each other as in Fig. 7.17, or if they have spacelike distance less than σ, then a flash X in A can be associated with a

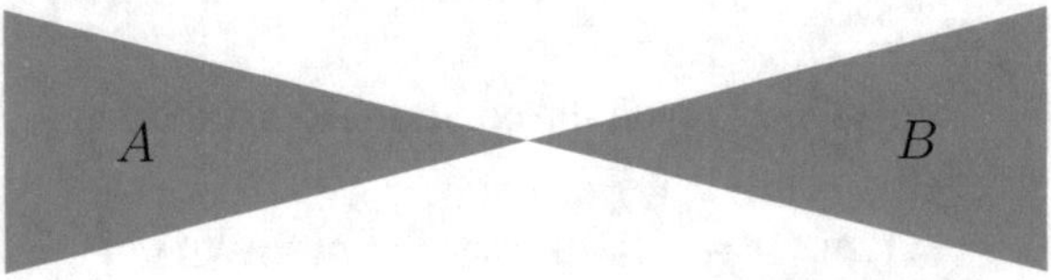

Fig. 7.17 Spacelike separated regions A, B are necessarily disjoint but can touch each other in the sense of having distance 0

Gaussian collapse factor $g_{\Sigma z}$ that reaches into B,[21] so the probability density (7.265) of X in A within distance σ of B may depend on the part of ψ_Σ with $x_1 \in B$ and thus on external fields in B. So in these cases, MPI is not exactly satisfied; in sum, it seems appropriate to say that *MPI is satisfied over distances greater than* σ.

Collapsed Wave Function

I now describe how to define, from a given pattern of flashes, a wave function ψ_Σ for every Cauchy surface (or, for that matter, hyperboloid) Σ, a wave function that incorporates, as Aharonov and Albert had suggested (see Sect. 7.7.1), all collapses associated with flashes in past(Σ). Since the flashes are random, ψ_Σ is random. Let $\Sigma_{ik} = \mathbb{H}_{|X_{ik}-X_{ik-1}|}(X_{ik-1})$, and suppose that for each i, n_i is the number of the last flash before Σ; then set

$$\psi_\Sigma = \mathcal{N}\Big[\bigotimes_{i=1}^{N} U^{\Sigma}_{1\Sigma_{in_i}}\big(g_{\Sigma_{in_i}X_{in_i}}U^{\Sigma_{in_i}}_{1\Sigma_{in_i-1}}\big)\cdots\big(g_{\Sigma_{i1}X_{i1}}U^{\Sigma_{i1}}_{1\Sigma_0}\big)\Big]\psi_0 \tag{7.286}$$

with $\mathcal{N}$ such that $\|\psi_\Sigma\|_\Sigma = 1$. Then the conditional distribution of the flashes in the future of Σ, given the flashes in the past of Σ, is

$$\mathbb{P}\Big(X_{ik} \in d^4x_{ik}\ \forall i\ \forall k > n_i \Big| X_{ik} = x_{ik}\ \forall i\ \forall k \le n_i \text{ and } X_{in_i+1} \in \text{future}(\Sigma)\Big)$$
$$= \tilde{\mathcal{N}}\Big\langle\psi_\Sigma\Big|G_{\Sigma,x_{1n_1}\ldots x_{Nn_N}}\Big(\prod_{i=1}^{N}\prod_{k>n_i} d^4x_{ik}\Big)\Big|\psi_\Sigma\Big\rangle, \tag{7.287}$$

where the notation $G_{\Sigma_0,x_{10}\ldots x_{N0}}$ makes the dependence of the N-particle POVM on the initial surface Σ_0 and the seed flashes x_{i0} explicit, the $\prod$ symbol means Cartesian product, and the normalizing factor is

$$\tilde{\mathcal{N}} = \Big\langle\psi_\Sigma\Big|G_{\Sigma,x_{1n_1}\ldots x_{Nn_N}}\Big(\text{future}(\Sigma)^{\{1\ldots N\}\times\mathbb{N}}\Big)\Big|\psi_\Sigma\Big\rangle^{-1}. \tag{7.288}$$

That is, x_{in_i} now takes the role of the seed flash for i. We can note that the equation (7.287) for the distribution *from a surface* Σ *onward* differs from the equation (7.279) *for the universe* by the occurrence of the factor $\tilde{\mathcal{N}}$. If $x_{in_i} \in \Sigma$ for all i, then $\tilde{\mathcal{N}} = 1$. In general, $1/\tilde{\mathcal{N}}$ is the probability that no flashes occur prior to Σ if the seed flashes are x_{in_i} and the initial wave function is ψ_Σ.

[21] In fact, no matter how distant B is from A, tails of the Gaussian factor may reach into B. Apart from the fact that these tails will be negligibly small for distances much larger than σ, the overlap with a distant B could be avoided entirely by replacing the Gaussian factor by some other shape with (strictly) compact support.

It also follows that in order to obtain from ψ_Σ the $\psi_{\Sigma'}$ on another surface $\Sigma' \subset \text{future}(\Sigma)$, we can use (7.286) *mutatis mutandis*,[22] i.e., with ψ_0 replaced by ψ_Σ, x_{in_i} inserted as the seed flashes, and the numbering of flashes starting anew.

Criticisms

Jones, Guaita, and Bassi (2019) [30] claimed that rGRWf fails to be relativistic, but I cannot follow their arguments. The central argument seems to be that if Σ is the surface $\{t = 0\}$ in one Lorentz frame and Σ' the surface $\{t = 0\}$ in another Lorentz frame, then they would say that Lorentz invariance demands the existence of an operator $W_\Sigma^{\Sigma'}$ that allows the computation of how the wave function transforms from one frame to the other, $\psi_{\Sigma'} = W_\Sigma^{\Sigma'}\psi_\Sigma$. Since, for given ψ_Σ, $\psi_{\Sigma'}$ depends on the flashes between Σ and Σ', there is no fixed operator $W_\Sigma^{\Sigma'}$. My reply is that demanding a fixed $W_\Sigma^{\Sigma'}$ goes way beyond Lorentz invariance, so the argument does not reveal a failure of Lorentz invariance. Keep in mind also that the theory is independent of any choice of Lorentz frame because it was defined without ever using or choosing a Lorentz frame.

Conway and Kochen (2006–2009) [14–16] as well as Esfeld and Gisin (2014) [25] claimed that rGRWf fails to be relativistic, and their objections seem to boil down to the following: If you wanted to simulate an rGRWf universe on a computer, you would use a Lorentz frame F and ask a random number generator for a random bit every time you need to make a random decision. But then events earlier in F influence those later in F, and not vice versa, so the relations between events prefer F over other frames. My reply is that their method of simulation breaks the Lorentz invariance, not the theory. As I explained in Property (vi) above, in nature, if governed by rGRWf, there is no fact about who influences whom; see [49] for further discussion.

7.8.3 Interacting Particles

I now give an outline of the model for interacting particles developed in [53]. The empirical predictions are again close to those of the standard quantum formalism. It is an open question how to set up a relativistic GRW theory with interaction for identical particles or a variable number of particles.

We assume that between the N-particle Hilbert spaces $\mathscr{H}_\Sigma$, unitary evolution operators $U_\Sigma^{\Sigma'}$ are given that incorporate interaction, and we assume that Σ, Σ' can be any set that is intersected exactly once by every timelike straight line (which includes Cauchy surfaces, hyperboloids, and surfaces consisting of pieces of hyperboloids). A collapse will again be implemented through multiplication by some function g on Σ^N, and the multiplication operator (which, if the position observable is given by a PVM P_Σ on Σ^N, should be defined as $\int_{\Sigma^N} P_\Sigma(dx)\, g(x)$) will again simply be denoted by g. We assume *interaction locality*, which can be

[22] = after having changed what must be changed (Latin).

formalized this way: For any set $A \subseteq \Sigma \cap \Sigma'$ in the overlap of two surfaces and any $i \in \{1, \ldots, N\}$,

$$P_{\Sigma'}\Big((\Sigma')^{i-1} \times A \times (\Sigma')^{N-i-1}\Big) = U_{\Sigma}^{\Sigma'} P_{\Sigma}\Big(\Sigma^{i-1} \times A \times \Sigma^{N-i-1}\Big) U_{\Sigma'}^{\Sigma}. \qquad (7.289)$$

We write again $\underline{x}$ for the collection of x_{ik} with $i \in \{1, \ldots, N\}$ numbering the particles and $k \in \{1, \ldots, n\}$ (or ultimately $k \in \mathbb{N}$) the flashes for each i, $d\underline{x} = \prod_{ik} d^4x_{ik}$, and

$$\Sigma_{ik} = \mathbb{H}_{|x_{ik}-x_{ik-1}|}(x_{ik-1}) \,. \qquad (7.290)$$

The joint distribution of the flashes is again of the form

$$\mathbb{P}(\underline{X} \in d\underline{x}) = \langle \psi_0 | D(\underline{x}) | \psi_0 \rangle \, d\underline{x} \,, \qquad (7.291)$$

D is again of the form

$$D(\underline{x}) = \Bigg(\lambda^{\nu} \prod_{i=1}^{N} \prod_{k=1}^{n_i} 1_{x_{ik} \in \text{future}(x_{ik-1})} e^{-\lambda |x_{ik}-x_{ik-1}|} \Bigg) L(\underline{x})^{\dagger} \, L(\underline{x}) \,, \qquad (7.292)$$

and L should be something like the product of the collapse operators C_{ik}, one for each flash x_{ik}, given by

$$C_{ik} = U_{\Sigma_{ik}}^{\Sigma_0} \, g_{\Sigma_{ik} x_{ik} i} \, U_{\Sigma_0}^{\Sigma_{ik}} \qquad (7.293)$$

with $g_{\Sigma x i}$ the function $g_{\Sigma x}$ applied to the i-th variable of ψ_{Σ}. Since the C_{ik} do not commute with each other, we need to specify the ordering of the factors in $L(\underline{x})$. The basic idea for the ordering is this: When, of two flashes x_{ik} and $x_{j\ell}$, x_{ik} lies in the (relativistic) past of $x_{j\ell}$, then the factor C_{ik} should appear to the right of $C_{j\ell}$. For two spacelike separated flashes $x_{ik} \times x_{j\ell}$, we would hope that C_{ik} commutes with $C_{j\ell}$. However, if the g's are Gaussian functions, they do not, because C_{ik} and $C_{j\ell}$ are multiplication operators on different surfaces, and the unitary evolution from one surface to the other must be expected to mix up directions in Hilbert space in a complicated way. Here is how we deal with that: we cut off the support of g; instead of Gaussians, we use other form factors that vanish exactly outside certain regions, and then we make sure that support($g_{j\ell}$) is spacelike from support(g_{ik}) when needed, with the consequence that $C_{j\ell}$ commutes with C_{ik}. To this end, we partition each hyperboloid Σ_{ik} (see Fig. 7.18) according to which parts ("3-cells") lie in the future or past of the other hyperboloids.

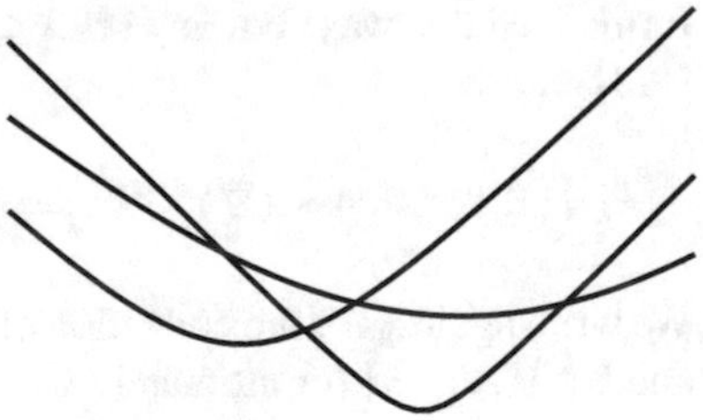

Fig. 7.18 Several hyperboloids in generic position cut space-time into regions called 4-cells and cut each other into pieces called 3-cells

In generic position, each out of n hyperboloids will be partitioned by the others into 2^{n-1} 3-cells. For each 3-cell A, we define the form factor g to be

$$g_{Az}(x) := 1_{z\in A}\, 1_{x\in A} \left\| \text{Gaussian}_{\Sigma x} 1_A \right\|_{\Sigma}^{-1} \text{Gaussian}_{\Sigma z}(x)\,. \tag{7.294}$$

For x far away from the boundary of A, this function still has a Gaussian shape, but for x and z close to the boundary, the shape differs from a Gaussian, although it is still bump-shaped around the center z (see [53] for more detail). If, for two 3-cells $A, A', A \times A'$, then the multiplication operator g_{Az} commutes with $g_{A'z'}$ for every $z \in A$ and $z' \in A'$.

Now we define the order of operators C_{ik} in $L(\underline{x})$ as follows. The hyperboloids also partition space-time into what I will call 4-cells. An *admissible sequence* is any ordering of the 4-cells such that whenever 4-cell C lies in the past of 4-cell C', C comes before C' in the sequence; see Fig. 7.19 for an example.

It can be shown [53] that admissible sequences always exist, that every admissible sequence crosses every 3-cell exactly once, and using interaction locality that the C_{ik} associated with 3-cells crossed in the same step commute. We can thus define, for any given admissible sequence, $L(\underline{x})$ to be the product of all C_{ik} *in the order in which the 3-cells get crossed in this admissible sequence*. It can then be shown using interaction locality that any two admissible sequences lead to the same operator $L(\underline{x})$ and that

$$\int d\underline{x}\, D(\underline{x}) = I\,, \tag{7.295}$$

which implies that $G(d\underline{x}) = D(\underline{x})\, d\underline{x}$ is a POVM. Construction complete.

7.8.4 Matter Density

In this section, I explain how to set up a relativistic GRW theory rGRWm with the *matter density ontology* (Tumulka 2007 [50], Bedingham et al. 2014 [3]). We use the wave function ψ_Σ of rGRWf, but not the flashes; the points x_{ik} thus retain the meaning of being the centers of the collapses but lose the meaning of being material points. Instead, as the primitive ontology, we use a matter density function $m(x^\nu)$,

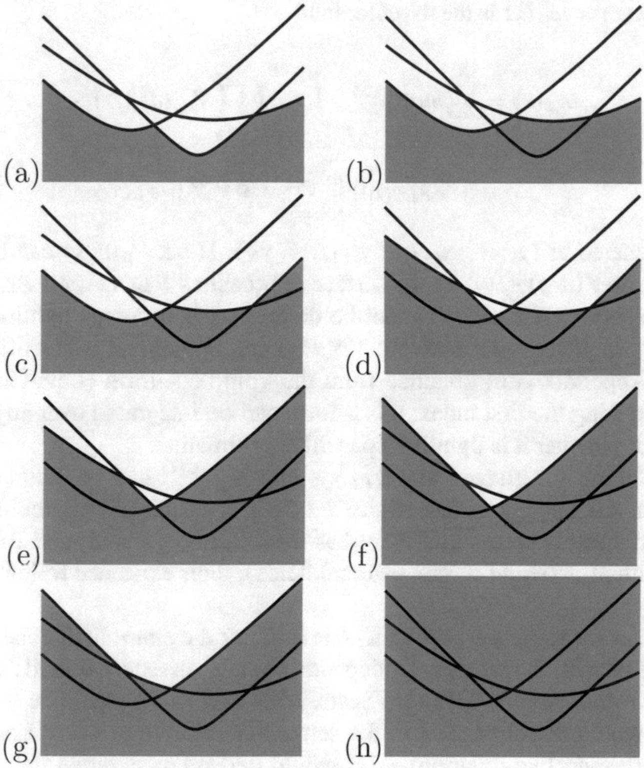

Fig. 7.19 Example of an admissible sequence of 4-cells: their future boundaries form a sequence of 3-surfaces (each consisting of 3-cells) that moves to the future

and we make it a law of rGRWm that

$$m(x) = \left\langle \psi_{PLC(x)} \middle| \mathcal{M}_{PLC(x)}(x) \middle| \psi_{PLC(x)} \right\rangle, \tag{7.296}$$

where $PLC(x)$ means the past light cone of x,

$$PLC(x) = \left\{ y \in M : y^0 \leq x^0 \text{ and } (y^\mu - x^\mu)(y_\mu - x_\mu) = 0 \right\}, \tag{7.297}$$

and $\mathcal{M}_{PLC(x)}(x)$ is the mass density operator at x acting in $\mathscr{H}_{PLC(x)}$. Here, $m(x)$ could be a scalar field or vector field $m_\mu(x)$ or tensor field $m_{\mu\nu}(x)$, depending on the appropriate nature of the $\mathcal{M}$ operators. Concretely, for N-particle Dirac wave functions, $m(x) = m_\mu(x)$ is the 4-vector field

$$m_\mu(x) = \sum_{i=1}^{N} m_i \eta_{\mu\mu_i} \int_{PLC(x)^{N-1}} \Big(\prod_{j\neq i} V_{\mu_j}(d^3y_j)\Big) \times \\ \times \; \overline{\psi}_{PLC(x)}[\gamma^{\mu_1} \otimes \cdots \otimes \gamma^{\mu_N}]\psi_{PLC(x)} \tag{7.298}$$

with ψ evaluated at $(y_1, \ldots, y_{i-1}, x, y_{i+1}, \ldots, y_N)$. Here, $V_\mu(d^3y)$ can be regarded as the limit of $V(d^3y)\, n_\mu(y)$ as the surface approaches $PLC(x)$; on $PLC(x)$ itself, $V(d^3y)$ vanishes, while $n_\mu(y)$ cannot be defined as it becomes lightlike and thus cannot be normalized. Alternatively, $V^\mu(d^3y)$ can be defined as the differential 3-form (see Appendix A.6) obtained from the volume 4-form (Levi-Civita tensor) $\varepsilon_{\mu\nu_1\nu_2\nu_3}$ by raising the first index; this 3-form can be integrated over any 3-surface, regardless of whether it is lightlike, spacelike, or timelike.

We are taking for granted that $\psi_{PLC(x)}$ and $U_\Sigma^{PLC(x)}$ can be defined, although $PLC(x)$ is not a Cauchy surface. Since $PLC(x)$ is a limit of hyperboloids (and of Cauchy surfaces), it seems plausible that they can be defined, and for the Dirac equation with $m > 0$ (and certain external fields), their existence follows from [52, Lemma 17].

The reason for using the *past* light cone and not the *future* light cone $FLC(x)$ in (7.297) and (7.298) is that $\psi_{FLC(x)}$ depends already on external fields at spacelike separation from x, so that MPI (see Sects. 7.5.2 and 7.8.2) would be violated, and the theory would be retrocausal in the sense that relative to every Lorentz frame (and to every spacelike foliation), $m(x)$ would depend on external fields with time coordinate later than x.[23]

It turns out [3] that this theory rGRWm makes predictions close to those of the standard quantum formalism and consistent with all present-day experiments. The key step for this conclusion is that if a quantum measurement takes place at some space-time point x_1 and x lies in the future of x_1, then the collapses associated with the quantum measurement have already occurred on $PLC(x)$ and are thus incorporated in $m(x)$.

[23] Since $PLC(x)$ and $FLC(x)$ are 3-surfaces that can be defined for any $x \in M$ from the metric alone, one might think of applying them in Bohmian mechanics and modifying $\mathrm{BM}_{\mathcal{F}}$ so as to use light cones instead of the leaves of the time foliation $\mathcal{F}$. Specifically, we might think of a law for dX_k^μ/ds whose right-hand side involves the wave function ϕ on either $PLC(X_k(s))$ or $FLC(X_k(s))$, that is, at the N points where the N world lines intersect $PLC(X_k(s))$, respectively, $FLC(X_k(s))$; such a law has been studied in Goldstein et al. (2003) [26]. If we use PLC, then the theory is local and thus cannot be empirically adequate. With FLC, the theory is retrocausal (which is perhaps not easy to accept) and nonlocal; in fact, it is interesting as an example of a nonlocal, relativistic particle theory without a preferred foliation. However, with either PLC or FLC, equivariance breaks down: there are no surfaces on which the configuration on which the configuration is $|\psi|^2$ distributed, so these theories do not make any discernible statistical predictions that could be compared to observations.

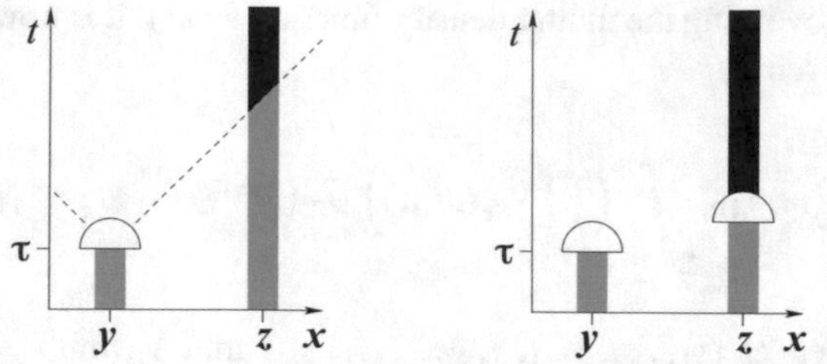

Fig. 7.20 Space-time diagram of $m(x)$ in the example. Semi-circles represent detectors; dashed lines, a light cone; and thick lines, the region where $m_1(x) = m_0$ (black) or $m_1(x) = m_0/2$ (gray). Left: with one detector. Right: with two detectors (Reprinted from [3] with permission by Springer Nature)

However, it is easy to mistakenly think that rGRWm makes wrong predictions; here is an example. Consider a single particle whose wave function at time 0 in some Lorentz frame consists of two equal-sized, well-localized packets at y and $z \in \mathbb{R}^3$ and that the detector is placed at y and gets activated at time τ shortly after. Then the detector has probability 1/2 to fire; suppose it does not fire (i.e., it "finds no particle"). Then, as shown in the left diagram of Fig. 7.20, the matter density associated with the particle will be $m(t, y) = 0$ for $t > \tau$, while at z, the matter density doubles only as soon as (t, z) lies in the future of (τ, y); put differently, the change in m at z is delayed at the speed of light.

This fact is unproblematical, but it might suggest that another detector at $(\tau+\varepsilon, z)$ (with $0 < \varepsilon \ll 1$) would have a probability of only 1/2 (instead of 1) to be triggered, which would be a big deviation from the quantum formalism. But actually, as shown in the right diagram of Fig. 7.20, the detector has probability 1 to be triggered because the attempted detection will cause collapses around $(\tau + \varepsilon, z)$, which are correlated with collapses around (τ, y) according to the stochastic evolution of ψ in such a way that (in the presence of the second detector!) $m(t, z)$ increases already at $t = \tau + \varepsilon$ instead of $t = \tau + |z - y|$. The upshot is that the empirical predictions of rGRWm for this experiment agree closely with the standard quantum formalism.

7.9 Other Approaches

7.9.1 Many-Worlds in Relativistic Space-Time

For Bohmian mechanics or GRW theory in relativistic space-time, we had to go through substantial mathematical considerations. In Everett's many-worlds theory SØ, of course, you say that no further math is needed once the time evolution of the wave function ϕ has been defined. In Schrödinger's many-worlds theory Sm, the fundamental definition of the theory in relativistic space-time, once we have ϕ, still requires a law governing the matter density function $m(x^\nu)$: it is now a 4-vector field $m_\mu(x^\nu)$ on M given by

$$m_\mu(x) = \sum_{i=1}^{N} m_i \eta_{\mu\mu_i} \int_{\Sigma^{N-1}} \Big(\prod_{k\neq i} V_{\mu_k}(d^3y_k)\Big) \overline{\psi}_\Sigma[\gamma^{\mu_1} \otimes \cdots \otimes \gamma^{\mu_N}]\psi_\Sigma \qquad (7.299)$$

with ψ evaluated at $(y_1, \ldots, y_{i-1}, x, y_{i+1}, \ldots, y_N)$ and $V_\mu(d^3x) = V(d^3x)\, n_\mu(x)$. This is the same expression as (7.298), except that $PLC(x)$ has been replaced by a Cauchy surface Σ. It turns out that this expression is independent of the choice of any Cauchy surface Σ passing through $x \in M$.

Here is a proof for non-interacting Dirac particles, showing that for each i, the integral is independent of Σ, even regardless of whether Σ passes through x: Let system a consist of particle i and system b of the other $N-1$ particles. Then the time evolution $U^{\Sigma'}_{N\Sigma} = U^{\Sigma'}_{a\Sigma} \otimes U^{\Sigma'}_{b\Sigma}$ factorizes; the reduced density matrix $\rho_a = \mathrm{tr}_b|\psi_\Sigma\rangle\langle\psi_\Sigma|$ does not change under unitaries acting on b, $\rho_a = \mathrm{tr}_b([I_a \otimes U_b]|\psi_\Sigma\rangle\langle\psi_\Sigma|[I_a \otimes U_b^\dagger])$; ρ_a defines a vector field j^μ on M because every 1-particle wave function ψ would, and ρ_a cannot be distinguished from a mixture of ψs; so, $j^\mu(x)$ is the value of the integral.

7.9.2 Wormholes as an Alternative?

Some physicists have developed thoughts in the direction that quantum nonlocality may have to do with an unusual space-time topology.[24] A "usual" space-time topology would be that of Minkowski space ($\mathbb{R}^4$) or perhaps $[0, \infty) \times \mathbb{S}^3$ as occurs in the cosmological models of Friedmann. Now general relativity allows for space-time to be curved and may in principle allow for a wormhole, i.e., a tube-shaped piece of space-time connecting two distant space-time regions. If we compare the actual geometry of a space-time $(\mathscr{M}, g)$ with a wormhole to a space-time $(\mathscr{M}', g')$ that approximates $(\mathscr{M}, g)$ but leaves out the wormhole, then two regions that are spacelike separated in $\mathscr{M}'$ may not be spacelike separated in $\mathscr{M}$ if there is a timelike curve through the wormhole connecting them but no timelike curve connecting them that does not pass through the wormhole. Now suppose that $\mathscr{M}'$ is a usual space-time such as Minkowski space or one of Friedmann's cosmological space-times and that the actual space-time $\mathscr{M}$ contains not just one but lots and lots of wormholes, say such that Paris and Tokyo are actually connected in every minute by millions of wormholes. If we imagine that many physical processes, in particular in the classical regime, do not "notice" the presence of the wormholes (i.e., are well approximated by corresponding processes in $\mathscr{M}'$), then we can imagine that

[24] Such thoughts may be inspired by the slogan "ER = EPR" of Maldacena and Susskind (2013) [35]. Here, "ER" stands for the Einstein-Rosen bridge (the first version of a wormhole), while "EPR" alludes to nonlocality.

physicists might mistake $\mathscr{M}'$ for the actual space-time geometry and regard regions as spacelike separated that in fact are not. Could that explain nonlocality?

If that proposal is right, then the laws of physics could be local after all; that is, it could be the case that influences cannot occur between spacelike separated regions, while we misjudge all the time which regions are spacelike separated because we ignore the wormholes and mistake the space-time geometry to be $\mathscr{M}'$.

However, let me point to a few limitations of the idea.

- If the nonlocal correlations of Bell experiments require wormholes, and if the wormholes have a certain size, then one should expect some finite-size effects leading to deviations from the predictions of quantum mechanics. The plausibility of the wormhole proposal may depend on whether such deviations can be empirically confirmed.
- The wormhole structure would essentially select a preferred foliation of $\mathscr{M}'$ as follows. We have discussed that in relativistic GRW theory, there is no fact about who influenced whom. In contrast, in the wormhole proposal, there is, because if two space-time points x, y that would normally (i.e., assuming $\mathscr{M}'$) be taken as spacelike separated are really timelike separated, then there is a fact about which one is in the past of the other. So for any two points $x, y \in \mathscr{M}'$, one should lie in the $\mathscr{M}$-past of the other (except for a set of space-time volume zero) because there could be nonlocal correlations between x and y. Thus, there is a foliation $\mathcal{F}$ of $\mathscr{M}'$ into $\mathscr{M}'$-spacelike 3-surfaces such that x lies in the $\mathscr{M}$-past of y whenever x lies earlier than y relative to $\mathcal{F}$. That is how the wormhole proposal singles out a preferred foliation of the macroscopic, effective space-time $\mathscr{M}'$.

  Let us compare the situation to that of Bohmian mechanics. Both involve a preferred foliation, so anybody who feels that Bohmian mechanics is not relativistic enough should have the same reservation about the wormhole proposal. At the same time, Bohmian mechanics has the advantage of being a full-fledged theory whose fundamental equations are known and correctly predict all phenomena of relativistic quantum mechanics, while the wormhole proposal is more of a program for future research. Moreover, Bohmian mechanics seems more parsimonious as it does not postulate a complicated, additional, so-far-unknown structure of many wormholes but instead uses only the wave function and the preferred foliation.
- For the wormhole idea to be convincing, one would need a really good explanation for why superluminal signaling is impossible. After all, if the influences violating Bell's inequality can propagate through wormholes, why could not signals? I do not see how such an explanation could work. For Bohmian mechanics, for comparison, we have clear-cut proofs that superluminal signaling is strictly impossible, while violations of Bell's inequality do arise.

To sum up, the wormhole idea does not seem promising.

Exercises

Exercise 7.1 can be found in Sect. 7.1.2, Exercise 7.2 in Sect. 7.2.1, Exercises 7.3 and 7.4 in Sect. 7.3.2, Exercise 7.5 in Sect. 7.3.6, Exercise 7.6 in Sect. 7.3.8, Exercise 7.7 in Sect. 7.3.9, Exercise 7.8 in Sect. 7.7.1, and Exercise 7.9 in Sect. 7.8.1.

Exercise 7.10 (Essay Question) Describe the concept of a Cauchy surface in simple terms.

Exercise 7.11 (Cauchy Surface) Draw several examples of Cauchy surfaces in 1 + 1-dimensional Minkowski space. Try to find examples that are as different from each other as possible.

Exercise 7.12 (Dirac Equation) Derive the barred Dirac equation

$$-i\partial_\mu \overline{\psi}\gamma^\mu = m\overline{\psi} \tag{7.300}$$

from the Dirac equation.

References

1. Y. Aharonov, D.Z. Albert, Can we make sense out of the measurement process in relativistic quantum mechanics?. Phys. Rev. D **24**, 359–371 (1981)
2. D. Bedingham, Relativistic state reduction dynamics. Found. Phys. **41**, 686–704 (2011). http://arxiv.org/abs/1003.2774
3. D. Bedingham, D. Dürr, G.C. Ghirardi, S. Goldstein, R. Tumulka, N. Zanghì, Matter density and relativistic models of wave function collapse. J. Stat. Phys. **154**, 623–631 (2014). http://arxiv.org/abs/1111.1425
4. J.S. Bell, Beables for quantum field theory. Phys. Rep. **137**, 49–54 (1986). Reprinted as chapter 19 of [6]. Also reprinted in: *Quantum Implications: Essays in Honour of David Bohm.* Ed. by F.D. Peat, B.J. Hiley (Routledge, London, 1987), p. 227
5. J.S. Bell, Are there Quantum Jumps?, in *Schrödinger. Centenary Celebration of a Polymath*, ed. by C.W. Kilmister (Cambridge University, Cambridge, 1987), pp. 41–52. Reprinted as chapter 22 of [6]
6. J.S. Bell. *Speakable and Unspeakable in Quantum Mechanics* (Cambridge University, Cambridge, 1987)
7. K. Berndl, D. Dürr, S. Goldstein, N. Zanghì, Nonlocality, Lorentz invariance, and Bohmian quantum theory. Phys. Rev. A **53**, 2062–2073 (1996). http://arxiv.org/abs/quant-ph/9510027
8. I. Bialynicki-Birula, On the Wave Function of the Photon. Acta Phys. Polon. **86**, 97–116 (1994)
9. D. Bohm, A Suggested Interpretation of the Quantum theory in terms of "Hidden" Variables II". Phys. Rev. **85**, 180–193 (1952)
10. D. Bohm, Comments on an Article of Takabayasi concerning the Formulation of Quantum Mechanics with Classical Pictures. Prog. Theor. Phys. **9**, 273–287 (1953)
11. D. Bohm, B.J. Hiley, *The Undivided Universe: An Ontological Interpretation of Quantum Theory* (Routledge, London, 1993)
12. S.M. Carroll, J. Chen, Does inflation provide natural initial conditions for the universe? Gen. Relativ. Gravit. **37**, 1671–1674 (2005). http://arxiv.org/abs/gr-qc/0505037. Reprinted in International Journal of Modern Physics D **14**, 2335–2340 (2005)

13. P.R. Chernoff, Essential Self-Adjointness of Powers of Generators of Hyperbolic Equations. J. Funct. Anal. **12**, 401–414 (1973)
14. J.H. Conway, S. Kochen, The Free Will Theorem. Found. Phys. **36**, 1441–1473 (2006). http://arxiv.org/abs/quant-ph/0604079
15. J.H. Conway, S. Kochen, Reply to Comments of Bassi, Ghirardi, and Tumulka on the Free Will Theorem, Found. Phys. **37**, 1643–1647 (2007). http://arxiv.org/abs/quant-ph/0701016
16. J.H. Conway, S. Kochen, The Strong Free Will Theorem. Not. Am. Math. Soc. **56**, 226–232 (2009). http://arxiv.org/abs/0807.3286
17. L. de Broglie, *Nonlinear Wave Mechanics* (Elsevier, Amsterdam, 1960)
18. D.-A. Deckert, F. Merkl, Dirac equation with external potential and initial data on Cauchy surfaces. J. Math. Phys. **55**, 122305. http://arxiv.org/abs/1404.1401 (2014)
19. J. Dimock, Dirac Quantum Fields on a Manifold. Trans. Am. Math. Soc. **269**, 133–147 (1982)
20. P.A.M. Dirac, The quantum theory of the electron. Proc. R. Soc. A **117**, 610–624 (1928)
21. F. Dowker, J. Henson, Spontaneous Collapse Models on a Lattice. J. Stat. Phys. **115**, 1327–1339 (2004). http://arxiv.org/abs/quant-ph/0209051
22. D. Dürr, S. Goldstein, K. Münch-Berndl, N. Zanghì, Hypersurface Bohm–Dirac Models. Phys. Rev. A **60**, 2729–2736 (1999). http://arxiv.org/abs/quant-ph/980107. Reprinted in [23]
23. D. Dürr, S. Goldstein, N. Zanghì, *Quantum Physics Without Quantum Philosophy* (Springer, Heidelberg, 2013)
24. D. Dürr, S. Goldstein, T. Norsen, W. Struyve, N. Zanghì, Can Bohmian mechanics be made relativistic? Proc. R. Soc. A **470**(2162), 20130699 (2014). http://arxiv.org/abs/1307.1714
25. M. Esfeld, N. Gisin, The GRW flash theory: a relativistic quantum ontology of matter in space-time? Philos. Sci. **81**, 248–264 (2014). http://arxiv.org/abs/1310.5308
26. S. Goldstein, R. Tumulka, Opposite arrows of time can reconcile relativity and nonlocality. Classical and Quantum Gravity **20**, 557–564 (2003). http://arxiv.org/abs/quant-ph/0105040
27. R.H. Good Jr., Particle aspect of the electromagnetic field equations. Phys. Rev. **105**(6), 1914–1919 (1957)
28. K.-E. Hellwig, K. Kraus, Formal description of measurements in local quantum field theory. Phys. Rev. D **1**, 566–571 (1970)
29. A. Ijjas, P.J. Steinhardt, A new kind of cyclic universe. Phys. Rev. B **795**, 666–672 (2019). http://arxiv.org/abs/1904.08022
30. C. Jones, T. Guaita, A. Bassi, On the (im)possibility of extending the GRW model to relativistic particles. Phys. Rev. A **103**, 042216 (2021). http://arxiv.org/abs/1907.02370
31. M.K.-H. Kiessling, A.S. Tahvildar-Zadeh, On the Quantum-Mechanics of a Single Photon, J. Math. Phys. **59**, 112302 (2018). http://arxiv.org/abs/1801.00268
32. M. Lienert, S. Petrat, R. Tumulka, *Multi-Time Wave Functions. An Introduction* (Springer, Heidelberg, 2020)
33. M. Lienert, R. Tumulka, Born's Rule for Arbitrary Cauchy Surfaces. Lett. Math. Phys. **110**, 753–804 (2020). http://arxiv.org/abs/1706.07074
34. S. Lill, R. Tumulka, Another Proof of Born's Rule on Arbitrary Cauchy Surfaces. Annales Henri Poincaré **23**, 1489–1524 (2022). http://arxiv.org/abs/2104.13861
35. J. Maldacena, L. Susskind, Cool horizons for entangled black holes, Fortschritte der Physik **61**, 781–811 (2013). http://arxiv.org/abs/1306.0533
36. G. Nimtz, A. Enders, H. Spieker, Photonic tunneling times. J. Phys. I **4**(4), 565–570 (1994)
37. P. Pearle, Toward a Relativistic Theory of Statevector Reduction, in *Sixty-Two Years of Uncertainty: Historical, Philosophical, and Physical Inquiries into the Foundations of Quantum Physics*, ed. by A.I. Miller (Plenum Press, New York, 1990), pp. 193–214
38. R. Penrose, W. Rindler. *Spinors and Space-time*, vol. 1 (Cambridge University, Cambridge, 1984)
39. S. Petrat, R. Tumulka, Multi-time wave functions for quantum field theory. Ann. Phys. **345**, 17–54 (2014). http://arxiv.org/abs/1309.0802
40. J. Schwinger, Quantum Electrodynamics. I. A Covariant Formulation. Phys. Rev. **74**(10), 1439–1461 (1948)
41. R.U. Sexl, H.K. Urbantke, *Relativity, Groups, Particles* (Springer, Heidelberg, 2001)

42. J.C. Slater, *Solid-State and Molecular Theory: A Scientific Biography* (Wiley, New York, 1975)
43. S. Teufel, R. Tumulka, Simple Proof for Global Existence of Bohmian Trajectories. Commun. Math. Phys. **258**, 349–365 (2005). http://arxiv.org/abs/math-ph/0406030
44. B. Thaller, *The Dirac Equation* (Springer, Berlin, 1992)
45. A. Tilloy, *Interacting Quantum Field Theories as Relativistic Statistical Field Theories of Local Beables*. Preprint. 2017. http://arxiv.org/abs/1702.06325
46. S. Tomonaga, On a relativistically invariant formulation of the quantum theory of wave fields. Prog. Theor. Phys. **1**(2), 27–42 (1946)
47. R. Tumulka, Closed 3-Forms and Random World Lines. Ph.D. thesis (Mathematics Institute, Ludwig-Maximilians-Universität, Munich, Germany 2001). http://edoc.ub.uni-muenchen.de/7/
48. R. Tumulka, A Relativistic Version of the Ghirardi–Rimini–Weber Model. J. Stat. Phys. **125**, 821–840 (2006). http://arxiv.org/abs/quant-ph/0406094
49. R. Tumulka, Comment on "The Free Will Theorem". Found. Phys. **37**, 186–197 (2007). http://arxiv.org/abs/quant-ph/0611283
50. R. Tumulka, The "Unromantic Pictures" of Quantum Theory. J. Phys. A Math. Theor. **40**, 3245–3273 (2007). http://arxiv.org/abs/quant-ph/0607124
51. R. Tumulka, A Kolmogorov Extension Theorem for POVMs. Lett. Math. Phys. **84**, 41–46 (2008). http://arxiv.org/abs/0710.3605
52. R. Tumulka, The point processes of the GRW theory of wave function collapse. Rev. Math. Phys. **21**, 155–227 (2009). http://arxiv.org/abs/0711.0035
53. R. Tumulka, A relativistic GRW flash process with interaction, in *Do Wave Functions Jump?*, ed. by V. Allori, A. Bassi, D. Dürr, N. Zanghì (Springer, Berlin, 2020), pp. 321–348. http://arxiv.org/abs/2002.00482
54. R. Tumulka, Boundary conditions that remove certain ultraviolet divergences. Symmetry **13**(4), 577 (2021)
55. R. Tumulka, *Positron Position Operators. I. A Natural Option*. Ann. Phys. **443**, 168988 (2022). http://arxiv.org/abs/2111.12304
56. H.G. Wells, *The Time Machine* (Heinemann, London, 1895)
57. E.P Wigner, Thirty Years of Knowing Einstein, in *Some Strangeness in the Proportion: A Centennial Symposium to Celebrate the Achievements of Albert Einstein*, ed. by H. Woolf (Addison-Wesley, Reading, MA, 1980)

8 Further Morals

Various morals can be drawn. I have already mentioned many throughout the book, for example, that "observables" should not be thought of as quantities but as classes of experiments; that the big divide about the foundations of quantum mechanics lies between the view that a good theory needs to describe a possible reality and the view that the attempt to describe reality is misguided, a relict of classical physics; that limitations to knowledge and control are inevitable; and that nonlocality is inevitable. But I have some more, as well as some remarks about the status of the foundations of quantum mechanics.

8.1 Controversy

One of them is that the foundations of quantum mechanics are more controversial than they should be. To many of the contentious issues, there is a surprisingly clear-cut conclusion. I think part of the controversy arises because some rather questionable assumptions, prejudices, and attitudes are firmly rooted in the traditional, orthodox view of quantum mechanics, and we would not naturally expect a majority view among scientists to go wrong. (Even if we know of historical examples where the majority of experts was wrong. Gauss did not publish his discovery of non-Euclidean geometry, apparently because he was afraid it would ruin his reputation among his contemporaries, which presumably it would have.)

8.2 Do We Need Ontology?

Sometimes students ask me a question along the following lines. In the end, all the different approaches to the quantum foundations agree about the quantum formalism, the rules for doing calculations. So, if I want to make calculations or want to make applications work, then I do not need to choose any particular approach. Then,

R. Tumulka, *Foundations of Quantum Mechanics*, Lecture Notes in Physics 1003,
https://doi.org/10.1007/978-3-031-09548-1_8

do we need to think about the possible approaches to the foundations? Do we need ontology?

Let me compare the question to, “Do we need mathematical proofs?” Do we need rigorous reasoning, or theorems, or definitions? Obviously, for many purposes, we do not. Just like we can drive a car without knowing how to build one, we can use many rules of mathematics without knowing how to prove them. And obviously, we do not always want to be fully precise when reasoning, as that would mean a burden and might turn our attention away from key aspects and toward details of little relevance. But sometimes we want to be precise. With too little precision, we might get confused and make mistakes. On top of that, even if we ourselves do not go through the proofs, we benefit if other people, maybe more specialized mathematicians, have figured out how to precisely define certain concepts, to precisely formulate certain theorems, and to precisely prove them.

The situation is similar for the foundational physical theories. To carry out an algorithm or recipe for calculating certain quantities of interest does not require understanding the justification of the algorithm or recipe. But some understanding helps with applying it properly, and too little understanding might lead to confusion or mistakes. On top of that, it is good to know that it *can* be justified.

And there is another component to this, in both the question about mathematics and the question about ontology. Some important questions only come up when you see the general definition of a mathematical concept or the general formulation of a fundamental physical law. Without thinking about the general definition, or the general theorem, or the general law, we would not ask the question, let alone be able to answer it.

8.3 What If Two Theories Are Empirically Equivalent?

“Empirical predictions” of a theory means the testable predictions for the outcomes of experiments. The empirical predictions of Bohmian mechanics and many-worlds theories agree exactly with those of orthodox quantum mechanics and very nearly with those of collapse theories such as GRW. Thus, an empirical test between these theories, an *experimentum crucis*, is impossible. They are *empirically equivalent*, and one says that the theory is empirically *underdetermined*. If several theories are empirically equivalent, then how much of a rational basis could we possibly have for preferring one over another?

A lot, actually. Already on theoretical grounds, one theory may be more plausible and convincing than another. Here are some examples:

- In Sect. 3.5.4, I described theoretical arguments against Bell’s second many-worlds theory (which of course was never seriously proposed). This theory is self-undermining: if we believed in it, we had to conclude that our belief is not justified.
- In classical mechanics, Newton’s law of gravity asserts that the gravitational force that a body exerts on another at distance r has magnitude r^{-2}. Consider

an alternative law according to which the magnitude is $r^{-2.0001}$. The difference is presumably hard to detect, so we do not actually have any empirical evidence against the alternative law, but still nobody would be inclined to believe that the alternative law is the correct law of nature. In fact, if empirical evidence showed the exponent *has* to be -2.0001, everyone would ask for a deeper explanation that does not involve the number -2.0001 in the fundamental laws.

- Let me turn to a more extreme type of example, so-called skeptical scenarios, such as the "brain in a vat" (Harman 1973 [4]). A vat is a kind of bathtub, and according to this hypothesis, your brain is presently not inside your body but is actually kept alive in a vat by some evil scientist. He uses some machines to supply oxygen and nutrients and an advanced computer connected to your nerves to provide simulated sensory data so that you do not notice you are in a vat. The data provided by the computer give you the illusion of being wherever you think you are right now and of reading this book. The point of this example is that, assuming the computer is good enough, you have no empirical way of distinguishing this situation from a normal one. If you decide to move your hand, the computer will read the signals of your motor nerves and include an appropriate motion of your hand in the simulation—the illusion is perfect. The brain-in-a-vat hypothesis is always a logical possibility; it is irrefutable and contradiction-free. But that does not mean that it is a serious possibility, or that it would be reasonable to believe this hypothesis, or that it would be reasonable to think the hypothesis is as likely as that your brain is in your body. The upshot is that

$$\textit{rational thinking is more than logic.} \tag{8.1}$$

Philosophers have long discussed skeptical scenarios. Here are two more: *solipsism* is the hypothesis that your mind alone is real, but the outside world is not, and no other person is; they all exist only in your mind; another hypothesis says that God created the universe one minute ago in just the physical state that it happened to be in one minute ago. In particular, everybody was created with memories of past events (those more than one minute ago) that actually never happened. Again, these hypotheses are logically possible and irrefutable, but not reasonable, and (8.1) applies again.

It seems hard to formulate clear, general principles for what is reasonable to believe. It may be easier to decide in concrete cases, that is, between concrete physical theories. There, the theoretical reasons should not be expected to be as compelling as for the brain-in-the-vat hypothesis, but they can very well give rational reasons for preferring one theory over an empirically equivalent one.

- Here is a somewhat artificial example of a Bohm-type theory that is empirically equivalent to the standard version of Bohmian mechanics but less convincing than it (Goldstein et al. 2005 [3]). According to the standard model of particle physics, there are various species of particles: electrons, muons, tauons, etc. Consider the theory that electrons have trajectories, but muons and tauons do not.

The muons and tauons appear as variables in the wave function ψ, but not in the actual configuration Q. For the Born distribution of Q, we consider not $\rho = |\psi|^2$ but a different, reduced distribution ρ_{red} obtained from $|\psi|^2$ by integrating out the positions of all muons and tauons. Likewise, the reduced current j_{red} is obtained from the usual current j by integrating out the muons and tauons, and as the equation of motion, we postulate

$$\frac{dQ}{dt} = \frac{j_{\text{red}}}{\rho_{\text{red}}}(t, Q_t)\,. \tag{8.2}$$

It can easily be shown that equivariance holds for ρ_{red}, so $Q_t \sim \rho_{\text{red}}(t, \cdot)$, which is the appropriate marginal of $|\psi_t|^2$. Since every macroscopic object contains electrons, its configuration will have probability distribution in agreement with $|\psi|^2$, so the theory makes exactly the same predictions for outcomes as standard Bohmian mechanics, in which all particles have trajectories.

At the same time, this artificial theory does not seem particularly convincing. Why should muons and tauons have such a different status than electrons? Would it not be simpler if all particles had trajectories? As Louis de Broglie remarked already in 1927 [1, p. 346]:

> It seems a little paradoxical to construct a configuration space with the coordinates of points that do not exist.

- In Sect. 7.3.7, I described theoretical arguments concerning the Bohm-type equation of motion (7.157) for a wave function governed by the Pauli equation versus (2.120).
- In Sect. 6.5.2, I described theoretical arguments concerning the field ontology versus the particle ontology for quantum field theory.
- In Appendix A.3, I describe Nelson's stochastic mechanics, a theory of quantum mechanics that provides a stochastic process $(Q_t)_{t\in\mathbb{R}}$ in configuration space $\mathbb{R}^{3N}$ so that $Q_t \sim |\psi_t|^2$ for every $t \in \mathbb{R}$. As I point out there, the theory is implausible for large values of the volatility parameter σ.
- If we change Bohm's equation of motion so that it is still of the form $d\boldsymbol{Q}_i/dt = \boldsymbol{j}_i/|\psi|^2$ but with a modified $\boldsymbol{j}_i$ differing from the standard expression (1.22) by addition of a term that is a curl and thus does not contribute to the continuity equation (1.21), then the modified theory will still make $|\psi|^2$ equivariant and thus entail exactly the same empirical predictions as Bohmian mechanics. Deotto and Ghirardi (1998) [2] have given expressions in ψ for such additional terms that will make the modified theory Galilean invariant. If we assume that every theory that is logically consistent and empirically adequate should have the same right to be taken seriously, then we have a problem here. However, if our goal is to guess the fundamental laws of physics, then the contrived, lengthy, unmotivated expressions considered by Deotto and Ghirardi are no serious competitor to Bohm's simple, elegant, and natural proposal.

- Even when two theories make slightly different predictions, and thus *can* in principle be tested against each other, such as Bohmian mechanics and GRW theories, we can ask about theoretical reasons for preferring one to the other. I would like to comment on some reasons concerning the comparison between Bohmian mechanics and GRW theories.
  - Some people say GRW theories are better than Bohmian mechanics or many-worlds because you can test them against standard quantum mechanics. I do not see why that would be a relevant reason.
  - Some people say it is good about GRW that the ontology is "ψ only." However, I do not think GRW∅ is satisfactory. On the contrary: much of the motivation for considering collapse theories starts out from the wish for a "ψ-only" theory, and once we realize that this promise cannot be kept, it subtracts from the appeal of collapse theories.
  - Some people say it is bad about GRW theory that it is more nonlocal than Bohmian mechanics, as GRW theory is nonlocal already in Einstein's boxes example (see Sects. 4.1.5 and 4.3.1), whereas Bohmian mechanics is not. I think this is not a strong reason.
  - It is an advantage of GRW theory over Bohmian mechanics that it does not require a preferred foliation of space-time. On the other hand, I would not dismiss Bohmian mechanics for that reason. We should weigh the different reasons in favor of one theory or another.
  - The particle ontology seems more natural than the flash or matter density ontology. Think of the de Broglie quote above. And let me tell you an anecdote.

    I was once sitting in Tim Maudlin's car while he was driving from New York to Princeton, and he asked me to explain the flash ontology, which I had published about, to his son, who was with us in the car and was perhaps 13 years old at the time. So I said something like, "It is the hypothesis that every electron and every quark, every particle in an atom, appears only very briefly and then disappears again for 100 million years. They just flash up for one instant and are gone again. They don't all flash up at the same time," I continued, "but each at a different time. Since macroscopic objects such as cars or trees consist of gigantic numbers of atoms, they still have billions of flashes every second, even though each particle shows up only once every 100 million years. While they are gone, only a kind of shadow of them exists in the wave function, which is a kind of field that keeps track of where they should appear next time."

    What this episode illustrates is how much more obvious, down-to-earth, and intuitive the particle ontology is than the flash ontology.
  - In addition, Bohmian mechanics is much simpler than GRW theories, not only conceptually but also mathematically.
- For different choices of the time foliation $\mathcal{F}$, $\text{BM}_{\mathcal{F}}$ can be regarded as different but empirically equivalent theories. After all, according to $\text{BM}_{\mathcal{F}}$, they lead to different trajectories but the same distribution of outcomes for any experiment. So, an empirical determination of $\mathcal{F}$ is out of the question. Then how much of

a rational basis can we have for believing in any particular foliation? Maybe we need to say that we just do not know which foliation it is, that it could be any. Then again, maybe the situation will change when we have a serious quantum gravity theory: maybe it will work only with certain foliations and not with others. Furthermore, it seems not unlikely that our space-time has an initial singularity, which can be regarded as a Cauchy surface Σ_0. In that case, first, it would seem to make the theory simpler if $\Sigma_0 \in \mathcal{F}$, and second, it would seem most natural that $\mathcal{F}$ consists of the surfaces of constant time-like distance from Σ_0, that is, $\mathcal{F} = \mathcal{F}_T$ in the notation of Sect. 7.5.4. $\mathcal{F}_T$ can be determined empirically, as it is approximately the foliation relative to which the cosmic microwave background (CMB) radiation is in thermal equilibrium. This foliation is also called the *rest frame* of the CMB; in the solar system, $\mathcal{F}_T$ is approximately flat, i.e., a Lorentz frame, relative to which the sun is moving at about 370 km/s in the direction of the constellation Leo.

8.4 Positivism and Realism

My high school physics textbook tried to give a positivistic definition of the concept of "force," presumably because the authors assumed that scientific style means positivistic style. So the definition tried to describe how to measure the force on an object S. The book said we should attach a spring to S and elongate it so much and in such a direction that S will move uniformly. That does not sound very practical. How do you apply it to a rolling billiard ball? It also has the disadvantage that it cannot be applied to planets, and even in cases in which it can be applied, it raises the worry whether the spring complies with Hooke's law, that is, whether we can trust that its elongation is proportional to the force.

For comparison, in the realist definition of Newtonian mechanics that I gave in Sect. 1.4.1, we could define the force exerted by particle j on particle i as simply the j-th terms on the right-hand side of the equation of motion (1.59).

Quite generally, operational statements (i.e., statements of the form "if we do X, then Y happens") tend to be complicated, even more so if we want them to be precise and to be true in generality. Let me elaborate a bit on this point.

A certain kind of operational statement we encountered was the law of operators (5.33) and the corresponding statement (5.163) about superoperators: they were made to express the effect of an experimenter's actions. However, this kind of statement already involves a lot of realism, as it talks about the apparatus wave function ϕ_A, the Hamiltonian $H_{S\cup A}$ fully characterizing the interaction between the system S and the apparatus A, and other things, as if they were known. If we want to formulate statements truly operationally, we could not refer to ϕ_A, and we would have to describe how to build the apparatus. Of course, this will increase the complexity of the statement phenomenally.

On top of that, it seems hopeless that the resulting statement could be *precise*. And on top of *that*, it will give us headaches if we want the statement to be always *true*. For example, consider the statement that a billiard ball will continue rolling in the same direction until it comes to rest or hits the cushion or hits another ball. Once you start thinking about which circumstances could lead to violations, you notice you should add clauses excluding the possibilities that the ball is spinning in addition to its rolling motion, that the billiard table is uneven, that a stream of air is blowing, that an earthquake occurs, that a truck driving by in the street causes vibrations disturbing the ball, that a player hits the ball with her queue, that a cat jumps on the table, that the roof breaks down, that the mass distribution within the ball is inhomogeneous, and more. You get the picture. Of course, all this is obvious to billiard players. But it makes the aim to formulate *laws* operationally seem hopeless. And often, it is the laws we are after.

8.5 If It Makes the Same Predictions, What Is It Good For?

I sometimes get asked what quantum theories without observers (QTWOs, such as Bohmian mechanics, GRW, and many-worlds) are good for, given that the predictions are the same as those of orthodox quantum mechanics (OQM). So let me make explicit what they are good for:

- Consider two theories making the same predictions: one clear, precise, and simple, the other unclear, vage, and incoherent. Which one is better? There is no doubt that Bohmian mechanics, GRW, and the many-worlds theories I described are clear, precise, and simple (each can be defined on less than a page). OQM is unclear (recall the Feynman quote: "Nobody understands quantum mechanics"), vague (it assumes the concepts of "observer" and "measurement" as given), and incoherent (because of the measurement problem).
- It is one of the goals of physics to find out how the world works. QTWOs provide possible answers; OQM does not.
- Bohmian mechanics is easier to learn than OQM.
- Bohmian mechanics has helped us find the right equations for absorbing boundary conditions (Sect. 5.2.4) and interior-boundary conditions (Sect. 6.4), among other things.
- QTWOs permit an analysis of quantum measurements, while they are taken as primitive and unanalyzable in OQM.
- We are often interested in issues that go beyond mere empirical predictions, such as non-locality, or how long it takes a particle to pass a potential barrier. The clear picture provided by QTWOs allows us to understand and deal with these issues.
- There are two meanings of "prediction": First, Alice describes an experimental setup and asks Bob to make a guess as to what the outcome of the observation (or the probabilities over outcomes) will be; Bob is not required to have principles, or rules, or to reason in a logical way, just to come up with the correct value; his guess will be called his "prediction." Second, Alice asks Bob to specify a complete set of laws of the universe; when Bob has submitted his set of laws, Alice will analyze what inhabitants of a hypothetical universe governed by these laws will observe in the experiment at hand, and the result will be called

Bob's "prediction." We may assume that Alice is a powerful mathematician with powerful computers and can carry out such an analysis even if it is too hard for Bob. For each of the two senses of "prediction," there are occasions when we want these kinds of prediction. The first one: when we send astronauts to Mars, we want to be confident about what is likely to happen during the trip if we construct the space ship in certain ways. It does not matter whether our prediction is based on fundamental laws or just on experience with previous flights, nor whether computations involved steps that mathematicians do not know how to justify rigorously; all that matters is that we can be confident the prediction is correct. The second one: when we want to find the fundamental laws of the universe or the correct explanation of certain phenomena, or when we want to compare two theories, then it will matter to us whether conclusions actually follow from the laws Bob stated.

8.6 Concluding Remarks

I hope to have helped the reader get a clearer understanding of quantum theory. But certainly, many questions remain open, the most profound one perhaps being "what can a clear theory of quantum field theory be like?" It can be frustrating that the full answers to our questions are not known and that perhaps they will not be found in our lifetimes. On the other hand, this means an opportunity for us and in particular for the younger generation: we have the opportunity to figure it all out. We can make progress. We live in an age of discoveries.

References

1. G. Bacciagaluppi, A. Valentini, *Quantum Theory at the Crossroads. Reconsidering the 1927 Solvay Conference* (Cambridge University, Cambridge, 2009). http://arxiv.org/abs/quant-ph/0609184
2. E. Deotto, G.C. Ghirardi, Bohmian mechanics revisited, in *Foundations of Physics*, vol. 28 (1998), pp. 1–30. http://arxiv.org/abs/quant-ph/9704021
3. S. Goldstein, J. Taylor, R. Tumulka, N. Zanghì, Are all particles real?, in *Studies in History and Philosophy of Modern Physics*, vol. 36 (2005), pp. 103–112. http://arxiv.org/abs/quant-ph/0404134
4. G. Harman, *Thought* (Princeton University, Princeton, 1973)

Appendix A

A.1 Topological View of the Symmetrization Postulate

This section is mathematically heavier. A modern view of the reasons behind the symmetrization postulate is based on the topology of the unordered configuration space ${}^N\mathbb{R}^3$ and goes back particularly to the work of J. Leinaas and J. Myrheim (1977) [18]. The symmetrization postulate is then a special case of a more general principle, according to which for any given topologically nontrivial manifold $\mathcal{Q}$, there are several quantum theories on $\mathcal{Q}$ corresponding to the one-dimensional unitary representations of the so-called fundamental group of $\mathcal{Q}$. This will be briefly summarized in this appendix. (There is also a vector bundle view of the symmetrization postulate, but that is a different story and will be told elsewhere [15, 18].)

A manifestation of the nontrivial topology of ${}^N\mathbb{R}^3$ is the fact that it is not *simply connected*. A topological space $\mathcal{Q}$ is said to be simply connected if every closed curve is *contractible*, i.e., can be continuously deformed into a point. A space that is not simply connected is also said to be *multiply connected*. For example, $\mathbb{R}^d$ is simply connected for every $d \geq 1$, whereas $\mathcal{Q} = \mathbb{R}^2 \setminus \{0\}$ is not: a curve encircling the origin cannot be contracted to a point without crossing the origin and thus leaving $\mathcal{Q}$. $\mathbb{R}^3 \setminus \{0\}$ is again simply connected because when we need to cross the origin, we can dodge it by going into the third dimension. But $\mathbb{R}^3$ without the z-axis is multiply connected. The sphere

$$\mathbb{S}^d = \{\boldsymbol{v} \in \mathbb{R}^{d+1} : |\boldsymbol{v}| = 1\} \tag{A.1}$$

is simply connected for $d \geq 2$. On a cylinder $\mathbb{R} \times \mathbb{S}^1$, closed curves that "go around the tube" cannot be contracted, whereas others can; in fact, a closed curve is contractible if and only if its so-called winding number is zero. (The winding number is the number of times, possibly negative, that the curve goes around the tube counterclockwise.) Closed curves are also called *loops*.

R. Tumulka, *Foundations of Quantum Mechanics*, Lecture Notes in Physics 1003,
https://doi.org/10.1007/978-3-031-09548-1

Example A.1 The following loop in ${}^N\mathbb{R}^3$ is not contractible: $q(t) = \{\boldsymbol{x}_1(t), \ldots, \boldsymbol{x}_N(t)\}$ for $0 \le t \le \pi$ with

$$\boldsymbol{x}_1(t) = (\cos t, \sin t, 0) \tag{A.2}$$

$$\boldsymbol{x}_2(t) = (-\cos t, -\sin t, 0) \tag{A.3}$$

$$\boldsymbol{x}_j(t) = \text{const.} \quad \forall j > 2\,, \tag{A.4}$$

say with $x_{j3} > 0$ so collisions cannot occur. It is a loop because $q(\pi) = q(0)$ as $\{\boldsymbol{e}_1, -\boldsymbol{e}_1\} = \{-\boldsymbol{e}_1, \boldsymbol{e}_1\}$.

A contraction (i.e., continuous deformation to a constant path) is impossible for the following reason. Every loop in ${}^N\mathbb{R}^3$, beginning and ending at (say) $y \in {}^N\mathbb{R}^3$, defines a permutation of y because the particles need to arrive in the same locations but may switch places. A continuous deformation will not change the permutation, as the permutation would have to jump in order to change. Therefore, loops with nontrivial permutation cannot be deformed into ones with trivial permutation, in particular cannot be contracted.

Example A.2 We show that $\mathbb{R}^{3,N}_{\neq}$ is simply connected. To begin with, $\mathbb{R}^{3N}$ is simply connected. Suppose an attempt to contract a loop in $\mathbb{R}^{3,N}_{\neq}$ intersects the collision set $\cup_{i<j}\Delta_{ij}$. Then we can dodge the intersection: Since a curve can be pulled past a point in $\mathbb{R}^3$ without intersecting it, and since Δ_{ij} has codimension 3, a curve can be pulled past Δ_{ij} in $\mathbb{R}^{3N}$ without intersecting it.

For a manifold $\mathcal{Q}$ that is not simply connected, one can algebraically characterize its way of being multiply connected by means of its *fundamental group*. It is defined as follows: Choose a point $q \in \mathcal{Q}$, and consider all loops that start and end at q. Two such loops are called *homotopic* if they can be continously deformed into each other. For example, a loop is homotopic to the constant path at q if and only if it is contractible. For example, two loops in the circle $\mathbb{S}^1$ are homotopic if and only if they have the same winding number. Homotopy is an equivalence relation; the set of equivalence classes $[g]$ of paths g is denoted $\pi_1(\mathcal{Q}, q)$ and becomes a group with the following operations. The group multiplication $[g][h] = [gh]$ is concatenation of the paths, i.e., the path obtained by first following h and then g, and the inverse of $[g]$ is obtained by following g in the opposite direction.

(If we replace g by a homotopic path g' and h by h', then the concatenation of h' and g' is homotopic to that of h and g, so that the product of the equivalence classes is independent of the choice of representative from each class. Concatenation is automatically associative (up to homotopy, where the relevant homotopy is re-parameterization). The neutral element of the group is the class of contractible loops. One can verify that the path obtained by first following g and then following it backward is contractible, thereby confirming that we have correctly identified the inverse element in the group.)

This group is called the first homotopy group or the fundamental group of $\mathcal{Q}$ based at q. For different choices q_1, q_2 of q, the groups $\pi_1(\mathcal{Q}, q_1)$, and $\pi_1(\mathcal{Q}, q_2)$ are isomorphic to each other, and any curve γ from q_1 to q_2 defines an isomorphism by first following γ backward, then any chosen loop starting at q_1, and then γ; this yields a loop starting at q_2. For example, the fundamental group of the circle $\mathbb{S}^1$ is, for any base point q, given by the additive group of the integers. In fact, the integer is the winding number, and when concatenating loops, then their winding numbers add. We report without proof that

Proposition A.1 *The fundamental group of* ${}^N\mathbb{R}^3$ *is, for any base point* q*, isomorphic to the permutation group* S_N.

Now we need to turn again to the forgetful mapping $\pi : \mathbb{R}^{3,N}_{\neq} \to {}^N\mathbb{R}^3$. A *diffeomorphism* is a bijective mapping that is smooth in both directions; π is locally a diffeomorphism: Every $\hat{q} \in \mathbb{R}^{3,N}_{\neq}$ has a neighborhood $U \subset \mathbb{R}^{3,N}_{\neq}$ such that π restricted to U is a diffeomorphism to its image $\pi(U)$ in ${}^N\mathbb{R}^3$. Even more, π is a covering map. A *covering map* is a smooth map $p : A \to B$ between manifolds such that for every $b \in B$, there exists an open neighborhood V of b such that $p^{-1}(V)$ is a union of disjoint open sets in A, each of which is mapped diffeomorphically onto V by p. For π this means that if a neighborhood V of an unordered configuration q is small enough, its pre-image $\pi^{-1}(V) = \{\hat{q} \in \mathbb{R}^{3,N}_{\neq} : \pi(\hat{q}) \in V\}$ consists of $N!$ disjoint neighborhoods, each one a neighborhood of one ordering of q.

Example A.3 The mapping $p : \mathbb{R} \to \mathbb{S}^1$ given by

$$p(\theta) = (\cos\theta, \sin\theta) \tag{A.5}$$

is a covering map. If we picture the real line as a helix above the circle (i.e., draw $\theta \in \mathbb{R}$ as the point $(\cos\theta, \sin\theta, \theta)$), then p is the projection to the $\{z = 0\}$ plane. For every interval on the circle of length less than 2π, the pre-image consists of an interval in $\mathbb{R}$ and all its translates by integer multiples of 2π.

The set $p^{-1}(q)$ is called the *covering fiber* of q. Relative to a given covering map $p : A \to B$, a *deck transformation* is a mapping $\varphi : A \to A$ is a diffeomorphism such that $p \circ \varphi = p$. That is, φ does not leave the covering fiber. For example, a deck transformation φ relative to $\pi : \mathbb{R}^{3,N}_{\neq} \to {}^N\mathbb{R}^3$ as in (6.10) must be such that at every ordered configuration $\hat{q}$, it can change the ordering but not the N points in $\mathbb{R}^3$ involved in $\hat{q}$. We report without proof that

Proposition A.2 *The deck transformations of* π *as in (6.10) are exactly the permutation mappings* $\varphi_\sigma(\boldsymbol{x}_1, \ldots, \boldsymbol{x}_N) = \sigma(\boldsymbol{x}_1, \ldots, \boldsymbol{x}_N)$ *with* $\sigma \in S_N$. *The deck transformations of* $p : \mathbb{R} \to \mathbb{S}^1$ *as in* (A.5) *are the mappings of the form* $\varphi_k(\theta) = \theta + 2\pi k$ *with* $k \in \mathbb{Z}$.

The deck transformations relative to a given covering map $p : A \to B$ form a group (the product is composition), called the *covering group*. If A is simply connected, then p is called a *universal covering* and A a *universal covering space*.

Proposition A.3 *For a universal covering, the covering group is isomorphic to the fundamental group of B at any $b \in B$. For every connected manifold B, there exist a universal covering space A and universal covering $p : A \to B$, and they are unique up to diffeomorphism (i.e., if $p : A \to B$ and $p' : A' \to B$ are both universal coverings, then there is a diffeomorphism $\chi : A \to A'$ such that $p' \circ \chi = p$).*

In the following, the universal covering space of B will be denoted $\widehat{B}$. Intuitively, $\widehat{B}$ is the "unfolding" of B that looks locally like B but removes the multiple connectedness. For example, the universal covering space of $\mathbb{S}^1$ is $\mathbb{R}$ and can be thought of as obtained by piecing together pieces of $\mathbb{S}^1$ in such a way that by going around the circle, you *do not* return to the same point. Instead, like on a spiral staircase, you arrive at the corresponding location on a different level.

Corollary A.1 *The universal covering space of ${}^N\mathbb{R}^3$ is $\mathbb{R}^{3,N}_{\neq}$, and the universal covering map is π.*

For identical particles, the wave function is defined on the universal covering space of the configuration space. Now we consider the general situation of that kind: the configuration space $\mathcal{Q}$ is a multiply connected Riemannian manifold, and ψ_t is defined on the universal covering space, $\psi_t : \widehat{\mathcal{Q}} \to \mathbb{C}$. We call the covering map $\pi : \widehat{\mathcal{Q}} \to \mathcal{Q}$; $\widehat{\mathcal{Q}}$ automatically becomes a Riemannian manifold, and the deck transformations are automatically isometries (i.e., preserve the metric). On every Riemannian manifold M, there is a natural way to define the Laplacian operator Δ and the gradient $\nabla\psi$ of a scalar function ψ; $\nabla\psi(x)$ is a complexified tangent vector at $x \in M$, $\nabla\psi(x) \in \mathbb{C}T_xM$. The relevant condition on ψ becomes particularly clear from a Bohmian perspective: we need that the velocity field v of the Bohmian equation of motion

$$\frac{dQ}{dt} = v^{\psi_t}(Q(t)) \tag{A.6}$$

is a vector field on $\mathcal{Q}$, but Bohm's formula for it,

$$\hat{v}^{\psi} = \frac{\hbar}{m}\mathrm{Im}\frac{\nabla\psi}{\psi}, \tag{A.7}$$

defines a vector field $\hat{v}$ on $\widehat{\mathcal{Q}}$. In order to be able to define

$$v(q) := D\pi|_{\hat{q}}(\hat{v}(\hat{q})) \tag{A.8}$$

in an unambiguous and consistent way, we need that $D\pi|_{\hat{q}}(\hat{v}(\hat{q}))$ is the same vector at every $\hat{q}$ in the covering fiber $\pi^{-1}(q)$. This will be the case if and only if

$$\hat{v}(\varphi(\hat{q})) = D\varphi|_{\hat{q}}(\hat{v}(\hat{q})) \tag{A.9}$$

for all $\hat{q}$ and all deck transformations φ. A natural sufficient condition on ψ ensuring (A.9) is

$$\psi(\varphi(\hat{q})) = \gamma_\varphi\, \psi(\hat{q})\,, \tag{A.10}$$

where γ_φ is a phase factor, a complex constant of modulus 1 (that depends on φ but not on $\hat{q}$) called a *topological factor*. Relation (A.10) means that the values of ψ at different points in the covering fiber are not independent of each other; it is called a *periodicity condition*. It entails that $\nabla\psi(\varphi(\hat{q})) = \gamma_\varphi\, D\varphi|_{\hat{q}}(\nabla\psi(\hat{q})$, so the factor γ_φ cancels out of (A.7), and (A.9) follows.

The periodicity condition (A.10) can only hold if

$$\gamma_{\varphi_1\circ\varphi_2}\, \psi(\hat{q}) = \psi(\varphi_1 \circ \varphi_2(\hat{q})) = \gamma_{\varphi_1}\, \psi(\varphi_2(\hat{q})) = \gamma_{\varphi_1}\, \gamma_{\varphi_2}\, \psi(\hat{q})\,, \tag{A.11}$$

so the gammas need to satisfy

$$\gamma_{\varphi_1\circ\varphi_2} = \gamma_{\varphi_1}\, \gamma_{\varphi_2}\,. \tag{A.12}$$

Together with $\gamma_{\mathrm{id}} = 1$, this means that the gammas form a one-dimensional *group representation* of the covering group. Since we assumed $|\gamma_\varphi| = 1$, it is in fact a *unitary* representation (by unitary 1×1 matrices). Such representations are also called *characters*. Since the covering group is isomorphic to the fundamental group of $\mathcal{Q}$, the characters of the covering group can be translated to the characters of the fundamental group. The upshot of the reasoning can be summarized in the following principle.

Character Quantization Principle (Leinaas and Myrheim (1977) [18], Dürr et al. (2007) [10]) *For quantum mechanics on a multiply-connected Riemannian manifold $\mathcal{Q}$, there are several possible types of wave functions, each corresponding to a character γ of the fundamental group $\pi_1(\mathcal{Q})$.*

Now let us apply this to the case of identical particles.

Proposition A.4 *For $N \geq 2$, the permutation group S_N (which is the fundamental group of ${}^N\mathbb{R}^3$) has exactly two characters: the trivial character $\gamma_{\varphi_\sigma} = 1$ and the alternating character $\gamma_{\varphi_\sigma} = (-)^\sigma$.*

For identical particles, the character quantization principle applied to the unordered configuration space yields that there are two possible theories of identical particles, one requiring

$$\psi(\sigma\hat{q}) = \psi(\hat{q}) \tag{A.13}$$

and one requiring

$$\psi(\sigma\hat{q}) = (-)^{\sigma}\, \psi(\hat{q})\,. \tag{A.14}$$

Obviously, (A.13) is a bosonic wave function and (A.14) a fermionic one. So, we obtain the boson-fermion alternative as a special case of the character quantization principle.

A.2 Ultraviolet Divergence Problem in Classical Electrodynamics

Here, I will say a bit more about the inconsistency of classical electrodynamics that I mentioned already in Sect. 7.1.6. If we write the relativistic equation of motion (7.23) for N classical point particles in Minkowski space-time involving the Lorentz force and the Maxwell equation (7.25) with source terms (7.26) given by the N particles, we run into a mathematical inconsistency because the Maxwell field diverges to ∞ where we need to evaluate for computing the Lorentz force. This inconsistency is often swept under the rug in textbooks. What are the possible ways out?

A straightforward and simple cure is to introduce N Maxwell fields, one for each particle, and postulate that the i-th field F_i feels only the source term stemming from the i-th particle, while the Lorentz force on the i-th particle involves all fields but the i-th. In this way, the divergent part has been removed from the equations, and a well-defined time evolution law for fields and particles has been obtained. This law may seem odd because we normally think of *one* Maxwell field, not of N Maxwell fields. On the other hand, we may be willing to tolerate some degree of oddness in a classical theory, also because we know that all classical theories are empirically wrong anyway. But it gets odder: As discovered by Deckert and Hartenstein (2016) [8], the well-defined equations with N fields develop shock fronts for generic initial data (discontinuities and Dirac delta functions along the future light cones of the points where the particle world lines intersect the $t = 0$ surface). We do not see such shock fronts empirically, and it does not seem believable that such shock fronts exist in reality (and would somehow remain unobservable).

There is an idea that takes care of both oddities at once: Liénard-Wiechert fields. The idea is that the field can be computed from the particle world line, so not all mathematical possibilities of initial data for F_i at $t = 0$ actually represent physical possibilities. There would be a formula for computing the field from the world line, and there is a natural and convincing formula for this due to Liénard and Wiechert, which expresses the field generated by a moving charge (and which I write down below). This picture means that the fields are not really dynamical degrees of freedom but that they are just a mathematical bookkeeping device encoding certain information about the world lines. It removes the oddness of N fields because it makes sense anyway to talk about the field generated by each particle. And the Liénard-Wiechert fields are free of shock fronts. On top of that, the Liénard-Wiechert fields also remove a third potential oddness: the Maxwell equation in Minkowski space-time allows waves coming from infinity, waves that were never generated by any charges but were included in the initial data. Liénard-Wiechert fields exclude waves coming from infinity.

However, there are also drawbacks with the Liénard-Wiechert fields. First, there is a 1-parameter family of them, as there are retarded Liénard-Wiechert fields, advanced Liénard-Wiechert fields, and convex combinations thereof; so we need to make a choice. Here are the formulas: For a point charge with electric charge e and time-like world line $Y^\mu(\tau)$ parametrized by proper time, the *retarded Liénard-Wiechert potential* at space-time point x is

$$A_\mu^{\text{ret}}(x) = \frac{e}{4\pi}\left(\frac{dY_\nu}{d\tau}(\tau_{\text{ret}})\big(x^\nu - Y^\nu(\tau_{\text{ret}})\big)\right)^{-1}\frac{dY_\mu}{d\tau}(\tau_{\text{ret}}), \tag{A.15}$$

where τ_{ret} is the unique parameter value for which $Y(\tau_{\text{ret}})$ lies on the past light cone of x. The *retarded Liénard-Wiechert field* is

$$F_{\mu\nu}^{\text{ret}} = \partial_\mu A_\nu^{\text{ret}} - \partial_\nu A_\mu^{\text{ret}}. \tag{A.16}$$

For the definition of the *advanced* Liénard-Wiechert potential and field, replace τ_{ret} by τ_{adv}, where τ_{adv} is the unique parameter value for which $Y(\tau_{\text{adv}})$ lies on the *future* light cone of x.

The theory using only the retarded Liénard-Wiechert fields breaks time reversal symmetry on the fundamental level, a symmetry respected by the Maxwell equation and all other serious classical theories. That would be odd. Note that in a theory with initial conditions for the fields, it would be less severe if fields in the distant past have different features (perhaps lower entropy) from those in the distant future; it is more severe if temporal asymmetry is not in the initial conditions but built into the fundamental dynamical laws. There is only one parameter choice that preserves time reversal invariance, and that is the $\frac{1}{2}$ retarded + $\frac{1}{2}$ advanced Liénard-Wiechert field, and the theory with this parameter choice is known as Wheeler-Feynman electrodynamics (Wheeler and Feynman 1945 [26], 1949 [27]).

However, and that is the second drawback, every Liénard-Wiechert field other than the retarded one (i.e., every one that involves any small dose of the advanced one) involves a kind of retrocausation (effects that precede the causes). This is not mathematically impossible, and I think it is interesting to explore and analyze such theories, but I would say it is the oddest of all the oddities we have faced so far in this section.

Another consideration comes from the fact that our space-time is not Minkowski space-time. Presumably it is roughly similar to one of the Friedmann space-times, it has a beginning in time, and its Cauchy surfaces have finite 3-volume. In such a space-time, Liénard-Wiechert fields are much less convincing than in Minkowski space-time: retarded Liénard-Wiechert fields, for example, vanish at spacelike separation from the end points of the world lines on the initial singular surface. (They also have shock fronts on the future light cones of these points, but that alone is not the point; the Liénard-Wiechert fields seem more fundamentally inappropriate here.) Moreover, in the relevant Friedmann space-times, there cannot be waves coming from infinity (because, technically speaking, the "past null infinity" is empty). One could say that there are waves coming out of the initial singularity (the big bang), but that does not seem very odd; after all, if particles can come out of the initial singularity, then why not radiation?

So let us go back to before we considered Liénard-Wiechert fields; let us go back to N-field theory with initial data for the fields. Also in Friedmann space-times, generic initial fields would develop shock fronts on the future light cones of the initial end points of world lines. On the other hand, these fronts lead to kinks in the world lines every time a world line crosses the future light cone of the initial end point of another, and it seems not impossible that such kinks might be unobservable; theoretical physicists might come to tolerate these kinks and these shock fronts even if they seem odd.

Then there is another aspect to consider. Since we observe the cosmic microwave background (CMB) radiation, which is approximately thermal, it seems like an obvious possibility that random radiation might also come out of the initial singularity (even though the CMB radiation does not directly come out of the initial singularity). Let me call that random primordial radiation (RPR). It does a particularly good job at removing our sense of oddness about radiation that was never generated by charges. And RPR might well contain so many shock fronts that the shock fronts on the future light cones of the initial end points of the world lines do not stand out any more. On the other hand, there seem to be difficulties with defining an appropriate probability distribution for RPR (an ultraviolet catastrophe) which might become better in quantum electrodynamics.

An alternative idea for obtaining a consistent classical, relativistic theory of point charges is to replace Maxwell's equations (7.25) by a different set of equations. Born and Infeld (1934) [4] have proposed a particular set of nonlinear equations, which however still has problems of its own; see Kiessling (2011) [17] for details and references.

To sum up, it is still not clear, at least to me, whether there exists a satisfactory theory of classical electrodynamics.

A.3 Nelson's Stochastic Mechanics

A theory similar to Bohmian mechanics was proposed by Edward Nelson (1932–2014) (1966 [20], 1985 [21]) and is known under the name *stochastic mechanics*; see also Goldstein (1987) [13]. The theory uses a particle ontology but replaces Bohm's deterministic equation of motion by a stochastic law of motion that can be written in the form

$$dQ_t = u^{\psi_t}(Q_t)\,dt + \sigma\,dW_t\,. \tag{A.17}$$

Before I explain this equation, let me say something about the solution $t \mapsto Q(t)$. It is a continuous curve in configuration space $\mathbb{R}^{3N}$, and it is a realization of a *stochastic process*, which means that random decisions are made during the motion, in fact continuously in time, so that $Q(t_1)$ does not fully determine $Q(t_2)$ at any $t_2 > t_1$. This process is designed in such a way that, at every time t,

$$Q(t) \sim |\psi_t|^2\,. \tag{A.18}$$

The type of motion of Q is called a *diffusion process*. The simplest and best known diffusion process is the *Wiener process* W_t. The Wiener process in 1d can be obtained as the limit $\Delta t \to 0$ of the following *random walk* X_t: In each time step Δt, let X_t move either upward or downward, each with probability 1/2, by the amount $\sqrt{\Delta t}$ (see Fig. A.1), so

$$X_{t+\Delta t} = X_t \pm \sqrt{\Delta t} \tag{A.19}$$

for $t \in \Delta t\mathbb{Z}$. For times between t and $t + \Delta t$, we may keep X constant (so it jumps at t and $t + \Delta t$) or define it to increase/decrease linearly (and thus be continuous but have kinks at t and $t + \Delta t$ as in Fig. A.1); both choices will converge to the same trajectory $t \mapsto W_t$ in the limit $\Delta t \to 0$. It turns out that the trajectory $t \mapsto W_t$ is everywhere continuous but nowhere differentiable; its velocity is, so to speak, always either $+\infty$ or $-\infty$; the trajectory is a very jagged curve reminiscent of the prices at the stock market.

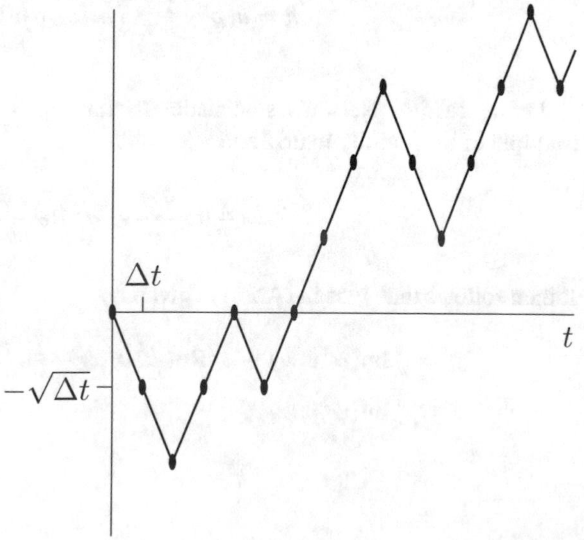

Fig. A.1 Realization of a random walk. The Wiener process is the limit $\Delta t \to 0$ of such a process

Now a diffusion process is a deformed Wiener process; it is the limit $\Delta t \to 0$ of a deformed random walk given by

$$X_{t+\Delta t} = X_t + u(t, X_t)\,\Delta t \pm \sigma(t, X_t)\,\sqrt{\Delta t}\,, \tag{A.20}$$

where the value $\sigma \geq 0$, called the *diffusion constant* or *volatility*, characterizes the strength of the random fluctuations and u, called the *drift*, represents a further contribution to the motion that would remain in the absence of randomness. (The σ has nothing to do with the collapse width of GRW theory.) A diffusion process can thus have a tendency to move in a certain direction and to fluctuate less in some regions than in others. The realizations of diffusion processes are still very jagged curves; in fact, stock prices are often modeled using diffusion processes. A diffusion process is characterized by its drift and volatility, and therefore a common notation for it is

$$dX_t = u(t, X_t)\,dt + \sigma(t, X_t)\,dW_t\,, \tag{A.21}$$

where $dX_t = X_{t+dt} - X_t$ and correspondingly dW_t for a Wiener process. The Wiener process and diffusion processes exist in any dimension d; then u becomes a d-dimensional vector and σ a $d \times d$ matrix. Equations of the type (A.21) are called *stochastic differential equations*. The probability density ρ_t of X_t evolves with time t according to the *Fokker-Planck equation*

$$\frac{\partial\rho}{\partial t} = -\sum_{i=1}^{d} \partial_i\big[u_i\rho\big] + \tfrac{1}{2}\sum_{i,j,k=1}^{d} \partial_i\partial_j\Big[\sigma_{ik}\sigma_{jk}\rho\Big] \tag{A.22}$$

with $\partial_i = \partial/\partial x_i$, a version of the continuity equation with probability current

$$j_i = u_i\rho - \tfrac{1}{2}\sum_{jk}\partial_j[\sigma_{ik}\sigma_{jk}\rho]. \tag{A.23}$$

In the case of Nelson's stochastic mechanics, σ is chosen to be a constant multiple of the identity matrix and

$$u^\psi = \frac{\hbar}{m}\mathrm{Im}\frac{\nabla\psi}{\psi} + \sigma^2\,\mathrm{Re}\frac{\nabla\psi}{\psi}\,. \tag{A.24}$$

It then follows that j_i as in (A.23) is given by

$$j_i = \frac{\hbar}{m}\mathrm{Im}(\psi^*\partial_i\psi) + \sigma^2\mathrm{Re}(\psi^*\partial_i\psi) - \tfrac{1}{2}\sigma^2\partial_i(\psi^*\psi) \tag{A.25}$$

$$= \frac{\hbar}{m}\mathrm{Im}(\psi^*\partial_i\psi)\,. \tag{A.26}$$

It now follows further from the Schrödinger equation and the Fokker-Planck equation (A.22) that $|\psi|^2$ is preserved, (A.18). As a consequence, the empirical predictions of stochastic mechanics agree with the quantum formalism. Furthermore, it solves the quantum measurement problem in the same way as Bohmian mechanics. Nelson proposed the value $\sigma = \sqrt{\hbar/m}$, but actually any value yields a possible theory, so stochastic mechanics provides a 1-parameter family of theories. For $\sigma = 0$, we obtain Bohmian mechanics, and in the limit $\sigma \to \infty$, the fluctuations become so extreme that Q_{t+dt} is independent of Q_t, so we obtain Bell's second many-world theory (see Sect. 3.5.4).

A remarkable fact is that Bohmian mechanics and stochastic mechanics are two theories that make *exactly* the same empirical predictions; they are *empirically equivalent*. Well, one could also say that orthodox quantum mechanics and Bohmian mechanics make exactly the same predictions; on the other hand, orthodox quantum mechanics is not a theory in the sense of providing a possible way the world may be, so perhaps the example of empirical equivalence provided by stochastic mechanics is a more serious one. This equivalence means that there is no experiment that could test one theory against the other.

This brings us once more to the question (as in Sect. 8.3) of how to decide between these two theories if there is no way of proving one of them wrong while keeping the other. Let me make a few remarks about theoretical grounds for a decision in this case. When we try to extend the theories to relativistic space-time, quantum electrodynamics, or quantum gravity, one theory might fare better than the other. But already in non-relativistic quantum mechnics, one might be simpler, more elegant, or more convincing than the other. For example, solutions of ODEs are mathematically a simpler concept than diffusion processes. Here is another example: In a macroscopic superposition such as Schrödinger's cat, the wave function is not exactly zero in configuration space between the regions corresponding to a live and a dead cat, and as a consequence for large diffusion constant σ, Nelson's configuration Q_t will fluctuate a lot (i.e., move around intensely and erratically) and also repeatedly pass through regions of small $|\psi|^2$; in fact, it will repeatedly switch from one packet to another and thus back and forth between a live and a dead cat. Resurrections are possible, even likely, even frequent, if σ is large enough. As discussed in Sect. 3.5.4 in the context of Bell's second many-worlds theory, we may find that hard to believe and conclude that smaller values of σ are more convincing than larger ones. In Bohmian mechanics, in fact, the configuration tends to move no more than necessary to preserve the $|\psi|^2$ distribution. These reasons contribute to why Bohmian mechanics seems more attractive than stochastic mechanics.

Nelson's motivation for stochastic mechanics was a different one. He defined a "stochastic derivative" of the non-differentiable trajectories, found that it satisfies a certain equation, and hoped that the process Q_t could be characterized without the use of wave functions and the Schrödinger equation. Nelson hoped that the Schrödinger equation would not have to be postulated but would somehow come out. However, these hopes did not materialize, and the only known way to make sense of stochastic mechanics is to assume, as in Bohmian mechanics, that the wave function exists as an independent object and guides the configuration. But then the theory has not much of an advantage over Bohmian mechanics.

A.4 Probability and Typicality in Bohmian Mechanics

A.4.1 Empirical Distributions in Bohmian Mechanics

The Law of Large Numbers

The *law of large numbers* is a theorem, or a family of theorems, in probability theory. The basic form asserts that after many independent trials of a random experiment, the relative frequency of each outcome z is close to the probability p_z of z in a trial. In more detail:

Law of Large Numbers in Probability Theory (Simple Version) *If $n \in \mathbb{N}$ is large, if $Z_1, \ldots, Z_n$ are independent random variables with values in $\{1, \ldots, r\}$, and if each has the same probability distribution $(p_1, \ldots, p_r)$ (i.e., $\mathbb{P}(Z_k = z) = p_z$), then the relative frequency of each z in the sample $(Z_1, \ldots, Z_n)$,*

$$F_z = \frac{\#\{k : Z_k = z\}}{n}, \tag{A.27}$$

is close to p_z with probability close to 1.[1]

The values $(p_1, \ldots, p_r)$ are called the *theoretical distribution*, the random values $(F_1, \ldots, F_r)$ are called the *empirical distribution*. We also write $n \gg 1$ for "n is large." So the law of large numbers can be paraphrased as, *for $n \gg 1$, the empirical distribution is close to the theoretical one with probability near 1*. An immediate consequence for Bohmian mechanics is this:

Law of Large Numbers in Bohmian Mechanics (Simple Version) *Suppose $n \gg 1$ systems are disentangled, each has wave function $\varphi : \mathbb{R}^d \to \mathbb{C}^k$ (so jointly they have wave function $\psi = \varphi^{\otimes n}$), they have configuration $X = (X_1, \ldots, X_n) \in (\mathbb{R}^d)^n$, $X \sim |\psi|^2$, the sets $A_1, \ldots, A_r$ form a partition of $\mathbb{R}^d$, $p_z := \int_{A_z} |\varphi|^2$, and Z_k is the cell containing X_k (i.e., $Z_k = z$ if and only if $X_k \in A_z$). Then the empirical distribution is close to $(p_1, \ldots, p_r)$, that is, to the $|\varphi|^2$ distribution, with probability near 1.*

Of course, when we use probabilities in practice, we use the law of large numbers all the time. For example, we determine whether a coin is fair or loaded by tossing

---

[1] The precise formulation of this statement says that for every $\varepsilon > 0$, every $\delta > 0$, every $r \in \mathbb{N}$, and every $(p_1, \ldots, p_r) \in [0, 1]^r$ with $p_1 + \ldots + p_r = 1$, there is $n_0 \in \mathbb{N}$ such that for all $n \geq n_0$, $\mathbb{P}\big(|F_z - p_z| < \varepsilon \,\forall z \in \{1, \ldots, r\}\big) \geq 1 - \delta$. It is furthermore possible to give an explicit expression for an n_0.

it many times. The same is true in Bohmian mechanics. For example, when we test the Born rule in the double-slit experiment, we compare the Born distribution to, say, the empirical distribution of $n = 70{,}000$ arrival points of Fig. 1.4.

Without a law of large numbers, the comparison would not be possible. Let me say an obvious thing explicitly: When we postulated that the initial configuration of the universe Q_0 is $|\Psi_0|^2$ distributed with Ψ_0 the initial wave function of the universe, and concluded that Q_t is $|\Psi_t|^2$ distributed, we were talking about a distribution over an ensemble of universes, all with the same wave function but different configurations. But only one universe is accessible to us, so the only distributions we ever see are *empirical* distributions.

If we are more careful about the law of large numbers, we notice that Tonomura's experiment, shown in Fig. 1.4, did not actually involve 70,000 non-interacting electrons at the same time, but rather only one electron at a time, which were detected on the screen at 70,000 different times. Maybe the experiment could even be done using the same electron again and again. And although it seems plausible that neither the different times nor using the same electron should change the empirical distribution, the question arises whether this actually follows from Bohm's equation of motion. So we want a more general version of the law of large numbers in Bohmian mechanics. The following one is an immediate consequence of equivariance and the law of large numbers from probability theory:

Law of Large Numbers in Bohmian Mechanics (Refined Version) *Suppose the wave function Ψ of a lab (including quantum objects and apparatus) splits repeatedly, at times $t_1 < \ldots < t_n$ with $n \gg 1$, into disjoint branches (as in Fig. 3.8), in such a way that*

1. *(no overlap) For $t \in [t_k, T]$ (with $T > t_n$ thought of as long after completion of the experiment),*

$$\Psi(t) = \sum_{z_1 \ldots z_k = 1}^{r} \Psi_{z_1 \ldots z_k}(t) \tag{A.28}$$

with $\Psi_{z_1 \ldots z_k}(t)$ supported (up to negligible tails) by sets $B_{z_1 \ldots z_k}(t) \subset \mathbb{R}^{3N}$ that are mutually disjoint, $B_{z'_1 \ldots z'_k}(t) \cap B_{z_1 \ldots z_k}(t) = \emptyset$ for $(z'_1 \ldots z'_k) \neq (z_1 \ldots z_k)$.

2. *(branching) For $t \in [t_{k+1}, T]$,*

$$\Psi_{z_1 \ldots z_k}(t) = \sum_{z_{k+1} = 1}^{r} \Psi_{z_1 \ldots z_{k+1}}(t)\,. \tag{A.29}$$

3. *(stay in branch) If $Q(t) \in B_{z_1 \ldots z_k}(t)$ for $t = t_k$ then also for all $t \in [t_k, T]$ (up to negligible probability).*
4. *(theoretical distribution) $\|\Psi_{z_1 \ldots z_{k+1}}(t_{k+1})\|^2 = p_{z_{k+1}} \, \|\Psi_{z_1 \ldots z_k}(t_{k+1})\|^2$.*

Then Q selects a branch, i.e., there are $Z_1, \ldots, Z_n \in \{1, \ldots, r\}$ such that $Q(t) \in B_{z_1 \ldots z_k}(t)$ for $t \in [t_k, T]$, and for this branch, the empirical distribution (A.27) *is close to the theoretical one,*

$$F_z \approx p_z \tag{A.30}$$

for all $z \in \{1, \ldots, r\}$, with probability near 1.

Again, the proof is trivial; what this statement does is make explicit certain sufficient ingredients for agreement between the empirical and the theoretical distribution: branching wave function and the trajectory staying in a branch (which is very plausible).[2] This kind of statement was presumably clear to Bohm in 1952 and, as a statement just about $|\Psi(t_n)|^2$, to Everett in 1957.

General Statement

There are several reasons why a further refined and more general statement is desirable. We may use different times for the experiments in different branches or do different experiments in different branches. Since the lab will actually not be isolated up to time T, we may want to use the wave function of the universe instead of the wave function of the lab; but then in some branches, the experiment does not even get carried out, so we may want to conditionalize on the event $\mathcal{M}$ that it does get carried out. More importantly, we may want to get rid of the issues of tails of wave packets and of whether Q will stay in a branch.

A solution in terms of conditional wave functions was developed by Dürr, Goldstein, and Zanghì (1992) [9]. To this end, let us separate the problem of deriving statements about the empirical distribution into two steps: first, statements about the joint theoretical distribution of positions of n objects (e.g., pointers) at various times and, second, conclusions from the joint theoretical distribution about properties that the empirical distribution will have with probability near 1. Since probability theory and statistics have standard results about the latter (of which the law of large numbers in probability theory is the first and simplest), let us focus on the first kind of statement. The following statement about the distribution is exact, regardless of whether n is large or not.

Theorem A.1 ([9, Sec. 10]) *Suppose that*

1. *On the space $\mathbb{R}^{3N}$ of initial configurations Q_0 of a Bohmian N-particle universe, functions $t_1(Q_0), \ldots, t_n(Q_0)$ are defined such that $t_1 \leq \ldots \leq t_n$.*

[2] Bohm's equation of motion tends to require as little motion as possible while maintaining equivariance. For comparison, Nelson's stochastic mechanics (see Appendix A.3) tends to move more; also, Bohm's equation of motion is the continuum limit of Bell's minimal jump rate (see Sect. 6.3.2), which in a sense leads to as few jumps as possible.

2. *For each* $k \in \{1, \ldots, n\}$, π_k *is a subset of* $\{1, \ldots, N\}$ *depending on* $Q_0 \in \mathbb{R}^{3N}$. *Set* $X_k = (\boldsymbol{Q}_i(t_k))_{i \in \pi_k}$ *and* $Y_k = (\boldsymbol{Q}_i(t_k))_{i \notin \pi_k}$.
3. *For every* $\pi \subset \{1, \ldots, N\}$, $t \geq 0$, *and* $Q_0 \in \mathbb{R}^{3N}$,

$$1_{\pi_k(Q_0)=\pi}\, 1_{t_k(Q_0)=t} = f\Big((\boldsymbol{Q}_i(t))_{i \notin \pi} \Big) =: f\big(Y_{\notin \pi}(t)\big) \tag{A.31}$$

for some function f; *that is, it can be read off from* $Y_{\notin \pi}(t)$ *whether* $\pi_k = \pi$ *and* $t_k = t$. *Whenever* $\pi_k = \pi$ *and* $t_k = t$, *let the conditional wave function of system* k *be*

$$\psi_{\pi_k, t_k}(x) = \Psi(t, x, Y_{\notin \pi}(t)) \,. \tag{A.32}$$

4. $\psi_{\pi_k, t_k}(x)$ *is deterministic,* $\psi_{\pi_k, t_k}(x) = \varphi_k(x)$ *for all* $k \in \{1, \ldots, n\}$ *and all* $Q_0 \in \mathbb{R}^{3N}$.
5. X_k *is encoded in* $Y_{k'}$ *for every* $k' > k$, *i.e.,* $X_k = f_{kk'}(Y_{k'})$ *for some function* $f_{kk'}$.
6. *The subset* $\mathcal{M} \subseteq \mathbb{R}^{3N}$ *is such that* $\int_{\mathcal{M}} |\Psi_0|^2 > 0$, *and for every* $k \in \{1, \ldots, n\}$,

$$1_{Q_0 \in \mathcal{M}} = f\Big((\boldsymbol{Q}_i(t_k(Q_0)))_{i \notin \pi_k(Q_0)} \Big) \tag{A.33}$$

for some function f; *that is, it can be read off from any* Y_k *whether* $Q_0 \in \mathcal{M}$.

Then, as a generalization of the fundamental conditional probability formula (5.57),

$$\mathbb{P}\big(X_k \in dx_k \big| X_1, \ldots, X_{k-1}, \mathcal{M}\big) = |\varphi_k(x_k)|^2 \, dx_k \tag{A.34}$$

for every $k \in \{1, \ldots, n\}$, *and therefore*

$$\mathbb{P}\Big(X_1 \in dx_1, \ldots, X_n \in dx_n \Big| \mathcal{M}\Big) = |\varphi_1(x_1)|^2 \cdots |\varphi_n(x_n)|^2 \, dx_1 \cdots dx_n \,. \tag{A.35}$$

A.4.2 Typicality

In Sect. 1.6.1, I said the initial configuration Q_0 of the universe *looks as if chosen randomly* with distribution $|\Psi_0|^2$, where Ψ_0 is the initial wave function of the universe. Here is some elucidation of this wording. As a first version, we may assume that Q_0 *gets* chosen randomly with distribution $|\Psi_0|^2$. That is how we would simulate a Bohmian universe on a computer. But in a sense, the statement is too strong: Suppose that a future (empirically adequate) theory of physics requires that Q_0 is exactly a particular point in $\mathbb{R}^{3N}$, say the unique solution of a certain equation. Then Q_0 would not at all *be random*, but it would still be completely possible that Q_0 *looks random.*

How can it be that something looks random if it is not random? Easily! Think, for example, of the decimal digits of π,

$$31415926535897932384626 43\ldots. \tag{A.36}$$

After the first few digits that most of us recognize, the sequence of digits looks random, although obviously it *is* not random. It is hard to give a sharp definition of what "looking random" means, but even without a sharp definition, we understand the meaning. As a substitute for a sharp definition, we could use criteria such as this: the relative frequency of each digit (such as 7) in the sequence converges to 1/10, the relative frequency of each pair of digits (such as 41) in the sequence converges to 1/100, and so on. Let me call this the frequency property. While it has not been proven that π has the frequency property, it seems very persuasive and is consistent with numerical evidence that it does. What can be proven, on the other hand, is that *almost all* numbers in (say) the interval [0, 10] have the frequency property, i.e., all except a set of measure 0; the rational numbers belong to the exceptions, of course. One says that a property is *typical* if all points except a set of very small measure have it; so the frequency property is typical in [0, 10].

There are typical properties that π does not have; for example, the property of x that there exists no algorithm that computes, upon input n, the first n decimal digits of x. (This property is typical because there are only countably many algorithms, and every countable set has measure 0.) Thus, there is an ambiguity as what the word "look" in "looking random" refers to. In addition, a sharp definition of which sequences of digits look random appears even less achieveable, less reasonable, and less desirable when we consider *finite* sequences. In practice, we get along well without a sharp definition.

For the application to Bohmian mechanics, relations such as $F_z \approx p_z$ define a sense of "looking like Born distributed"; then the various laws of large numbers tell us that the empirical observations will obey $F_z \approx p_z$ "with probability near 1" if we think of Q_0 as randomly chosen. This means, in other words, that the set S_0 of Q_0s for which, at the end of our experiment, $F_z \approx p_z$ with certain error bounds (i.e., the set S_0 for which our experiment confirms Born's rule) is such that

$$\int_{S_0} |\Psi_0|^2 \approx 1\,. \tag{A.37}$$

Put differently, the property "$F_z \approx p_z$" is typical. Further relevant typical properties include that repeated runs of the double-slit experiment look like *independent* $|\varphi|^2$-distributed random points or, more generally, that the outcomes of experiments look random with the distribution as given by the quantum formalism or may concern macroscopic behavior such as that at a certain age of the universe, hydrogen atoms form (or, at another age, stars and galaxies).

So, what we assume about Q_0 is that $Q_0 \in S_0$ for the relevant typical properties. And that is ultimately meant by saying that Q_0 is typical relative to $|\Psi_0|^2$ or that Q_0 looks as if randomly chosen with distribution $|\Psi_0|^2$.

A.4.3 The Explanation of Quantum Equilibrium

The $|\varphi|^2$ distribution is also called the *quantum equilibrium distribution* because it is equivariant under the time evolution of Q_t, in analogy to the *thermal equilibrium distributions* in statistical mechanics, which are invariant under the time evolution of the phase point. There is a crucial disanalogy, however, between the two: while some of the relevant systems in nature are in thermal equilibrium and others are not, there is no evidence of any system in nature that, if we take Bohm's equation of motion for granted, is not in quantum equilibrium. Such a system would violate Born's rule and would thus provide an empirical falsification (i.e., refutation) of the quantum formalism. No such system is known to date.

Some authors, specifically Bohm (1952b) [2] and Valentini and co-authors (e.g., [25]), had the thought that the quantum equilibrium distribution $\rho = |\varphi|^2$ requires an explanation. Why, of all the infinitely many possible ρ functions, should we find exactly $|\varphi|^2$? They thought that a satisfactory explanation would be provided by a mechanism that would push an arbitrary ρ closer and closer to $|\varphi|^2$. And such a mechanism exists indeed: if the Bohmian motion is sufficiently chaotic, which is presumably the case for sufficiently complicated systems and wave functions, then ρ_t as transported by the Bohmian motion tends to get closer over time to $|\varphi_t|^2$. The relevant chaoticity property of a dynamical system is called *mixing*; it is a stronger variant of being *ergodic* and has been studied particularly in statistical mechanics in connection with invariant distributions instead of equivariant distributions. Valentini and Westman (2005) [25] have provided numerical simulations illustrating how, for some model systems of a single Bohmian particle in a potential and some model wave functions φ, rather arbitrary initial density functions ρ_0 evolve toward $|\varphi_t|^2$ over time.

And yet, something is wrong with this explanation of quantum equilibrium; I call it the *illusion of justifying equilibrium*. To explain it, let me begin with the Maxwellian distribution of velocities. As James Clerk Maxwell (1831–1879) discovered in 1860, the velocities $\boldsymbol{v}$ of molecules in a gas in thermal equilibrium are distributed with density

$$\rho(\boldsymbol{v}) = \mathcal{N}\, e^{-\boldsymbol{v}^2/2\sigma^2}, \tag{A.38}$$

a Gaussian distribution with normalizing factor $\mathcal{N}$ and width $\sigma = \sqrt{kT/m}$ (where $k = 1.38 \times 10^{-23}$ J/K is the Boltzmann constant, T the absolute temperature, and m the mass of a molecule). Ludwig Boltzmann gave the following explanation of this distribution in the framework of Newtonian mechanics: Consider the phase space of the gas; let us assume that the gas consists of N molecules and that each molecule is a point particle, so a phase point corresponds to listing the position $\boldsymbol{x} \in \Lambda \subset \mathbb{R}^3$ and velocity $\boldsymbol{v} \in \mathbb{R}^3$ of each molecule, where Λ is the available volume (the gas container). So, the phase space is $(\Lambda \times \mathbb{R}^3)^N$. Suppose it is an ideal gas, so we neglect the interaction energy between molecules. Then the total energy is $E = \sum_{k=1}^{N} \frac{m}{2} \boldsymbol{v}_k^2$, and the surface of constant energy is $\Lambda^N \times \mathbb{S}_E$, where $\mathbb{S}_E$ is the sphere in $\mathbb{R}^{3N}$ of radius $\sqrt{2E/m}$ around the origin. Let us drop the position variables since they are not relevant. Boltzmann found the following law of large numbers: *If $N \gg 1$ and we choose a point $v = (\boldsymbol{v}_1, \ldots, \boldsymbol{v}_N)$ randomly on $\mathbb{S}_E$ with uniform probability distribution, then the empirical distribution of $\boldsymbol{v}_1, \ldots, \boldsymbol{v}_N$ in $\mathbb{R}^3$ is close to the Maxwellian distribution* (A.38) *with probability near 1.*

So why, of all the infinitely many possible ρ functions, should we find exactly the Maxwellian density (A.38)? Because the Maxwellian, although it may appear like a very special choice, is not special at all! It is what comes out when there is no particular reason forcing ρ to be any particular function! This becomes visible from the more fundamental point of view, that of phase space: The Maxwellian comes out for the overwhelming majority of phase points v. It is the empirical distribution that is typical among the phase points.

Now let us return to Bohmian mechanics; here, the more fundamental point of view is that of the configuration space of the universe. The question is not, "What was the initial ρ?" It is, "What was the initial Q?" Some initial configurations Q_0 of the universe will lead to empirical observations in agreement with Born's rule (say, those in S_0), and others will not (those in $S_0^c = \mathbb{R}^{3N} \setminus S_0$). So the basic fact behind quantum equilibrium is that

$$Q_0 \in S_0 \,. \tag{A.39}$$

And what is the explanation of that? The various versions of the law of large numbers in Bohmian mechanics tell us that if Q_0 is chosen randomly with $|\Psi_0|^2$ distribution, then it will lie in S_0 with probability near 1. Put differently, if we measure sizes of sets R in configuration space by $\mu(R) = \int_R |\Psi_0|^2$, then

$$\text{most configurations lie in } S_0. \tag{A.40}$$

That is, it is typical to lie in S_0.

This brings us to the question, "Why should we quantify the size of a set R by $\mu(R)$?" There are two kinds of answer to this, corresponding to two pictures of Bohmian mechanics, or to two ways of understanding Bohmian mechanics, that I have mentioned already briefly. The first, which I will call the *random picture*, can be expressed in a creation myth: God tells the demiurgs to choose Q_0 randomly with $|\Psi_0|^2$ distribution, and the demiurgs say, "Will do." In the second, which I will call the *uniqueness picture*, a future fundamental physical theory will involve principles that determine Q_0 exactly. (Further pictures in between the two are conceivable.) In the random picture, μ is the right measure by virtue of a law of nature that demands $Q_0 \sim \mu$. Still in the random picture, this does not mean that the measure is completely arbitrary; just like the Schrödinger equation and Bohm's equation of motion are not any old equations but ones that fit together, μ fits together with both because it is time independent by virtue of equivariance; it does not single out any particular time. The law "$Q_0 \sim \mu$" is a particularly natural law that makes the theory particularly simple and elegant. Now the uniqueness picture: it sounds more speculative but makes visible why the random picture makes assumptions that are perhaps stronger than necessary. For compatibility of this future theory with Bohmian mechanics, we only need that Q_0 looks like a μ-typical point. And if it does, then the *reason* why it does is presumably just *the absence of reasons to the contrary*. For comparison, if π has the frequency property (which is presumably the case), then the *reason* why it has this property is the *absence of reasons* for any simple pattern in its decimal expansion; while for every rational number there is a reason for a simple pattern (i.e., periodicity) in its decimal expansion, π presumably has no special relation to the decimal system (or the place-value system in any other base instead of 10), and that presumably makes its decimal sequence look chaotic or random. Likewise with Q_0: if it is not specially selected for making the empirical distribution F_z of outcomes of experiments look a particular way, then F_z should look the same as for "any old" initial configuration; and if the principles selecting Q_0 are not particularly related to the time $t = 0$ but refer to the entire trajectory, then the sense of typicality of Q_0 should not refer to any particular time, i.e., it should be equivariant; and the natural equivariant measure is μ.

So then, what is the justification of μ? It is either, in the random picture, that in the most natural possible law for the distribution of Q_0, this distribution is μ; or, in the uniqueness picture, that in the absence of reasons to the contrary and of reasons preferring a particular time, Q_0 should be μ-typical. And what is the explanation of quantum equilibrium? That μ-most Q_0 imply quantum equilibrium at all times.

Let us compare that to classical statistical mechanics. Once the number N of particles and a particular value E for the energy of the universe have been selected, the natural measure on the energy surface Γ_E in phase space $\mathbb{R}^{6N}$ is the Liouville measure λ_E (which can roughly be thought of as uniform, although it is in general not the same as surface area). So we might expect the initial phase point X_0 to be λ_E-typical. However, a λ_E-typical X_0 would be in thermal equilibrium and would remain in thermal equilibrium for approximately the next 10^N years. Now we know that our universe is not in thermal equilibrium, for example, because some subsystems are not in thermal equilibrium, for example, because the temperature is not the same in all places. So, our universe cannot be λ_E-typical. This is a fact.

Is it a fact that refutes typicality reasoning? No, it is a fact that calls for an explanation. It calls for discovering a deeper reason why our universe has an atypical trajectory (assuming for the sake of the argument that it is governed by Newtonian mechanics). If a Newtonian universe is in thermal equilibrium for the first 10^N years of its existence, then no further explanation is needed. That is the default possibility, it is what we would expect in the absence of reasons to the contrary. The atypicality of our universe is unexpected and points to the existence of a deeper reason behind this behavior. The correct explanation has not been reliably identified yet; it is still an open question; but it has been suggested that this deeper reason may involve a further law of nature that has been termed the *past hypothesis*; see, e.g., Goldstein et al. (2020) [16, Sec. 5.7] for discussion and references.

For Bohmian mechanics, this means in parallel that if we saw evidence of quantum non-equilibrium, for example, if we saw empirical distributions $|\varphi|^4$ or $|\varphi|^0$, then that would be unexpected. Then we would be surprised, and in need of an explanation, we would have a problem, an open question. It is a crucial difference between statistical mechanics and Bohmian mechanics that whereas even everyday evidence contradicts the thought that the whole universe could be in thermal equilibrium, all evidence available is compatible with the possibility that quantum equilibrium holds always and everywhere in the universe. This is the default possibility that requires no further explanation. And this is why it is an illusion to think that quantum equilibrium needs to be justified through mixing or other ways of pushing ρ toward $|\psi|^2$.

The explanation of quantum equilibrium in terms of typicality that I just described is due to Dürr, Goldstein, and Zanghì (1992) [9]. As a last remark, also for stochastic processes such as Bell's jump process we can regard the probability measure over the histories of the universe as a typicality measure.

A.4.4 Historical Notes

Here are a few more remarks about the history of Bohmian mechanics.

Quantum Potential

Bohm had a curious way of writing his equation of motion. When you take the time derivative of (1.77), you obtain that

$$m_k \frac{d^2 \boldsymbol{Q}_k}{dt^2} = -\nabla_k \big(V + V_Q\big)(t, Q(t)) \tag{A.41}$$

with V the potential in the Schrödinger equation and the function

$$V_Q := -\sum_{j=1}^{N} \frac{\hbar^2}{2m_j} \frac{\nabla_j^2 |\psi|}{|\psi|} \tag{A.42}$$

called the "quantum potential."

Proof Expressing ψ in terms of modulus and phase, $\psi = Re^{iS/\hbar}$, the Schrödinger equation implies that[3]

$$\frac{\partial R^2}{\partial t} = -\sum_{k=1}^{N} \nabla_k \cdot \left(R^2 \tfrac{1}{m_k} \nabla_k S\right) \tag{A.43a}$$

$$\frac{\partial S}{\partial t} = -\sum_{k=1}^{N} \frac{(\nabla_k S)^2}{2m_k} - V - V_Q\,. \tag{A.43b}$$

Writing Bohm's equation of motion as $d\boldsymbol{Q}_k/dt = \frac{1}{m_k}\nabla_k S$, we obtain that

$$m_k \frac{d^2 \boldsymbol{Q}_k}{dt^2} = \frac{d}{dt}\nabla_k S \tag{A.44a}$$

$$= \left(\frac{\partial}{\partial t} + \sum_{j=1}^{N} \tfrac{1}{m_j} \nabla_j S \cdot \nabla_j\right) \nabla_k S \tag{A.44b}$$

$$= -\nabla_k \sum_{j=1}^{N} \frac{(\nabla_j S)^2}{2m_j} - \nabla_k (V + V_Q) + \sum_{j=1}^{N} \tfrac{1}{m_j} \nabla_j S \cdot \nabla_j \nabla_k S \tag{A.44c}$$

$$= -\nabla_k (V + V_Q)\,. \tag{A.44d}$$

□

The remarkable thing about Eq. (A.41) is that it looks like Newton's equation of motion, only with an additional potential V_Q.

Bohm declared [1, 3] that the equation of motion was (A.41) but demanded in addition that the initial conditions obey the "constraint condition"

$$\frac{d\boldsymbol{Q}_k}{dt} = \tfrac{1}{m_k} \nabla_k S(t, Q(t)) \tag{A.45}$$

[3] However, the two Eqs. (A.43) together are *not equivalent* to the Schrödinger equation, for several reasons: (1) S is not defined where $\psi = 0$. (2) PDEs need boundary conditions at places where the unknown function is not defined, or else they do not define a unique time evolution. (3) Even on the set M where $\psi \neq 0$, S is only defined up to addition of an integer multiple of $2\pi\hbar$; in particular, it may be impossible to define $S : M \to \mathbb{R}$ in a continuous way on M, as a jump by $2\pi\hbar$ may be necessary (e.g., think of $\psi(x, y, z) = (x + iy)\exp(-x^2 - y^2 - z^2) = e^{i\varphi}\sin(\theta)\, r \exp(-r^2)$); as a consequence, ∇S is locally but not globally a gradient, and the curve integral of ∇S along a closed curve surrounding a point with $\psi = 0$ must be an integer multiple of $2\pi\hbar$ (which, by the way, would be an unmotivated requirement if we wanted to think of (A.43) as more fundamental than the Schrödinger equation).

at $t = 0$. This is actually equivalent to using (A.45) as the equation of motion, for the following reason. We have verified that every solution of (A.45) is also a solution of (A.41); since the solution of (A.41) is unique once the initial configuration $Q(t = 0)$ and the first derivative $(dQ/dt)(t = 0)$ are given, a solution with initial condition obeying (A.45) at $t = 0$ must agree with the solution of (A.45) with initial condition $Q(t = 0)$. So, Bohm's prescription leads to the same trajectories as (A.45) as a first-order ODE.

Of course, Bohm's prescription is an artificial, intransparent, and unnecessarily complicated way of characterizing the solutions of (A.45). What is worse, giving Eq. (A.41) the status of a law of nature makes the theory seem unnatural, complicated, and contrived in view of the unmotivated formula (A.42) for V_Q. In contrast, (A.45) is a simple formula. Even more obviously natural is the version (7.130) for the Dirac equation, $dX^\mu/ds \propto j^\mu$, with $j^\mu = \overline{\psi}\gamma^\mu\psi$ the most obvious way of defining a 4-vector field from a Dirac wave function ψ. That is why I did not use Bohm's original formulation (A.41). The quantum potential is still of interest, though, when one studies the classical limit of Bohmian mechanics, as that involves the question under which conditions the Bohmian trajectories are close to Newtonian trajectories, and that is related to the question under which conditions the derivatives of V_Q are small.

"And Then Throws It Away"

Bohm had discovered his law of motion in 1951 independently of de Broglie and Rosen, who had discovered equivalent equations in 1926 and 1945, respectively. About whether de Broglie or he should have the priority for discovering Bohmian mechanics, Bohm wrote in a letter to Pauli dated October 1951 [22, p. 390]:

> If one man finds a diamond and then throws it away because he falsely concludes that it is a valueless stone, and if this stone is later found by another man who recognize[s] its true value, would you not say that the stone belongs to the second man? I think the same applies to this interpretation of the quantum theory.

Ironically, Bohm abandoned Bohmian mechanics around 1957 (see, e.g., Freire Jr. 2019 [12]). Publicly, he kept advocating it, but rather half-heartedly. It is my impression that Bohm's reasons for abandoning Bohmian mechanics were these:

1. He thought that the probability distribution $\rho = |\psi|^2$ could be justified only if a stochastic law of motion was introduced that would drive ρ toward $|\psi|^2$.
2. The formula for the quantum potential looked arbitrary, and that seemed unsatisfactory.
3. He thought that Bohmian mechanics was incompatible with relativity.

From what I said earlier in this Appendix and in Sect. 7.5.4, readers will be aware that I do not regard any of these reasons as valid.

Interestingly, that was not the end of the story. Decades later, in the 1980s, Bohm returned to liking his equation of motion, in particular after his students Chris Dewdney and Chris Philippidis became enthusiastic about it and created a diagram similar to Fig. 1.5. Some years later, he even wrote (jointly with Basil Hiley) a book [3] about it, which appeared in 1993 after Bohm's death. He thought that the first of the three reservations mentioned above was resolved through mixing, but the other two he perhaps never regarded as resolved.

A.5 Philosophical Topics

A.5.1 Free Will

Free will comes up in the foundations of quantum mechanics in three places: In connection with whether a fundamental physical theory could be deterministic; as an assumption in Bell's proof of nonlocality; and in connection with a theorem of Conway and Kochen's that they call the "free will theorem." I will briefly discuss each topic in turn.

Free Will and Determinism

It would seem that any deterministic theory of physics (such as Newtonian or Bohmian mechanics) excludes "fundamental" free will. After all, according to such theories, our decisions would have been pre-determined by the initial state of the universe, or its state at any time since. For example, suppose I had pasta for lunch today; from determinism I must conclude it was never actually possible for me to have anything else for lunch today, although I had the clear impression before lunch of having a choice between several meals. This situation might suggest that deterministic theories are wrong and even absurd. So, determinism seems to conflict with common sense. But upon more careful thinking, I find it hard to point to where exactly determinism would conflict with common sense, and I therefore suspect that there is no real conflict between determinism and a reasonable understanding of free will. Let me explain.

When a computer is playing chess, it follows a deterministic algorithm, and each of its moves is pre-determined, given the earlier moves of the opponent. When we play against a computer, we usually do not exploit this determinism for predicting its moves (say, by running the same algorithm on another computer); instead, we think about which moves would be good moves for the computer and which advantages or disadvantages for the computer would follow from certain moves. That is, when playing against a computer, we treat the computer like a human opponent; more precisely, we treat it as if it had free will. We consider all possible moves the computer could make and analyze their consequences; if a move seems advantageous for the computer, we prepare for the situation that the computer will make this move. This suggests that free will may be a way of treating an agent, and not so much a property of the agent. (Here, "agent" means any being that acts.)

Maybe a computer, when playing chess, should be regarded as endowed with a kind of "effective free will" that is more relevant than "fundamental free will."

When we make a decision, we begin by thinking through our options of action and their consequences. For doing this, we have to think of each option as possible, of the future as not fixed yet, and of our will as free. After all, we try to analyze what would happen if we chose a certain option. Suppose that on a particular historical date at a particular time t I chose a particular move M in a game of chess. Shortly before t, when considering my options, I did not know which move I would make. Still, it is a fact that at t I made the move M. But this fact does not remove the need for my thinking through the options to arrive at this decision. Even if my move was pre-determined, this fact did not allow me to know before t which move I would make. And anybody predicting my move by simulating the processes in my brain would also effectively repeat all of my considerations about which move is better than others and why. By the way, if one of the moves actually is objectively better than the others, then it would seem most natural and obvious that my move could be predicted.

I can also mention that I sometimes wish I had no choice, when I am facing a tough decision. Then physical determinism (if true) and a conflict between free will and determinism would seem to come in handy. Alas, determinism is of no use then! It will remain true that the decision in the end depends on the thought process leading to it. The fact that ultimately I will make the decision that I will make is a mere tautology—a statement without content.

It is also striking that my thought process before a chess move is pretty similar to what a computer is doing when playing chess. In the old days before computers, many people assumed that the human mind cannot be a machine, as machines cannot be capable of thinking. Nowadays, with artificial intelligence and chess computers, it is obvious that machines can carry out various types of thinking. That should reduce the perceived conflict between free will and determinism.

Some people feel that in a deterministic world, without fundamental free will, there could not be any moral responsibility. Also for this claim, I do not see a compelling case. In a deterministic world, it could still be true that some actions are evil, and such moral truths could play a role in our considerations during decision-making. You might think that if brains work deterministically, then they cannot recognize moral truths. Then again, you would have just as much reason to think the same about factual truths or mathematical truths, but obviously computers can arrive at mathematical truths, so it is not clear that there is any real problem.

Consider also stochasticism instead of determinism. By that, I mean physical theories according to which the course of events is fundamentally random, so that nature makes random decisions in the middle of the world history. GRW theory can be thought of in this way (see Sect. 3.3). Now it seems that people who think that free will conflicts with determinism should also regard free will as impossible in a stochastic world. After all, the random events are not under my control, and there does not seem to be any room for my will to influence them. If my decisions are random. they would appear no more free than in a deterministic world—random is not the same as free. Now it is widely agreed in science that human brains, like all systems under usual conditions on Earth, are governed by the quantum formalism—the empirical rules consisting of the Schrödinger equation, the Born rule, and the projection postulate. In the situation of a brain making a decision, these rules will fix, from the initial wave function of the brain, the probability distribution over the possible decision outcomes. So any version of quantum mechanics will have to agree that our decisions are either deterministic or stochastic.

The upshot of this discussion is that, as far as I can tell, there is no real conflict between free will and determinism. But there is one more relevant aspect to this question, the existence of "free variables," which I will discuss now in connection with Bell's proof of nonlocality.

Free Will and Nonlocality

Let me come back to the computer playing chess and us treating it as having free will. Also in common physical theories, there is a way of treating a system as having free will, and that is by the concept of an *external field*. Although a physical theory needs to do no more than account for our universe, essentially, all known theories do more as they also provide equations for treating subsystems of the universe and in these equations allow for arbitrary external fields. (Maybe the fields have to be differentiable and bounded, but that is not a relevant limitation to their arbitrariness.) For example, we can put an arbitrary potential in Newton's equation of motion (1.58) or the Schrödinger equation (1.1). The system $\mathscr{S}$ described by these equations is not treated as free, but another system $\mathscr{S}'$ is a system interacting with $\mathscr{S}$ in a way represented by the external field. That is, known theories have a modular structure, where subsystems are the modules that can be treated separately. (A generic differential equation for the phase point of the universe would not have this structure.) These theories thus answer how the system $\mathscr{S}$ will react to any choice of external fields, with the consequence that the external fields can be regarded as *free variables* (i.e., variables whose values can be chosen at free will).

These free variables play a role in the proof of nonlocality (see Sect. 4.2). They appear as the parameters $\boldsymbol{\alpha}$, $\boldsymbol{\beta}$ chosen by the experimenters (say, as the field of the Stern-Gerlach magnets) and correspond to Alice and Bob's choice of observables. The proof assumes that the values of these variables cannot be foreseen and that nature can react to any of their values. In Bell's 1976 proof (see Sect. 4.2.3), this is expressed mathematically as the fact that the distribution $\rho(\lambda)$ does not depend on $\boldsymbol{\alpha}$ or $\boldsymbol{\beta}$. If the two particles "knew" in advance which way the magnets are oriented, then no contradiction between the observed outcomes and locality would arise. Since in a deterministic world Alice and Bob's decisions are pre-determined, it may seem like the argument has no force on deterministic theories. But Bohmian mechanics illustrates that it does: Bohmian mechanics has an answer to how the particles will react to *any* external field. Thus, despite being deterministic, Bohmian mechanics satisfies the assumption of free variables.

One can imagine "super-deterministic" theories that would correctly reproduce the whole world history but do not have the relevant modular structure and violate the assumption of free variables, so that Alice and Bob's later decisions are already "built into" the state and behavior of the two particles. Such a theory seems unsatisfactory, for three reasons. First, it goes against our intuitions about free will and is at odds with common sense. Second, it does not seem like a satisfactory explanation of the statistics of outcomes of Bell's experiment to say that no influence is superluminal, but the particles knew all along which parameters Alice and Bob would choose. Third, such a theory will presumably involve contrived dynamics or complicated initial conditions.

For the nonlocality proof, the difference between "random" and "free" does not matter. The proof will go through as well if the parameters $\boldsymbol{\alpha}, \boldsymbol{\beta}$ are chosen randomly, independently of λ.

Conway and Kochen's "Free Will Theorem"

John Horton Conway (1937–2020) and Simon Kochen (2006 [6], 2009 [7]) published a theorem they called the "free will theorem" (see also Tumulka 2007 [24] and Goldstein et al. 2010 [14]). Their proof is really just another proof of Bell's nonlocality theorem, but Conway and Kochen disagreed with this assessment of mine and insisted that their theorem has further philosophical implications. Here is what is going on. They had the view that a relativistic theory has to satisfy microscopic parameter indepedence, or MPI. (This assumption is not made explicit in their mathematical discussion but introduced vaguely in their prose.) It is not hard to see from Bell's theorem (and was known long before Conway and Kochen's work) that deterministic theories agreeing with the quantum mechanical statistics of outcomes, such as Bohmian mechanics, must sometimes violate MPI. Thus, Conway and Kochen concluded the true theory of our universe cannot be deterministic. (This, of course, depends on whether one insists on MPI.) As a further step, Conway and Kochen also claimed to have excluded stochastic theories. Their argument is that any stochastic theory can be simulated by a deterministic theory if sufficiently many random bits are provided in advance. However, this step contains a mistake, as illustrated by relativistic GRW theory, which obeys MPI. Indeed, the conversion of a stochastic theory obeying MPI into a deterministic theory will break MPI because for the simulation, a choice must be made about which pre-existing random bits to use for which random decision, and thus about the order of random decisions, or the temporal order of (spacelike separated) events. Put differently, a stochastic theory can have fundamental random events (such as flashes) that are correlated without specifying who influenced whom, and that feature is lost when all randomness is realized at the beginning of the universe.

What Conway and Kochen meant by "free will" is that the true theory of the universe be neither deterministic nor stochastic. Unfortunately, no example of such a theory is presently known.

A.5.2 Causation

> All philosophers, of every school, imagine that causation is one of the fundamental axioms or postulates of science, yet, oddly enough, in advanced sciences such as gravitational astronomy, the word "cause" never occurs.
>
> Bertrand Russell (1912) [23, p. 1]

In previous chapters, we introduced various theories that are meant as fundamental physical theories—as governing a hypothetical universe: Newtonian mechanics, Bohmian mechanics, GRW theories, etc. You may find it striking that the definitions of these theories did not mention the concepts of *cause* and *effect*. Instead, the definitions specified laws concerning how much matter is where in space at what time.

In this section, I elaborate a bit on why that is so, why causation is not a fundamental concept, and why fundamental physical theories *should not* talk about causation in their definition.

Causation has to do with how *agents* (acting beings) consider the world. Let us consider an example: the spaceship *Challenger* exploded on January 28, 1986, about a minute after takeoff. When we ask about the cause or causes of this disaster, we have in mind that the engineers who built the rockets had plans for what the different parts of the vessel were *supposed* to do at certain stages of the flight, and we want to know how the actual course of events deviated from the intended one and how the course of events would have been different if we had prepared this or that differently. By the last question, we mean in what way future flights must be prepared differently to avoid further disasters. (Later investigations concluded that a so-called O-ring, made of rubber and sealing a joint of the rocket, was not as elastic as supposed to be, and this was related to the freezing temperatures before the launch, temperatures for which the O-rings were not designed.)

For such considerations, it is not needed that the physical laws assume or imply a fundamental concept or category of causation or that they talk at all about causes. It is not needed either that the physical laws be deterministic, although determinism often makes it easier to give a precise meaning to counterfactual statements ("if we had done X, which we did not, then Y would have happened"). It is relevant, however, that the physical laws have the modular structure allowing for external fields or free variables that I mentioned in Appendix A.5.1.

In fact, when, in a modularly structured theory, we limit ourselves to regarding a term in the equations as an external field or a free variable, we can give a precise meaning to the concepts of cause and effect, with the external field playing the role of the cause and all other variables that of the effect. Other than that, however, "causation" is a higher-level concept related to the perspective of agents, much like, e.g., the concept of "message." Sometimes authors suggested that the orbit of the moon is caused by the gravitational pull of Earth or that the position and velocity of the moon at time t were caused by its position and velocity at an earlier time t_0. To me, such statements seem to overly extend the concept of cause.

A.5.3 The Mind-Body Problem

Suppose a professor gives her students the following homework exercise: Write a computer program that will make the computer see the color red. It seems obvious that this is an impossible assignment and that no program whatsoever will make the computer see red. After all, if a digital camera is connected to the computer, then red light (i.e., light with a wave length of around 800 nanometers) will be translated into pixels with a certain color code. But having the information that the pixels have the color code that corresponds to 800 nanometers and the English word "red" is not the same as seeing red.

At this point, the most surprising aspect is perhaps that human brains see red. It is clear that the physical processes inside the brain could serve for processing information, but how and why would that lead to seeing red? A variant of the question is to explain our experience of seeing red: Why do we see red? That is, it seems like a conceivable alternative possibility that our brains, like the computer, have the information available that our eyes have detected light of the wave length commonly called red, but that nothing *looks red* to the brain. Why is that not what happens? Why do we have, in addition, this experience of color that is hard to convey in words?

This question is very difficult to answer, apparently even impossible. If somebody described the complete physical processes in the brain for processing data from the eye, a key part of the question would remain: Why would these processes be associated with the experience of seeing red? If we knew the exact trajectories (or wave function or other variables) of all electrons and quarks in the brain, why would these trajectories be associated with an experience of seeing red? It seems that there is no logical reason for that, while there might be laws of nature that ensure that certain trajectories or wave functions are associated with certain experiences. The apparent fact that the experience of red does not follow logically from the physical facts is known as the *mind-body problem*. It was known to Gottfried Wilhelm Leibniz (1646–1716)[4] and presumably already to Democritus of Abdera (around 460 BCE–around 370 BCE);[5] among contemporary authors, Thomas Nagel (1974) [19] and David Chalmers (1996) [5] have given detailed descriptions and discussions of it. The same situation arises not only with colors but also, for example, with flavors, and not just perceptions but also various feelings; together, these experiences are called *conscious experiences* or *qualia*. (Here, "conscious" has nothing to do with "self-aware.") In contrast, life (i.e., the existence and functioning of living beings) is just a consequence of physical facts. It remains unclear what kind of laws govern conscious experiences, but it seems inevitable to conclude that not all facts in the world follow logically from the physical facts. Fortunately, we can deduce from physical theories their observable consequences without having to deduce the conscious experience of the observer, as we can deduce where pointers point and take for granted that the observers' conscious experience of pointer positions usually agrees with the pointers' actual positions.

[4] It is perhaps not a coincidence that Leibniz had been working on the artificial intelligence of his day, a mechanical calculator.

[5] It is perhaps not a coincidence that Democritus had co-proposed a fundamental physical theory, according to which particles of nonzero size and various shapes (called "atoms") fly around, rotate, and collide.

A.6 Differential Forms

I give a brief introduction to the concept of a differential form. It is mainly relevant to *relativistic* calculations.

Often, the concept of integration $\int f(x)\,\mu(dx)$ involves an *integrand* (a function f) and a concept of *volume* or *measure*. The concept of a differential form, in contrast, describes an integrand that brings its own measure already built in, that is, it describes something like the product $f\mu$ of integrand and measure. This is useful when we do not already have a concept of volume or when we have several ones and want to switch between them. And it is particularly relevant to integrating over surfaces (submanifolds). For example, when we want to switch between the Euclidean metric and the Minkowski metric on $\mathbb{R}^4$, then the two are associated with the same concept of 4d volume, but with different concepts of surface area or 3d volume of 3d sets.

Definition A.1 A *differential form of degree* $k \in \{0, 1, 2, \ldots\}$, or k*-form* for short, is an anti-symmetric tensor field with k lower indices,

$$T_{\mu_1 \ldots \mu_k} = (-)^\sigma \, T_{\mu_{\sigma(1)} \ldots \mu_{\sigma(k)}} \tag{A.46}$$

for any permutation $\sigma \in S_k$ and $(-)^\sigma$ the sign of σ.

Key Fact *If T is a k-form on $\Omega = \mathbb{R}^d$ (or a manifold Ω of dimension d) and Σ is a k-dimensional subset of Ω, then it is well defined what the integral $\int_\Sigma T$ of T over Σ means (except that the integral might be infinite or might have infinite positive and negative contributions).*

The point is that T contains all the information needed for integration over Σ, independently of any notion of volume or metric in Ω. Heuristically, this point can be understood by decomposing Σ into infinitesimal k-dimensional parallelepipeds; for an infinitesimal parallelepiped in Euclidean space spanned by the vectors $dx_1^\mu, \ldots, dx_k^\mu$, its k-dimensional volume is given by $T_{\mu_1 \ldots \mu_k}\, dx_1^{\mu_1} \cdots dx_k^{\mu_k}$ for some anti-symmetric T; thus, also $f(x)$ times the k-dimensional volume is of this form for an x-dependent T.

Note that for the standard d-dimensional volume in d-dimensional Euclidean (or Minkowski) space, $T_{\mu_1 \ldots \mu_k}\, dx_1^{\mu_1} \cdots dx_k^{\mu_k}$ is, up to a factor, nothing but the determinant of the matrix with columns $dx_1^\mu, \ldots, dx_k^\mu$. The corresponding tensor field is the *Levi-Civita tensor* or *epsilon tensor* whose components in any ONB (respectively, Lorentz frame) are

$$\varepsilon_{\mu_1 \ldots \mu_d} = \begin{cases} 0 & \text{if } \mu_i = \mu_j \text{ for some } i \neq j \\ (-)^\sigma & \text{if } \mu_i = \sigma(i) \text{ for a permutation } \sigma \in S_d. \end{cases} \tag{A.47}$$

By the way, we note that a 0-form is just a function $\Omega \to \mathbb{R}$ and a 1-form is a covector field, while for $k > d$ every k-form is 0.

Definition A.2 The *exterior derivative* of a k-form T is the $k+1$-form dT equal to $k+1$ times the anti-symmetrization of $\partial_{\mu_1} T_{\mu_2 \ldots \mu_{k+1}}$. In another notation, $dT = \nabla \wedge T$.

Example A.4 The electromagnetic 4-vector potential A_μ is a 1-form on Minkowski space, and the field tensor $F_{\mu\nu} = \partial_\mu A_\nu - \partial_\nu A_\mu$ is the exterior derivative of A, $F = dA$.

Key Fact *On a manifold, different curvilinear coordinate systems have different derivative operators* ∂_μ*, but they all lead to the same exterior derivative* dT*. On the tangent bundle, different connections have different covariant derivative operators, but they all lead to the same exterior derivative* dT*.*

Another key fact is the generalization of the Ostrogradski-Gauss integral theorem (divergence theorem) to differential forms, usually called the *Stokes integral theorem for differential forms*:

$$\int_\Sigma dT = \int_{\partial\Sigma} T \,. \tag{A.48}$$

References

1. D. Bohm, A suggested interpretation of the quantum theory in terms of "hidden" variables I. Phys. Rev. **85**, 166–179 (1952)
2. D. Bohm, A suggested interpretation of the quantum theory in terms of "hidden" variables II. Phys. Rev. **85**, 180–193 (1952)
3. D. Bohm, B.J. Hiley, *The Undivided Universe. An Ontological Interpretation of Quantum Theory* (Routledge, London, 1993)
4. M. Born, L. Infeld, Foundation of the new field theory. Proc. R. Soc. Lond. A **144**, 425–451 (1934)
5. D. Chalmers, *The Conscious Mind* (Oxford University Press, Oxford, 1996)
6. J.H. Conway, S. Kochen, The free will theorem. Found. Phys. **36**, 1441–1473 (2006). http://arxiv.org/abs/quant-ph/0604079
7. J.H. Conway, S. Kochen, The strong free will theorem. Notices Amer. Math. Soc. **56**, 226–232 (2009). http://arxiv.org/abs/0807.3286
8. D.-A. Deckert, V. Hartenstein, On the initial value formulation of classical electrodynamics. J. Phys. A Math. Theor. **49**, 445202 (2016). http://arxiv.org/abs/1602.04685
9. D. Dürr, S. Goldstein, N. Zanghì, Quantum equilibrium and the origin of absolute uncertainty. J. Statist. Phys. **67**, 843–907 (1992). http://arxiv.org/abs/quant-ph/0308039. Reprinted in [11]
10. D. Dürr, S. Goldstein, J. Taylor, R. Tumulka, N. Zanghì, Quantum mechanics in multiply-connected spaces. J. Phys. A Math. Theor. **40**, 2997–3031 (2007). http://arxiv.org/abs/quant-ph/0506173
11. D. Dürr, S. Goldstein, N. Zanghì, *Quantum Physics Without Quantum Philosophy* (Springer, Heidelberg, 2013)

12. O. Freire Jr, *David Bohm: A Life Dedicated to Understanding the Quantum World* (Springer, New York, 2019)
13. S. Goldstein, Stochastic mechanics and quantum theory. J. Statist. Phys. **47**, 645–667 (1987)
14. S. Goldstein, D.V. Tausk, R. Tumulka, N. Zanghì, What does the free will theorem actually prove? Notices Amer. Math. Soc. **57**, 1451–1453 (2010). http://arxiv.org/abs/0905.4641
15. S. Goldstein, J. Taylor, R. Tumulka, N. Zanghì, Fermionic wave functions on unordered configurations (2014). http://arxiv.org/abs/1403.3705
16. S. Goldstein, J.L. Lebowitz, R. Tumulka, N. Zanghì, Gibbs and Boltzmann entropy in classical and quantum mechanics, in *Statistical Mechanics and Scientific Explanation*, ed. by V. Allori (World Scientific, Singapore, 2020), pp. 519–581. http://arxiv.org/abs/1903.11870
17. M.K.-H. Kiessling, Some uniqueness results for stationary solutions to the Maxwell-Born-Infeld field equations and their physical consequences. Phys. Lett. A **375**, 3925–3930 (2011). http://arxiv.org/abs/1107.2333
18. J.M. Leinaas, J. Myrheim, On the theory of identical particles. Il Nuovo Cimento **37B**, 1–23 (1977)
19. Th. Nagel, What is it like to be a bat? Philos. Rev. **83**, 435–450 (1974)
20. E. Nelson, Derivation of the Schrödinger equation from Newtonian mechanics. Phys. Rev. **150**, 1079 (1966)
21. E. Nelson, *Quantum Fluctuations* (Princeton University Press, Princeton, 1985)
22. W. Pauli, *Wissenschaftlicher Briefwechsel mit Bohr, Heisenberg, Einstein und anderen, Band IV Teil I 1950–1952*, ed. by K. von Meyenn (Springer, Berlin, 1996)
23. B. Russell, On the notion of cause. Proc. Aristotelian Soc. **13**, 1–26 (1912)
24. R. Tumulka, Comment on "The Free Will Theorem". Found. Phys. **37**, 186–197 (2007). http://arxiv.org/abs/quant-ph/0611283
25. A. Valentini, H. Westman, Dynamical origin of quantum probabilities. Proc. R. Soc. Lond. A **461**, 253–272 (2005). http://arxiv.org/abs/quant-ph/0403034
26. J.A. Wheeler, R.P. Feynman, Interaction with the absorber as the mechanism of radiation. Rev. Modern Phys. **17**, 157–181 (1945)
27. J.A. Wheeler, R.P. Feynman, Classical electrodynamics in terms of direct interparticle action. Rev. Modern Phys. **21**, 425–433 (1949)

Index

R. Tumulka, *Foundations of Quantum Mechanics*, Lecture Notes in Physics 1003,
https://doi.org/10.1007/978-3-031-09548-1

Printed in the United States
by Baker & Taylor Publisher Services